Louise Willcox

D1506052

CARPENTRY

Gaspar J. Lewis

DELMAR PUBLISHERS INC.

This book is dedicated to my wife, Lorraine,
for her help, encouragement, and understanding.

Cover photos

Home plan 3714-1 as seen in *Homes for the 80's* **plan book. Courtesy of Home Building Plan Service, 2235 N.E. Sandy Blvd., Portland, OR 97232**

Hand woodworking tools. Courtesy of Garret Wade Company

Portable power tools. Courtesy of Milwaukee Electric Tool Corporation

For information, address Delmar Publishers Inc.,
2 Computer Drive West, Box 15-015
Albany, New York 12212

COPYRIGHT © 1984
BY DELMAR PUBLISHERS INC.

All rights reserved. No part of this work covered by the copyright hereon may be reproduced or used in any form or by any means — graphic, electronic, or mechanical, including photocopying, recording, taping, or information storage and retrieval systems — without written permission of the publisher.

10 9 8 7 6 5 4 3

LIBRARY OF CONGRESS CATALOG CARD NUMBER: 83-71049
ISBN: 0-8273-1800-6

Printed in the United States of America
Published simultaneously in Canada
by Nelson Canada,
A Division of International Thomson Limited

CONTENTS

PREFACE

Most carpentry and building construction books are one of two types: textbooks that present general information intended to familiarize the reader with building parts; and reference books that provide specific data, but little educational material. *Carpentry* provides the student with both. The principles of building construction are presented in an easy-to-understand format. Throughout the text the duties of carpenters are integrated with the explanations of building construction. Unlike most textbooks, *Carpentry* includes detailed step-by-step procedures for each of the tasks carpenters are expected to perform.

The book is divided into four sections. Section One discusses the basic tools and materials. Section Two deals with rough carpentry including building layout, concrete forms, scaffolding, framing, and insulation. Section Three explains how to apply exterior finish such as windows, doors, siding, cornices, and roofing. Section Four presents information on the application of interior finish such as drywall, paneling, ceiling tile, window and door trim, baseboard, stair finish, and cabinets.

Each unit of the text begins with a list of objectives that explain exactly what tasks the student is expected to accomplish. These objectives help guide and evaluate student progress through the text. In order to meet the objectives, the student must effectively learn the skills presented in each unit.

Material is presented through the use of information, illustrations, and step-by-step procedures. This three-fold approach gives the student many opportunities to grasp the material. After reading the explanations, students gain practical, hands-on experience by following step-by-step procedures that are highly illustrated, making the tasks easier to understand and accomplish. More than 900 drawings and photographs show modern construction details and methods. New terms appear in italics and are defined both on first use in the unit and in the glossary at the end of the text. The unit review questions measure students' knowledge of the material.

Additional study questions and suggested activities are provided in a study guide. The suggested activities may be used to supplement the text material by giving students extra work to generalize their knowledge in the same area or may be used to complement the material presented in the unit.

In writing this book, the author has drawn on his more than thirty years experience as a carpenter. Each unit was carefully developed through research and task analysis. The manuscript for the text was reviewed for technical accuracy and educational approach by carpentry instructors in vocational schools, community colleges, and apprentice training programs.

Although space prohibits naming everyone who has contributed to *Carpentry,* the author especially appreciates the reviews by: R.A. Wigman, Skyline High School in Dallas, Texas; James L. Barnes, Spotswood Senior High School in Penn Laird, Virginia; Odelle Grose, Forsythe Technical Institute in Winston-Salem, North Carolina; Silas Bruner, Harrison High School in Colorado Springs, Colorado; Wes Boydston, State University College in Oswego, New York; James Lee Morris, Georgia Southern College in Statesboro, Georgia; and Robert Simonds, Orange Coast College in Costa Mesa, California.

SPECIAL THANKS

The author wishes to thank the administration, faculty, and students of the Construction Trades Technology of Pinellas Vocational Technical Institute for their cooperation.

The author wishes also to express special thanks to the following:

Lorraine Lewis, his wife, for typing the original manuscript.

Robert Morency for photographic assistance.

NOTICE TO THE READER

Publisher does not warrant or guarantee any of the products described herein or perform any independent analysis in connection with any of the product information contained herein. Publisher does not assume, and expressly disclaims, any obligation to obtain and include information other than that provided to it by the manufacturer.

The reader is expressly warned to consider and adopt all safety precautions that might be indicated by the activities described herein and to avoid all potential hazards. By following the instructions contained herein, the reader willingly assumes all risks in connection with such instructions.

The publisher makes no representations or warranties of any kind, including but not limited to, the warranties of fitness for particular purpose or merchantability, nor are any such representations implied with respect to the material set forth herein, and the publisher takes no responsibility with respect to such material. The publisher shall not be liable for any special, consequential or exemplary damages resulting, in whole or in part, from the readers' use of, or reliance upon, this material.

INTRODUCTION

The carpenter constructs and repairs structures and their parts using wood, plywood, and other building materials by laying out, cutting, fitting, and fastening the materials to erect the framework and apply the finish.

The majority of workers in the construction industry are carpenters. They are the first trade workers on the job, laying out excavation and building lines. They take part in every phase of the construction, working below or at ground level or at great heights. They are the last to leave the job when the key is put in the lock.

SPECIALIZATION

In large cities, where there is a great volume of construction, carpenters tend to specialize in one area of the trade. They may be specialists in framing and are called rough carpenters. Rough carpentry does not mean that the workmanship is crude. Just as much care is taken in the rough work as in any other work. Rough carpentry is that which will be covered eventually by the finish work or dismantled, as in the case of concrete form construction. Finish carpenters specialize in applying exterior and interior finish, sometimes called trim. Some other specialties are building concrete forms, laying finish flooring, stair building, applying gypsum board, roofing, insulation, and installing acoustical ceilings.

In smaller communities, where the volume of construction is lighter, carpenters perform tasks in all areas of the trade from the rough to the finish. The general carpenter needs a more complete knowledge of the trade than the specialist does.

REQUIREMENTS

Carpenters need to know how to use and maintain hand and power tools. They need to know the kinds, grades, and characteristics of the materials with which they work — how each can be cut, shaped, and most satisfactorily joined. Carpenters must be familiar with the many different fasteners available and wisely choose the most appropriate on each occasion.

They must know how to lay out and frame floors, walls, stairs, and roofs. They must know how to install windows and doors, and how to apply numerous kinds of exterior and interior finish. Carpenters must use good judgment to decide on proper procedures to do the job at hand in the most efficient and safe manner.

Carpenters must be in good physical condition because much of the work is done by hand and sometimes requires great exertion. They must lift large sheets of plywood, heavy wood timbers, and bundles of roof shingles; they also have to climb ladders and scaffolds.

An attitude of care and concern for the job and for fellow workers is vital. The carpenter must feel that nothing less than a first-class job is acceptable.

TRAINING

Vocational training in carpentry is offered in many high schools for those who become seriously interested at an early age. For those who have completed high school, carpentry training programs are offered at many post-secondary vocational schools and community colleges. Most of these programs train students for duties up to and including that of foreman. Because there is so much more to learn, vocational school training should then be followed by an apprenticeship. Indentured apprenticeship training programs (usually four years in length) are offered by the United Brotherhood of Joiners and Carpenters of America and the Associated Builders and Contractors, among others. Usually these organizations give apprenticeship credit for completion of previous vocational school training in carpentry.

Starting pay for the apprentice carpenter is usually a percentage of the journeyman carpenter's pay with periodic increases throughout the apprenticeship. The apprentice becomes a journeyman carpenter when the training is completed.

Fig. I-A. Many opportunities lie ahead for the apprentice carpenter. *(Courtesy of Brotherhood of Carpenters and Joiners of America)*

Fig. I-B The journeyman carpenter

Many opportunities exist for the journeyman carpenter. Advancement depends on skill, productivity, and ingenuity, among other characteristics. Carpentry foremen, construction superintendents, and general contractors usually rise from the ranks of the journeyman carpenter. Many who started as apprentice carpenters now operate their own construction firms.

SUMMARY

Carpentry is a trade in which there is a great deal of self-satisfaction, pride, and dignity associated with the work. It is an ancient and honorable trade considered to be of key importance in the building industry.

Skilled carpenters who have labored to the best of their ability can take pride in their workmanship whether the job was a rough concrete form or the finest finish in an elaborate staircase. At the end of each working day, carpenters can stand back and actually see the results of their labor. As the years roll by, the buildings that carpenters' hands had a part in creating still can be viewed with pride throughout the community.

Fig. I-C Construction superintendents usually rise from the ranks of the carpenter. *(Courtesy of Brotherhood of Carpenters and Joiners of America)*

SECTION
1
Tools and Materials

WOOD 1

OBJECTIVES

After completing this unit, the student will be able to

- *name the parts of a tree trunk and state their function.*
- *describe methods of cutting the log into lumber.*
- *explain moisture content at various stages of seasoning, tell how wood shrinks and describe some common lumber defects.*
- *define hardwood and softwood, give examples of some common kinds, and tell their characteristics.*
- *state the grades and sizes of lumber and compute board measure.*

The carpenter works with wood more than any other material and must understand its characteristics in order to use it intelligently. Wood is a remarkable substance. It can be cut, shaped, or bent in just about any form. There are many kinds that vary in strength, color, grain, and texture. With proper care, wood will last indefinitely. It is a material with beauty and warmth that has thousands of uses and is one of our greatest natural resources. With wise conservation practices, wood will always be in abundant supply.

STRUCTURE AND GROWTH

Wood is made up of many hollow cells held together by a natural substance called *lignin*. The size, shape, and arrangement of these cells determine the strength, weight, and other properties of wood. Tree growth takes place in the *cambium layer* which is just inside the protective shield of the tree called the bark, Figure 1-1. The roots absorb water which passes upward through the *sapwood* to the leaves where it is combined with carbon dioxide from the air. Sunlight causes these materials to change into food which is then carried down and distributed toward the center of the trunk through the *medullary rays*. As the tree grows outward from the pith (center) the inner cells become inactive and turn to *heartwood*. Heartwood is the central part of the tree and is usually darker in color and more durable than sapwood. The heart of redwood, for instance, is extremely resistant to decay and

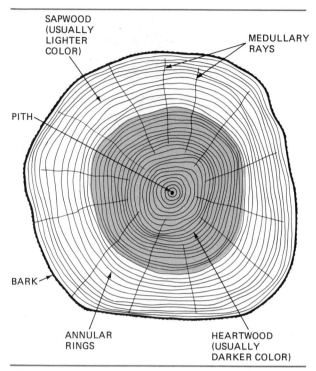

Fig. 1-1 Cross section of a tree *(Courtesy of Western Wood Products Association)*

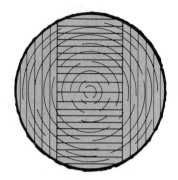

Fig. 1-2 a. Plain-sawed lumber b. Surface of plain-sawed lumber *(Courtesy of California Redwood Association)*

is extensively used for outdoor furniture, patios, and exterior siding. Used for the same purposes, its sapwood decays more quickly.

Each spring and summer, a tree adds new layers to its trunk. Wood growth in the spring is rapid, rather porous, and light in color. In summer, the tree growth is slower, more dense, and darker, forming distinct rings. Because these rings are formed each year, they are called *annular rings.* By counting the dark rings, the age of a tree can be determined. By studying the width of the rings, periods of abundant rainfall and sunshine or periods of slow growth can be determined. Some trees, like Douglas fir, grow rapidly to great heights and have very wide and pronounced annular rings. Mahogany, which grows in a tropical climate where the weather is more constant, has annular rings that are not so contrasting and sometimes are hardly visible.

LUMBER

Wood becomes what is known as *rough lumber* when the log is sawed into usable pieces at the sawmill. The operator of the saw, called a *sawyer,* decides the best way to cut the log.

PLAIN-SAWED LUMBER

A common way of cutting lumber is called the *plain-sawed* method in which the log is cut tangent to the annular rings. This method produces a distinctive grain pattern on the wide surface, Figure 1-2. This method of sawing is the least expensive and produces greater widths. However, plain-sawed lumber shrinks more in drying and it warps easily.

QUARTER-SAWED LUMBER

Another method of cutting the log, called *quarter-sawing,* produces pieces in which the annular rings are at or almost at right angles to the wide surface. Quarter-sawed lumber has less tendency to warp and shrinks less and more evenly when dried. This type lumber is durable because the wear is on the end of the annular rings and is frequently used for flooring.

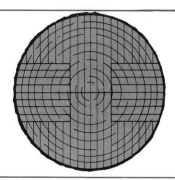

Fig. 1-3 a. Quarter-sawed lumber b. Surface of quarter-sawed lumber *(Courtesy of California Redwood Association)*

A distinctive and desirable grain pattern is produced in some woods, such as oak because the lumber is sawed along the length of the medullary rays. Quarter-sawed lumber is sometimes called vertical-grained, Figure 1-3.

COMBINATION SAWING

Most logs are cut into a combination of plain-sawed and quarter-sawed lumber. The sawyer must use great skill in determining how to cut the log with as little waste as possible in the shortest amount of time to get the desired amount and kinds of lumber, Figure 1-4.

MOISTURE CONTENT AND SHRINKAGE

When a tree is first cut down, it contains a great amount of water. Lumber, when first cut from the log, is called green lumber and is very heavy because, in some cases, over 50 percent of its weight is water.

Green lumber cannot be used in construction because it will eventually dry to the same moisture content as the surrounding air. As green lumber dries, it shrinks considerably and unequally in size. The use of green lumber in construction results in cracked ceilings and walls, squeaking floors, sticking doors, and many other problems caused by shrinking, warping, and twisting of the lumber as it dries.

Green lumber is also subject to decay. Decay is caused by fungi (low forms of plant life) that feed on wood. *Decay will not occur unless wood moisture content is in excess of 19 percent. Wood construction maintained at mois-ture content of less than 20 percent will not decay.* It is important that lumber be protected to prevent the entrance of moisture.

MOISTURE CONTENT

The moisture content (M.C.) of lumber is expressed as a percentage of its total weight. Therefore, a moisture content of 50 percent indicates that one-half the weight of the wood is water. Lumber used for framing and exterior finish should not exceed 19 percent M.C., preferably 15 percent M.C. For interior finish, a M.C. of 10 to 12 percent is recommended.

Green lumber has water in the hollow part of the wood cells and also in the cell walls. When wood starts to dry, the water in the cell cavities, called *free water,* is first removed. When all of

Fig. 1-4 Combination-sawed lumber *(Courtesy of Western Wood Products Association)*

the free water is gone, the wood has reached the *fiber-saturation point;* about one-third of the remaining weight is water. No noticeable shrinkage of wood takes place up to this point.

As wood continues to dry, the water in the walls of the cell is removed and the wood starts to shrink. It shrinks considerably from the size at its fiber saturation point of approximately 30 percent M.C. to the desired percent M.C. suitable for construction. Lumber at this stage is called *dry* or *seasoned.*

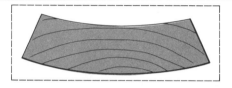

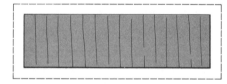

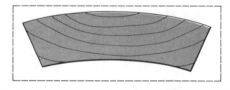

Fig. 1-5 Lumber shrinks in the direction of the annular rings.

Fig. 1-6 Drying lumber in the air *(Courtesy of Western Wood Products Association)*

It is important to understand not only that wood shrinks as it dries, but also how it shrinks. So little shrinkage occurs along the length of lumber that it is not considered. Most of the shrinkage is along the annular rings with more shrinkage taking place on the longer rings when viewing lumber in cross section, Figure 1-5. A cross section of quarter-sawed lumber shows annular rings of equal length. Therefore, although the piece shrinks somewhat, it shrinks evenly with no warp. Wood warps as it dries according to the way it was cut from the tree.

When the moisture content of lumber reaches that of the surrounding air (about 10 to 12 percent M.C.) it is at an *equilibrium moisture content.* At this point, lumber shrinks or swells only slightly with changes of moisture in the air.

DRYING LUMBER

Lumber is either *air-dried* or *kiln-dried.* In air-drying the lumber is stacked in piles with spacers, which are called *stickers,* between each layer to permit air to circulate through the pile, Figure 1-6. Kiln-dried lumber is stacked in the same manner, but is dried in huge kilns (ovens) in which the drying method is carefully controlled, Figure 1-7. Kiln-drying has the advantage of drying lumber in a shorter period of time. Air-drying may take 4 to 6 months depending on the thickness of the lumber and the weather. An advantage of kiln-dried lumber is that it can be brought to a lower moisture content than air-dried lumber. Air-dried lumber cannot be

Fig. 1-7 Drying lumber in a kiln *(Courtesy of Western Wood Products Association)*

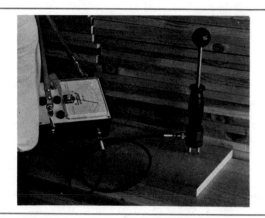

Fig. 1-8 Moisture meter *(Courtesy of Moisture Register Company)*

brought to a lower than equilibrium moisture content. Lumber with low moisture content (8 to 10 percent) is necessary for cabinet work.

The moisture content of lumber is determined by the use of a *moisture meter,* Figure 1-8. Points on the ends of the wires of the meter are driven into the wood, and the moisture content is read on the meter.

When lumber is sufficiently dry, it is brought to the planer mill where it is surfaced and uniformly sized. Most construction lumber is surfaced on four sides (S4S), although some may be surfaced on only two sides to the required thickness (S2S).

LUMBER DEFECTS

A defect in lumber is any fault that detracts from its appearance, function, or strength. One type of defect is called a *warp.* Warps are caused by drying lumber too fast, by careless handling and storage, and by surfacing the lumber before it is thoroughly dry. Warps are classified as cups, bows, crooks, and twists, Figure 1-9.

Splits in the end of lumber running lengthwise and across the annular rings are called *checks,* Figure 1-10. These are caused by faster drying of the end than of the rest of the stock. Checks can be prevented to a degree by sealing the ends of lumber with paint, wax, or other material during the drying period. Cracks that run parallel to and between the annular rings are called *shakes* and may be caused by storm or other damage to the tree.

The *pith* is the spongy center of the tree and if possible should not be used. *Knots* are cross sections of branches in the trunk of the tree. Knots are not necessarily defects unless they are loose or weaken the piece. *Pitch pockets*

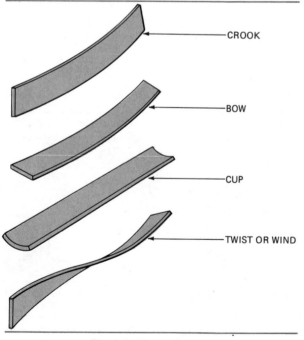

CROOK

BOW

CUP

TWIST OR WIND

Fig. 1-9 Kinds of warps

Fig. 1-10 A severe check in the end of a piece of lumber

are small cavities that hold pitch which usually runs out of the cavity. A *wane* is bark on the edge of lumber or the surface from which the bark has fallen. *Pecky* wood has small grooves or channels running with the grain. This is common in cypress. Pecky cypress is often used as an interior wall paneling when that effect is desired. Some other defects are *stains, decay,* and *wormholes.*

HARDWOODS AND SOFTWOODS

Woods are classified as either hardwood or softwood. There are different methods of distinguishing these woods, but the most common method is that hardwood comes from decidious trees that shed their leaves each year. Softwood is cut from conifers or cone bearing trees, commonly known as evergreens. In this method of classifying wood, some of the softwoods may actually be harder than the hardwoods. For instance, fir, a softwood, is harder and stronger than basswood, a hardwood. There are other methods of classifying hardwoods and softwoods, but this method is the one most widely used.

Some common hardwoods are ash, birch, cherry, hickory, maple, mahogany, oak, and walnut. Some common softwoods are pine, fir, hemlock, spruce, cedar, cypress, and redwood.

Wood may also be divided into two groups according to cell structure. Open-grained lumber has large cells which show tiny openings or pores in the surface. In order to obtain a smooth finish these pores must be filled with a specially prepared wood filler. Examples of open-grained wood are oak, mahogany, and walnut. All softwoods are close-grained. Some close-grained hardwoods are birch, cherry, maple, and poplar. See Figure 1-11 for common kinds of hardwoods and softwoods and their characteristics.

LUMBER GRADES AND SIZES

Lumber grades and sizes are established by wood products associations of which many wood mills are members. Member wood mills are closely supervised by the associations to assure that association standards are maintained.

The grade stamp of the association is assurance that lumber grade standards have been met.

SOFTWOOD GRADES

One of the largest softwood associations is the Western Wood Products Association which grades lumber in three categories: *boards* (under 2" thick), *dimension* (2" to 4" thick), and *timbers* (5" and thicker). These three categories are further classified according to strength and appearance as shown in Figure 1-12.

Member mills use the association grade stamp to indicate strict quality control. A typical grade stamp is shown in Figure 1-13 and shows the association trademark, the mill number, and lumber grade, the species of wood, and the moisture content.

HARDWOOD GRADES

Hardwood grades are established by the National Hardwood Lumber Association. *First and Seconds (FAS)* is the best grade of hardwood and must yield about 85 percent clear cutting. Each piece must be at least 6 inches wide and 8 feet long. The next best grade is called *select*. For this the minimum width is 4 inches, and the minimum length is 6 feet. *No. 1 common* allows even narrower widths and shorter lengths, with about 65 percent clear cutting.

LUMBER SIZES

Lumber that comes directly from the saw mill is close in size to what it is called (nominal size). There are slight variations to its nominal size because of the heavy machinery used to cut the log into lumber. When rough lumber is planed, it is reduced in thickness and width to standard and uniform sizes. Its nominal size does not change even though the actual size does. Therefore, when surfaced, although a piece may be called a 2x4, its actual size is 1 1/2 inches by 3 1/2 inches. The same applies to all surfaced lumber; the nominal size and the actual size are not the same.

SOFTWOODS

KIND	COLOR	GRAIN	HARDNESS	STRENGTH	WORK-ABILITY	ELASTICITY	DECAY RESISTANCE	USES	OTHER
Red Cedar	Dark Reddish Brown	Close Medium	Soft	Low	Easy	Poor	Very High	Exterior	Cedar Odor
Cypress	Orange Tan	Close Medium	Soft to Medium	Medium	Medium	Medium	Very High	Exterior	
Fir	Yellow to Orange Brown	Close Coarse	Medium to Hard	High	Hard	Medium	Medium	Framing Millwork Plywood	
Ponderosa Pine	White with Brown Grain	Close Coarse	Medium	Medium	Medium	Poor	Low	Millwork Trim	Pine Odor
Sugar Pine	Creamy White	Close Fine	Soft	Low	Easy	Poor	Low	Patternmaking Millwork	Large Clear Pieces
Western White Pine	Brownish White	Close Medium	Soft to Medium	Low	Medium	Poor	Low	Millwork Trim	
Southern Yellow Pine	Yellow Brown	Close Coarse	Soft to Hard	High	Hard	Medium	Medium	Framing Plywood	Much Pitch
Redwood	Reddish Brown	Close Medium	Soft	Low	Easy	Poor	Very High	Exterior	Light Sapwood
Spruce	Cream to Tan	Close Medium	Medium	Medium	Medium	Poor	Low	Siding Subflooring	Spruce Odor

Fig. 1-11 Common softwood and hardwood characteristics

8

HARDWOODS

KIND	COLOR	GRAIN	HARDNESS	STRENGTH	WORK-ABILITY	ELASTICITY	DECAY RESISTANCE	USES	OTHER
Ash	Light Tan	Open Coarse	Hard	High	Hard	Very High	Low	Tool Handles Oars Baseball Bats	
Basswood	Creamy White	Close Fine	Soft	Low	Easy	Low	Low	Drawing Bds Veneer Core	Imparts No Taste Or Odor
Beech	Light Brown	Close Medium	Hard	High	Medium	Medium	Low	Food Containers Furniture	
Birch	Light Brown	Close Fine	Hard	High	Medium	Medium	Low	Furniture Veneers	
Cherry	Lt. Reddish Brown	Close Fine	Medium	High	Medium	High	Medium	Furniture	
Hickory	Light Tan	Open Medium	Hard	High	Hard	Very High	Low	Tool Handles Diving Boards	
Lauan	Lt. Reddish Brown	Open Medium	Soft	Low	Easy	Low	Low	Veneers Paneling	
Mahogany	Russet Brown	Open Fine	Medium	Medium	Excellent	Medium	High	Quality Furniture	
Maple	Light Tan	Close Medium	Hard	High	Hard	Medium	Low	Furniture Flooring	
Oak	Light Brown	Open Coarse	Hard	High	Hard	Very High	Medium	Flooring Boats	
Poplar	Greenish Yellow	Close Fine	Medium Soft	Medium Low	Easy	Low	Low	Furniture Veneer Core	
Teak	Honey	Open Medium	Medium	High	Excellent	High	Very High	Furniture Boat Trim	Heavy Oily
Walnut	Dark Brown	Open Fine	Medium	High	Excellent	High	High	High Quality Furniture	

Fig. 1-11 (Continued)

Grade Selector Charts
Boards

APPEARANCE GRADES	**SELECTS**	B & BETTER (IWP—SUPREME)* C SELECT (IWP—CHOICE) D SELECT (IWP—QUALITY)	
	FINISH	SUPERIOR PRIME E	
	PANELING	CLEAR (ANY SELECT OR FINISH GRADE) NO. 2 COMMON SELECTED FOR KNOTTY PANELING NO. 3 COMMON SELECTED FOR KNOTTY PANELING	
	SIDING (BEVEL, BUNGALOW)	SUPERIOR PRIME	
	BOARDS SHEATHING & FORM LUMBER	NO. 1 COMMON (IWP—COLONIAL) NO. 2 COMMON (IWP—STERLING) NO. 3 COMMON (IWP—STANDARD) NO. 4 COMMON (IWP—UTILITY) NO. 5 COMMON (IWP—INDUSTRIAL) **ALTERNATE BOARD GRADES** SELECT MERCHANTABLE CONSTRUCTION STANDARD UTILITY ECONOMY	

SPECIFICATION CHECK LIST
- ☐ Grades listed in order of quality.
- ☐ Include all species suited to project.
- ☐ Specify lowest grade that will satisfy job requirement.
- ☐ Specify surface texture desired.
- ☐ Specify moisture content suited to project.
- ☐ Specify Ⓦ grade stamp. For finish and exposed pieces, specify stamp on back or ends.

Western Red Cedar

FINISH PANELING AND CEILING	CLEAR HEART A B
BEVEL SIDING	CLEAR — V.G. HEART A — BEVEL SIDING B — BEVEL SIDING C — BEVEL SIDING

Idaho White Pine carries its own comparable grade designations.

Dimension/All Species 2″ to 4″ thick (also applies to finger-jointed stock)

LIGHT FRAMING 2″ to 4″ Thick 2″ to 4″ Wide	CONSTRUCTION STANDARD UTILITY	This category for use where high strength values are **NOT** required; such as studs, plates, sills, cripples, blocking, etc.
STUDS 2″ to 4″ Thick 2″ to 6″ Wide 10′ and Shorter	STUD	An optional all-purpose grade limited to 10 feet and shorter. Characteristics affecting strength and stiffness values are limited so that the "Stud" grade is suitable for all stud uses, including load bearing walls.
STRUCTURAL LIGHT FRAMING 2″ to 4″ Thick 2″ to 4″ Wide	SELECT STRUCTURAL NO. 1 NO. 2 NO. 3	These grades are designed to fit those engineering applications where higher bending strength ratios are needed in light framing sizes. Typical uses would be for trusses, concrete pier wall forms, etc.
STRUCTURAL JOISTS & PLANKS 2″ to 4″ Thick 5″ and Wider	SELECT STRUCTURAL NO. 1 NO. 2 NO. 3	These grades are designed especially to fit in engineering applications for lumber five inches and wider, such as joists, rafters and general framing uses.

Timbers 5″ and thicker

BEAMS & STRINGERS 5″ and thicker Width more than 2″ greater than thickness	SELECT STRUCTURAL NO. 1 NO. 2** NO. 3**	**POSTS & TIMBERS** 5″ x 5″ and larger Width not more than 2″ greater than thickness	SELECT STRUCTURAL NO. 1 NO. 2** NO. 3**	

**Design values are not assigned.

Fig. 1-12 Softwood lumber grades *(Courtesy of Western Wood Products Association)*

Standard Lumber Sizes / Nominal, Dressed, Based on WWPA Rules

Product	Description	Nominal Size		Dressed Dimensions		
				Thicknesses and Widths In.		
		Thickness In.	Width In.	Surfaced Dry	Surfaced Unseasoned	Lengths Ft.
DIMENSION	S4S .... Other surface combinations are available. See "Abbreviations" below.	2 3 4	2 3 4 5 6 8 10 12 Over 12	1-1/2 2-1/2 3-1/2 4-1/2 5-1/2 7-1/4 9-1/4 11-1/4 Off 3/4	1-9/16 2-9/16 3-9/16 4-5/8 5-5/8 7-1/2 9-1/2 11-1/2 Off 1/2	6 ft. and longer in multiples of 1'
SCAFFOLD PLANK	Rough Full Sawn or S4S	1¼ & Thicker	8 and Wider	Same	Same	6 ft. and longer in multiples of 1'
TIMBERS	Rough or S4S	5 and Larger		Thickness In. ½ Off Nominal	Width In.	6 ft. and longer in multiples of 1'

Product	Description	Nominal Size		Dressed Dimensions		
		Thickness In.	Width In.	Thickness In.	Width In.	Lengths Ft.
DECKING	2″ Single T&G	2	5 6 8 10 12	1½	4 5 6¾ 8¾ 10¾	6 ft. and longer in multiples of 1'
	3″ and 4″ Double T&G	3 4	6	2½ 3½	5¼	
FLOORING	(D & M), (S2S & CM)............	3/8 1/2 5/8 1 1¼ 1½	2 3 4 5 6	5/16 7/16 9/16 3/4 1 1¼	1⅛ 2⅛ 3⅛ 4⅛ 5⅛	4 ft. and longer in multiples of 1'
CEILING AND PARTITION	(S2S & CM)	3/8 1/2 5/8 3/4	3 4 5 6	5/16 7/16 9/16 11/16	2⅛ 3⅛ 4⅛ 5⅛	4 ft. and longer in multiples of 1'
FACTORY AND SHOP LUMBER	S2S	1 (4/4) 1¼ (5/4) 1½ (6/4) 1¾ (7/4) 2 (8/4) 2½ (10/4) 3 (12/4) 4 (16/4)	5 and wider except (4″ and wider in 4/4 No. 1 Shop and 4/4 No. 2 Shop)	25/32 (4/4) 1 5/32 (5/4) 1 13/32 (6/4) 1 19/32 (7/4) 1 13/16 (8/4) 2⅜ (10/4) 2¾ (12/4) 3¾ (16/4)	Usually sold random width	4 ft. and longer in multiples of 1'

ABBREVIATIONS

Abbreviated descriptions appearing in the size table are explained below.
S1S — Surfaced one side.
S2S — Surfaced two sides.

S4S — Surfaced four sides.
S1S1E — Surfaced one side, one edge.
S1S2E — Surfaced one side, two edges.
CM — Center matched.

D & M — Dressed and matched.
T & G — Tongue and grooved.
Rough Full Sawn — Unsurfaced green lumber cut to full specified size.

Product Classification

	thickness in.	width in.		thickness in.	width in.
board lumber	1″	2″ or more	beams & stringers	5″ and thicker	more than 2″ greater than thickness
light framing	2″ to 4″	2″ to 4″	posts & timbers	5″ x 5″ and larger	not more than 2″ greater than thickness
studs	2″ to 4″	2″ to 6″ 10′ and shorter	decking	2″ to 4″	4″ to 12″ wide
structural light framing	2″ to 4″	2″ to 4″	siding		thickness expressed by dimension of butt edge
structural joists & planks	2″ to 4″	5″ and wider	mouldings		size at thickest and widest points

Lengths of lumber generally are 6 feet and longer in multiples of 2′

Nailing Diagram

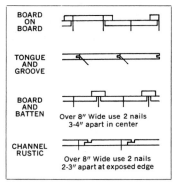

BOARD ON BOARD

TONGUE AND GROOVE

BOARD AND BATTEN — Over 8″ Wide use 2 nails 3-4″ apart in center

CHANNEL RUSTIC — Over 8″ Wide use 2 nails 2-3″ apart at exposed edge

Fig. 1-12 (Continued)

Fig. 1-13 Typical softwood lumber grade stamp *(Courtesy of Western Wood Products Association)*

Figure 1-14 shows the nominal size and actual size of the thickness and width of softwood lumber. Figure 1-15 shows only the nominal and actual thickness of hardwoods because they usually are surfaced on only two sides.

BOARD MEASURE

Softwood lumber is usually purchased by specifying the thickness, width, number of pieces, and length — in addition to the grade. Often, when no particular lengths are required, the thickness, width, and total number of linear feet (length in feet) is ordered. The length of the pieces then may vary and are called *random* lengths. Another method of purchasing softwood lumber is by specifying the thickness, width, and total number of *board feet.* Lumber purchased in this manner may also contain random lengths.

Hardwood lumber is purchased by specifying the grade, thickness, and total number of board feet. Large quantities of both softwood and hardwood lumber are priced and sold by the board foot.

A board foot is a measure of lumber. It is equivalent to a piece 1 inch thick, 12 inches wide, and 1 foot long. A piece of lumber 1 inch thick and 6 inches wide must be two feet long to equal one board foot. A piece two inches thick has twice as many board feet as a piece 1 inch thick of the same width and length.

To calculate the number of board feet use the formula:

number of pieces X thickness in inches X width in inches X length in feet divided by 12 = number of board feet

For example: 16 pieces of 2"x4"x8' equal 85 1/3 board feet.

$$\frac{\overset{4}{\cancel{16}} \times 2 \times 4 \times 8}{\underset{3}{\cancel{12}}} = \frac{256}{3} = 85 \text{ } 1/3 \text{ bd ft}$$

THICKNESS (INCHES)		WIDTH (INCHES)	
NOMINAL	ACTUAL	NOMINAL	ACTUAL
1	3/4	2	1 1/2
1 1/4	1	3	2 1/2
1 1/2	1 1/4	4	3 1/2
2	1 1/2	5	4 1/2
2 1/2	2	6	5 1/2
3	2 1/2	8	7 1/4
3 1/2	3	10	9 1/4
4	3 1/2	12	11 1/4

Fig. 1-14 Softwood lumber sizes

THICKNESS (INCHES)	
NOMINAL	ACTUAL
1/2	5/16
3/4	9/16
1	13/16
1 1/4	1 1/16
1 1/2	1 5/16
2	1 3/4
3	2 3/4
4	3 3/4

Fig. 1-15 Hardwood lumber sizes

REVIEW QUESTIONS

Select the most appropriate answer.

1. The center of a tree is called the
   a. heartwood.   c. pith.
   b. lignin.   d. sapwood.

2. New wood cells of a tree are formed in the
   a. heartwood.   c. medullary rays.
   b. bark.   d. cambium layer.

3. Tree growth is faster in the
   a. spring.   c. fall
   b. summer.   d. winter.

4. Quarter-sawed lumber is sometimes called
   a. tangent grained.   c. vertical grained.
   b. slash-sawed.   d. plain-sawed.

5. Lumber is called "green lumber" when
   a. it is stained by fungi.
   b. the tree is still standing.
   c. it is first cut from the log.
   d. it has reached equilibrium moisture content.

6. Wood will not decay unless its moisture content is in excess of
   a. 15 percent.   c. 25 percent.
   b. 19 percent.   d. 30 percent.

7. When all of the free water in the cell cavities of wood is removed and before water
   is removed from cell walls, lumber is at what is called
   a. fiber saturation point.
   b. 30 percent moisture content.
   c. equilibrium moisture content.
   d. shrinkage commencement.

8. Air-dried lumber cannot be dried less than
   a. 8 to 10 percent moisture content.
   b. equilibrium moisture content.
   c. 25 to 30 percent moisture content.
   d. its fiber saturation point.

9. A commonly used and abundant softwood is
   a. ash.   c. basswood.
   b. fir.   d. birch.

10. One of the woods extremely resistant to decay is
    a. pine.   c. cypress.
    b. spruce.   d. hemlock.

SHEET MATERIALS

OBJECTIVES

After completing this unit, the student will be able to describe the composition, kinds, sizes, grades, and several uses of

- *plywood and composite panels*
- *particleboard*
- *waferboard*
- *strand board*
- *fiberboards*
- *gypsum board*
- *plastic laminates*
- *polyethylene film, asphalt felt, building paper, and sheet metal*

Sheet materials are man-made products. In some cases, the tree has been taken apart and its contents have been redistributed into sheet or panel form. The sheets are widely used in the construction industry. The sheets are also used in the aircraft, automobile, and boat-building industries, as well as in the making of road signs, furniture, and cabinets. Each of the sheet forms mentioned in the text have different characteristics that determine their use.

With the use of sheet materials, construction progresses at a faster rate because a greater area is covered in a shorter period of time, Figure 2-1. The use of sheet material, in certain cases, presents a more attractive appearance and gives more protection to the surface than does the use of solid lumber. It is important to know the various kinds and uses of various sheet materials in order to use them to the best advantage.

Fig. 2-1 Sheet material covers a greater area in a shorter period of time.
(Courtesy of American Plywood Association)

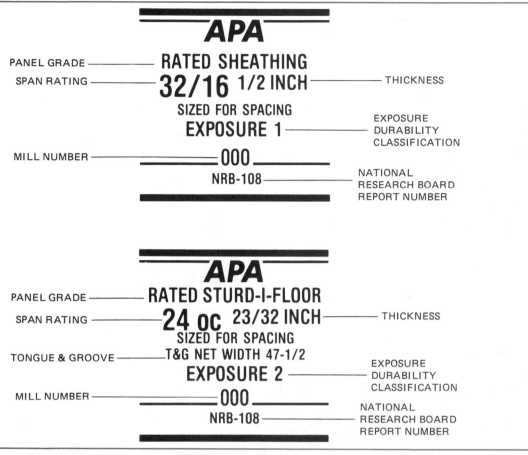

Fig. 2-2 The grade stamp is assurance of a high-quality performance-rated panel.
(Courtesy of American Plywood Association)

Many mills belong to associations that inspect, test, and allow the mill to stamp the product to certify it conforms to government and industrial standards. The grade stamp assures the consumer that the product has met the rigid quality and performance requirements of the association.

APA RATED PANELS

The trademarks of the largest association of this type, The American Plywood Association (APA), appear only on products manufactured by APA member mills, Figure 2-2. This association is concerned not only with quality supervision and testing of *plywood* (cross-laminated wood veneer) but also of *composites* (veneer faces bonded to reconstituted wood cores) and nonveneered panels including *waferboard, oriented strand board,* and certain specific classes of *particleboard*. These sheets are called APA Rated Panels, Figure 2-3.

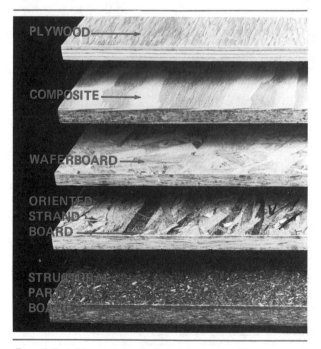

Fig. 2-3 APA performance-rated panels *(Courtesy of American Plywood Association)*

PLYWOOD

One of the most extensively used sheet materials is plywood. Plywood is a sandwich of wood. Most plywood panels are made up of sheets of veneer (thin pieces) called *plies.* These

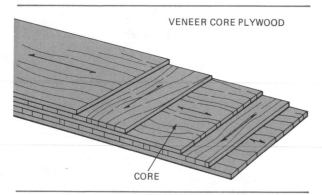

VENEER CORE PLYWOOD

CORE

Fig. 2-4 Plywood construction

Fig. 2-5 The veneer is peeled from the log like unwinding paper from a roll. *(Courtesy of American Plywood Association)*

Fig. 2-6 Gluing and assembling plywood veneers into panels *(Courtesy of American Plywood Association)*

plies are bonded under pressure with glue to form a very strong panel. The plies are glued together so that the grain of each layer is at right angles to the next one. This cross-graining results in a sheet that is as strong or stronger than the wood it is made from. Plywood usually contains an odd number of plies so that the grain on both sides of the sheet runs in the same direction, Figure 2-4. Softwood plywood is commonly made with three, five, or seven plies. Because of its construction, plywood resists shrinking and is more stable with changes of humidity. Plywood is less likely to warp and does not split or check like solid wood.

Manufacture of Veneer Core Plywood. Specially selected ''peeler logs'' are mounted on a huge lathe in which the log is rotated against a sharp knife. As the log turns, a thin layer is peeled off like unwinding paper from a roll, Figure 2-5. The entire log is used. The small remaining spindles are utilized for making other wood products.

The long ribbon of veneer is then cut into desired widths, sorted, and dried to a moisture

VENEER GRADES

N — Smooth surface ''natural finish'' veneer. Select, all heartwood or all sapwood. Free of open defects. Allows not more than 6 repairs, wood only, per 4x8 panel, made parallel to grain and well matched for grain and color.

A — Smooth, paintable. Not more than 18 neatly made repairs, boat, sled, or router type, and parallel to grain, permitted. May be used for natural finish in less demanding applications.

B — Solid surface. Shims, circular repair plugs and tight knots to 1 inch across grain permitted. Some minor splits permitted.

C Plugged — Improved C veneer with splits limited to 1/8 inch width and knotholes and borer holes limited to 1/4 x 1/2 inch. Admits some broken grain. Synthetic repairs permitted.

C — Tight knots to 1-1/2 inch. Knotholes to 1 inch across grain and some to 1-1/2 inch if total width of knots and knotholes is within specified limits. Synthetic or wood repairs. Discoloration and sanding defects that do not impair strength permitted. Limited splits allowed. Stitching permitted.

D — Knots and knotholes to 2-1/2 inch width across grain and 1/2 inch larger within specified limits. Limited splits allowed. Stitching permitted. Limited to Interior, Exposure 1 and Exposure 2 panels.

Fig. 2-7 Veneer grades *(Courtesy of American Plywood Association)*

content of 5 percent. After drying, the veneers are fed through glue spreaders that coat them with a uniform thickness. The veneers are then assembled to make panels, Figure 2-6. Large presses bond the assembly under controlled heat and pressure. From the presses, the panels are either left unsanded, touch-sanded or smooth-sanded, cut to size, inspected, and stamped.

Veneer Grades. In declining order, the letters *N, A, B, C plugged, C,* and *D* are used to indicate the quality of panel veneers. Two letters are found in the grade-stamp of veneered panels. One letter indicates the quality of one face, while the other letter indicates the quality of the opposite face. The exact description of these letter grades is shown in Figure 2-7. Panels with B-grade or better veneer faces are always sanded smooth. Some panels, such as APA Rated Sheathing are unsanded because their intended use does not require sanding. Other panels used for such purposes as subflooring and underlayment, require only a touch-sanding to make the panel thickness more uniform.

Strength Grades. Softwood veneers are made of many different kinds of wood. These woods are classified in groups according to their strength, Figure 2-8. Group 1 is the strongest. Douglas fir is in Group 1 and is used to make most of the softwood plywood. The group number is also shown in the grade-stamp.

CLASSIFICATION OF SPECIES				
Group 1	**Group 2**	**Group 3**	**Group 4**	**Group 5**

Apitong
Beech,
  American
Birch
  Sweet
  Yellow
Douglas
  Fir 1[a]
Kapur
Keruing
Larch,
  Western
Maple, Sugar
Pine
  Caribbean
  Ocote
Pine, South
  Loblolly
  Longleaf
  Shortleaf
  Slash
Tanoak
	Cedar, Port
	  Orford
	Cypress
	Douglas
	  Fir 2[a]
	Fir
	  Balsam
	  California
	  Red
	  Grand
	  Noble
	  Pacific
	  Silver
	  White
	Hemlock,
	  Western
	Lauan
	  Almon
	  Bagtikan
	  Mayapis
	  Red
	  Tangile
	  White
	Maple, Black
	Mengkulang
	Meranti,
	  Red[b]
	Mersawa
	Pine
	  Pond
	  Red
	  Virginia
	  Western
	  White
	Spruce
	  Black
	  Red
	  Sitka
	Sweetgum
	Tamarack
	Yellow-
	  Poplar
		Alder, Red
		Birch, Paper
		Cedar, Alaska
		Fir,
		  Subalpine
		Hemlock,
		  Eastern
		Maple
		  Bigleaf
		Pine
		  Jack
		  Lodgepole
		  Ponderosa
		  Spruce
		Redwood
		Spruce
		  Engelmann
		  White
			Aspen
			  Bigtooth
			  Quaking
			Cativo
			Cedar
			  Incense
			  Western
			  Red
			Cottonwood
			  Eastern
			  Black
			  (Western
			  Poplar)
			Pine
			  Eastern
			  White
			  Sugar
				Basswood
				Poplar,
				  Balsam

(a) Douglas Fir from trees grown in the states of Washington, Oregon, California, Idaho, Montana, Wyoming, and the Canadian Provinces of Alberta and British Columbia shall be classed as Douglas Fir No. 1. Douglas Fir from trees grown in the states of Nevada, Utah, Colorado, Arizona and New Mexico shall be classed as Douglas Fir No. 2.

(b) Red Meranti shall be limited to species having a specific gravity of 0.41 or more based on green volume and oven dry weight.

Fig. 2-8 Plywood is classified in groups according to strength and stiffness.
(Courtesy of American Plywood Association)

BRUSHED

Brushed or relief-grain surfaces accent the natural grain pattern to create striking textured surfaces. Generally available in 11/32", 3/8", 1/2", 19/32" and 5/8" thicknesses. Available in redwood, Douglas fir, cedar, and other species.

KERFED ROUGH-SAWN

Rough-sawn surface with narrow grooves providing a distinctive effect. Long edges shiplapped for continuous pattern. Grooves are typically 4" o.c. Also available with grooves in multiples of 2" o.c. Generally available in 11/32", 3/8", 1/2", 19/32" and 5/8" thicknesses. Depth of kerfgroove varies with panel thickness.

APA TEXTURE 1-11

Special 303 Siding panel with shiplapped edges and parallel grooves 1/4" deep, 3/8" wide; grooves 4" or 8" o.c. are standard. Other spacings sometimes available are 2", 6" and 12" o.c., check local availability. T 1-11 is generally available in 19/32" and 5/8" thicknesses. Also available with scratch-sanded, overlaid, rough-sawn, brushed and other surfaces. Available in Douglas fir, cedar, redwood, southern pine, other species.

ROUGH-SAWN

Manufactured with a slight, rough-sawn texture running across panel. Available without grooves, or with grooves of various styles; in lap sidings, as well as in panel form. Generally available in 11/32", 3/8", 1/2", 19/32" and 5/8" thicknesses. Rough-sawn also available in Texture 1-11, reverse board-and-batten (5/8" thick), channel groove (3/8" thick), and V-groove (1/2" or 5/8" thick). Available in Douglas fir, redwood, cedar, southern pine, other species.

CHANNEL GROOVE

Shallow grooves typically 1/16" deep, 3/8" wide, cut into faces of 3/8" thick panels, 4" or 8" o.c. Other groove spacings available. Shiplapped for continuous patterns. Generally available in surface patterns and textures similar to Texture 1-11 and in 11/32", 3/8" and 1/2" thicknesses. Available in redwood, Douglas fir, cedar, southern pine and other species.

REVERSE BOARD-AND-BATTEN

Deep, wide grooves cut into brushed, roughsawn, coarse sanded or other textured surfaces. Grooves about 1/4" deep, 1" to 1-1/2" wide, spaced 8", 12" or 16" o.c. with panel thickness of 19/32" and 5/8". Provides deep, sharp shadow lines. Long edges shiplapped for continuous pattern. Available in redwood, cedar, Douglas fir, southern pine and other species.

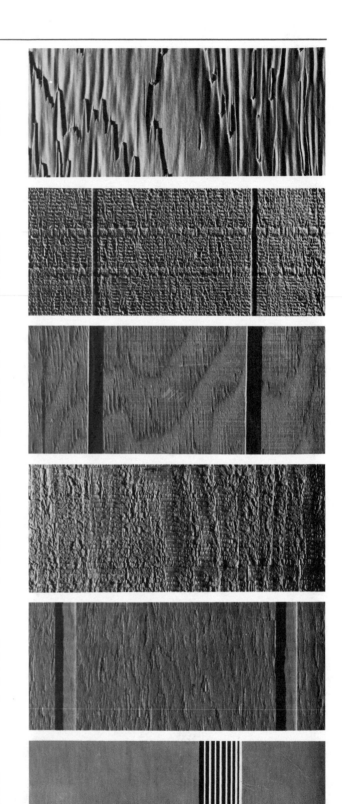

Fig. 2-9 303 Siding surface patterns and textures *(Courtesy of American Plywood Association)*

Fig. 2-10 Applying texture 1-11 siding

APA 303 SIDING

APA 303 plywood siding products are manufactured in a number of special surface treatments such as V-groove, channel groove, striated, brushed, rough sawed and texture-embossed, Figures 2-9 and 2-10. This product is divided into four basic classes: special series 303, 303-6, 303-18, and 303-30. Each class is further divided into grades according to the number of repairs permitted, Figure 2-11.

PARTICLEBOARD

Particleboard is a reconstituted wood panel made of wood flakes, chips, sawdust, and planer shavings, Figure 2-12. These wood particles are mixed with an adhesive, formed in a mat, and pressed into sheet form. The kind, size, and arrangement of the wood particles determine the quality of the board.

The highest quality particleboard is made up of large wood flakes in the center. The flakes become gradually smaller toward the surfaces where the finer particles are found. This type of construction results in an extremely hard board with a very smooth surface. Softer and lower quality boards contain the same size particles throughout. These boards usually have a rougher

		Patches	
303 SIDING FACE GRADES			
Class	Grade[1]	Wood	Synthetic
Special Series 303	303-OC[2][3] 303-OL[4] 303-NR[5] 303-SR[6]	Not permitted Not applicable Not permitted Not permitted	Not permitted for overlays Not permitted Permitted as natural-defect shape only
303-6	303-6-W 303-6-S 303-6-S/W	Limit 6 Not permitted Limit 6 — any combination	Not permitted Limit 6
303-18	303-18-W 303-18-S 303-18-S/W	Limit 18 Not permitted Limit 18 — any combination	Not permitted Limit 18
303-30	303-30-W 303-30-S 303-30-S/W	Limit 30 Not permitted Limit 30 — any combination	Not permitted Limit 30

(1) Limitations on grade characteristics are based on 4 ft. x 8 ft. panel size. Limits on other sizes vary in proportion. All panels except 303-NR allow restricted minor repairs such as shims. These and such other face appearance characteristics as knots, knotholes, splits, etc., are limited by both size and number in accordance with panel grades, 303-OC being most restrictive and 303-30 being least. Multiple repairs are permitted only on 303-18 and 303-30 panels. Patch size is restricted on all panel grades. For additional information, including recommendations, see APA Product Guide: 303 Plywood Siding, E300.

(2) Check local availability
(3) "Clear"
(4) "Overlaid" (e.g. Medium Density Overlay siding)
(5) "Natural Rustic"
(6) "Synthetic Rustic"

Fig. 2-11 APA 303 Siding face grades *(Courtesy of American Plywood Association)*

Fig. 2-12 Particleboard is made from wood flakes, shavings, resins, and waxes. *(Courtesy of Duraflake Division, Williamette Industries, Inc.)*

surface texture. In addition to the size, kind, and arrangement of the particles, the quality of the board is determined by the method of manufacture.

The quality of particleboard is indicated by its density (hardness) which ranges from 28 to 55 lbs., per cubic foot. Rated particleboard panels may be used for wall sheathing, roof sheathing and subflooring, if so designated. Non-structural particleboard is used in the construction industry as an underlayment for finish floors, for the construction of kitchen cabinets and countertops, and for the core of veneer doors and similar panels, Figure 2-13. *Duraflake, Novoply,* and *Tuf-Flake* are some brand names for particleboard.

WAFERBOARD

Another sheet material currently being rated for structural applications is *waferboard.* Commonly called *Aspenite,* it is a nonveneered panel of compressed waferlike wood particles or flakes randomly or directionally placed. It is manufactured in a manner similar to that used for particleboard and is now being used for

sheathing and subflooring, Figure 2-14. Non-rated waferboard can be used for the same purposes as nonrated particleboard.

ORIENTED STRAND BOARD

Oriented strand board is nonveneered panel composed of small strand-like wood pieces arranged in layers (usually three to five) with each layer oriented at right angles to the others. This construction makes a stronger board, has uses similar to other reconstituted panels, and is also rated for performance standards in construction applications.

SPAN RATINGS

The span rating in the grade stamp on APA Rated Sheathing appears as two numbers separated by a slash, such as 32/16 or 48/24. The left hand number denotes the maximum recommended spacing of supports when the panel is used for roof sheathing. The long dimension of the panel is placed across three or more supports. The right hand number indicates the maximum

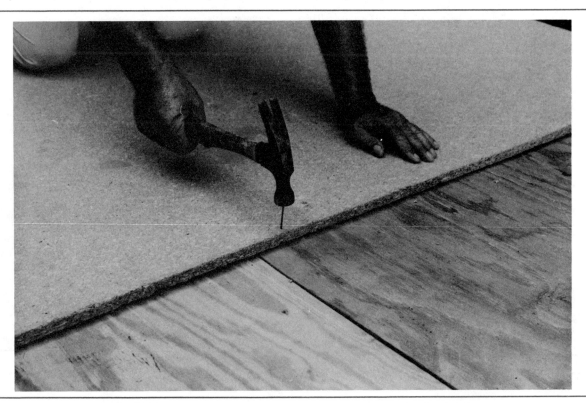

Fig. 2-13 Particleboard is used extensively as underlayment for finish floors.

Fig. 2-14 Waferboard being used for wall sheathing *(Photo by Robert Morency)*

recommended spacing of supports when the panel is used for subflooring. The long dimension of the panel is placed across three or more supports. A panel marked 32/16, for example, may be used for roof sheathing over rafters not more than 32 inches on center or for subflooring over joists not more than 16 inches on center.

The span ratings on APA Rated Sturd-I-Floor and 303 Siding appear as a single number. APA Rated Sturd-I-Floor panels are designed specifically for combined subfloor-underlayment applications and are manufactured with span ratings of 16, 20, 24, and 48 inches. The span rating is the maximum spacing of joists based on the application of the panel with the long dimension across three or more joists.

APA 303 sidings are produced with span ratings of 16 and 24 inches. The rating applies to vertical installation of the panel. All 303 siding panels may be applied horizontally direct to studs 16 to 24 inches on center provided horizontal joints are blocked.

EXPOSURE DURABILITY

APA performance-rated panels can be manufactured in three exposure durability classifica-
tions: *Exterior, Exposure 1,* and *Exposure 2.* Panels marked Exterior are designed for continuous exposure to the weather or moisture. Panels marked as Exposure 1 are intended for use for protected construction applications where ability to resist moisture during construction delays is required. Panels marked Exposure 2 are designed for use when moderate delays in providing protection from the weather is expected. The exposure durability of a panel may be found in the grade-stamp.

Plywood is produced in two basic types. *Exterior* type has 100 percent waterproof glueline. *Interior* type may be manufactured with exterior, intermediate, or interior glue. *Interior type* panels with exterior glue are suitable for application where ability to resist moisture during long construction delays is required. However, because the lower grade of veneer permitted for backs and inner plies of interior type panels may affect glueline performance, only Exterior type plywood should be used for permanent exposure to moisture.

To select the size, kind, and grade of APA performance-rated panels and other APA panel products, consult the selection guides in Figure 2-15. Three guides are furnished for your use.

GUIDE TO APA PERFORMANCE-RATED PANELS[1] [2]

	Grade Designation	Description & Common Uses	Typical Trademarks	Most Common Thicknesses (in.)				
				5/16	3/8	1/2	5/8	3/4
PROTECTED OR INTERIOR USE	**APA RATED SHEATHING EXP 1 or 2**	Specially designed for subflooring and wall and roof sheathing, but can also be used for a broad range of other construction and industrial applications. Can be manufactured as conventional veneered plywood, as a composite, or as a nonveneered panel. For special engineered applications, including high load requirements and certain industrial uses, veneered panels conforming to PS 1 may be required. Specify Exposure 1 when construction delays are anticipated.	APA RATED SHEATHING 32/16 1/2 INCH SIZED FOR SPACING EXPOSURE 1 000 PS 1-74 C-D INT/EXT GLUE NRB-108	●	●	●	●	●
	APA STRUCTURAL I & II RATED SHEATHING EXP 1	Unsanded all-veneer PS 1 plywood grades for use where strength properties are of maximum importance: structural diaphragms, box beams, gusset plates, stressed-skin panels, containers, pallet bins. Made only with exterior glue (Exposure 1). STRUCTURAL I more commonly available. (3)	APA RATED SHEATHING STRUCTURAL I 24/0 3/8 INCH SIZED FOR SPACING EXPOSURE 1 000 PS 1-74 C-D INT/EXT GLUE NRB-108	●	●	●	●	●
	APA RATED STURD-I-FLOOR EXP 1 or 2	For combination subfloor-underlayment. Provides smooth surface for application of resilient floor covering and possesses high concentrated and impact load resistance. Can be manufactured as conventional veneered plywood, as a composite, or as a nonveneered panel. Available square edge or tongue-and-groove. Specify Exposure 1 when construction delays are anticipated.	APA RATED STURD-I-FLOOR 24 oc 23/32 INCH SIZED FOR SPACING T&G NET WIDTH 47-1/2 EXPOSURE 1 000 INT/EXT GLUE NRB-108 FHA-UM-66				● 19/32	● 23/32
	APA RATED STURD-I-FLOOR 48 oc (2-4-1) EXP 1	For combination subfloor-underlayment on 32- and 48-inch spans and for heavy timber roof construction. Provides smooth surface for application of resilient floor coverings and possesses high concentrated and impact load resistance. Manufactured only as conventional veneered plywood and only with exterior glue (Exposure 1). Available square edge or tongue-and-groove.	APA RATED STURD-I-FLOOR 48 oc 1-1/8 INCH SIZED FOR SPACING EXPOSURE 1 T&G 000 INT/EXT GLUE NRB-108 FHA-UM-66	1-1/8				
EXTERIOR USE	**APA RATED SHEATHING EXT**	Exterior sheathing panel subflooring and wall and roof sheathing, siding on service and farm buildings, crating, pallets, pallet bins, cable reels, etc. Manufactured as conventional veneered plywood.	APA RATED SHEATHING 48/24 3/4 INCH SIZED FOR SPACING EXTERIOR 000 PS 1-74 C-C NRB-108	●	●	●	●	●
	APA STRUCTURE-I & II RATED SHEATHING EXT	For engineered applications in construction and industry where fully waterproof panels are required. Manufactured only as conventional veneered PS 1 plywood. Unsanded. STRUCTURAL I more commonly available. (3)	APA RATED SHEATHING STRUCTURAL I 24/0 3/8 INCH SIZED FOR SPACING EXTERIOR 000 PS 1-74 C-C NRB-108	●	●	●	●	
	APA RATED STURD-I-FLOOR EXT	For combination subfloor-underlayment, under resilient floor coverings where severe moisture conditions may be present, as in balcony decks. Possesses high concentrated and impact load resistance. Manufactured only as conventional veneered plywood. Available square edge or tongue-and-groove.	APA RATED STURD-I-FLOOR 20 oc 19/32 INCH SIZED FOR SPACING EXTERIOR 000 NRB-108 FHA-UM-66				● 19/32	● 23/32

(1) Specific grades, thicknesses, constructions and exposure durability classifications may be in limited supply in some areas. Check with your supplier before specifying.
(2) Specify Performance-Rated Panels by thickness and Span Rating.
(3) All plies in STRUCTURAL I panels are special improved grades and limited to Group 1 species. All plies in STRUCTURAL II panels are special improved grades and limited to Group 1, 2, or 3 species.

Fig. 2-15 Guides for panel selection *(Courtesy of American Plywood Association)*

GUIDE TO OTHER APA PANEL PRODUCTS/PROTECTED OR INTERIOR USE[1]

Grade Designation	Description & Common Uses	Typical Trademarks	Veneer Grade			Most Common Thicknesses (in.)					
			Face	Inner Plies	Back	1/4	5/16	3/8	1/2	5/8	3/4
APA N-N, N-A, N-B INT	Cabinet quality. For natural finish furniture, cabinet doors, built-ins, etc. Special order items. (2)	N-N G-1 INT·APA PS1-74 000	N	C	N,A, or B						●
APA N-D INT	For natural finish paneling. Special order item. (2)	N-D G-2 INT·APA PS1-74 000	N	D	D	●					
APA A-A INT	For applications with both sides on view: built-ins, cabinets, furniture, partitions. Smooth face, suitable for painting. (2)	A-A G-1 INT·APA PS1-74 000	A	D	A	●		●	●	●	●
APA A-B INT	Use where appearance of one side is less important but where two solid surfaces are necessary. (2)	A-B G-1 INT·APA PS1-74 000	A	D	B	●		●	●	●	●
APA A-D INT	Use where appearance of only one side is important: paneling, built-ins, shelving, partitions, flow racks. (2)	APA A-D GROUP 1 INTERIOR 000 PS 1-74 EXTERIOR GLUE	A	D	D	●		●	●	●	●
APA B-B INT	Utility panel with two solid sides. Permits circular plugs. (2)	B-B G-2 INT·APA PS1-74 000	B	D	B	●		●	●	●	●
APA B-D INT	Utility panel with one solid side. Good for backing, sides of built-ins, industry shelving, slip sheets, separator boards, bins. (2)	APA B-D GROUP 2 INTERIOR 000 PS 1-74 EXTERIOR GLUE	B	D	D	●		●	●	●	●
APA UNDERLAYMENT INT	For application over structural subfloor. Provides smooth surface for application of resilient floor coverings. Touch-sanded. Also available with exterior glue. (3)	APA UNDERLAYMENT GROUP 1 INTERIOR 000 PS 1-74 EXTERIOR GLUE	C Plgd.	C & D	D				●	● 19/32	● 23/32
APA C-D PLUGGED INT	For built-ins, wall and ceiling tile backing, cable reels, walkways, separator boards. Not a substitute for UNDERLAYMENT or STURD-I-FLOOR as it lacks their indentation resistance. Touch-sanded. Also made with exterior glue. (3)	APA C-D PLUGGED GROUP 2 INTERIOR 000 PS 1-74 EXTERIOR GLUE	C Plgd.	D	D				●	● 19/32	● 23/32
APA DECORATIVE INT	Rough-sawn, brushed, grooved, or striated faces. For paneling, interior accent walls, built-ins, counter facing, display exhibits. (5)	APA DECORATIVE GROUP 4 INTERIOR 000	C or btr.	D	D		●	●	●	●	
APA PLYRON INT	Hardboard face on both sides. For counter-tops, shelving, cabinet doors, flooring. Faces tempered, untempered, smooth or screened.	PLYRON ·INT·APA 000	C & D						●	●	●

(1) Specific grades and thicknesses may be in limited supply in some areas. Check with your supplier before specifying.
(2) Sanded both sides.
(3) Can also be manufactured in STRUCTURAL I (all plies limited to Group 1 species) and STRUCTURAL II (all plies limited to Group 1, 2, or 3 species).
(4) C or better for 5 plies. C Plugged or better for 3 and 4 plies.
(5) Can also be made by some manufacturers in Exterior for exterior siding, gable ends, fences, and other exterior applications. Use recommendations for Exterior Decorative panels vary with the particular product. Check with the manufacturer for specific application recommendations.

Fig. 2-15 (Continued)

23

GUIDE TO OTHER APA PANEL PRODUCTS/EXTERIOR USE[1]

Grade Designation	Description & Common Uses	Typical Trademarks	Veneer Grade			Most Common Thicknesses (in.)					
			Face	Inner Plies	Back	1/4	5/16	3/8	1/2	5/8	3/4
APA A-A EXT	Use where appearance of both sides is important: fences, built-ins, signs, boats, cabinets, commercial refrigerators, shipping containers, tote boxes, tanks, ducts. (2) (3)	A-A G-1 EXT-APA PS1-74 000	A	C	A	●		●	●	●	●
APA A-B EXT	Use where the appearance of one side is less important. (2) (3)	A-B G-1 EXT-APA PS1-74 000	A	C	B	●		●	●	●	●
APA A-C EXT	Use where the appearance of only side is important: soffits, fences, structural uses, boxcar and truck linings, farm buildings, tanks, trays, commercial refrigerators. (2) (3)	APA A-C GROUP 1 EXTERIOR 000 PS1-74	A	C	C			●	●	●	●
APA B-B EXT	Utility panel with solid faces. (2) (3)	B-B G-2 EXT-APA PS1-74 000	B	C	B	●		●	●	●	●
APA B-C EXT	Utility panel for farm service and work buildings, boxcar and truck linings, containers, tanks, agricultural equipment. Also as a base for exterior coatings for walls, roofs. (2) (3)	APA B-C GROUP 1 EXTERIOR 000 PS1-74	B	C	C	●		●	●	●	●
APA UNDERLAYMENT C-C PLUGGED EXT	For application over structural subfloor. Provides smooth surface for application of resilent floor coverings where severe moisture conditions may be present. Touch-sanded. (3)	APA UNDERLAYMENT C-C PLUGGED GROUP 2 EXTERIOR 000 PS1-74	C Plgd.	C	C				●	● 19/32	● 23/32
APA C-C PLUGGED EXT	For use as tile backing where severe moisture conditions exist. For refrigerated or controlled atmosphere rooms, pallet fruit bins, tanks, boxcar and truck floors and linings, open soffits. Touch-sanded. (3)	APA C-C PLUGGED GROUP 2 EXTERIOR 000 PS1-74	C Plgd.	C	C				●	● 19/32	● 23/32
APA HDO EXT	High Density Overlay. Has a hard semi-opaque resin-fiber overlay both faces. Abrasion resistant. For concrete forms, cabinets, countertops, signs, tanks. Also available with skid-resistant screen-grid surface. (3)	HDO A-A G-1 EXT-APA PS1-74 000	A or B	C or C Pldg.	A or B			●	●	●	●
APA MDO EXT	Medium Density Overlay. Smooth, opaque, resin-fiber overlay one or both faces. Ideal base for paint, both indoors and outdoors. Also available as a 303 Siding. (3)	MDO B-B G-2 EXT-APA PS1-74 000	B	C	B or C			●	●	●	●
APA MARINE EXT	Ideal for boat hulls. Made only with Douglas fir or western larch. Special solid jointed core construction. Subject to special limitations on core gaps and number of face repairs. Also available with HDO or MDO faces.	MARINE A-A EXT-APA PS1-74 000	A or B	B	A or B	●		●	●	●	●
APA PLYRON EXT	Hardboard faces both sides, tempered, smooth or screened.	PLYRON EXT-APA 000		C					●	●	●
APA 303 SIDING EXT	Proprietary plywood products for exterior siding, fencing, etc. Special surface treatment such as a V-groove, channel groove, striated, brushed, rough-sawn and texture-embossed (MDO). Stud spacing (Span Rating) and face grade classification indicated in trademark.	APA 303 SIDING 18-S/W 24 OC 23/32 INCH GROUP 1 EXTERIOR 000 PS1-74 FHA UM-64	(4)	C	C			● 11/32	● 15/32	● 19/32	
APA T 1-11 EXT	Special 303 panel having grooves 1/4" deep, 3/8" wide, spaced 4" or 8" o.c. Other spacing optional. Edges shiplapped. Available unsanded, textured and MDO.	APA 303 SIDING 6-S/W 16 OC 19/32 INCH GROUP 2 EXTERIOR 000 T-1-11 PS1-74 FHA-UM-64	C or btr.	C	C					● 19/32	
APA B-B PLYFORM CLASS I and CLASS II EXT	Concrete form grades with high reuse factor. Sanded both sides and mill-oiled unless otherwise specified. Special restrictions on species. Class I panels are stiffest, strongest and most commonly available. Also available in HDO for very smooth concrete finish, in STRUCTURAL I (all plies limited to Group 1 species) and with special overlays.	APA PLYFORM B-B CLASS I EXTERIOR 000 PS1-74	B	C	B					●	●

Fig. 2-15 (Continued)

HARDWOOD FACE VENEER

Plywood is also available with hardwood face veneers. Beautifully grained hardwoods are sometimes matched in a number of ways to produce interesting face designs. Hardwood plywood is used in the interior of buildings for such things as wall paneling, built-in cabinets, and fixtures. Much of this kind of plywood is manufactured with a lumber core, Figure 2-16.

FIBERBOARDS

Fiberboards are manufactured as *high density, medium density* and *low density* boards.

HARDBOARDS

High density fiberboards are called hardboards and are commonly known by the trademark, *Masonite*. The hardboard industry makes almost a complete use of the great natural resource of wood by utilizing the wood chips and board trimmings, which were once considered waste, in addition to using logs in the manufacture of hardboards.

Wood chips are reduced to fibers and water is added to make a soupy pulp. The soupy pulp flows onto a traveling mesh screen where water is drawn off to form a mat. The mat is then pressed under heat to weld the wood fibers back together by utilizing lignin, the natural adhesive in wood.

Some panels are *tempered* (coated with oil and baked to increase hardness, strength, and water resistance). Carbide-tipped saws trim the panels to standard sizes.

Sizes of Hardboard. The most popular thicknesses of hardboard range from 1/8 inch to 3/8 inch. The most popular sheet size is 4 feet by 8 feet, although sheets may be ordered in practically any size.

Classes and Kinds of Hardboard. Hardboard is available in five different classes: tempered, standard, service-tempered, service, and industrialite, Figure 2-17. It may be obtained smooth-one-side (S1S) or smooth-two-sides

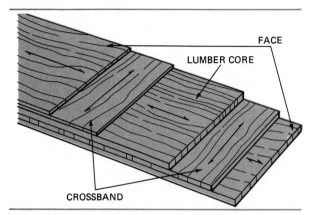

Fig. 2-16 Construction of lumber-core plywood

Class	Surface	Nominal thickness
		inch
	S1S	1/12
1 Tempered	S1S and S2S	1/10 1/8 3/16 1/4 5/16 3/8
2 Standard	S1S and S2S	1/12 1/10 1/8 3/16 1/4 5/16 3/8
3 Service-tempered	S1S and S2S	1/8 3/16 1/4 3/8
4 Service	S1S and S2S	1/8 3/16 1/4 3/8 7/16 1/2
	S2S	5/8 11/16 3/4 13/16 7/8 1 1-1/8
5 Industrialite	S1S and S2S	3/8 7/16 1/2
	S2S	5/8 11/16 3/4 13/16 7/8 1 1-1/8

Fig. 2-17 Kinds and thicknesses of hardboard (*Courtesy of Masonite Corporation*)

(S2S). Hardboard is available in many forms, such as perforated, grooved, and striated.

USES OF HARDBOARD

Hardboard may be used inside or outside. It is widely used for exterior siding and interior wall paneling. It is also used extensively for cabinet backs and drawer bottoms. It can be used wherever a dense, hard panel is required, Figure 2-18.

Because it is a wood-base product, hardboard can be sawed, routed, shaped, and drilled with standard woodworking tools. It can be securely fastened with glue, screws, staples, or nails.

Hardboard products are manufactured to comply with U.S. Department of Commerce Voluntary Product Standards. Most hardboard producers belong to the American Hardboard Association.

MEDIUM DENSITY FIBERBOARD

Commonly called *MDF,* medium density fiberboard is known by such names as *Medite, Baraboard* and *Fibrepine* — among others. It is manufactured in a manner similar to that used to make hardboard except that the fibers are not pressed so tightly together. The refined fiber produces a fine-textured and homogenous board

with an exceptionally smooth surface. Densities range from 28 to 65 pounds. It is available in thicknesses ranging from 3/16 inch to 1 1/2 inches and comes in widths of 4 feet and 5 feet. Lengths run from 6 feet to 18 feet.

MDF may be used for case goods, drawer parts, kitchen cabinets, cabinet doors, signs, and for some interior wall finish.

LOW DENSITY FIBERBOARD

Low density fiberboard is called *softboard.* Common brand names include *Temlok, Celotex,* and *Baracore.* Softboard is very light and contains many tiny air spaces because the particles are not compressed tightly.

The most common thicknesses range from 1/2 inch to 1 inch. The most common sheet size is 4 feet by 8 feet, although many sizes are available.

USES OF SOFTBOARD

Because of its lightness, softboard panels are used primarily for insulating or sound control purposes. They are used extensively as decorative ceiling panels in suspended ceilings and as ceiling tiles, Figure 2-19. Much use is made of softboard for exterior wall sheathing. This type may be coated or impregnated with

Fig. 2-18 Hardboard is widely used for exterior siding and interior paneling. *(Courtesy of Masonite Corporation)*

Fig. 2-19 Softboards are used extensively for decorative ceiling panels. *(Courtesy of Armstrong Company)*

asphalt to protect it from moisture during construction.

Because of its softness, panels can be easily cut with a knife, hand saw, or power saw. They cannot be hand-planed with any satisfactory results. Wide-headed nails, staples, or adhesive are used to fasten softboards in place.

GYPSUM BOARD

Gypsum board is made from gypsum rock, which contains a mineral that will not burn. Gypsum boards make fire-resistant construction; most types must be protected from moisture, however.

Gypsum rock is heated to form a powder and is mixed with water. The mixture is then encased in heavy paper to form a dry stiff board. The paper is folded along the long edges; the ends are cut square and finished smooth. Some types of gypsum board are made with a water-resistant core and covering for use around showers and bathtubs.

Common thicknesses are 3/8 inch, 1/2 inch, and 5/8 inch. *Wallboard* (another name for gypsum board) comes in a four-foot width only and is available in 8-, 9-, 10-, 12-, and 14-foot lengths. Edges may be rounded or tapered.

USES OF GYPSUM BOARD

Primary use of gypsum board is for interior wall and ceiling covering. (See Unit 22 — Dry-wall.) Used in this manner, it is commonly called *dry-wall construction,* Figure 2-20. Gypsum board is also used extensively as a base for plaster and is commonly called *rock lath.* Special types of gypsum board are used on the exterior for roof and wall covering.

PLASTIC LAMINATES

Plastic laminate is a very thin, tough material used primarily as a decorative covering for kitchen and bath cabinets and countertops, Figure 2-21. The material is commonly known as "formica" although Formica is a registered trademark of that company. Plastic laminates are manufactured by a number of companies under

Fig. 2-20 Gypsum wallboard provides high fire-resistant wall and ceiling covering. *(Courtesy of Georgia Pacific Corporation)*

Fig. 2-21 Plastic laminates are used primarily to cover cabinets and countertops. *(Courtesy of Rockwell, International)*

such names as Micarta, Melamite, Pionite, Wilson Art, and others. They are manufactured in a great number of colors and designs, including many wood-grain patterns.

A base of several layers of resin-impregnated kraft paper, a saturated printed sheet, and a transparent top sheet are bonded together under extremely high pressure and temperature. This process produces a decorative sheet of very thin, hard, smooth material which is highly resistant to wear and scratching. It is unharmed by boiling water, fruit juices, alcohol, oil, grease, and most household cleaners. It withstands heat up to 275°F.

Plastic laminates are available in widths of 24 inches to 60 inches. Lengths range from 6 feet to 12 feet. Actual sizes are 3/8 inch to 1 inch greater than the nominal size.

The most widely used laminate is called general purpose grade and is 1/16 inch thick. Vertical surface type is 0.032 inch thick and is used only on vertical surfaces. A thinner type is called a ''backer'' or cabinet liner. This type is used to cover the insides of cabinets or the backs of cabinet doors.

POLYETHYLENE

Polyethylene is a thin plastic film and is used as a vapor barrier to prevent the penetration of moisture. It is used on the inside of exterior walls before the wall finish is applied, Figure 2-22. It is also used under concrete slab floors and in crawl spaces beneath the building. It is available in thicknesses of 2 to 4 mils and comes in rolls of various widths and lengths. A common tradename is Visqueen.

BUILDING PAPER AND ASPHALT FELT

Building paper comes in rolls 36 inches wide and is used mainly under exterior siding and finish floors. The paper is fairly heavy and is treated to give it more strength. It is porous, so it is used when the passage of vapor through it is not a consideration.

Fig. 2-22 Polyethylene film applied to the inside surface of an exterior wall acts as a vapor barrier. *(Photo by Robert Morency)*

Asphalt felt is similar to building paper except it is saturated with asphalt to give it more resistance to moisture. It also comes in rolls 36 inches wide. The amount contained in the roll depends on the weight of the felt. Felts are available in 15-, 30-, and 60-pound weights, per hundred square feet.

Asphalt felts are used beneath roof coverings, with some heavier felts used as roof covering alone, Figure 2-23. It is also used anywhere a moisture-resistant paper covering is desired.

METALS

Much metal is used in the construction industry for flashing. *Flashing* is thin sheet metal that is placed around chimneys, windows, and doors; and in roof valleys, and other places to stop the entrance of water in the opening, Figure 2-24.

Fig. 2-23 Asphalt felt being applied to a roof deck

Fig. 2-24 Flashing is usually thin metal placed at inter- sections to prevent the entrance of water. *(Photo by Robert Morency)*

Common metals used are aluminum, zinc, and copper. Copper is the longest lasting of all three.

The metal usually comes in rolls of varying widths, from 6 to 12 inches and upward. Some- times asphalt felt is used for flashing instead of metal.

OTHER

There are many more sheet and other ma- terials used in the construction industry other than those mentioned in this and following units. It is recommended that the student study *Sweet's Architectural File,* a well-known refer- ence for architects published annually by McGraw-Hill Publishing Company, to become better acquainted with the thousands of building material products on the market.

REVIEW QUESTIONS

Select the most appropriate answer.

1. Plywood always contains
   a. three layers.
   b. four layers.
   c. an even number of layers.
   d. an odd number of layers.

2. Most of the softwood plywood is made of
   a. cedar.
   b. fir.
   c. pine.
   d. spruce.

3. The best appearing face veneer of a softwood plywood panel is indicated by the letter
   a. A
   b. B
   c. E
   d. N

4. Which is a good selection of plywood for exterior wall sheathing?
   a. APA structural rated sheathing EXP1
   b. APA A-C EXT
   c. APA Rated Sheathing Exp 1 or 2
   d. CD, Plugged, EXT

5. A number "0" on the right of the slash in the Identification Index of Plywood means
   a. there are no knots in the inner plies.
   b. it should not be used for subfloor.
   c. it should not be used for roof sheathing.
   d. it should not be used for any structural purpose.

6. Particleboard not rated as structural may be used for
   a. underlayment.
   b. subflooring.
   c. wall sheathing.
   d. roof sheathing.

7. Hardboard may be used in
   a. interior applications only.
   b. exterior and interior applications.
   c. applications protected from moisture.
   d. cabinet and furniture work only.

8. Much of the softboard is used in the construction industry for
   a. underlayment for wall to wall rugs.
   b. roof covering.
   c. decorative ceiling panels.
   d. interior wall finish.

9. One of the most important characteristics of gypsum board is its
   a. strength.
   b. fire resistance.
   c. moisture resistance.
   d. appearance.

10. Plastic laminates are widely used in construction for
    a. cabinets and countertops.
    b. exterior wall finish.
    c. insulation and sound control.
    d. vapor barriers.

FASTENERS 3

OBJECTIVES

After completing this unit, the student will be able to name and describe commonly used fasteners:

- *Nails*
- *Screws*
- *Bolts and lag screws*
- *Solid and hollow wall anchors*
- *Adhesives*

Many kinds of fasteners are used in the construction industry. It is important to know the best kind and size to use for each job.

NAILS

There are hundreds of different kinds of nails, Figure 3-1. They differ according to shape, material, coating, and in other ways. Nails are made of aluminum, brass, copper, steel, and other metals. Different coatings are applied to reduce corrosion, increase holding power, and for appearance. Some nails are also hardened so that they can be driven into masonry.

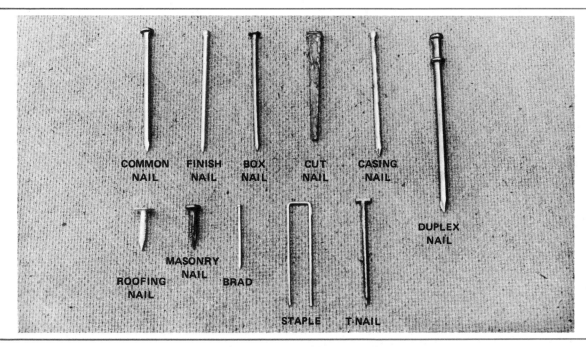

Fig. 3-1 Kinds of nails

NAIL SIZES

Nails are sized according to the *penny* system. The symbol for penny is *d.* For instance, a sixpenny nail is written as 6d and is two inches long. In the penny system the shortest nail is 2d and is one inch long. The longest nail is 60d and is six inches long, Figure 3-2.

The diameter of the nail is called its gauge. The gauge of a nail depends on its kind and length. Long nails, 16d and over, are called *spikes.* Nails shorter than one inch and longer than six inches are listed by inches and fractions of an inch.

KINDS OF NAILS

Wire nails are made from long rolls of metal wire and are the most used nail in the trade. *Cut nails* are wedge-shaped and stamped from thin sheets of metal.

Common nails are made of wire, are fairly thick, and have a medium-sized head. They have a pointed end and a smooth shank except for a barbed section just under the head, which increases the holding power of the nails.

Box nails are similar to common nails, except that they are thinner. Because of their small gauge, they can be used close to the edge and the end with less danger of splitting the wood. Many box nails are coated to increase their holding power and are commonly used to fasten underlayment.

Finish nails are thin with a very small head and are used mostly to fasten interior trim. The small head is sunk into the wood with a nail set and covered with a filler. The small head of the finish nail does not detract from the appearance of a job as much as a nail with a larger head would.

Casing nails are similar to finish nails. Many carpenters prefer them to fasten exterior finish. The head is cone-shaped and slightly larger than the finish nail. The shank is the same gauge as the common nail.

Brads are small finishing nails. They are sized according to their length in inches and gauge. Usual lengths are from 1/2 inch to 1 1/2 inches and gauges run from #14 to #20. The higher the gauge number, the thinner the brad.

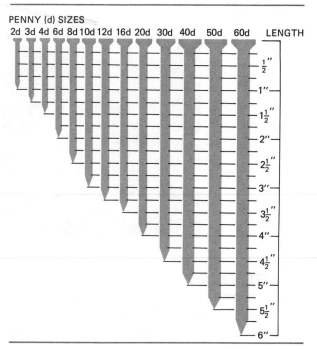

Fig. 3-2 The penny system of nail sizes

Brads are used for fastening thin material, such as small moulding.

Roofing nails are short nails of fairly heavy gauge and have wide, round heads. They are used for such purposes as fastening asphalt shingles, asphalt felt, and softboard wall sheathing. The large head holds thin or soft material more securely. Some roofing nails are made from aluminum. Steel roofing nails are galvanized (zinc coated) or cadmium plated to retard rusting. The shank is usually barbed to increase holding power.

Masonry nails and *flooring nails* are often cut nails. These steel nails are cut from sheets and hardened to prevent bending when being driven into hard material. The cut nail has a blunt point which tends to prevent splitting when it is driven into hardwood. Some masonry and flooring nails have round shanks of different designs to hold better.

Tacks are small pointed nails with a head. Sometimes used to fasten lightweight material such as wire screen, they are available in sizes 1 to 24. The larger the number is, the longer the tack.

Staples are U-shaped fasteners. They may be purchased to be driven individually with a hammer. However, widely used in the industry

Fig. 3-3 Using a power stapler to fasten roof sheathing plywood
(Courtesy of American Plywood Association)

are staples that come glued together in rows to be driven by hand or power staplers, Figure 3-3. Staple lengths generally are from 1/2 inch to 2 inches. Widths of staples usually run from 3/8 inch to 1 inch. The common gauge sizes range from 14 to 16. There are many other sizes for special work. Make sure the staples fit the machine being used. Follow the manufacturer's instructions for the operation of the stapler.

T-nails are specially designed nails to be used in nailing machines. These nails are used for rapid fastening of subfloors, underlayment, and wall and roof sheathing.

NAILING TECHNIQUES

Hold the hammer handle firmly, close to the end, and hit the head of the nail squarely. Rub the face of the hammer head with sandpaper or on a rough surface to clean it and to help prevent it from glancing off the nailhead. As a general rule, use nails three times longer than the thickness of the material to be fastened.

Toenailing is the technique of driving nails at an angle to fasten the end of one piece to another, Figure 3-4. Start the nail about 3/4 inch to 1 inch from the end and at an angle of about 30° from vertical. To avoid splitting the stock do not toenail too close to the edge.

Drive finish nails almost home and then set the nail below the surface with a nail set if necessary to prevent hammer marks on the surface. Finish nails are set at least 1/8 inch deep so that the filler used to cover the head will not fall out.

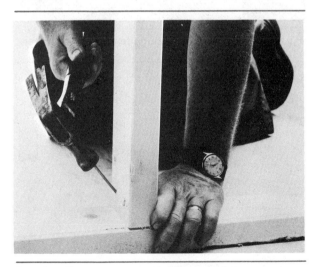

Fig. 3-4 Toenailing is the technique of driving nails at an angle.

In hardwood, or close to edges, drill a hole slightly smaller than the nail shank to prevent the wood from splitting or the nail from bending. A little paraffin applied to the nail shank makes driving the nail easier.

When nailing along the length of a piece, stagger the nails from edge to edge. This technique provides greater strength. Splitting the wood is also less likely than if nailing in a straight line.

SPECIAL PURPOSE NAILS

There are many kinds of nails designed for special purposes. Some of these are used to install prefinished plywood paneling, exterior asbestos siding, gypsum board, and many other kinds of material.

On temporary structures, such as wood scaffolding, the *duplex nail* is used. The lower head assures that the piece is fastened tightly. The projecting upper head is used to pry the nail out when the structure is dismantled, Figure 3-5.

SCREWS

Wood screws are used when greater holding power is needed and when the work being fastened must at times be removed. For example, door hinges must be applied with screws. Stop beads that hold the sash in a window frame are fastened with screws to permit removal of the sash when necessary. However, screws cost more than nails and require more time to drive. When ordering screws, specify the length, gauge, type of head, coating, kind of metal, and screwdriver slot.

KINDS OF SCREWS

The shape of the screwhead and screwdriver slot determine the kind of screw used. Three of

Fig. 3-5 Duplex nails are used on temporary structures.

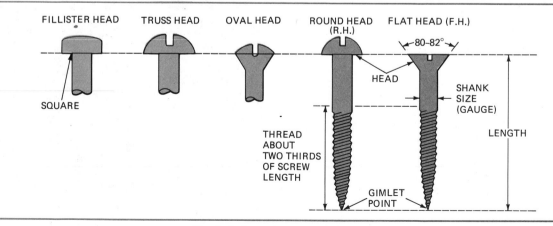

Fig. 3-6 Common kinds of screws

34

the most common shapes of screwheads are the *flat head, round head* and *oval head,* Figure 3-6. Other shapes include the *pan head, truss head* and *fillister head.*

Screwheads made with a straight single slot are called *common screws. Phillips head screws* have a crossed slot. There are many other types of screwdriver slots, each with a different name, Figure 3-7.

The pointed end of a screw is called a *gimlet point.* The threaded section is called the *thread.* The smooth section between the head and thread is called the *shank.*

Wood screws are threaded, in most cases, only part way to the head. Sheet metal screws have a deeper thread all the way up to the head. Sheet metal screws are recommended for fastening hardboard and particleboard because of their deeper thread. Another type screw, used with power screwdrivers, is the *self-drilling screw.* This screw has a cutting edge on its point to eliminate predrilling a hole.

Many other screws are available that are designed for special purposes. Like nails, screws come in a variety of metals and coatings. Steel nails and screws with no coating are called *bright.*

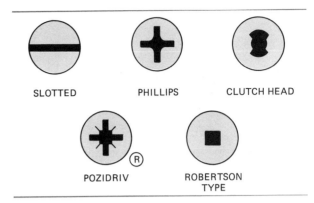

Fig. 3-7 Common kinds of screw slots

SCREW SIZES

Screws are made in many different sizes. Usual lengths range from 1/4 inch to 4 inches. Gauges run from 0 to 24, Figure 3-8. The higher the gauge number, the greater the diameter of the screw. Screw lengths are not available in every gauge. The lower gauge numbers (thinner screws) are for shorter screws. Higher gauge numbers are for longer screws. Screw lengths are measured from the point to that part of the head that sets flush with the wood when fastened.

SCREWDRIVING TECHNIQUES

If possible, select screws so that two-thirds of their length penetrates the piece into which

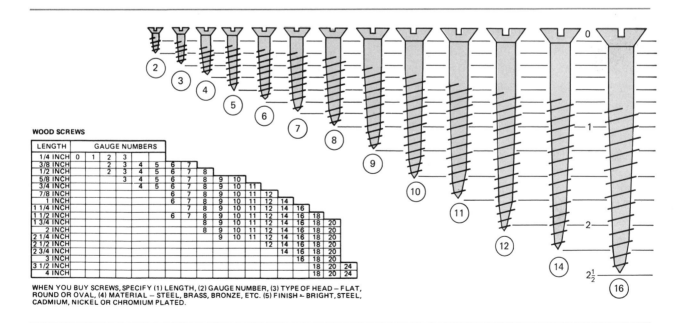

WOOD SCREWS

LENGTH	GAUGE NUMBERS														
1/4 INCH	0	1	2	3										
3/8 INCH			2	3	4	5	6	7						
1/2 INCH			2	3	4	5	6	7	8					
5/8 INCH				3	4	5	6	7	8	9	10			
3/4 INCH					4	5	6	7	8	9	10	11		
7/8 INCH							6	7	8	9	10	11	12	
1 INCH							6	7	8	9	10	11	12	14
1 1/4 INCH								7	8	9	10	11	12	14	16
1 1/2 INCH							6	7	8	9	10	11	12	14	16	18
1 3/4 INCH									8	9	10	11	12	14	16	18	20
2 INCH									8	9	10	11	12	14	16	18	20
2 1/4 INCH										9	10	11	12	14	16	18	20
2 1/2 INCH													12	14	16	18	20
2 3/4 INCH														14	16	18	20
3 INCH															16	18	20
3 1/2 INCH																18	20	24
4 INCH																18	20	24

WHEN YOU BUY SCREWS, SPECIFY (1) LENGTH, (2) GAUGE NUMBER, (3) TYPE OF HEAD – FLAT, ROUND OR OVAL, (4) MATERIAL – STEEL, BRASS, BRONZE, ETC. (5) FINISH – BRIGHT, STEEL, CADMIUM, NICKEL OR CHROMIUM PLATED.

Fig. 3-8 Wood screw sizes

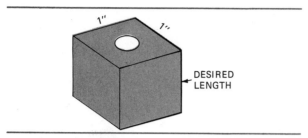

Fig. 3-9 Simple drill stop made from piece of 1x1 stock, drilled to desired length

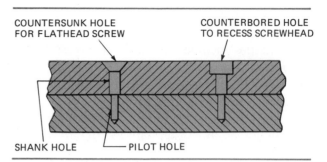

Fig. 3-10 Countersunk and counterbored holes

the screws are gripping. If the material to be fastened is thick, the screw may be set below the surface by *counterboring* to gain additional penetration without resorting to a longer screw.

Before driving a screw, two holes must be drilled. One hole for the shank is called the *shank hole.* The other hole, for the thread, is called the *pilot hole.*

To select a drill for the shank hole, hold the drill bit against the shank and determine by eye if it is the same size as the shank. For the pilot hole, hold the drill bit against the threaded portion of the screw so that the drill covers the solid center section, but leaves the threads visible.

Select drills with great care. Smaller drills may be used for a pilot hole in softwoods. However, in hardwoods if the pilot hole is too small, difficulty may be encountered in driving the screw. Also, if too much pressure is applied when driving the screw, the head may be twisted off. This is particularly true when driving screws of soft metal, such as aluminum or brass. Remember that if the pilot hole is too large, the screw will not grip. Special drills or attachments for drills are available so that shank hole, pilot hole, countersinking, and counterboring can be done in one operation.

Always drill holes for screws, whether in softwood or hardwood. Not only will driving be easier, but the screw will be driven straight. Drill the pilot hole deep enough so that the screw will not bottom. When there is danger of drilling through the material, use a stop on the drill. A simple stop can be made by drilling a hole lengthwise through a piece of 1x1 stock cut to the desired length and inserted on the drill, as shown in Figure 3-9. Rub some wax (paraffin) on the

threads of the screw to make driving easier. Many carpenters carry a small piece of candle in their tool box for this and other purposes.

COUNTERSINKING AND COUNTERBORING

In addition to shank and pilot holes being drilled, oval and flat head screws must be countersunk (set flush with the surface). A special tool, called a countersink, is used to make the recess for the screwhead.

To set the screwhead below the surface, bore the counterbored hole first to the desired depth. The diameter of the hole is equal to or slightly larger than the diameter of the screwhead. Next, drill the shank hole; then, the pilot hole, Figure 3-10. The center would be lost for the counterbored hole if the smaller holes were drilled first.

Screwdrivers should fit snugly, without play, into the slot of the screw being driven. The screwdriver tip should not be wider than the screwhead, nor should it be too narrow, Figure 3-11. Exert firm pressure downward while turning the screwdriver. Seat the head firmly. When seated, the screw slot should look the same as before the screw was driven — with no burred edges to the slot.

Spiral screwdrivers (refer to Unit 4 Hand Tools) save much energy when many screws are to be driven. When using spiral screwdrivers, take care not to slip off the screw head. This can be avoided by twisting the handle while pushing down when the screw nears its seat.

Slipping may also be avoided when using a bit brace and screwdriver bit by using the

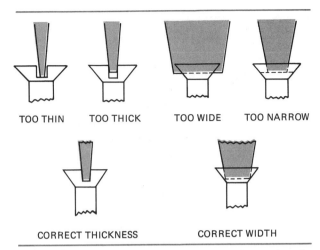

TOO THIN   TOO THICK   TOO WIDE   TOO NARROW

CORRECT THICKNESS   CORRECT WIDTH

Fig. 3-11 Select the correct size screwdriver for the screw being driven.

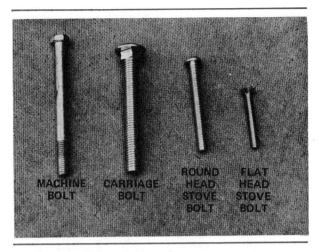

MACHINE BOLT   CARRIAGE BOLT   ROUND HEAD STOVE BOLT   FLAT HEAD STOVE BOLT

Fig. 3-12 Kinds of commonly used bolts

ratchet on the bit brace. When the screw is almost driven home, ratchet the bit brace and pull the handle about one-eighth of a turn each time until the screw is well seated.

BOLTS

Commonly used bolts are the carriage, machine, and stove bolts, Figure 3-12. As with nails and screws, they are available in different kinds of metal and coatings. Most of the bolts used in construction are made of steel. To retard rusting, galvanized or cadmium plated bolts are used.

The *carriage bolt* has a square section under its oval head. The square section is embedded in wood and prevents the bolt from turning as the nut is tightened.

The *machine bolt* has a square or hexagonal (six-sided) head. This is held with a wrench to keep the bolt from turning as the nut is tightened.

Stove bolts have either round or flat heads with a screwdriver slot. They are usually threaded all the way up to the head.

BOLT SIZES

Bolt sizes are specified by the diameter and length of the bolt. Carriage and machine bolts range from 3/4 inch to 20 inches in length and from 3/16 inch to 3/4 inch in diameter. Stove bolts are small in comparison to other bolts. They commonly come in lengths from 3/8 inch

to 6 inches and from 1/8 inch to 3/8 inch in diameter.

ANCHOR BOLTS

Anchor bolts, shown in Figure 3-13, are frequently used to fasten wood sills to the top of foundation walls. Usually 1/2 inch in diameter, they have a long shank with a hooked end which is embedded in the fresh concrete. Instead of a hooked end, some types have a large flat washer.

BOLT FASTENING TECHNIQUES

Drill holes for bolts the same diameter as the bolt. Use washers under the head (except for carriage bolts) and under the nut to prevent the nut from cutting into the wood and to distribute the pressure over a wider area. Apply a little light lubricating oil to the threads before the nut is turned on. Use wrenches of the correct size to tighten the bolt.

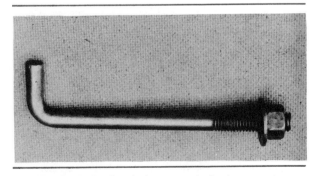

Fig. 3-13 Anchor bolts are set in fresh concrete.

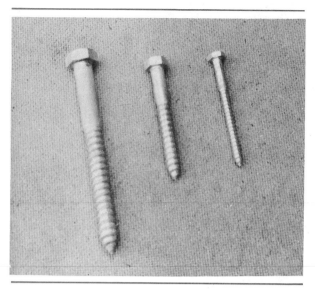

Fig. 3-14 Lag screws are large screws with a square or hex head.

LAG SCREWS

Lag screws, Figure 3-14, are similar to wood screws except that they are larger and have a square head designed to be turned with a wrench instead of a screwdriver.

Lag screws are sized by their diameter and length. Diameters range from 1/4 inch to 1 inch, with lengths from 1 inch to 12 inches and up.

This fastener is used when great holding power is needed to join heavy parts and where a bolt cannot be used. Shank and pilot holes to receive lag screws are drilled in the same manner as for wood screws. Place a flat washer under the head to prevent the head from digging into the wood as the lag screw is tightened down. Ap-

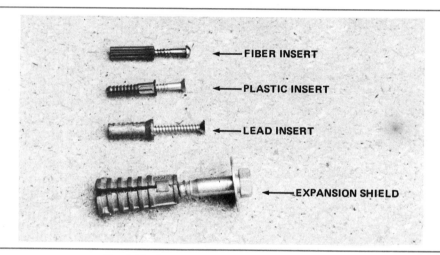

Fig. 3-15 Commonly used solid wall anchors

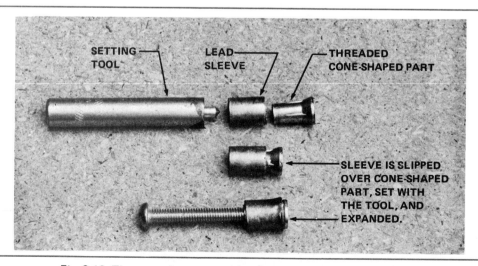

Fig. 3-16 The expansion screw anchor is commonly called a tamp-in.

ply a little wax to the threads to turn the screw easier and to prevent twisting the head off.

ANCHORS

Anchors are used to fasten parts to solid masonry or to hollow walls and ceilings of various materials. There are hundreds of different types available. Those most commonly used are discussed here.

SOLID WALL ANCHORS

Expansion shields take either lag screws or machine bolts. The shield is a split sleeve of soft metal, usually lead, which is inserted in a hole drilled in the wall. As the fastener is threaded in, the shield expands tightly and securely in the drilled hole. Expansion shields are usually used to fasten heavy objects to solid masonry walls, Figure 3-15.

To fasten lighter objects to solid masonry, *lead, plastic,* and *fiber inserts* are commonly used. These inserts have an unthreaded hole into which a wood screw is driven. The insert is placed into the proper-size drilled hole. As the screw is turned, the threads of the screw cut into the soft material of the insert and cause it to expand and tighten in the drilled hole.

The *lead expansion screw anchor,* commonly called a tamp-in, Figure 3-16, takes a stove bolt and consists of two parts. A lead sleeve slides over a cone-shaped piece containing threads for the bolt. With a special setting punch that comes with the set, the lead sleeve is driven over the cone-shaped piece to expand the sleeve and hold it securely in the hole.

Fig. 3-17 Toggle bolt

To drill holes in masonry, use a carbide-tipped bit in a power drill.

HOLLOW WALL FASTENERS

Toggle bolts, Figure 3-17, have a wing-type end or a tumble head. The wing end is fitted with springs which causes the toggle to open as it passes through the hole. The tumble-type toggle falls in a vertical position when passed through the hole. A hole must be drilled large enough for the toggle of the bolt to slip through. A disadvantage of using toggle bolts is that, if removed, the toggle falls off inside the wall.

Star expansion anchors are commonly called *molly screws,* Figure 3-18. The anchor consists of a shield and a stove bolt. The unit is inserted in the drilled hole and prongs on the shield are tapped into the surface of the wall. These prongs prevent the shield from turning while the anchor bolt is tightened. Tightening the bolt expands the shield against the inside of the wall. The bolt is then removed, inserted through the part to be attached, and then screwed back into the shield. The advantage of using this type anchor is that the fastened part can be removed and replaced without losing the shield inside the wall.

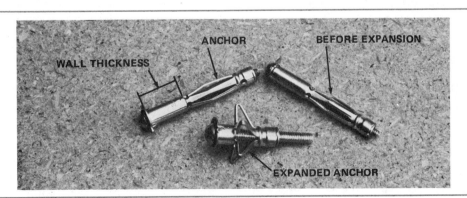

Fig. 3-18 Star expansion anchors are commonly called molly screws.

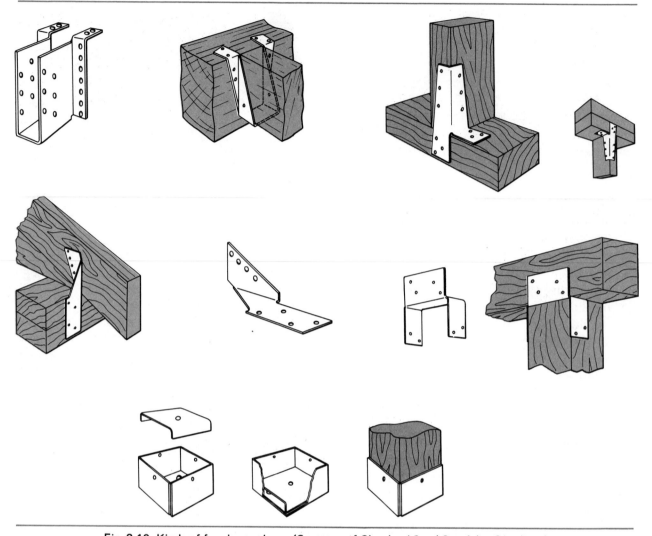

Fig. 3-19 Kinds of framing anchors *(Courtesy of Cleveland Steel Specialty Company)*

These anchors are available in many different sizes and are designed for different wall thicknesses. When ordering, specify the wall thickness, diameter, and length. Drill a hole size specified on the anchor.

MISCELLANEOUS FASTENERS

Many other types of fasteners are used in the construction industry. *Framing anchors* are pieces of metal bent to various shapes according to their purpose. They have punched holes through which nails are driven, Figure 3-19. Some of these anchors are designed to fasten the tops and bottoms of wood posts, to hold roof rafters to walls, to secure the ends of floor joists, and to splice two members in a straight line.

Corrugated fasteners, Figure 3-20, are thin metal pieces used to hold butt or miter joints of small pieces together. They are used on light rough work or where the fastener will not be exposed.

Dowels are round pieces of hardwood ranging in diameter from 1/4 inch to 1 inch, and up. They are used most frequently to strengthen joints in cabinet work.

ADHESIVES

The carpenter seldom uses any glue in the frame or exterior finish. Glue is used on some joints and other parts of the interior finish work. A number of *mastics* (heavy, pasty-type adhesives) are used throughout the construction.

GLUE

Most of the glue used by the carpenter is the so-called white glue or yellow glue. The white glue is *polyvinyl acetate;* the yellow glue is *aliphatic resin.* Neither type is resistant to moisture. Both types are fast setting; joints should be made quickly after applying the glue. The yellow glue is said to be a little faster setting and stronger than the white glue. Both kinds are available under a number of different trade names.

When a moisture-resistant glue is needed, *plastic resin* glue is used. This type comes in powder form and is mixed with water to the consistency of heavy cream. The glue sets slowly, and the joint must remain under pressure for eight hours or more, depending on the temperature.

MASTICS

Mastics are used on subflooring and wall paneling and as a base for ceiling and floor tiles. Mastics are usually so thick that they are spread with a trowel. Usually the trowels used have notched edges to allow the desired amount of adhesive to remain on the surface. The depth of the notches determines the amount of adhesive left on the surface. It is important to use a trowel with the correct notch depth.

Asphalt cement is commonly used on roofing to cement down shingle tabs or roof flashing. Special mastics are made to bond rubber baseboard, floor tile, ceramic tile, drywall, wall paneling, subflooring, and many other materials. These mastics are discussed in later units in which they apply. Mastics come in cans or are available in cartridges which are placed in *guns* (holders) to apply the mastic more easily, Figure 3-21.

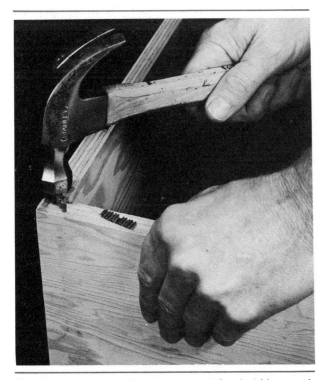

Fig. 3-20 Corrugated fasteners are used to hold butt and miter joints on light work. *(Courtesy of American Plywood Association)*

Fig. 3-21 Applying a mastic-type adhesive with a gun *(Courtesy of Masonite Corporation)*

REVIEW QUESTIONS

Select the most appropriate answer.

1. The length of an eightpenny nail is
   a. 1 1/2 inches.
   b. 2 inches.
   c. 2 1/2 inches.
   d. 3 inches.

2. Fasteners coated with zinc to retard rusting are called
   a. coated.
   b. dipped.
   c. electroplated.
   d. galvanized.

3. Brads are
   a. types of screws.
   b. small box nails.
   c. small finishing nails.
   d. kinds of stove bolts.

4. To help prevent the hammer from glancing off a nailhead
   a. choke up on the hammer handle.
   b. drill holes for the nails.
   c. rub the hammer face with sandpaper.
   d. start the nail square with the work.

5. Many carpenters prefer to use casing nails for
   a. interior finish.
   b. exterior finish.
   c. door casings.
   d. roof shingles.

6. The blunt point on the end of a cut nail helps
   a. to drive the nail straight.
   b. prevent splitting the wood.
   c. hold the fastened material more securely.
   d. to start the nail in the material.

7. On temporary structures such as wood scaffolding and concrete forms, use
   a. common nails.
   b. duplex nails.
   c. galvanized nails.
   d. spikes.

8. As a general rule, how should the length of a nail compare to the thickness of the material being fastened?
   a. The same as
   b. Twice as long
   c. 2 1/2 times as long
   d. 3 times as long

9. What part of the length of a wood screw should penetrate the piece which it is gripping?
   a. one-third
   b. one-half
   c. two-thirds
   d. three-quarters

10. The first hole to be made when counterboring for screws is the
    a. counterbored hole.
    b. pilot hole.
    c. shank hole.
    d. any of these.

HAND TOOLS 4

OBJECTIVES

After completing this unit, the student will be able to

- *identify and describe the hand tools that are commonly used by the carpenter.*

- *tell how and why each of these tools is used.*

- *sharpen chisels, plane irons, handsaws, auger bits and twist drills.*

Even though there has been a great increase in the use of power tools, there are still many occasions when hand tools must be used. There may be times when it is easier and faster to use hand tools rather than power tools.

Knowing how to choose the proper hand tool, how to use it skillfully, and how to keep it in good condition is essential to the carpentry trade. A student should not underestimate the importance of hand tools and neglect training in their use and care.

Buy tools of good quality. Inferior quality tools sometimes cannot be brought to a sharp edge and dull rapidly. Keep the hand tools sharp and in good condition. If they get wet on the job, dry them as soon as possible and coat them with a little light oil to prevent them from rusting.

In most cases, carpenters are expected to have their own hand tools on the job and keep them in working condition. The contractor usually provides the necessary power tools needed to do the work.

LAYOUT TOOLS

Many layout tools must be used by the carpenter. They are used, among other things, to measure distances, lay out lines and angles, test for depths of cuts, and to set vertical and level pieces.

RULES

Rules used in construction are divided into feet, inches, and usually 16ths of an inch. The

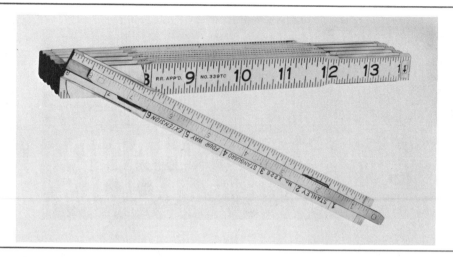

Fig. 4-1 Six-foot folding rule *(Courtesy of Stanley Tools)*

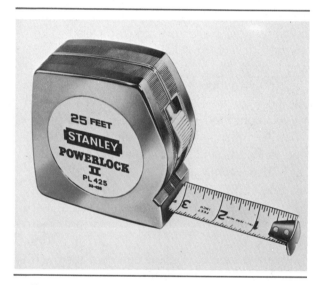

Fig. 4-2 Pocket tape *(Courtesy of Stanley Tools)*

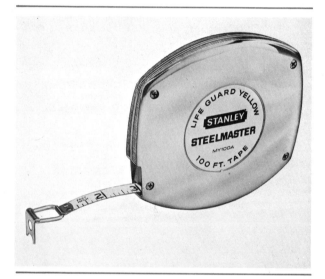

Fig. 4-3 Steel Tape *(Courtesy of Stanley Tools)*

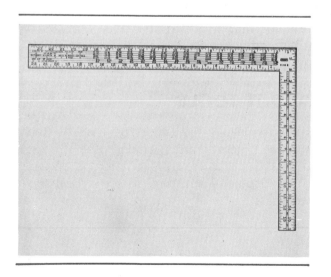

Fig. 4-4 Framing square *(Courtesy of Stanley Tools)*

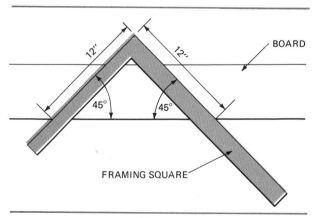

Fig. 4-5 Laying out a 45° angle with the framing square

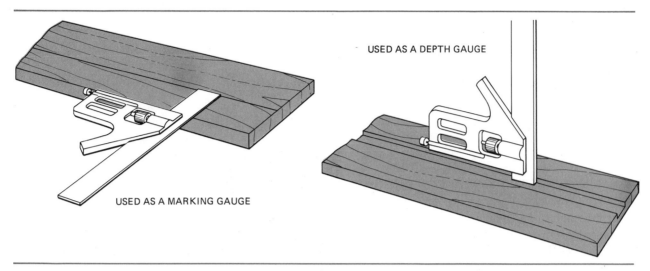

USED AS A DEPTH GAUGE

USED AS A MARKING GAUGE

Fig. 4-6 Combination square *(Courtesy of Stanley Tools)*

ability to read a rule quickly and accurately is important.

Some carpenters prefer the *6-foot folding rule,* Figure 4-1, while others use the *pocket tape,* Figure 4-2. The folding rule sometimes has a metal extension on one end for taking inside measurements. The user must oil the joints of the folding rule occasionally to prevent it from breaking when opening and closing it. Pocket tapes are available in 6- to 25-foot lengths and have an adjustable hook on the end for taking outside and inside measurements.

Steel tapes, Figure 4-3, of 50- and 100-foot lengths are commonly used to lay out longer measurements. Most of the rules and tapes used by the carpenter have clearly marked increments of 16 inches to help in laying out spaced framing members. Use *graphite* lubricant on the sides of

the steel tape for easy operation. Oil will make it bind.

The *framing square,* Figure 4-4, is used for squaring across wide pieces, laying out 45° angles (as shown in Figure 4-5), bridging, stairs, and rafters. This tool has many other uses in the field of carpentry.

The *combination square,* Figure 4-6, is used to lay out and check 90° and 45° angles on smaller stock. It is also used with a pencil as a marking gauge to draw lines parallel to the edge of a board. It can be used to test the depth of mortises, grooves, and dadoes. The *sliding T-bevel,* Figure 4-7, is used in a way similar to that used with a combination square to lay out or test any angle.

LEVELS

In construction, the term *level* is used to indicate *horizontal.* The term *plumb* is used to express *vertical.*

The *carpenter's level,* Figure 4-8, is a hand tool used to test both level and plumb surfaces.

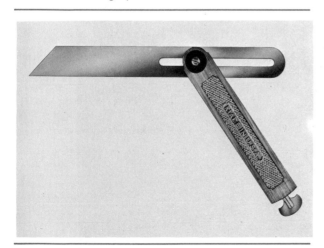

Fig. 4-7 Sliding T-bevel *(Courtesy of Stanley Tools)*

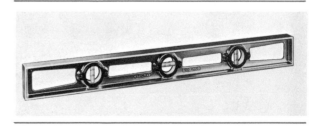

Fig. 4-8 Carpenter's level
(Courtesy of Stanley Tools)

Fig. 4-9 Line level *(Courtesy of Stanley Tools)*

Accurate use of the level depends on accurate reading. The air bubble in the slightly crowned glass tube of the level must be exactly centered between the lines marked on the tube. The pair of tubes located in the center are used to test for level. The pairs located at each end are used to test for plumb.

Levels are made of wood or metal, usually aluminum. Most carpenters prefer the metal type. The levels usually come in lengths of 24 to 48 inches. **Note: Care must be taken not to drop the level because dropping could break the glass or disturb the accuracy of the level.**

The *line level,* Figure 4-9, consists of one glass tube encased in a metal sleeve with hooks on each end. The hooks are attached to a stretched line and the bubble centered for approximate levelness. Care must be taken that the level be attached to the center of the suspended line because the weight of the level causes the line to sag.

The *plumb bob,* Figure 4-10, is very accurate and is used frequently for testing and establishing plumb lines. Suspended from a line,

Fig. 4-10 Plumb bob *(Courtesy of Stanley Tools)*

the plumb bob hangs absolutely vertical when it stops swinging. However, it is difficult to use outside when the wind is blowing.

Heavy plumb bobs are easier to use than lighter ones. Some plumb bobs are hollowed out and filled with mercury (a very heavy liquid) to increase the weight without enlarging the size.

CHALK LINES

Long straight lines are laid out by using a *chalk line.* A line coated with chalk dust is stretched tightly between two points and snapped against the surface, Figure 4-11. The

Fig. 4-11 Snapping a chalk line

Fig. 4-12 Chalk-line reel *(Courtesy of Stanley Tools)*

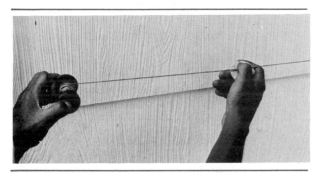

Fig. 4-13 Chalking a line with cake chalk

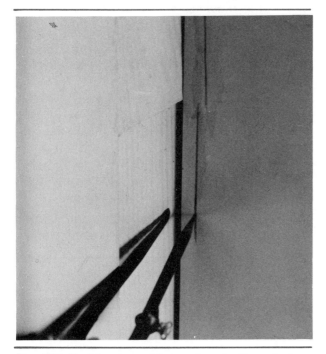

Fig. 4-14 Scribing is marking a piece to fit against an irregular surface.

chalk dust is dislodged from the line and remains on the surface.

Widely used at present is the *chalk-line reel,* Figure 4-12. The reel is filled with chalk dust and refilled when necessary. Some carpenters still prefer the *spool and awl* in which the line is chalked by hand, Figure 4-13. The cake of half-round chalk is rubbed over the line in a fashion that prevents the line from cutting into the cake. At the same time, the line is unwound by holding the spool ends between the middle finger and the thumb of the other hand. Chalk is available in colors of white, blue, yellow and maroon; blue is used most.

Keep the line from getting wet. A wet line is practically useless. Use cotton line to hold chalk dust better. Use braided cotton for extra strength and to prevent unraveling. Make sure lines are stretched tight. Sight long lines by eye for straightness, especially when snapping lines on vertical surfaces to make sure there is no sag in the line. Another method of marking long lines is to hold the center of the line to the deck and snap both ends. Be sure to sight the line for straightness before it is snapped.

Dividers are used to lay out circles and arcs, and to space off equal distances. This tool is also used for scribing and is often called a scriber. *Scribing* is marking the edge of a board to be fitted against an irregular surface, Figure 4-14. For easier and more accurate scribing, heat and bend the end of the solid metal leg outward, Figure 4-15. Use pencils with hard lead which keep their points longer.

HEAT AND BEND THE LEG OUTWARD FOR EASIER AND MORE ACCURATE SCRIBING.

Fig. 4-15 Dividers *(Courtesy of Stanley Tools)*

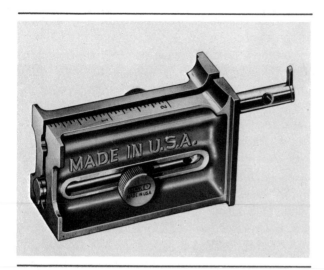

Fig. 4-16 Butt gauge *(Courtesy of Stanley Tools)*

Fig. 4-17 Butt marker *(Courtesy of Stanley Tools)*

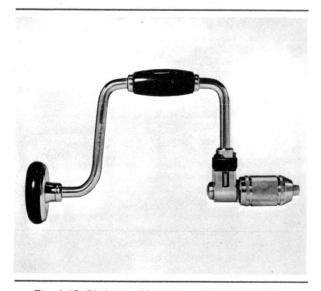

Fig. 4-18 Bit brace *(Courtesy of Stanley Tools)*

BUTT GAUGES

The butt gauge, Figure 4-16, is used to lay out the gain (cutout) for butt hinges. Butt hinges are commonly used on swinging doors. Butt markers, Figure 4-17, are available in three sizes and are often used to mark hinge gains.

PENCILS

Two types of pencils are used by carpenters. For rough work, a heavy, flat-sided, hard lead pencil is used. For finish work, a 3H or 4H drawing pencil is preferred. The hard lead in these pencils helps them hold their point longer. Do not draw lines any heavier than necessary.

BORING TOOLS

The carpenter is often required to cut holes in wood and metal. Boring tools include those which actually do the cutting and those used to turn the cutting tool.

BIT BRACE

The bit brace, Figure 4-18, is used to hold and turn auger bits to bore holes in wood. It may also be used to hold screwdriver bits to drive large screws. Its size is determined by its *sweep* (the diameter of the circle made by its handle). Sizes range from 8 to 12 inches. Most bit braces come with a ratchet which can be used where there is not enough room to make a complete turn of the handle.

Auger bits, Figure 4-19, are available with fast or slow *feed screws.* As the bit is turned, the *spurs* score the circle in advance of the *cutting lips.* The cutting lips lift the chip up and through the twist of the bit. Care must be taken not to strike any nails or other objects that might reduce the length of the spurs.

A full set of auger bits ranges in sizes from 1/4 inch to 1 inch, graduated in sixteenths of an

Fig. 4-19 Auger bit *(Courtesy of Stanley Tools)*

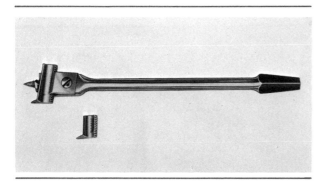

Fig. 4-20 Expansive bit *(Courtesy of Stanley Tools)*

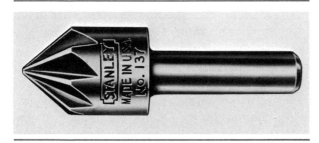

Fig. 4-21 Hand drill *(Courtesy of Stanley Tools)*

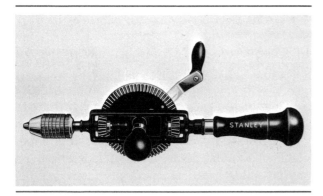

Fig. 4-22 Countersink *(Courtesy of Stanley Tools)*

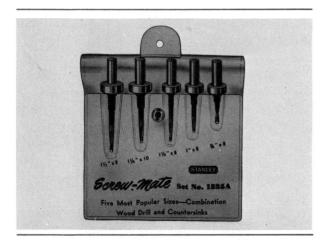

Fig. 4-23 Combination drill and countersink *(Courtesy of Stanley Tools)*

inch. The bit size is designated by the number of sixteenths of an inch. For instance, a #12 bit will bore a 3/4-inch diameter hole.

EXPANSIVE BIT

To bore holes over 1 inch in diameter, the carpenter usually carries an expansive bit, Figure 4-20. With interchangeable cutters, holes up to 3 inches in diameter may be bored.

HAND DRILL AND TWIST DRILLS

Making holes in metal or small holes in wood is called *drilling.* The hand drill, Figure 4-21, is used to turn *twist drills* to make holes in wood that are 1/4 inch or less in diameter. It is a very useful tool when a few holes are to be drilled or no power is available.

Twist drills ranging in size from 1/16 inch to 1/4 inch in increments of 1/64 inch satisfy the carpenter's needs. These drills are particularly useful for drilling holes for screws.

The *countersink,* Figure 4-22, may be turned in a hand drill. It forms a recess for a flat head screw to set flush with the surface of the material in which it is driven.

Combination drills and countersinks, Figure 4-23, are used to drill shank and pilot holes for screws and countersink in one operation.

EDGE CUTTING TOOLS

Chisels, planes, tin snips, hatchets, and knives are edge-cutting tools. The carpenter must have an understanding of the types available and the various features of each.

WOOD CHISELS

The wood chisel, Figure 4-24, is used to cut recesses in wood for such things as door hinges and locksets and to make joints.

Fig. 4-24 Wood chisel *(Courtesy of Stanley Tools)*

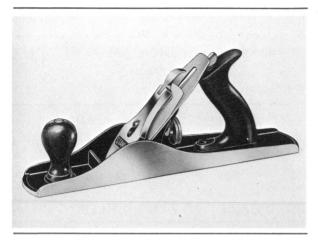

Fig. 4-25 Bench plane *(Courtesy of Stanley Tools)*

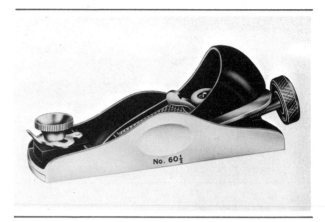

Fig. 4-26 Block plane *(Courtesy of Stanley Tools)*

Chisels are sized according to the width of the blade and are available in widths of 1/8 inch to 2 inches. Most carpenters can do their work with a set consisting of chisels that are 1/4, 1/2, 3/4, 1, and 1 1/2 inches in size.

Firmer chisels have long, thick blades and are used on heavy framing. Butt chisels are short, with a thinner blade and are preferred for finish work.

Improper use of chisels has caused many accidents. When using chisels, keep both hands back of the cutting edge at all times.

BENCH PLANES

Bench planes, Figure 4-25, come in a variety of sizes. They are used for smoothing rough surfaces and to bring work down to the desired size. Large planes are used on long work; small planes are used on short work.

Bench planes are given names according to the length. Beginning with the longest they are the jointer, fore, jack, and smooth. The jack plane is 14 inches long and is considered a general purpose plane to do all-around work. Most carpenters carry the jack plane in their tool box. Many have a jointer plane also to do long work, such as fitting doors. There is no need to carry a complete set of bench planes.

BLOCK PLANES

Block planes, Figure 4-26, are small planes designed to be held in one hand. They are used to smooth the edges of short pieces and for trimming end grain to make fine joints.

Block planes are available with their blades set at a high angle or at a low angle. Most carpenters prefer the low angle block plane.

Other types of planes such as the rabbet plane, the router plane, and the spokeshave are available, but are not generally used by the carpenter.

SNIPS

Straight tin snips, Figure 4-27, are generally used to cut straight lines on thin metal, such as roof flashing and metal roof edging. In some parts of the country, straight snips are used to trim asphalt shingles at each end of the roof.

Aviation snips, Figure 4-28, are available for straight cutting and for left and right curved cuts in thin metal. If the carpenter carries the straight snips just mentioned, only the right and the left cutting aviation snips are necessary to do the work encountered.

HATCHETS

For wood shingling of side walls and roofs among other purposes, the shingling hatchet, Figure 4-29, is used. In addition to the shingling hatchet, many carpenters carry a slightly heavier hatchet for such uses as pointing stakes or otherwise tapering rough stock.

Caution: When using hatchets for driving fasteners, watch out for the sharp edge of the backswing.

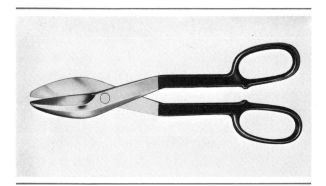

Fig. 4-27 Straight snips *(Courtesy of Stanley Tools)*

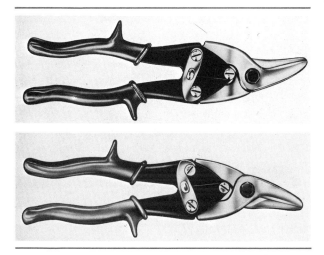

Fig. 4-28 Right- and left-cutting aviation snips *(Courtesy of Stanley Tools)*

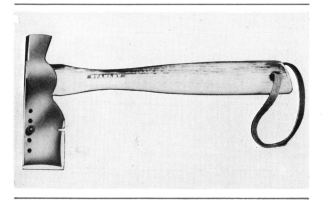

Fig. 4-29 Shingling hatchet *(Courtesy of Stanley Tools)*

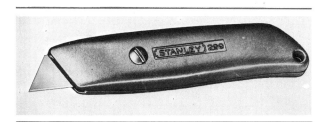

Fig. 4-30 Utility knife *(Courtesy of Stanley Tools)*

KNIVES

Carpenters usually have a jackknife of good quality. The jackknife is used mostly for sharpening pencils and for laying out some types of finish hardware, such as door hinges. The jackknife is used for laying out this type of work because a finer line can be obtained with it than with a pencil.

The *utility knife,* Figure 4-30, is a tool frequently used for such things as cutting gypsum board and softboards. Extra cutting blades are carried inside the handle and can be used to replace dull ones.

TOOTH-CUTTING TOOLS

To cut material to size, the use of some type of saw is required. There are many types available; each is designed for a particular purpose.

HANDSAWS

The handsaw, Figure 4-31, is necessary to do the carpenter's general work. Two *crosscut saws* are needed. A 7-point or 8-point (number of tooth points to the inch) saw is used to cut framing or rough work. An 11-point or 12-point saw is used for fine cross cuts on finish work.

One *ripsaw,* to cut with the grain of wood, is needed. Usually a 5 1/2-point ripsaw is used. Figure 4-32 shows the cutting action of crosscut saws and ripsaws.

COMPASS OR KEYHOLE SAW

The compass saw, Figure 4-33, is used to make circular cuts in wood. The keyhole saw is

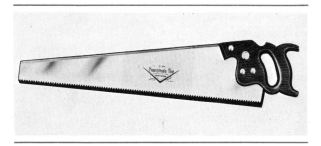

Fig. 4-31 Handsaw *(Courtesy of Stanley Tools)*

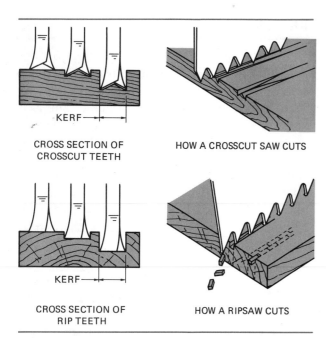

KERF

CROSS SECTION OF
CROSSCUT TEETH

HOW A CROSSCUT SAW CUTS

KERF

CROSS SECTION OF
RIP TEETH

HOW A RIPSAW CUTS

Fig. 4-32 Cutting action of hand rip and crosscut saws
(Courtesy of Disston Saw Company)

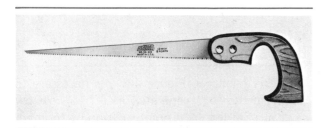

Fig. 4-33 Compass saw *(Courtesy of Stanley Tools)*

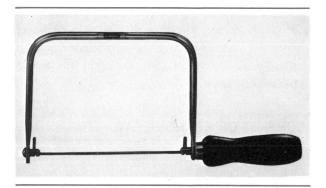

Fig. 4-34 Coping saw *(Courtesy of Stanley Tools)*

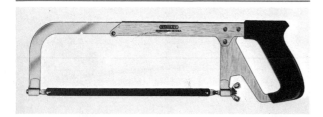

Fig. 4-35 Hacksaw *(Courtesy of Stanley Tools)*

similar to the compass saw except its blade is narrower. To start the saw cut, usually a hole needs to be bored.

COPING SAW

The coping saw, Figure 4-34, is used primarily to make certain joints between moldings, called coped joints. A *coped joint* is made by fitting the end of a molding against the face of a similar piece. The coping saw is also used to make small, irregular curved cuts.

HACKSAW

Hacksaws, Figure 4-35, are used to saw metal. Hacksaw blades are available with 18, 24, and 32 points to the inch. Coarse-toothed blades are for fast cutting in thick metal. Fine-toothed blades are used to cut thin metal. Make sure that blades are installed with the teeth pointing away from the handle.

WALLBOARD SAW

The wallboard saw, Figure 4-36, is designed especially for cutting gypsum board. The point is sharpened to make self-starting cuts for electric outlets, pipes, and other projections.

ASSEMBLING AND DISMANTLING TOOLS

In this classification are those tools used to drive screws, nails, and other fasteners. Tools used to hold or dismantle workpieces are also included in this category.

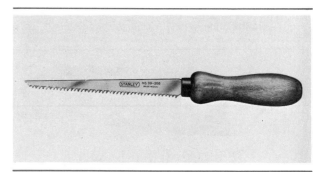

Fig. 4-36 Wallboard saw *(Courtesy of Stanley Tools)*

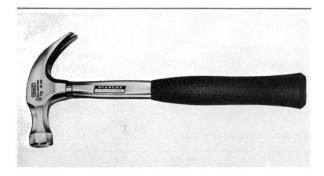

Fig. 4-37 Curved claw hammer *(Courtesy of Stanley Tools)*

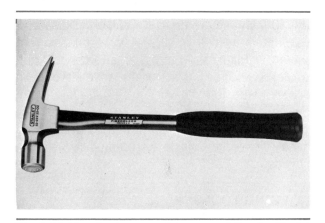

Fig. 4-38 Framing hammer *(Courtesy of Stanley Tools)*

Fig. 4-39 Nail set *(Courtesy of Stanley Tools)*

Fig. 4-40 Common screwdriver *(Courtesy of Stanley Tools)*

Fig. 4-41 Phillips screwdriver *(Courtesy of Stanley Tools)*

Fig. 4-42 Spiral screwdriver *(Courtesy of Stanley Tools)*

HAMMERS

The hammer is available in a number of styles and weights. The claws may be straight or curved. Weights range from 7 to 22 ounces. Most popular for general work is the 16-ounce, *curved claw hammer,* Figure 4-37. For rough work, the *framing hammer,* Figure 4-38, is often preferred. This hammer weighs 22 ounces and has a long handle for extra driving power.

NAIL SETS

Nail sets, Figure 4-39, are used to set nail heads below the surface. The most common sizes are 1/32, 2/32, and 3/32 inch. The size refers to the diameter of the tip. The surface of the tip is concave to prevent it from slipping off the nail head. If the tip becomes flattened, the nail set has lost its usefulness.

Note: Do not set hardened nails or hit the tip of the nail set with a hammer as this will cause the tip to flatten.

SCREWDRIVERS

Screwdrivers come in many styles and sizes. The size is determined by the length of the blade. *Common* screwdrivers, Figure 4-40, have a straight tip. *Phillips* screwdrivers have a cross-shaped tip, Figure 4-41. Carpenters should have access to a number of screwdrivers because of the different sizes and styles of wood screws. In addition, *spiral screwdrivers,* with interchangeable tips, Figure 4-42, may be desired to drive many screws at a faster rate with greater ease.

Screwdriver bits, Figure 4-43, are used with the bit brace to drive large screws with less effort. These bits are available in a number of different sizes to accommodate a variety of screws.

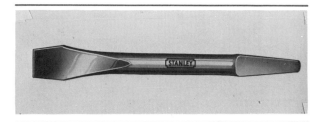

Fig. 4-43 Screwdriver bits *(Courtesy of Stanley Tools)*

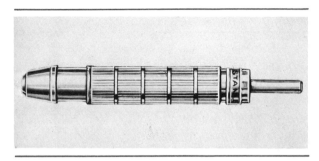

Fig. 4-44 Self-centering screw hole punch *(Courtesy of Stanley Tools)*

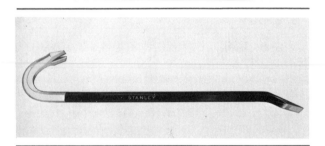

Fig. 4-45 Wrecking bar *(Courtesy of Stanley Tools)*

Fig. 4-46 Pry bar *(Courtesy of Stanley Tools)*

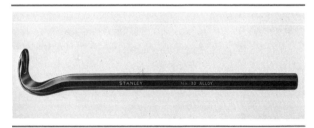

Fig. 4-47 Nail claw *(Courtesy of Stanley Tools)*

Fig. 4-48 Flooring chisel *(Courtesy of Stanley Tools)*

The *self-centering punch,* Figure 4-44, is used to center-punch screw holes on butt hinges and similar hardware. Its beveled end centers the punch on the countersunk hole of the hinge. The spring-loaded punch is lightly tapped with a hammer.

BARS AND PULLERS

The *wrecking bar,* Figure 4-45, is used to withdraw spikes and for prying purposes when dismantling parts of the structure. They are available in lengths from 12 to 36 inches, with the 30-inch size preferred for construction work. Carpenters need a small *pry bar* similar to that shown in Figure 4-46 to pry small work and pull small nails. To extract nails that have been driven home (all the way in) a *nail claw* (commonly called a *cat's paw*) is used, Figure 4-47. The *flooring chisel,* Figure 4-48, is often quite handy to wedge objects for holding or fastening purposes.

HOLDING TOOLS

To turn nuts, lag screws, bolts, and other objects, an *adjustable wrench* is often used, Figure 4-49. The wrenches are sized by their overall length. The 10-inch adjustable wrench is the one most widely used.

For extracting, turning, and holding objects, a pair of *pliers* such as shown in Figure 4-50, is often used. Small *C clamps,* Figure 4-51, are very useful to hold objects together while they are being fastened or to hold temporary guides.

SCRAPING AND SMOOTHING TOOLS

Scraping and smoothing tools include not only hand scrapers, but putty knives, tool

Fig. 4-49 Adjustable wrench *(Courtesy of Stanley Tools)*

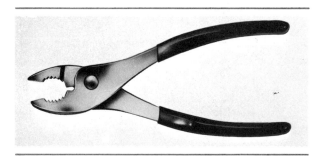

Fig. 4-50 Combination pliers *(Courtesy of Stanley Tools)*

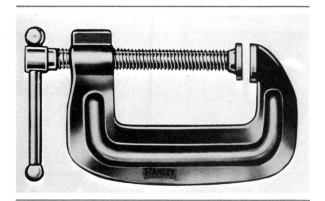

Fig. 4-51 C clamp *(Courtesy of Stanley Tools)*

Fig. 4-52 Hand scraper *(Courtesy of Stanley Tools)*

Fig. 4-53 Putty knife *(Courtesy of Stanley Tools)*

sharpening stones, files, and rasps. The carpenter must be able to care for and use all of these tools properly.

SCRAPER

The *hand scraper,* Figure 4-52, is used to remove old paint, dried glue, and pencil and crayon marks from the wood surface. The blades are reversible and replaceable. They also can be sharpened by filing against the cutting edge on the original bevel. **Note: Care should be taken not to hollow out the center of the blade; otherwise, the edges will dig into the wood.**

PUTTY KNIFE

The *putty knife,* Figure 4-53, is used to apply filler to cover nail heads and for similar purposes.

FILES AND RASPS

Files and rasps, Figure 4-54, are commonly used for smoothing parts that are difficult to smooth with other tools. There are a variety of shapes, sizes, and cuts. Generally *files* are used when little stock is to be removed and may be used to smooth metal. Files are also used for sharpening some tools such as saws and scrapers. *Rasps* are used on wood when considerable stock is to be smoothed.

Files and rasps should not be used without handles.

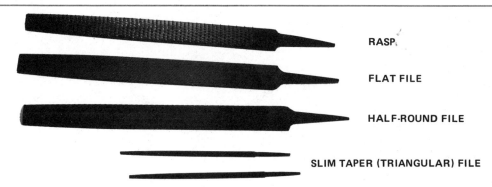

RASP

FLAT FILE

HALF-ROUND FILE

SLIM TAPER (TRIANGULAR) FILE

Fig. 4-54 Files and rasps

Fig. 4-55 Oilstone

OILSTONES

Oilstones, Figure 4-55, are used for *whetting* (sharpening) plane blades, chisels, and other tools. These stones are purchased in fine, medium, or coarse grits. Some stones are combination stones having a coarse grit on one side and a fine grit on the other side. The finer the grit of an oilstone, the keener the edge that can be obtained.

Oilstones should be kept in wood boxes to prevent breakage. To keep the stone clean and sharp, use a liberal amount of very light oil with every whetting.

OTHER TOOLS AND EQUIPMENT

In addition to the tools already mentioned in this unit, carpenters must have hard hats, safety goggles and shoes with steel toe caps, Figure 4-56.

Other hand tools that are used by the carpenter, but normally supplied by the contractor, include the *builder's level* (see Unit 8 Building Layout) and the *miter box,* Figure 4-57.

The miter box is used to cut angles on finish lumber. The angle cuts made to join pieces together are called miters. The miter box saw should only be used in the miter box.

Fig. 4-56 The carpenter needs a hard hat and safety glasses.

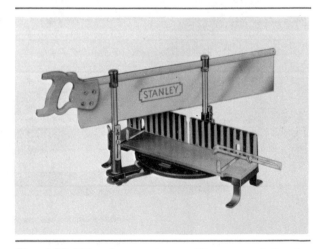

Fig. 4-57 Miter box and saw *(Courtesy of Stanley Tools)*

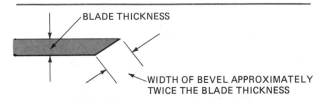

Fig. 4-58 Bevel width of wood chisel blade

SHARPENING HAND TOOLS

WOOD CHISELS

The bevel of the wood chisel is shaped by a grinding wheel. The bevel should have a concave surface which is called a hollow grind. In order to obtain a hollow grind, use a grinding attachment.

Use safety goggles or otherwise protect the eyes when grinding.

If a grinding attachment is not available, hold the chisel by hand on the tool rest at the proper angle. A general rule of thumb is that the width of the bevel is approximately twice the thickness of the chisel blade, Figure 4-58. Let the index finger of the hand holding the chisel ride against the outside edge of the tool rest as the chisel is moved back and forth across the revolving wheel. Never move the position of the guiding finger until grinding is complete. Cool the chisel by dipping it in water frequently to prevent overheating the metal. Grind the chisel until an edge is formed, Figure 4-59.

Grinding does not sharpen the tool, it only shapes it. To produce a keen edge, the chisel must be whetted on an oilstone. Hold the chisel on a well-oiled stone so that the bevel rests on it. Move the chisel back and forth across the stone for a few strokes. Then, make a few strokes with the flat side of the chisel held absolutely flat on the stone to avoid producing a bevel on the flat side. Continue whetting in this manner until as keen an edge as possible is obtained. To obtain a keener edge, repeat the procedure on a finer stone.

Chisels do not have to be ground each time they need sharpening. Grinding is necessary only when the bevel has lost its concave shape or the edge is badly nicked.

PLANES

Plane irons are ground and sharpened in the same manner as wood chisels. The slightest bevel on the flat side of the plane iron may cause a poor fit between it and the plane-iron cap and make planing difficult.

KNIFE

A knife should never be ground. Only whetting is required to bring a dull edge to its original condition. Take strokes on both sides of the blade against the cutting edge as if to peel off a thin slice from the top of the stone, Figure 4-60.

Fig. 4-59 Grinding a wood chisel

Fig. 4-60 Whetting a jackknife

57

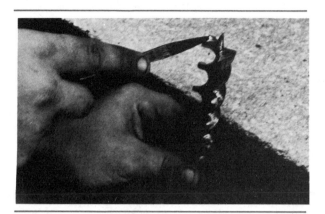

Fig. 4-61 Sharpening an auger bit

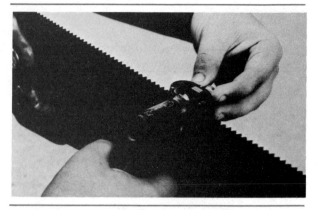

Fig. 4-62 Setting a handsaw

AUGER BITS

Auger bits are sharpened with an auger bit file. File across the inside of the spurs only enough to bring them to a sharp edge. Excessive filing of the spurs reduces their height, making the bit useless. File the cutting lips on the top side and on the original bevel until the edge is sharp, Figure 4-61.

HANDSAWS

Many carpenters prefer to send their handsaws to sharpening shops for reconditioning. Because of the expense and inconvenience, other carpenters prefer to sharpen their own.

The first step in sharpening a handsaw is called *jointing*. This is done by running a flat mill file over the tops of the teeth to bring them to uniform height. Do not joint any more than necessary; this avoids excessive filing later on. Jointing produces flats on the teeth which are quite visible.

Saws are next *set* with the use of a handsaw set, Figure 4-62. Setting is bending the teeth alternately right and left — providing clearance to keep the blade from binding in the cut. The saw set is adjusted to correspond with the number of points of the saw for the correct amount of set. Jointing and setting are not always necessary. If the teeth are straight and have sufficient set, these steps may be eliminated.

The last step is filing the teeth. Triangular files of the proper size are used. Ripsaws are filed at a 90° angle to the face with the file tilted to the correct angle, Figure 4-63. Every

tooth is filed from the same side. File each tooth until the shiny flat top disappears. Any more filing only makes the tooth smaller than it should be.

To file crosscut saws, hold the file at about 60° to the face of the blade and tilted to the original angle. File every other tooth from one side and the remaining teeth from the other side. Remember to stop filing when the shiny tip of the saw tooth disappears. This will produce teeth of uniform height.

HATCHETS AND SNIPS

Hatchets are whetted on an oilstone to obtain a keen edge. If the edge is nicked, it may be ground to remove the nick, but must be sharpened with an oilstone.

Snips do not dull rapidly. When sharpening is needed, it is best to take them to a professional sharpener.

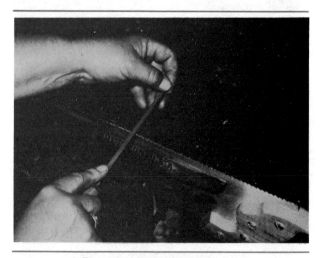

Fig. 4-63 Filing a hand ripsaw

REVIEW QUESTIONS

Select the most appropriate answer.

1. For easy operation of steel tapes, apply
   a. graphite.   c. wax.
   b. oil.   d. silicon.

2. In construction, the term plumb means
   a. horizontal.   c. straight.
   b. level.   d. vertical.

3. The most preferred chalkline is made of
   a. cotton.   c. nylon.
   b. linen.   d. manila.

4. A number 8 auger bit will bore a hole
   a. 1/4 inch in diameter.
   b. 1/2 inch in diameter.
   c. 3/4 inch in diameter.
   d. 1 inch in diameter.

5. Countersinks are tools used to
   a. make sink cut-outs in countertops.
   b. drive large screws.
   c. form recesses for flat head screws.
   d. set nails below the surface.

6. The longest of the bench planes is called a
   a. fore plane.   c. jointer plane.
   b. jack plane.   d. smooth plane.

7. For rough work, the carpenter uses a crosscut saw with
   a. 5 1/2 points.   c. 11 to 12 points.
   b. 7 to 8 points.   d. 16 to 18 points.

8. The kind of joint made by cutting one piece of molding to fit against the irregular
   face of a similar piece is made by using a
   a. chisel.   c. compass saw.
   b. coping saw.   d. countersink.

9. To prevent a nail set from slipping off the nail head, its tip is
   a. flat and smooth.   c. concave.
   b. convex.   d. checkered.

10. When sharpening handsaws, file only
    a. two strokes on each tooth.
    b. with very light pressure.
    c. until the shiny tip disappears.
    d. until the teeth are even.

PORTABLE POWER TOOLS 5

OBJECTIVES

After completing this unit, the student will be able to

- *state general safety rules for operating portable power tools.*
- *identify, describe, and safely use the following portable power tools: circular saws, saber saws, reciprocating saws, drills, hammer-drills, screwdrivers, planes, routers, sanders, staplers and nailers, and powder-actuated drivers.*

Portable power tools are widely used in the construction industry. They make the carpenter's work easier and faster with much less effort. However, they can cause serious injury if operated improperly. Most accidents are caused by a complete disregard of common sense safety precautions. Be aware of the dangers of the tool and take steps to avoid these dangers.

GENERAL SAFETY RULES

- **Have a complete understanding of the tool before attempting to operate it.**
- **Electrical shock has caused many fatal accidents. Make sure the tool is properly grounded.**
- **Do not allow your attention to be distracted when operating power tools.**
- **Never use a power tool that has a worn or frayed cord.**
- **Use the proper size extension cord.**
- **Be careful to place extension cords where they will not be damaged, cut, caught in the tool, or tripped on.**
- **Unplug tools when making adjustments or changing cutters.**
- **Do not wear loose clothing that might become caught in the tool.**
- **Wear safety goggles with any operation when particles are flying.**
- **Make sure the material being worked on is securely held.**

- **Use sharp tools only.**

- **Use all safety guards supplied with the tool.**

- **Stay alert and develop an attitude of care and concern for yourself and your fellow worker.**

ELECTRIC CIRCULAR SAW

Commonly called the *skilsaw,* the portable electric circular saw, Figure 5-1, is used by the carpenter more than any other portable power tool. The circular saw blade is driven by an electric motor. The saw has a base which rests on the work to be cut. A handle with a trigger switch is provided for the operator to control the tool. The base may be tilted for making beveled cuts. The saw is adjustable for depth of cut. A retractable safety guard is provided over the blade, extending from under the base.

Saws are manufactured in many styles and sizes. They are available from very light to heavy duty models ranging from 1/6 hp to 1 1/2 hp. The size is determined by the diameter of the blade which ranges from 4 1/2 inches to 12 inches. The handle and switch may be located on the top or in back. The blade may be driven directly by the motor or through a worm gear. These saws cut on the upstroke. The saw blade rotates upward through the material at the point where the teeth make contact. Guides are provided to follow the lines to be cut. Make sure the saw blade is installed with the teeth in the correct direction. The teeth of the saw blade projecting below the base should be pointing away from the operator.

SAW BLADES

Saw blades are available in a great number of styles. The shape and number of teeth per inch determine their cutting action. The chisel tooth combination blade is used for both crosscut and rip work and is the blade most frequently used for general carpentry.

Fine tooth blades are available for splinter-free cutting of thin plywood and other laminates. Carbide-tipped blades are used when cutting

Fig. 5-1 Using the portable electric circular saw to cut rafters *(Courtesy of Skil Corporation)*

material that would dull ordinary blades quickly and for longer life in general work.

MAKING CUTS

Safe and efficient cutting follows an established method:

- Make sure the work is securely held and that the waste will fall clear and not bind the saw blade.

- Adjust the depth of cut so that the blade just cuts through the work. Never expose the blade any more than is necessary, Figure 5-2.

- Make sure the guard operates properly. Be aware that the guard may possibly stick in the open position. *Never* wedge the guard back in an open position.

Fig. 5-2 The blade of the saw is adjusted for depth only enough to cut through the work.

- Mark the stock, put on safety goggles, and rest the forward end of the base on the work. With the blade clear of the material, start the saw.

- Advance the saw into the work, while observing the line to be followed. With the saw cut in the waste, cut as close to the line as possible for a short distance. Then observe the guide and follow the line by the guide on the saw.

- Follow the line closely. Any deviation from the line may cause the saw to bind and kick back. In case the saw does bind, stop the motor and then bring the saw back to where it will run free. Continue following the line closely.

- Nearing the end of the cut, the guide will go off the work. Guide the saw by observing the line at the saw blade and finish the cut. Let the waste drop clear, release the switch, and keep the saw clear of the body until the saw blade has completely stopped.

- When cutting across stock at an angle, it may be necessary to retract the guard by hand at the start of the cut. A handle is provided for this purpose, Figure 5-3. Release the handle after the cut has been started and continue as above.

- Compound miter cuts may be made by cutting across the stock at an angle with the base tilted.

Fig. 5-3 Retracting the guard of the portable electric saw by hand

Fig. 5-4 Making a plunge cut with the portable circular saw

MAKING PLUNGE CUTS

Many times it is necessary to make internal cuts in the material such as for sinks in countertops or openings in floors and walls. To make these cuts with a portable electric circular saw, the saw must be plunged:

- Accurately lay out the cut to be made and adjust the saw for depth of cut.

- Wearing safety goggles, hold the guard open and tilt the saw up with the front edge of the base resting on the work. Have the saw blade over, and in line with, the cut to be made, Figure 5-4.

- Make sure the teeth of the blade are clear of the work and start the saw. Lower the blade into the work, following the line carefully, until the base rests on the material. Advance the saw into the corner. Release the switch and wait until the saw stops before removing it from the cut.

- Proceed in like manner to cut the other sides. Cut into the corners with a handsaw or saber saw.

SABER SAW

The *saber saw*, Figure 5-5, is sometimes called a *jigsaw* or *bayonet saw* and is widely used

to make curved cuts. The teeth of the blade point upward when installed, so consequently the saw cuts on the upstroke. To produce a splinter-free cut on the face side, it is best to cut on the side opposite the face of the work, if possible.

There are many styles and varieties. The length of the stroke along with the capacity of the motor determines its size. Strokes range from 1/2 inch to 1 inch with the longest stroke being the best for faster and easier cutting. The base of the saw may be tilted to make bevel cuts.

Fig. 5-5 The saber saw
(Courtesy of Rockwell, International)

63

METAL CUTTING BLADES

Part/Item Code No.	Length	Blade Type	Teeth Per Inch	Applications
 24085	3 5/8″	tapered wavy set	12	1/8 - 5/16 ferrous and non-ferrous metals.
24086	3 5/8″	tapered wavy set	21	1/16 - 5/32 ferrous and non-ferrous metals.
24087	3 5/8″	tapered wavy set	36	Up to 1/16 ferrous and non-ferrous metals.

WOOD CUTTING BLADES

Part/Item Code No.	Length	Blade Type	Teeth Per Inch	Applications
 22644	3 5/8″	set and ground teeth (1/4″ wide)	6	Fast and efficient scroll cutting in all woods, plywoods and hardboard.
22652	3 5/8″	set and ground teeth (3/8″ wide)	6	Fast and efficient straight line cutting in all wood, plywood and hardboard.
 22650	3 5/8″	ground teeth, taper ground body	10	Smooth finish scroll cutting in wood, plywood, hardboard, plastic laminates and fiberglass.
22651	4″	ground teeth, taper ground body	6	Smooth finish straight cutting in wood, plywood, hardboard, plastic laminates and fiberglass.

MISCELLANEOUS APPLICATIONS

Part/Item Code No.	Length	Blade Type	Teeth Per Inch	Applications
 22649	3 5/8″	skip tooth (1/4″ wide)	8	Fast scroll cutting in fiberglass, tough abrasive materials, asbestos, 1/8 - 3/4″ aluminum and other non-ferrous metals, synthetics up to 3/4″ and plasterboard.
24084	3 5/8″	skip tooth (5/16″ wide)	8	Straight cutting in fiberglass, tough abrasive materials, asbestos, 1/8 - 3/4″ aluminum and other non-ferrous metals, synthetics up to 3/4″ and plasterboard.
71941	3 5/8″	set tooth (5/16″ wide)	12	General purpose metal, wood, plastic, cutting 1/8 - 1/4″; steel 1/8 - 1/2″, non-ferrous metals.

WOOD AND COMPOSITION CUTTING BLADES

Part/Item Code No.	Length	Blade Type	Teeth Per Inch	Applications
 22645	3 5/8″	ground teeth, taper ground body	12	Smooth finish straight cutting in thin wood, metal backed doors, plastic laminate counter tops, hard synthetics and composite materials.
 24089	3 5/8″	ground teeth, taper ground body	12	Smooth finish scroll cutting in thin wood, metal backed doors, plastic laminate counter tops, hard synthetics and composite materials.

Fig. 5-6 Saber saw blade selection guide *(Courtesy of Skil Corporation)*

Many blades are available for fine or coarse cutting in wood or metal, Figure 5-6. Wood cutting blades have teeth with from 6 to 12 points to the inch.

MAKING CUTS

Follow an established procedure:

- Outline the cut to be made and secure the work.

- Using safety goggles, hold the base of the saw firmly on the work and, with the blade clear, pull the trigger.

- Push the saw into the work, following the line closely. Make the saw cut into the waste and cut as close to the line as possible without removing it. Keep the saw moving forward, holding the base down firmly on the work. Turn the saw as necessary in order to follow the line to be cut. Feeding the saw into the work as fast as possible, but not forcing it, finish the cut. Keep the saw clear of the body until it has stopped.

MAKING PLUNGE CUTS

Plunge cuts are made with the sabre saw in a manner similar to that used with the circular saw:

- Tilt the saw up on the forward end of its base with the blade in line and clear of the work, Figure 5-7.

- Start the motor, holding the base steady. Very gradually lower the saw until the blade penetrates the work and the base rests firmly on it.

- Cut along the line into the corner. Back up for about an inch, turn the corner, and cut the other side into the corner. Continue in this manner until all the sides of the opening are cut.

- Turn the saw around and cut in the opposite direction to cut out the corners.

RECIPROCATING SAW

The reciprocating saw, Figure 5-8, sometimes called a *sawzall,* is used primarily for roughing-in work. This work consists of cutting holes and openings for such things as pipes, heating and cooling ducts, and roof vents. It can be likened to a powered compass saw.

Most models have a variable speed of from 0 to 2400 strokes per minute. For greater cutting

Fig. 5-7 Making an internal cut by plunging the saber saw

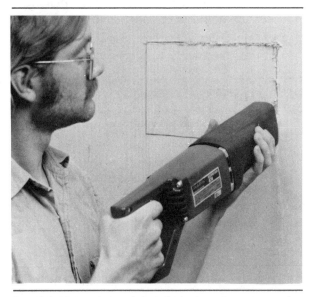

Fig. 5-8 Using the reciprocating saw to cut an opening *(Courtesy of Skil Corporation)*

speed, some models may be switched to an orbital cutting action. Common blade lengths run from 4 to 12 inches long and are available for practically every type of sawing purpose.

USING THE RECIPROCATING SAW

The reciprocating saw is used in a manner similar to that used with the saber saw. The difference is that the reciprocating saw is heavier and more powerful; therefore, it can be used better to cut through rough, thick material, such as walls when remodeling. With a long blade it can be used to cut flush with a floor or the side of a stud.

To use the saw, lines must be laid out and followed. The base or shoe of the saw is held firmly against the work whenever possible.

MAKING PLUNGE CUTS

Plunge cuts are made in a manner opposite to that used with a saber saw. Check for location of electrical wiring and plumbing before making the cut.

* Tilt the saw *down* on the bottom edge of its shoe.
* Hold the blade above the layout line and clear of the work.
* Start the saw and gradually lift the handle upward while keeping the shoe firmly on the work. Let the blade slowly penetrate the work and continue the cut.
* Remove the blade from the work only after it has completely stopped.

DRILLS, HAMMER-DRILLS, SCREWDRIVERS

Drills are manufactured in a great number of styles and sizes, Figure 5-9. The size of an electric drill is determined by the capacity of the chuck. The *chuck* is that part of the drill that holds the cutting tool. The most popular sizes are 1/4 inch, 3/8 inch, and 1/2 inch. The drill has an encased motor that drives the chuck. The lightweight drills usually have a pistol grip type handle, and heavier drills usually have a double handle.

Twist drills are used in the electric drills to make small holes in wood or metal. For larger holes in wood, a variety of wood-cutting bits are used, Figure 5-10.

> CAUTION: Never use bits with threaded center points in electric drills when boring deep holes, because the threaded center will draw the bit into the work — making it difficult to withdraw the bit and possibly causing the operator to lose control of the drill.

USING PORTABLE ELECTRIC DRILLS

* Select the proper size bit or twist drill and insert it. Tighten the chuck with the chuck key.
* For accuracy, holes in wood are center punched. Holes in metal must be center punched because the drill will wander off center.

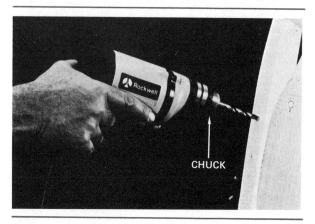

Fig. 5-9 Portable electric drill *(Courtesy of Rockwell, International)*

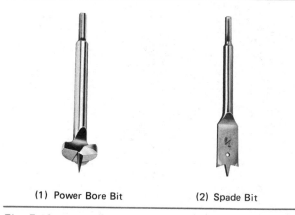

(1) Power Bore Bit   (2) Spade Bit

Fig. 5-10 Wood-boring bits used in power drills *(Courtesy of Stanley Tools)*

- Hold small pieces securely by clamping or other means. Place the bit on the center of the hole to be drilled, hold the drill at the desired angle, and start the motor.

- Apply pressure as required and drill into the stock, being careful not to wobble the drill.

CAUTION: Remove the bit from the hole frequently to clear the chips. Failure to do this may result in the drill binding and twisting from the operator's hands. Be ready to instantly release the trigger switch if the drill does bind.

HAMMER-DRILLS

Hammer-drills, Figure 5-11, are similar to other drills except they can be changed to hammer as they drill, quickly making holes in concrete or other masonry. Some models deliver as much as 19,000 hammer blows per minute. Most popular are the 3/8-inch size and the 1/2-inch size. A depth stop is usually attached to the side of the hammer-drill. It can be converted to a conventional drill by a quick-change mechanism. Most models have a variable speed of from 0 to 2500 rpm.

The hammer-drill has the same chuck and is used in the same manner as conventional drills. Carbide-tipped bits are used for drilling into masonry.

ELECTRIC SCREWDRIVERS

Electric screwdrivers, Figure 5-12, are similar in appearance to light-duty drills except for the chuck. Most screwdrivers have a pistol-type grip for one hand operation with a variable speed and reversing switch.

The chuck is made to receive special screwdriver bits of various shapes and sizes. Two

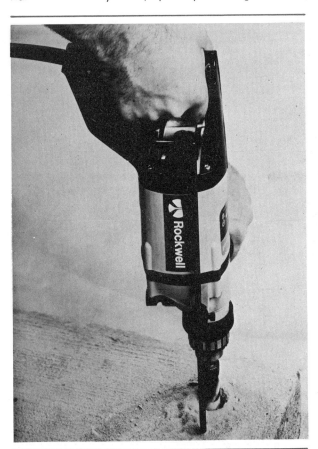

Fig. 5-11 The hammer-drill is used to make holes in concrete. *(Courtesy of Rockwell, International)*

Fig. 5-12 The portable electric screwdriver *(Courtesy of Rockwell, International)*

Fig. 5-13 Using the portable electric jointer plane
(Courtesy of Skil Corporation)

Fig. 5-14 Using the portable electric block plane
requires the use of only one hand.

popular models have a positive clutch or an adjustable clutch.

USING THE ELECTRIC SCREWDRIVER

In the positive clutch model, the screw is turned when the trigger is pulled and pressure is applied to the screw. The amount of tightening is dependent on the operator. As soon as pressure is released, the screwdriver stops turning. This model is preferred when fastening different kinds of materials or driving different size screws.

For uniform fastening using identical screws in the same type material, the adjustable clutch model is used. The clutch may be adjusted for different amounts of torque. The screwdriver automatically stops turning when the set torque is reached.

Accessories are available for driving nuts as well as for driving screws. The screwdrivers are extensively used for fastening gypsum board to walls and ceilings with screws.

POWER PLANES

The two basic types of available power planes are the *jointer plane,* Figure 5-13, and the *block plane,* Figure 5-14. The jointer plane (about 18 inches) is much longer than the block plane (about 7 inches). The block plane is operated with one hand, while the jointer plane takes two hands for operation.

The jointer plane is used on large jobs such as fitting doors in openings. The block plane is used to smooth the edges and ends of small pieces.

The planes are similar in construction. An electric motor powers a spiral cutting blade which measures 3 inches wide in the jointer plane and 1 3/4 inches wide in the block plane. The base has an adjustment for depth of cut. A side guide permits square or beveled planing.

OPERATING POWER PLANES

CAUTION: **Extreme care must be taken when operating power planes. Their high-speed cutters are unguarded.**

- Set the side guide to the desired angle and adjust the depth of cut.

- Hold the toe (front) firmly on the work with the plane cutter clear.

- Start the motor and with a steady, even pressure make the cut through the work. Guide the angle of the cut by holding the guide against the side of the stock. Apply pressure to the toe of the plane at the beginning of the cut and to the heel (back) at the end of the cut to prevent tipping the plane over the ends of the work.

- Keep the tool clear of the body until it has completely stopped.

ELECTRIC ROUTERS

One of the most versatile portable tools used in the construction industry is the *router,* Figure 5-15. It is available in many models

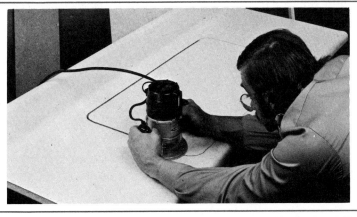

Fig. 5-15 Using the portable electric router to make a sink cutout *(Courtesy of Skil Corporation)*

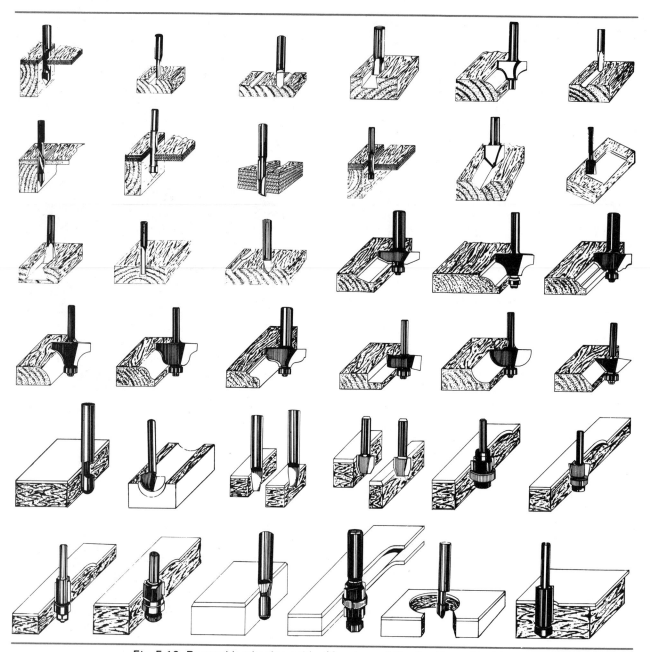

Fig. 5-16 Router bit selection guide *(Courtesy of Stanley Power Tools)*

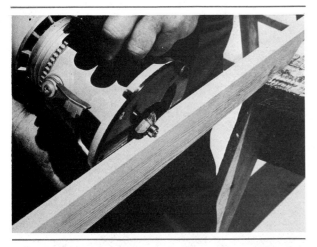

Fig. 5-17 Using a router bit with a pilot for a guide

Fig. 5-18 Using a straightedge to guide the router

ranging from 1/4 hp to 2 1/2 hp with speeds of from 18,000 rpm to 27,000 rpm. These tools have the high-speed motors that are necessary to produce clean, smooth-cut edges.

The motor powers a chuck in which cutting bits of various sizes and shapes are held, Figure 5-16. An adjustable base is provided to control the depth of cut. A trigger or toggle switch controls the motor.

The router is used to make many different cuts such as grooves, dadoes, rabbets, and dovetails. It is also used to shape edges and make cutouts such as for sinks in countertops. It is extensively used with accessories to cut in hinges for doors.

GUIDING THE ROUTER

Controlling the sideways motion of the router is accomplished by the following methods:

- By using a router bit with a pilot (guide), Figure 5-17.

- By guiding the edge of the router base against a straightedge, Figure 5-18.

- By using a guide attached to the base of the router, Figure 5-19.

Fig. 5-19 A guide attached to the base of the router controls the sideways motion.
(Courtesy of Black & Decker)

71

Fig. 5-20 Guiding the router by means of a template and template guide *(Courtesy of Rockwell, International)*

Fig. 5-21 Freehand routing
(Courtesy of Black & Decker)

- By using a template (pattern) with template guides attached to the base of the router, Figure 5-20.
- By freehand routing in which the sideways motion is controlled by the operator, Figure 5-21.

USING THE ROUTER

Before beginning to use the router, make sure power is disconnected. Follow the method outlined:

- Select the correct bit for the type of cut to be made.
- Insert the bit in the chuck. Make sure the chuck grabs at least 1/2 inch of the bit. Adjust the depth of cut.
- Control the sideways motion of the router by one of the methods previously described.
- Clamp the work securely in position and plug in the cord.

- Lay the base of the router on the work with the router bit clear of the work and start the motor.
- Advance the bit into the cut, pulling the router in a direction that is against the rotation of the bit. On outside edges and ends, the router is moved counterclockwise around the piece. When making internal cuts, the router is moved in a clockwise direction.
- Finish the cut, keeping the router clear of the body until it has stopped.

Be aware that the router bit is unguarded.

ELECTRIC SANDERS

The *belt sander,* Figure 5-22, is used frequently for sanding cabinetwork and interior finish. The size of the belt determines the size of the sander. Belt widths range from 2 to 4 inches and belt lengths from 21 to 24 inches. The 3 by 21-inch belt sander is a popular, lightweight model. Some sanders have a bag to collect the sanding dust. Sanding belts are available in grits from very fine to very coarse.

Belts are usually installed by retracting the forward roller. There is also an adjustment on the forward roller to keep the sanding belt centered. An arrow is stamped on the inside of the sanding belt to indicate the direction in which the belt should run. The sanding belt should run with its lap (where it is joined) and not against it.

USING THE BELT SANDER

- Secure the work to be sanded and make sure the belt is centered on the rollers and tracking properly.

- Start the machine and place the pad of the sander flat on the work. Pull the sander back and lift it clear of the work at the end of the backward stroke.

- Bring the sander forward and continue sanding, using a skimming motion. Sanding in this manner prevents overheating the

sander, belt, and material being sanded. It allows debris to be cleared from the work, and the operator to see what has been done.

- Do not sand in one spot too long. Be careful not to tilt the sander in any direction. Always sand with the pad flat on the work. Do not exert excessive pressure. The weight of the sander is enough. Always sand with the grain to produce a smooth finish. Make sure the sander has stopped before setting it down.

FINISHING SANDER

The finishing sander, Figure 5-23, is used for the final sanding of interior work. The machines are manufactured in many styles and are classified according to their motion.

Finishing sanders either have an orbital (circular) motion, a straight line (back and forth) motion, or a combination of motions controlled by a switch. The orbital motion has a faster cutting action, but leaves scratches across the grain. The straight line motion cuts slower, but produces a scratch-free surface.

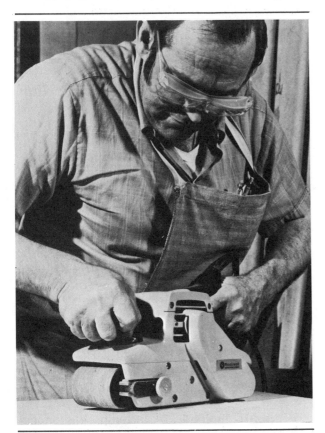

Fig. 5-22 Using the portable electric belt sander *(Courtesy of Rockwell, International)*

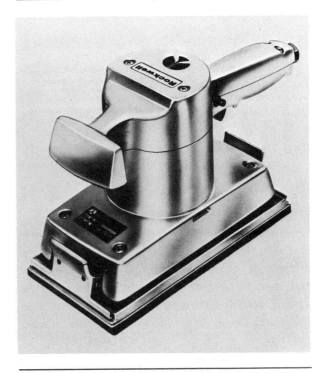

Fig. 5-23 The portable electric finishing sander *(Courtesy of Rockwell, International)*

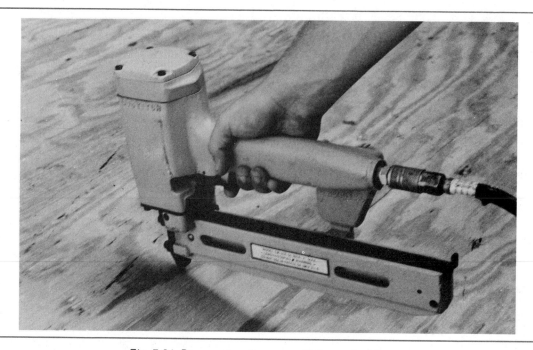

Fig. 5-24 Pneumatic nailers are widely used to fasten
subflooring and wall and roof sheathing.

Most sanders take one-third or one-half of a sheet of sandpaper. It is usually attached to the pad by some type of friction or spring device.

To use the finishing sander, follow the established procedure:

• Select the desired grit sandpaper and attach it to the pad, making sure it is tight. A loose sheet will tear easily.

• Start the motor and sand the surface evenly, *slowly* pushing and pulling the sander with the grain. Do not use excessive pressure because this may overload the machine and burn out the motor. Always hold the sander flat on its pad.

STAPLERS AND NAILERS

Staplers and nailers, Figure 5-24, are used widely for quick fastening of subfloors and of wall and roof sheathing, among other things. These tools are generally pneumatic (operated by compressed air) and are commonly called *stapling or nailing guns.* All are equipped with a cartridge or magazine which is loaded with specially designed fasteners. The fasteners are available in many sizes.

USING STAPLERS AND NAILERS

Because of the many different designs and sizes of staplers and nailers, you should study the manufacturer's directions and follow them carefully:

• Load the magazine with the desired size staplers or nails.

• Point nailer at ground (or away from oneself).

• Connect the air supply to the tool. Make sure there is an oiler in the air supply line with adequate oil to keep the gun lubricated during operation. Use the recommended air pressure.

• Press the trigger and tap the nose of the gun to the work. When the trigger is depressed, a fastener is driven each time the nose of the gun is tapped to the work.

• Upon completion of fastening, disconnect the air supply. Never leave an unattended gun with the air supply connected.

CAUTION: Always keep the gun pointed toward the work. Never point it at other workers or fire a staple except into the work. A serious injury can be the result of horseplay with the tool.

POWDER-ACTUATED DRIVERS

Powder-actuated drivers, Figure 5-25, are used to drive specially designed pins into masonry or steel. They are used in a manner similar to firing a gun. Powder charges of various strengths drive the pin when detonated.

Drivepins are available in a great variety of sizes. Three styles are commonly used. The *headed* type is used for fastening material. The *threaded* type is used to bolt an object after the pin is driven. The *eyelet* type is used when attachments are to be made with wire.

The powder charges are color-coded according to strength. Learn the color codes for immediate recognition of the strength of the charge.

Because of the danger involved in operating these guns, many states require certification of the operator. Certificates may be obtained from the manufacturer's representative after a brief training course.

USING POWDER-ACTUATED DRIVERS

Study the manufacturer's directions for safe and proper use of the gun:

- Make sure the drivepin will not penetrate completely through the material into which it is driven. This has been the cause of fatal accidents.

- To prevent ricochet hazard, make sure the recommended shield is in place on the nose of the gun. A number of different style shields are available for special fastening jobs.

- Select the proper fastener for the job. Consult the manufacturer's drivepin selection chart to determine the correct fastener size and style.

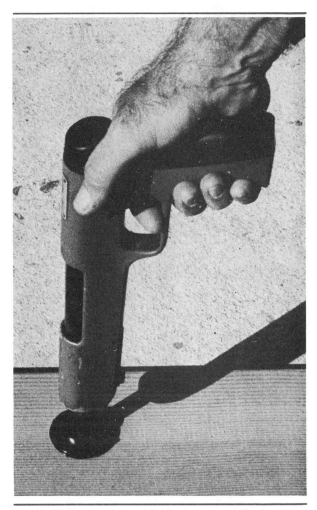

Fig. 5-25 Powder-actuated drivers are used for fastening into masonry or steel.

- Select a powder charge of necessary strength. Always use the weakest charge that will do the job. Load the driver with the pin first and the cartridge second.

- Keep the tool pointed at the work. Wear safety goggles, press hard against the work surface, and pull the trigger. The resulting explosion drives the pin. Eject the spent cartridge.

CAUTION: If the gun does not fire, hold it against the work surface for at least 30 seconds. Then remove the cartridge according to the manufacturer's directions. Do not attempt to pry out the cartridge with a knife or screwdriver because most of the cartridges are rim-fired.

REVIEW QUESTIONS

Select the most appropriate answer.

1. To prevent electrical shock, make sure the tool is properly
   a. connected.   c. insulated.
   b. grounded.   d. wired.

2. The guard of the portable electric saw should never be
   a. lubricated.   c. retracted by hand.
   b. adjusted.   d. wedged open.

3. The saber saw is used primarily for making
   a. curved cuts.   c. internal cuts.
   b. holes.   d. straight cuts.

4. The reciprocating saw can be compared most to a
   a. hammer-drill.   c. saber saw.
   b. powered compass saw.   d. vibrating sander.

5. The size of an electric drill is determined by its
   a. ampere rating.   c. weight.
   b. horsepower.   d. chuck capacity.

6. In power drills, never use bits with
   a. a double twist.   c. a single spur.
   b. no center.   d. a threaded center.

7. To prevent a drill bit from binding in a hole
   a. clear chips frequently.
   b. drill at a slow speed.
   c. drill at a high speed.
   d. use bits with a twist.

8. The amount of screw tightening in the positive clutch model electric screwdriver
   is dependent on the
   a. adjustment of clutch torque.
   b. release of pressure.
   c. size of the screw being driven.
   d. ampere rating of the screwdriver.

9. The portable electric jointer plane is well-suited to
   a. nose the edge of closet shelves.
   b. plane miter joints to fit.
   c. trim end grain of stair treads.
   d. plane doors to fit in openings.

10. When using the router to make shaped edges and ends, the router is pulled in a
    a. direction with the grain.
    b. clockwise direction.
    c. counterclockwise direction.
    d. direction with the rotation of the bit.

RADIAL ARM AND TABLE SAW

6

OBJECTIVES

After completing this unit, the student will be able to

- *describe different types of circular saw blades and select the proper blade for the job at hand.*
- *sharpen high-speed steel circular saw blades.*
- *describe, adjust, and operate the radial arm saw and the table saw safely to cross-cut and rip lumber.*
- *operate the radial arm and table saws safely to make miters, compound miters, dadoes, grooves, and rabbets.*
- *operate the table saw safely to taper rip and to make cove cuts.*

In most cases, the radial arm saw and the table saw are the only stationary power wood-working tools available to the carpenter on the job site. Sometimes both are available, and at other times, either one or the other, or none, are provided. Ordinarily, the radial arm saw is brought to the site when construction begins and is used for cutting framing members. The table saw is provided when the finish work begins.

The student must be able to operate these tools in a safe and efficient manner. Both tools use circular saw blades and to help insure safe and efficient operation, it is important to know which type to select for a particular use.

CIRCULAR SAW BLADES

The more teeth there are in a saw blade, the smoother the cut will be; however, the stock must be fed more slowly. A coarse tooth saw blade leaves a rough edge, but the stock may be fed faster by the operator. If the feed is too slow, the blade may overheat which causes it to lose its shape and wobble. The saw will then start to bind in the cut and possibly cause a kickback. The same results occur from trying to cut material with a dull blade. Therefore, always use a sharp blade, use fine tooth blades for cutting thin dry material, and coarse tooth blades for cutting heavy, rough lumber.

Saw blades are classified into two general groups. In one group, the saw blade is made with high-speed steel. In the other group, a steel blade has *tungsten carbide cutting tips.* The carbide-tipped blades stay sharp longer than steel ones and are used for cutting material that contains adhesives or other foreign material such as particleboard, plywood, hardboards, and plastics.

In both groups of saw blades, the number and shape of the teeth vary to give different cutting action according to the kind, size, and condition of the material to be cut.

HIGH SPEED STEEL CIRCULAR SAW BLADES

Steel blades are classified as *rip, crosscut,* or *combination* blades. They may sometimes be given other names such as plywood or panel saws, but they are still in one of the three classifications just named.

Ripsaw blade. The ripsaw blade, Figure 6-1, usually (but not always) has less teeth than crosscut or combination blades. Every tooth of the ripsaw is filed or ground at right angles to the face of the blade. This produces teeth with

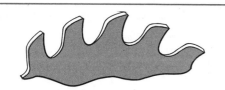

Fig. 6-1 The teeth of a rip saw blade have square-edge cutting tips.

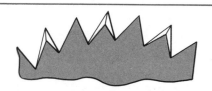

Fig. 6-2 The teeth of a crosscut saw blade have beveled sides and pointed tips.

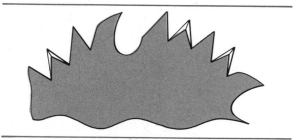

Fig. 6-3 The teeth of a combination blade

a cutting edge all the way across the tip of the tooth. Look for this cutting edge to determine a ripsaw.

Use a ripsaw for cutting solid lumber with the grain when a smooth edge is not necessary. Also use a ripsaw when cutting unseasoned lumber and lumber of heavy dimension with the grain.

Crosscut saw blade. The teeth of a crosscut circular saw blade are shaped similar to those of the crosscut handsaw, Figure 6-2. The sides of the teeth are alternately filed or ground on a bevel. This produces teeth that come up to points instead of edges. Look for these beveled sides and points on the teeth to determine a crosscut blade.

The shape slices across the wood fibers smoothly and is an ideal blade for cutting across the grain of solid lumber. Crosscut blades also cut plywood satisfactorily with little splintering of the cut edge.

Combination blade. The combination blade is used when a variety of ripping and crosscutting is to be done. This type blade eliminates the need of changing blades for different operations of crosscutting and ripping.

There are several types of combination blades. One type has groups of teeth around its circumference, Figure 6-3. The leading tooth in each group is a rip tooth while the ones following are crosscut teeth.

The *chisel-point* combination blade, Figure 6-4, is probably used more than any other blade in the construction industry. It is generally available with from 20 to 40 teeth per blade depending on the size of the blade. Every tooth is the same shape. Something like a ripsaw, the teeth are given a slight alternate bevel to produce combination crosscut and rip cutting action.

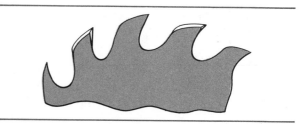

Fig. 6-4 The teeth of a chisel-point combination blade

PROVIDING CLEARANCE IN THE SAW CUT

All saw blades must have some provision for clearance in the saw cut. Without this clearance the saw blade, no matter how sharp, will bind in the saw cut. The teeth of the blades just described are set in a manner similar to that used for handsaws to provide this clearance. Special circular saw blade setting tools are needed for this purpose. Small circular saw blades may be set with the same type of saw set used for hand saws. However, larger blades need special saw-setting tools.

Another type of blade provides this clearance by being taper ground. A *taper ground blade* is thicker at the tips of the teeth and is ground thinner towards its center, Figure 6-5. Its teeth have no set. The taper of the blade provides the clearance in the saw cut. A taper ground blade can be determined by the ridge of thicker metal around the arbor hole.

The reason for using a taper ground blade is to obtain an extremely smooth edge. Taper ground blades usually are a combination type and are called *planer* blades. Use planer blades only on straight, dry lumber of relatively small dimension. Taper ground blades do not provide much clearance, and the slightest twist in the stock will cause it to bind against the blade.

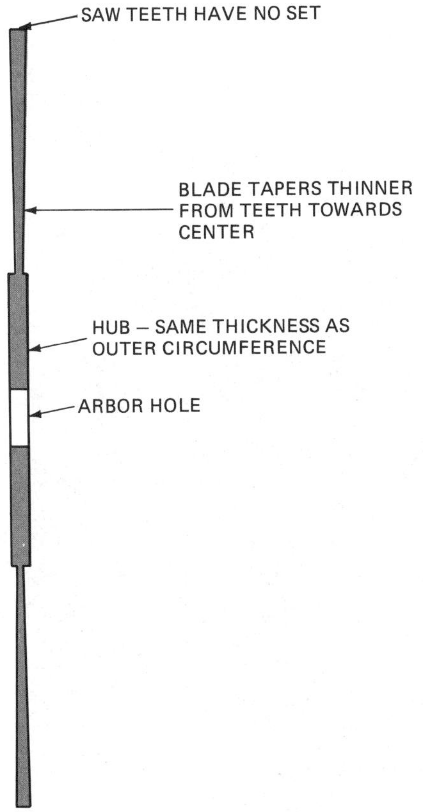

CARBIDE-TIPPED BLADES

Although carbide-tipped blades are shaped in many ways for various purposes, three main styles of teeth are commonly used, Figure 6-6. The *square grind* is similar to the rip teeth in a steel blade and is used primarily to cut solid wood with the grain. It can also be used on composition boards when the quality of the edge is not important.

The *alternate top bevel grind* is used with excellent results for crosscutting solid lumber and also for cutting plywood, hardboard, particleboard, and fiberboard.

The *triple chip grind* is designed for cutting brittle material without splintering or chipping the surface. It is particularly useful for cutting an extremely smooth edge on plastic laminated material. It can also be used like a planer blade to produce a smooth edge on straight, dry lumber of small dimension.

SAW TEETH HAVE NO SET

BLADE TAPERS THINNER FROM TEETH TOWARDS CENTER

HUB — SAME THICKNESS AS OUTER CIRCUMFERENCE

ARBOR HOLE

Fig. 6-5 Cross section of taper ground circular saw planer blade

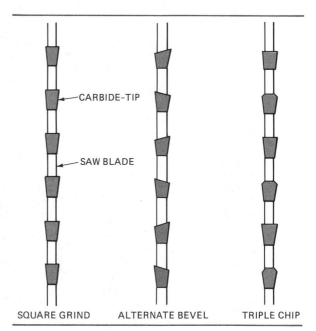

SQUARE GRIND   ALTERNATE BEVEL   TRIPLE CHIP

Fig. 6-6 Main styles of carbide-tipped saw blades — top view

Carbide-tipped saws also come with a combination of teeth. The leading tooth in each set is square ground, while the following teeth are ground at alternate bevels. Carbide-tipped teeth are not set. The carbide tips are slightly thicker than the saw blade and, therefore, provide clearance for the blade in the saw cut.

REMOVING AND REPLACING CIRCULAR SAW BLADES

Saw arbor shafts may have a right- or left-hand thread for the nut depending on which side of the blade it is located. No matter what direction the arbor shaft is threaded, the arbor nut is loosened in the same direction as the saw blade rotates, Figure 6-7. The arbor nut is always tightened against the rotation of the saw blade. Saws are designed in this manner to prevent the arbor nut from loosening during operation.

Make sure power is disconnected before changing saw blades.

SHARPENING CIRCULAR SAW BLADES

Carbide-tipped blades are sent out to shops that have specialized equipment for sharpening. Usually high-speed steel blades are also sent to sharpening shops. However, in an emergency, it is sometimes necessary to bring these blades back into condition on the job.

Jointing. To sharpen a steel circular saw blade, it first must be jointed. With the blade on the table saw, lower it until the teeth are below the surface of the table. Hold an oilstone over the teeth of the blade, start the machine and **slowly** raise the blade until the teeth come in contact with the stone, Figure 6-8. Stop the machine and inspect the teeth to see if all or most of the teeth have been touched by the stone. Do not joint any more than is necessary.

Gumming. With the blade still mounted on the table saw and the power disconnected, raise the blade until the point of a pencil held flat on the table and against the side of the blade is just below the deepest gullet (the valley between teeth). Rotate the blade by hand so that a

Fig. 6-7 The saw arbor nut is loosened in the same direction that the blade rotates

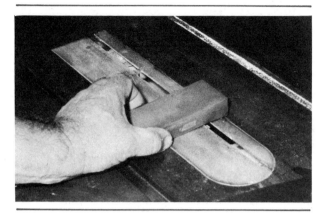

Fig. 6-8 Jointing a circular saw blade

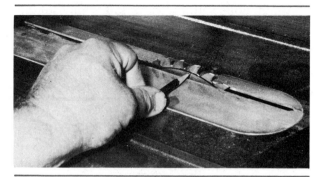

Fig. 6-9 Marking the saw blade for gumming

pencil mark is made around it, Figure 6-9. This will be a guide for grinding the gullets to their proper depth (gumming).

Remove the blade from the machine and grind the gullets using a bench grinder with a narrow round-edge grinding wheel, Figure 6-10. Grind down to the pencil mark, maintaining the original shape of the gullet. Be careful not to overheat the teeth. If necessary, the gullets may be filed instead of ground.

Use eye protection when operating the grinder.

Setting. The teeth of the blade are set by using a circular saw set. The saw teeth are laid on a beveled anvil so that about the top half of the tooth is bent outward. This is done by striking the side of the tooth with a punch. Set the teeth alternately. Crosscut and chisel point teeth are set with their cutting points bent outward.

Filing. For filing hold the saw blade in a circular saw vise. The teeth are filed in the same manner as for handsaws. Try to maintain a uniform shape. Use flat or triangular files as necessary. File only until the shiny tip of the tooth disappears, Figure 6-11.

OPERATING THE RADIAL ARM AND TABLE SAW

When operating stationary power tools, follow the general safety rules that apply for portable power tools as outlined in Unit 5. In

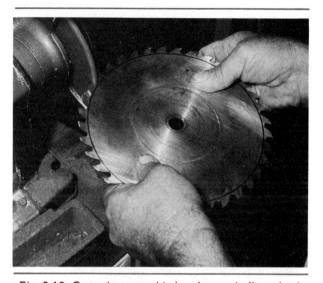

Fig. 6-10 Gumming a saw blade using a grinding wheel

Fig. 6-11 File saw teeth only until the shiny tips disappear

addition, follow the specific safety precautions outlined in this unit which are given when the various operations are described.

RADIAL ARM SAW

The radial arm saw, Figure 6-12, is ideally suited for crosscutting operations. The stock remains stationary while the saw moves across it. Long lengths are easily cut to shorter lengths. The cut is made above the work and all layout lines are clearly visible. The radial arm saw may also be used for ripping operations. In many cases, this saw is brought to the job site when construction begins.

The size of the radial arm saw is determined by the diameter of the largest blade that can be used. The arm of the saw moves horizontally in a complete circle, Figure 6-13. The motor unit tilts to any desired angle and also rotates in a complete circle. The depth of cut is controlled by raising or lowering the arm. This flexibility allows practically any kind of cut to be made.

CROSSCUTTING

For straight crosscutting, make sure the arm is at right angles to the fence. Adjust the depth of cut so that the teeth of the saw blade are about 1/16 inch below the surface of the table. With the saw all the way back and all guards in place, hold the stock against the fence and make the cut by bringing the saw forward and cutting to the layout line, Figure 6-14. When crosscutting stock thicker than the capacity of the saw, cut through half the thickness, turn the stock over and make another cut.

When holding the stock on the left side of the blade, hold with the left hand and pull the saw with the right hand. When holding the stock on the right side of the blade, hold with the right hand and pull the saw with the left hand. Do not cross the hands over each other. When the cut is complete, return the saw to the starting position and turn off the power.

CAUTION: Do not pull the saw quickly through the work. Forcing the saw may cause it to jam in the work or ride over it, possibly causing injury to the operator.

Fig. 6-12 The radial arm saw is well-suited for crosscutting operations. *(Courtesy of DeWalt)*

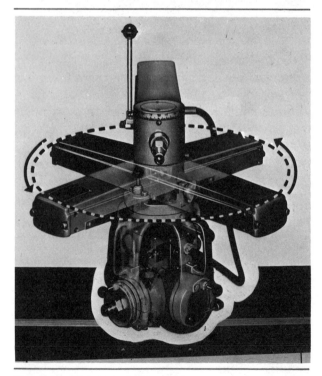

Fig. 6-13 The arm of the radial saw moves horizontally in a complete circle. *(Courtesy of Rockwell, International)*

Fig. 6-14 Straight crosscutting, using the radial arm saw *(Courtesy of Rockwell, International)*

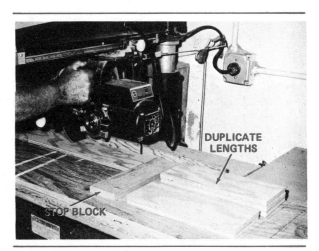

Fig. 6-15 Cutting duplicate lengths

CUTTING IDENTICAL LENGTHS

When many pieces of the same length need to be cut, clamp or fasten a stop block to the table in the desired location. Cut one end of the stock, slide the cut end against the stop block, and cut the other end, Figure 6-15. Continue in this manner until the desired number of pieces are cut. Be careful not to bring the stock against the stop block with too much force. This may move the stop block, resulting in pieces of unequal length. Also be careful that no sawdust or wood chips are trapped between the stock and the stop block. This results in pieces shorter than desired. A stop block with its end rabbeted helps prevent sawdust or chips being trapped between the stop and the piece being cut.

> **CAUTION: Replace the fence when saw cuts in it become too wide or too close together. Wide spaces in the fence allow small pieces to shoot out toward the rear. The pieces can bounce back and cause injury to the operator.**

BEVEL CROSSCUTTING

Bevel crosscutting is done in the same manner as straight crosscutting except that the motor unit is tilted to the desired angle. The *bevel locking pin* locks the motor unit into a 90°, 45°, or 0° angle position. The saw is held at any other angle by the *bevel clamp* only.

MITERING

To cut a *miter* (an angle cut across the stock), the arm of the saw is rotated to the right or left to the desired angle. Usually the arm is swung to the right for most miter cuts. The *miter latch* locks the arm in the 90° or 45° angle position. For any other angle, the arm is held in position by the *miter clamp* only. Miter cuts are made in the same manner as straight crosscutting, Figure 6-16.

Fig. 6-16 Making miter cuts

Fig. 6-17 Using a mitering jig on the radial arm saw

Mitering Jig. Miters may also be cut with the use of a *mitering jig.* One type of mitering jig consists of a piece of plywood to which strips are fastened at a 45° angle to its edge. The plywood is clamped to the radial arm table surface. The pieces are cut by holding them at a 45° angle while the radial arm saw is in its straight crosscutting position, Figure 6-17.

Using this jig allows right and left miters, such as for door and window casings, to be cut without swinging the arm of the saw. Mitering jigs may also be constructed to cut miters other than 45°.

Making Compound Miters. A compound miter cut is a combination of a miter and a bevel cut. Compound miters are frequently made on some types of roof rafters. The compound miter cut is made by adjusting the arm and also tilting the motor to the desired angles, Figure 6-18.

RIPPING OPERATIONS

For ripping operations, the arm is locked in the straight crosscutting position at right angles to the fence. The motor unit is rotated horizontally to either the in-rip or the out-rip position. The out-rip position is used to rip wider material. Lock the saw carriage in the desired position by

tightening the *rip lock* against the side of the arm. Lower the motor until the saw teeth are just below the table surface.

Adjust the safety guard so that the in-feed end almost touches the material to be cut. This is done to help prevent sawdust from flying at the operator. Lower the kickback assembly so that the kickback fingers are about 1/8 inch lower than the material. Using the fence as a guide, feed it evenly into the saw blade, Figure 6-19.

Fig. 6-18 Making a compound miter cut

Fig. 6-19 Using the radial arm saw to rip lumber (Note direction of feed in relation to blade rotation.)

CAUTION: Do not feed stock from the kickback side of the guard. Feed the stock against the rotation of the blade. Feeding stock in the wrong direction may cause it to be pulled from the operator's hands and through the saw with great force, causing serious injury to anyone in its path.

Do not force the stock into the saw, but feed it with a continuous motion. Use a push stick when ripping narrow stock.

Bevel ripping is done in the same manner as straight ripping except that the blade is tilted to the desired angle.

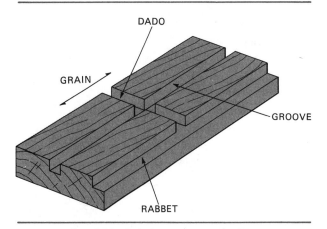

Fig. 6-20 A dado, groove, and rabbet

DADOING, GROOVING, AND RABBETING

A *dado* is a wide cut, part way through the thickness of the stock and across the grain. A *groove* is the same kind of cut, but with the grain. When a corner of the edge or end is cut away, it is called a *rabbet*, Figure 6-20.

To make these cuts, a *dado head* is commonly used. One type of dado head consists of two outside circular saw blades with several *chippers* of different thicknesses placed in between, Figure 6-21. Most dado heads make cuts of from 1/4 inch to 13/16 inch wide. Wider cuts are obtained by making two or more passes through the saw.

Fig. 6-21 The dado head consists of two blades and several chippers of different thicknesses.

When installing this type of dado head, make sure the tips of the chippers are clear and not against the side of the blade. Chipper tips are made wider (swaged) than the body of the chipper to assure a clean cut.

Another type is the one-unit adjustable dado head, Figure 6-22. This type can be adjusted without removing it from the saw arbor. Both types are available in high-speed steel or carbide-tipped.

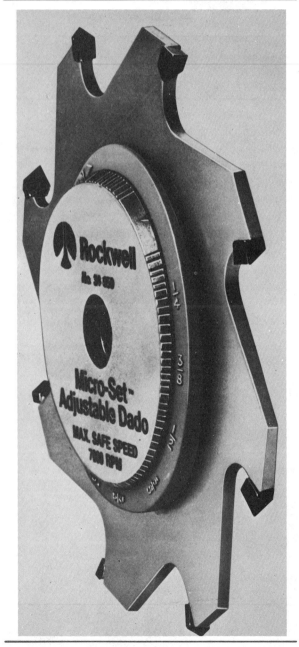

Fig. 6-22 The one-unit adjustable dado head *(Courtesy of Rockwell, International)*

Make dadoes and end rabbets in the same manner as for crosscutting. Edge rabbets and grooves are made with the saw in the ripping position.

TABLE SAW

The table saw is one of the most frequently used woodworking power tools, Figure 6-23. In many cases, the table saw is brought to the job site when the interior finish work begins. Many different operations are performed using the table saw and many different jigs are constructed to aid the operations. Common table saw operations are discussed in this unit. However, ingenuity helps the operator make the best use of jigs and devise ways to make cuts in the most efficient and practical manner.

The size of the saw is determined by the diameter of the circular saw blade, which may measure up to 16 inches or more. The most commonly used table saw on the construction site is the 10-inch model.

The blade is adjusted for depth of cut and tilted up to 45° by means of handwheels. A *rip fence* guides the work during ripping operations. A *miter gauge* is used to guide the work when crosscutting. The miter gauge slides in grooves in the table surface and may be adjusted to cut miters.

Fig. 6-23 The 10-inch table saw is frequently used on the job site. *(Courtesy of Rockwell, International)*

CAUTION: Many different kinds of saw blade guards are used. Many operations cannot be performed without removing the guard. Extreme caution should be used if operating without a guard. However, use the guard whenever possible.

RIPPING OPERATIONS

For ripping operations, the table saw is easier to use and safer than the radial arm saw. To rip stock to width, measure from the fence to the point of a saw tooth set closest to the fence the desired distance. Lock the fence in place. Adjust the height of the blade to about 1/4 inch above the stock to be cut. With the stock clear of the blade, turn on the power.

CAUTION: Stand to the left of the blade. Never stand directly in back of the saw blade. Make sure no one else is in line with the saw blade in case of kickback.

Hold the stock against the fence with the left hand and push the stock forward with the right hand, Figure 6-24. Hold the stock on the end. As the end approaches the saw blade, let it slip through the left hand. Remove the left hand from the work and push the end all the way through the saw blade with the right hand if the stock is of sufficient width (at least 5 inches wide). If the stock is not this wide, use a push stick. The left hand should never be allowed to come up to or beyond the saw blade when ripping.

CAUTION: Make sure the stock is pushed all the way through the saw blade. Leaving the cut stock between the fence and a running saw blade may cause a kickback injuring anyone in its path. Use a push stick when ripping narrow pieces. Do not pick small waste pieces or strips from the saw table when the saw is running. Remove them with a stick or wait until the saw has stopped.

Always use the rip fence for ripping operations. Never make freehand cuts.

Fig. 6-24 Using the table saw to rip lumber

BEVEL RIPPING

Ripping on a bevel is done in the same manner as straight ripping except that the blade is tilted. Whenever possible, place the fence on the side that the blade tilts away from.

TAPER RIPPING

Tapered pieces (one end narrower than the other) are made on the table saw with the use of a taper ripping jig. The jig consists of a wide board with the length and amount of taper cut out of one edge. The straight edge is held against the rip fence and the stock to be tapered is held in the cutout of the jig, Figure 6-25. The taper is cut by holding the stock in the jig as both are passed through the blade.

By using taper ripping jigs, small tapers such as wedges for concrete forms or large tapers such as rafter tails may be cut according to the design of the jig. A handle on the jig makes it safer to use. Also, if the jig is made the same thickness as the stock to be cut, then the cutout in the jig can be covered with a thin strip of wood to prevent the stock from flying back.

MAKING COVE CUTS

At times it may be necessary to make a cove (concave) cut, such as when cutting a short length of a large size cove molding to match an existing piece. The table saw is used to make these cuts by feeding the stock across the saw blade at an angle. A number of light cuts must

Fig. 6-25 Using a taper ripping jig to cut identical wedges (Guard is removed for clarity.)

be made because of the side stress placed on the saw blade.

A straightedge clamped to the table surface guides the stock. The angle at which the straightedge is clamped determines the radius of the cove, Figure 6-26.

RABBETING AND GROOVING

Making rabbets and narrow grooves may be done using a single saw blade in a ripping fashion.

Fig. 6-26 Making a cove cut with the table saw

To make a rabbet, usually two settings of the rip fence and two passes through the saw blade are required. For narrow grooves, one or more passes are required, moving the rip fence slightly with each pass until the desired width of groove is obtained.

To make these cuts with fewer passes, a dado head is used.

CROSSCUTTING OPERATIONS

For most crosscutting operations, the miter gauge is used. To cut stock to length with square ends, first check the miter gauge for accuracy by holding a framing square against it and the side of the saw blade. Usually the miter gauge is operated in the left-hand groove. The right-hand groove is used only when it is more convenient.

Square one end of the stock by holding the work firmly against the miter gauge with one hand while pushing the miter gauge forward with the other hand. Measure from the squared end the desired distance and mark on the front edge of the stock. Repeat the procedure, cutting to the layout line, Figure 6-27.

CUTTING DUPLICATE LENGTHS

When a number of duplicate pieces need to be cut, first square one end of each piece. Clamp

Fig. 6-27 Using the miter gauge for a guide

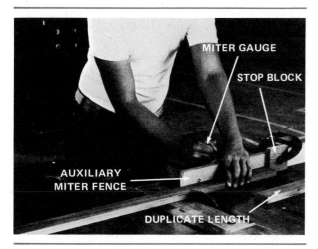

Fig. 6-28 Cutting duplicate lengths using a stop block on an auxiliary fence of the miter gauge (Guard is removed for clarity.)

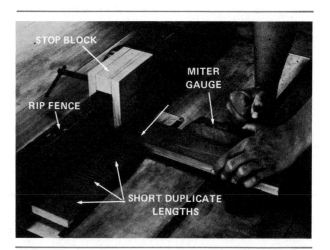

Fig. 6-29 Using the rip fence as a stop to cut short duplicate lengths (Guard is removed for clarity.)

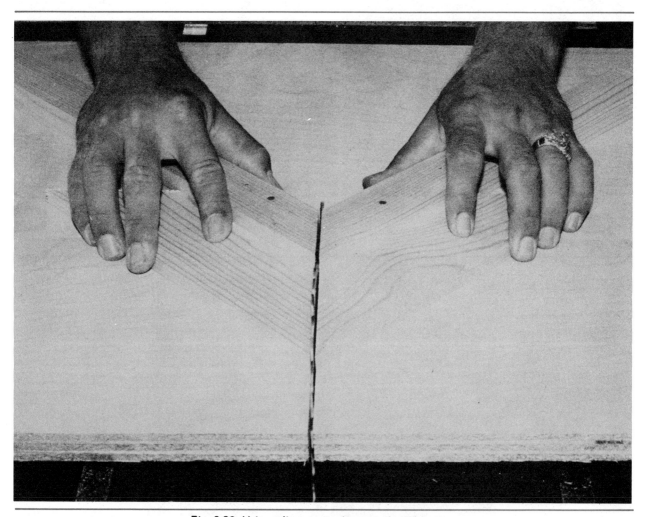

Fig. 6-30 Using a jig to cut miters on the table saw
(Guard is removed for clarity.)

a stop to an auxiliary wood fence installed on the miter gauge. Place the square end of the stock against the stop block and make the cut. Slide the remaining stock across the table until its end comes in contact with the stop block and make another cut. Continue in this manner until the desired number of pieces are cut, Figure 6-28.

When short pieces of identical length are to be cut, a stop block is clamped to the rip fence. The stop is clamped just to the operator's side of the blade so that once the cut is made, there is clearance between the piece and the rip fence.

The fence is adjusted so that the distance between the face of the stop block and saw blade is the desired length. Square one end of the stock, slide the squared end against the stop block, and make a cut. Continue making cuts in this manner until the desired number of pieces are obtained, Figure 6-29.

MITERING

Flat miters are cut in the same manner as cutting square ends except the miter gauge is turned to the desired angle. Edge miters are made by adjusting the miter gauge to a square position and making the cut with the blade tilted to the desired angle.

Compound miters are cut with the miter gauge turned and the blade tilted to the desired angles.

A mitering jig similar to that used on the radial arm saw can be used with the table saw. Two strips of wood are attached to the bottom side of the jig to slide in both miter gauge grooves. Use of such a jig eliminates turning the miter gauge each time for left- and right-hand miters, Figure 6-30.

DADOING

Dadoing is done in the same manner as crosscutting except with the use of a dado head.

TABLE SAW AIDS

Feather boards are useful aids to hold work against the fence and also down on the table surface during ripping operations, Figure 6-31. Feather boards are pieces of wood with one end cut at a 45° angle. Saw cuts are made in this end about 1/4 inch apart to give the end some spring and allow it to apply pressure to the piece being ripped, Figure 6-32.

Another useful aid is an auxiliary rip fence. A straight piece of 3/4-inch plywood about 12 inches wide and as long as the ripping fence is screwed or bolted to the metal fence.

When cuts must be made close to the fence, the use of a wood fence prevents the saw blade from cutting into the metal fence. Also, the additional height provided by the auxiliary wood fence gives a broader surface to steady wide work when its edge is being cut. The auxiliary wood fence also enables feather boards to be clamped to it to hold work down on the table surface.

Fig. 6-31 Feather boards are useful aids to hold work during table saw operations.

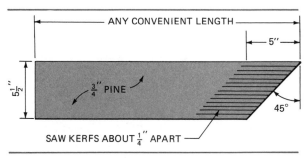

Fig. 6-32 Feather board design

REVIEW QUESTIONS

Select the most appropriate answer.

1. To cut lumber with the grain use a
   a. carbide-tipped blade.
   b. ripsaw blade.
   c. high-speed steel blade.
   d. blade with alternate beveled teeth.

2. The most frequently used blade in general carpentry is
   a. the combination planer blade.
   b. the chisel-point combination blade.
   c. square-grind carbide-tipped blade.
   d. ripsaw blade.

3. A circular saw blade that has no set in its teeth is
   a. the combination blade.   c. ripsaw blade.
   b. crosscut blade.   d. taper ground blade.

4. The alternate top bevel grind carbide-tipped circular saw blade is used with excellent results to
   a. rip solid lumber.
   b. crosscut solid lumber.
   c. cut in either direction of the grain.
   d. cut green lumber of heavy dimension.

5. Saw arbor nuts are loosened
   a. clockwise.   c. counterclockwise.
   b. with the rotation of the blade. d. against the rotation of the blade.

6. When using the radial arm saw for ripping operations, the in-feed end of the blade guard is lowered close to the work to
   a. prevent kickback.
   b. protect the operator from the saw blade.
   c. prevent sawdust from flying out.
   d. hold the work down on the table surface.

7. The table saw guide used for cutting with the grain is called a
   a. rip fence.   c. tilting arbor.
   b. miter gauge.   d. ripping jig.

8. When using the table saw for bevel ripping, the fence is placed on the
   a. right side.   c. left side.
   b. side away from the blade tilt. d. side toward the blade tilt.

9. A dado is a wide cut part way through the thickness of the material and
   a. across the grain.   c. in either direction of the grain.
   b. with the grain.   d. close to the edge.

10. When using the table saw for ripping operations
   a. hold the stock by its end no matter how long it is.
   b. hold long pieces of lumber in the middle to balance them.
   c. get help if the pieces are too long.
   d. use a support in the center of long pieces to prevent sagging.

BLUEPRINT READING 7

OBJECTIVES

After completing this unit, the student will be able to

- *read a scale rule.*
- *identify types of lines and read dimensions.*
- *identify and read plans, such as plot plans, foundation plans, floor plans and framing plans.*
- *identify and read exterior and interior elevations.*
- *identify, locate, and read sections and details.*
- *interpret window and door schedules.*
- *identify symbols, conventions, and abbreviations used on working drawings.*
- *define and state the purpose of specifications.*

Architects are persons who design buildings. The designs are made on paper through which light can pass, so that many copies can be made from the original drawing.

BLUEPRINTS

Real *blueprints* are copies that have a deep blue background with white lines. Other kinds of copies have blue, black, or sepia lines on a white background. Regardless of the color of the lines or background these copies are commonly called blueprints. Carpenters must be able to read and understand the combination of lines, dimensions, symbols, and notations on the drawings to be able to build exactly as the architect has designed the construction, Figure 7-1. No deviation from the blueprints may be made without the approval of the architect.

Fig. 7-1 The carpenter must be able to read blueprints.

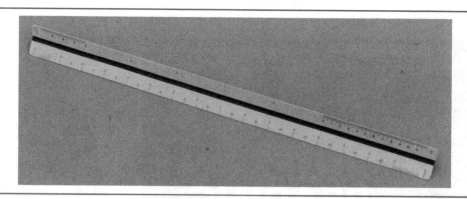

Fig. 7-2 Architectural scale rule

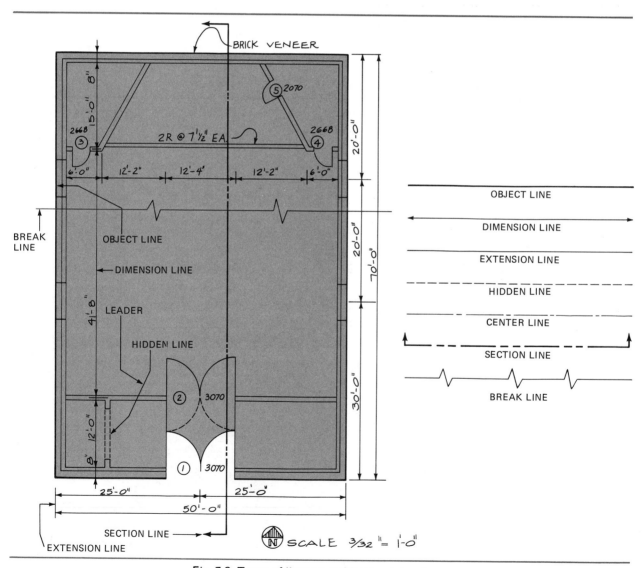

Fig. 7-3 Types of lines on architectural drawings

SCALE

Because it would be inconvenient and impractical to make full-sized drawings of a building, they are made *to scale.* This means each line in the drawing is reduced proportionally to a size that clearly shows the information, yet is not too large to handle conveniently. The scale to which the drawing is made is stated on the drawing.

THE SCALE RULE

In order to draw to scale, the architect uses a *scale rule,* Figure 7-2. A common type has a triangular shape. Each of the three faces of the rule contains two sets of graduations; one set along each edge. Except for one set which is divided into inches and sixteenths of an inch, the others are paired so that two scales can be used, depending on which end of the rule is used. The scales are paired as follows:

3″ = 1′-0″ and 1 1/2″ = 1′-0″
1″ = 1′-0″ and 1/2″ = 1′-0″
3/4″ = 1′-0″ and 3/8″ = 1′-0″
1/4″ = 1′-0″ and 1/8″ = 1′-0″
3/16″ = 1′-0″ and 3/32″ = 1′-0″

COMMONLY USED SCALES

The most commonly used scale found on blueprints is 1/4 inch equals 1 foot. This is indicated as 1/4″ = 1′-0″, and is often referred to as a "quarter-inch scale." This means that every quarter inch on the drawing will equal one foot on the building.

When drawing a building showing its location on a lot, the architect may use a scale of 1/16 inch equals 1 foot in order to reduce the size of the drawing to fit it on the paper.

To show certain details more clearly, larger scales of 1 1/2″ = 1′-0″ or 3″ = 1′-0″ are used. Complicated details may be drawn full size or half size.

Other scales are used when appropriate. Views showing the elevation of exterior or interior walls are often drawn at 1/2″ = 1′-0″ or 3/4″ = 1′-0″.

Drawing blueprints to scale is important because the building and its parts are shown in true proportion making it easier for the builder to visualize the construction. However, *the use of a scale rule to determine a dimension should be a last resort.* Dimensions on blueprints should be determined either by reading the dimension or by calculating it by adding and subtracting other dimensions. The use of a scale rule to determine a dimension results in inaccuracies.

TYPES OF LINES

In order to make the drawing clear and readable, some lines are made thick, while others are fine or broken, Figure 7-3.

- Those lines which outline the object being viewed are broad solid lines called *object lines.*

- Lines hidden below a surface are shown with short, fine, uniform dashes. *Hidden lines* are used only when necessary so that the drawing does not become confusing to read.

- *Centerlines* are indicated by a fine long dash, then a short dash, then a long dash, and so on. They are valuable to show the centers of doors, windows, partitions, and similar parts of the construction.

- A *section line* is a broad line consisting of a long dash followed by two short dashes. At its ends are arrows showing the direction in which the cross section is viewed. It actually is a reference line and is numbered or lettered to indicate the location of a cross-sectional view of part of the building.

- A *break line* is used when a detail is terminated in the drawing, but the object actually continues. It is also used to shorten a drawing when the construction remains the same.

- A *dimension line* is a fine solid line used to indicate the location, length, width, or thickness of an object.

- *Extension* lines are fine solid lines projecting from an object to show where dimension lines end.

- A *leader line* is a fine solid line pointing to an object from a notation.

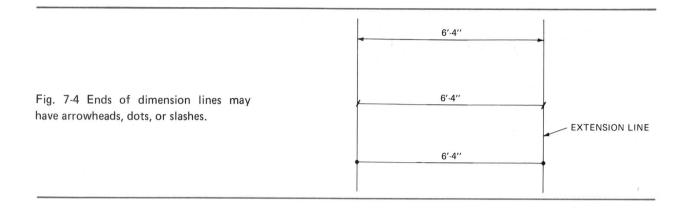

Fig. 7-4 Ends of dimension lines may have arrowheads, dots, or slashes.

DIMENSIONING

Dimension lines on a blueprint are generally drawn as continuous lines with the dimension appearing above the line. As the drawing is being viewed, all dimensions on vertical lines should appear above the line when the print is rotated one-quarter of a turn clockwise. Extension lines are drawn from the object so that the end of a dimension line is clear. Dimension lines may have arrowheads, slashes, or dots at their ends, as shown in Figure 7-4. When the space is too small to permit dimensions to be shown clearly, they may be shown as seen in Figure 7-5.

KINDS OF DIMENSIONS

Dimensions on blueprints are given in feet and inches, such as 3'-6", 4'-8", and 13'-7". A dash separates the foot measurement from the inch measurement. When the dimension is a whole number of feet with no inches, the dimension is written with zero inches, as 14'-0". Dimensions of one foot and under are given in inches, as 10", 8". Dimensions in fractions of an inch are shown as 1'-0 1/2" or 2'-3 3/4".

MODULAR MEASURE

In recent years, *modular measurement* has been used extensively. A grid with a unit of four inches is used in designing buildings, Figure 7-6. The idea is to draw the plans to use material manufactured to fit the grid spaces. Drawing plans to a modular measure enables the builder to use manufactured component parts with less waste — 4x8 sheet materials and manufactured wall, floor and roof sections that fit together with greater precision, for example. The spacing of framing members and the location and size of windows and doors adhering to the concept of modular measurement cuts down cost and conserves materials.

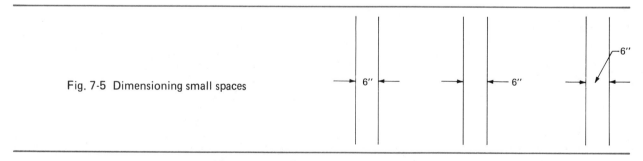

Fig. 7-5 Dimensioning small spaces

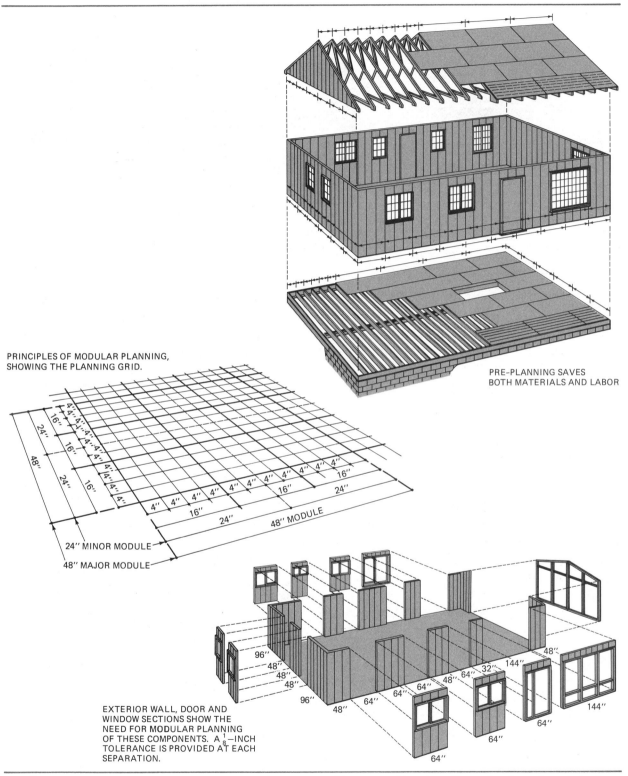

PRINCIPLES OF MODULAR PLANNING, SHOWING THE PLANNING GRID.

PRE-PLANNING SAVES BOTH MATERIALS AND LABOR

24" MINOR MODULE

48" MAJOR MODULE

EXTERIOR WALL, DOOR AND WINDOW SECTIONS SHOW THE NEED FOR MODULAR PLANNING OF THESE COMPONENTS. A $\frac{1}{8}$—INCH TOLERANCE IS PROVIDED AT EACH SEPARATION.

Fig. 7-6 Modular measurement uses a grid of four inches. *(Courtesy of Southern Forest Products Association)*

ELEVATION C

Fig. 7-7 A perspective drawing

KINDS OF DRAWINGS

A number of different views are required in order for the carpenter to build exactly as the architect has designed the construction. These views are called *perspectives*, *plans*, *elevations*, *sections*, and *details*. When put together, they constitute a set of prints.

PERSPECTIVES

A perspective rendering is a drawing that resembles a photograph of the building, Figure 7-7. The architect shows the house from a desirable vantage point to display its most interesting features. Walks, streets, shrubs, trees, and even shadows are drawn. Many times, the original drawing is colored for greater appeal. Perspectives have little information that the builder can use and are usually used to interest buyers.

PLANS

Plans are views looking from the top downward. In other words, they are bird's-eye views. There are many different kinds of plans.

- The *floor plan*, one of the most important to the builder, is a view as though a horizontal cut were made about 4 to 5 feet above the floor and the material above the cut removed, Figure 7-8. Among other things, the floor plan shows the location of walls, windows, and doors. Floor plans are given titles such as *first floor* or *second floor*, for example.

- The *plot plan*, Figure 7-9, shows information about the lot, the location of the building on the lot, the size and location of walks and driveways, and may include the kind and location of trees and shrubs. The contour of the finished grade may also be included.

- The *foundation plan*, Figure 7-10, shows the shape and dimensions of the foundation, among other things.

- *Framing plans*, not always found in a set of prints, include floor framing and roof framing. They show the direction and spacing of the framing members, Figure 7-11.

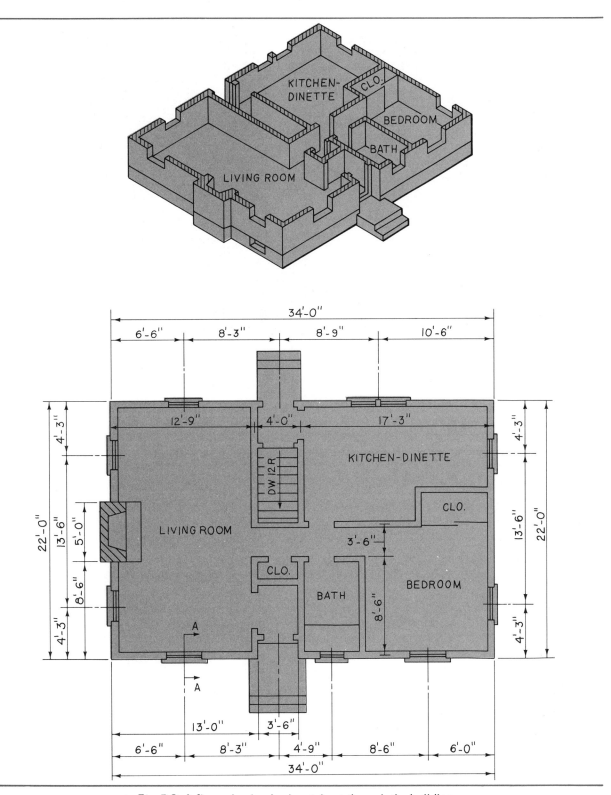

Fig. 7-8 A floor plan is a horizontal cut through the building.

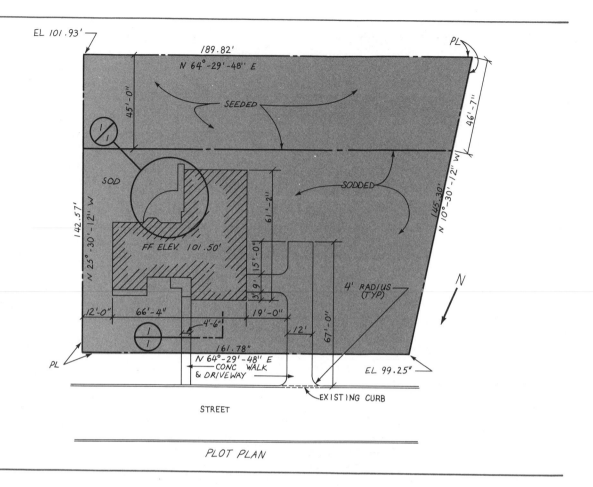

Fig. 7-9 A plot plan *(Courtesy of Pinellas Voc-Tech Institute)*

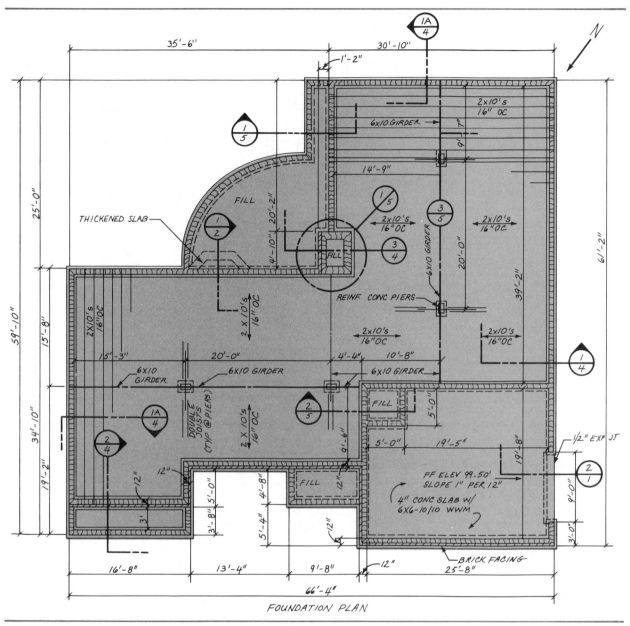

Fig. 7-10 A foundation plan *(Courtesy of Pinellas Voc-Tech Institute)*

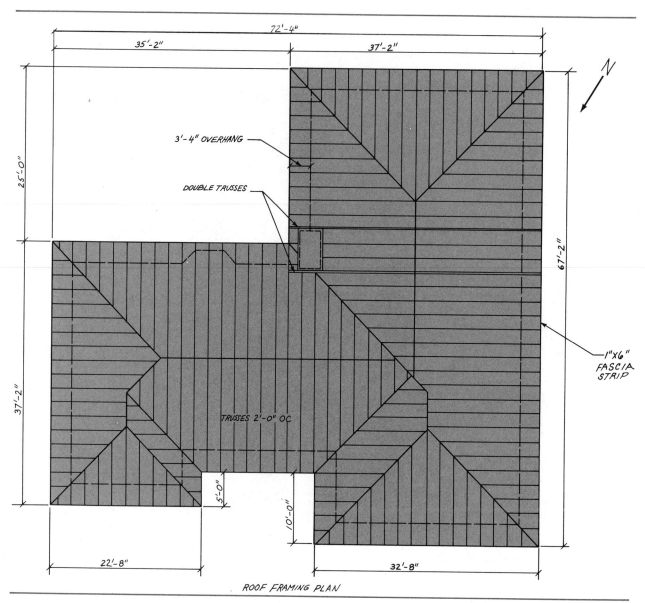

Fig. 7-11 A roof framing plan *(Courtesy of Pinellas Voc-Tech Institute)*

ELEVATIONS

Elevations are those views that show height. Most common elevations are exterior elevations that show the sides of a complete building, Figure 7-12. These are termed right elevation, left elevation, front elevation, and rear elevation. Sometimes they are titled according to the direction in which they are facing — North or South, for example.

Interior elevations show the height of certain interior walls. Most common of these are kitchen and bathroom wall elevations that show the design and size of cabinets built upon the wall, Figure 7-13. Other walls that have interesting features such as a special type of wall finish or storage cabinets may require an elevation drawing.

Occasionally found in some sets of prints are *framing elevations*. Similar to framing plans, they show the spacing, location and sizes of wall framing members.

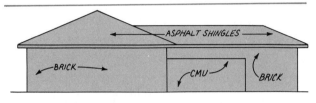

LEFT ELEVATION

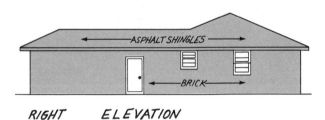

RIGHT ELEVATION

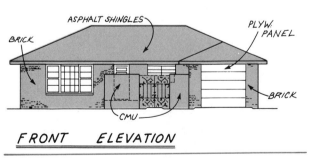

REAR ELEVATION

FRONT ELEVATION

Fig. 7-12 Elevations
(Courtesy of Pinellas Voc-Tech Institute)

SECTIONS

A section is a view showing a vertical cut through all or part of a construction, Figure 7-14. A section reference line found in the plans or elevations has arrows showing the direction from which the section drawing is viewed. In order to aid in locating the section drawing, the reference line may be designated as 1/5 or 3/4, for example, as shown in Figure 7-10. One number indicates the page of the blueprint on which the section drawing can be found. The other number is used to pinpoint the section drawing in case more than one is drawn on the same page.

DETAILS

To make parts of the construction more clear, the architect usually includes details in a set of prints. A small part is drawn at a very large scale for more clarity. A detail may be even drawn full-size and may be a plan view, elevation, section, or perspective. For instance, in order to show the construction more clearly, the architect may draw a section at a large scale through part of the foundation or cornice, Figure 7-15.

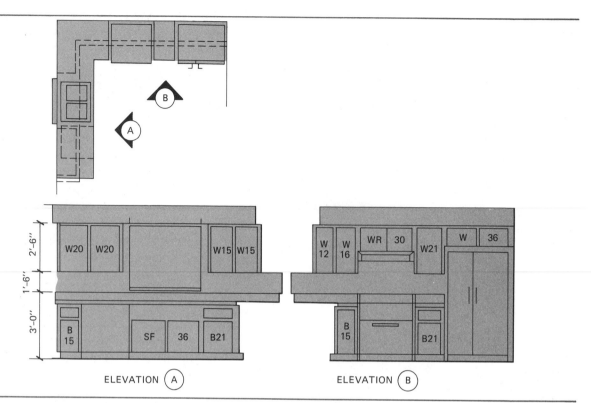

Fig. 7-13 Interior wall elevations *(Courtesy of Pinellas Voc-Tech Institute)*

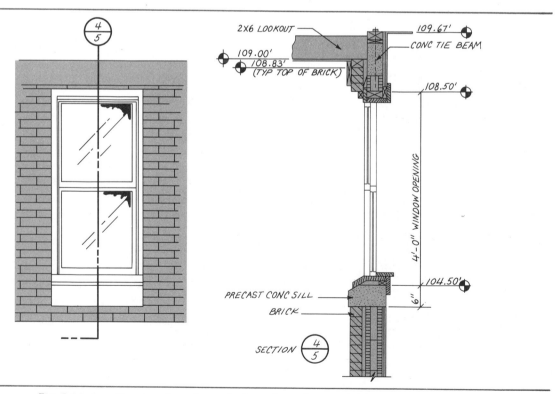

Fig. 7-14 A section is a view of a vertical cut through part of the construction.
(Courtesy of Pinellas Voc-Tech Institute)

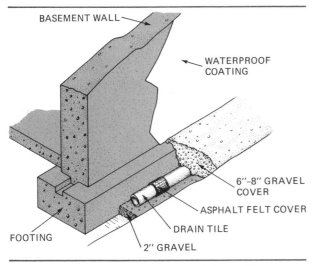

Fig. 7-15 Detail of a foundation wall

OTHER DRAWINGS

Some other drawings found in a set of prints are those relating to electrical work, plumbing, and heating and ventilating. Although the carpenter need not be able to read these plans, a general understanding is helpful because the trades often work together and a spirit of cooperation exists when each knows the problems and responsibilities of the other.

VERBAL INFORMATION

Besides drawings, printed instructions are included in a set of prints.

SCHEDULES

One kind of instruction is called a window and door schedule, Figure 7-16. This gives in-

WINDOW SCHEDULE (SIZE OF OPENING FOR FRAME)				
TYPE	HEIGHT	WIDTH	STYLE	MATERIAL
1	4^0	2^0	CASEMENT	WOOD
2	6^0	4^0	CASEMENT	WOOD
3	6^0	3^0	CASEMENT	WOOD
4	3^0	4^0	CASEMENT	WOOD
5	5^0	3^0	CASEMENT	WOOD
6	6^0	8^0	CASEMENT	WOOD
7	4^0	3^0	CASEMENT	WOOD

DOOR SCHEDULE			
TYPE	SIZE	STYLE	MATERIAL
A	$2^8 \times 6^8$	H.C.	WOOD
B	$(2)2^0 \times 6^8$	H.C. DOUBLE (LOUVER)	WOOD
C	$(2)2^0 \times 6^8$	S.C. DOUBLE (SWINGING)	WOOD
D	$(2)2^0 \times 6^8$	S.C. FRENCH DOORS	WOOD & GLASS
E	$3^0 \times 6^8$	S.C.	WOOD
F	$3^0 \times 6^8$	H.C.	WOOD
G	$5^0 \times 6^8$	DOUBLE SLIDING GLASS	GLASS & ALUM.
H	$3^0 \times 6^8$	S.C. (ONE — LIGHT)	WOOD

Fig. 7-16 A typical window and door schedule *(Courtesy of Pinellas Voc-Tech Institute)*

FINISH SCHEDULE							
ROOM	WALLS	DADO	BASE	FLOOR	CEILING	CORNICE	REMARKS
LIV. RM.	PLASTER	CANVAS	WOOD	OAK	PLASTER	WOOD	BOOKCASE
DIN. RM	"	"	"	"	"	PICT. MLDG	CLIPBD.
KITCHEN	"	4'-0" TILE	TILE	LINOLEUM	"	—	—
HALL	"	—	WOOD	OAK	"	WOOD	SEE DTL.
ENTRY	"	—	"	"	"	—	—

Fig. 7-17 A typical finish schedule

formation about the location, size and kind of windows and doors to be installed in the building. Each of the different units is given a number or letter. A corresponding number or letter is found on the floor plan to show the location of the unit.

A *finish schedule*, Figure 7-17, may also be included in a set of prints. This schedule gives information on the kind of finish material to be used on the floors, walls, and ceilings of the individual rooms.

SPECIFICATIONS

Specifications are written by the architect to give information about the job. They supplement the working drawings with information about the work to be done; material, equipment and fixtures to be used; and how they are applied. If there is a conflict between prints and specifications, the specifications take precedence.

A sample paragraph taken from specifications for a residence is shown in Figure 7-18.

On simpler plans, notations made on the same sheets as the drawings may take the place of specifications. The ability to read the notations and also the specifications accurately is essential to conforming to the architect's design.

SYMBOLS

Drawings are made even more clear by the use of symbols. However, symbols are not used unless it is necessary for clarity. They are used to represent various materials in construction. A different symbol for the same material may be used in a plan than is used in an elevation. Plumbing fixtures and electrical outlets and devices are shown on prints by the use of symbols. In order to read blueprints, the most commonly used symbols must be learned. Figure 7-19 shows some commonly used symbols.

SECTION 6A
FRAMING

6A-1 All framing lumber for partitions, ceilings and roof shall be No. 2 yellow pine spaced 16" O.C. maximum, roof sheathing shall be 1/2" plywood. All wood in contact with masonry shall be pressure treated. Anchor plates to masonry with 1/2" anchor bolts 6'-0" O.C.

(END OF SECTION 6A)

Fig. 7-18 Sample paragraph from a set of specifications for a residence
(Courtesy of Pinellas Voc-Tech Institute)

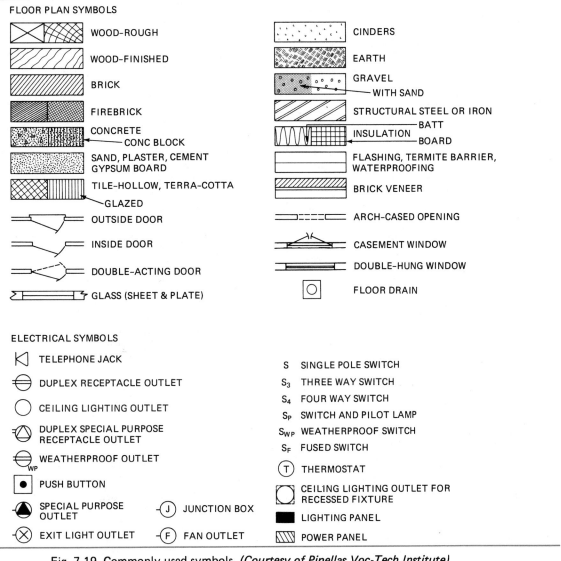

Fig. 7-19 Commonly used symbols *(Courtesy of Pinellas Voc-Tech Institute)*

ABBREVIATIONS

Architects find it necessary to use abbreviations on drawings to conserve space. Only capital letters, such as DR for door, are used. Abbreviations, which make an actual word, such as FIN. for finish, are usually followed by a period. A list of commonly used abbreviations is found in Figure 7-20.

READING BLUEPRINTS

Somewhere on each sheet in a set of blueprints is a block giving such information as description of the building, owners name, architect's name and address, date of the drawing, the sheet number, and number of pages in the set. Under each drawing on a sheet is given the kind of drawing and the scale at which it is drawn.

THE FLOOR PLAN

From the first floor plan, the *overall dimensions* of the building can be read. In a wood frame, the dimensions are usually to the outside face of studs. Walls of masonry are dimensioned to their outside face. Brick veneer

107

Access Panel . . . . .AP	Dressed and Matched . . . . .D & M	Plate . . . . .PL
Acoustic . . . . .ACST	Dryer. . . . . .D	Plate Glass . . . . .PL GL
Acoustical Tile . . . . .AT	Electric Panel . . . . .EP	Platform . . . . .PLAT
Aggregate . . . . .AGGR	End to End . . . . .E to E	Plumbing. . . . . .PLBG
Air Conditioning . . . . . .AIR COND	Excavate . . . . .EXC	Plywood . . . . .PLY
Aluminum . . . . .AL	Expansion Joint. . . . . .EXP JT	Porch . . . . .P
Anchor Bolt . . . . .AB	Exterior . . . . .EXT	Precast . . . . .PRCST
Angle L	Finish. . . . . .FIN	Prefabricated . . . . .PREFAB
Apartment . . . . .APT	Finished Floor. . . . . .FIN FL	Pull Switch. . . . . .PS
Approximate . . . . .APPROX	Firebrick . . . . .FBRK	Quarry Tile Floor . . . . .QTF
Architectural . . . . .ARCH	Fireplace . . . . .FP	Radiator . . . . .RAD
Area . . . . .A	Fireproof. . . . . .FPRF	Random . . . . .RDM
Area Drain. . . . . .AD	Fixture . . . . .FIX	Range. . . . . .R
Asbestos . . . . .ASB	Flashing . . . . .FL	Recessed . . . . .REC
Asbestos Board . . . . .AB	Floor . . . . .FL	Refrigerator . . . . .REF
Asphalt. . . . . .ASPH	Floor Drain. . . . . .FD	Register . . . . .REG
Asphalt Tile . . . . .AT	Flooring . . . . .FLG	Reinforce or Reinforcing. . . . .REINF
Basement. . . . . .BSMT	Fluorescent . . . . .FLUOR	Revision . . . . .REV
Bathroom . . . . .B	Flush. . . . . .FL	Riser . . . . .R
Bathtub . . . . .BT	Footing. . . . . .FTG	Roof . . . . .RF
Beam . . . . .BM	Foundation . . . . .FND	Roof Drain. . . . . .RD
Bearing Plate. . . . . .BRG PL	Frame . . . . .FR	Room. . . . . .RM or R
Bedroom . . . . .BR	Full Size . . . . .FS	Rough . . . . .RGH
Blocking . . . . .BLKG	Furring . . . . .FUR	Rough Opening . . . . .RO
Blueprint. . . . . .BP	Galvanized Iron . . . . .GI	Rubber Tile . . . . .R TILE
Boiler. . . . . .BLR	Garage . . . . .GAR	Scale . . . . .SC
Book Shelves . . . . .BK SH	Gas . . . . .G	Schedule . . . . .SCH
Brass . . . . .BRS	Glass . . . . .GL	Screen . . . . .SCR
Brick . . . . .BRK	Glass Block . . . . .GL BL	Scuttle . . . . .S
Bronze . . . . .BRZ	Grille . . . . .G	Section . . . . .SECT
Broom Closet . . . . .BC	Gypsum . . . . .GYP	Select. . . . . .SEL
Building . . . . .BLDG	Hardware. . . . . .HDW	Service . . . . .SERV
Building Line . . . . .BL	Hollow Metal Door . . . . .HMD	Sewer. . . . . .SEW
Cabinet. . . . . .CAB	Hose Bib . . . . .HB	Sheathing . . . . .SHTHG
Calking. . . . . .CLKG	Hot Air. . . . . .HA	Sheet . . . . .SH
Casing . . . . .CSG	Hot Water . . . . .HW	Shelf and Rod . . . . .SH & RD
Cast Iron CI	Hot Water Heater . . . . .HWH	Shelving . . . . .SHELV
Cast Stone . . . . .CS	I Beam I	Shower . . . . .SH
Catch Basin . . . . .CB	Inside Diameter . . . . .ID	Sill Cock . . . . .SC
Cellar. . . . . .CEL	Insulation . . . . .INS	Single Strength Glass . . . . . . .SSG
Cement. . . . . .CEM	Interior. . . . . .INT	Sink. . . . . .SK or S
Cement Asbestos Board . . . CEM AB	Iron. I	Soil Pipe . . . . .SP
Cement Floor . . . . .CEM FL	Jamb . . . . .JB	Specification. . . . . .SPEC
Cement Mortar . . . . . . CEM MORT	Kitchen. . . . . .K	Square Feet . . . . .SQ FT
Center . . . . .CTR	Landing . . . . .LDG	Stained . . . . .STN
Center to Center . . . . .C to C	Lath . . . . .LTH	Stairs . . . . .ST
Center Line . . . . .or CL	Laundry . . . . .LAU	Stairway . . . . .STWY
Center Matched . . . . .CM	Laundry Tray . . . . .LT	Standard . . . . .STD
Ceramic . . . . .CER	Lavatory . . . . .LAV	Steel . . . . .ST or STL
Channel . . . . .CHAN	Leader L	Steel Sash . . . . .SS
Cinder Block. . . . . .CIN BL	Length . . . . .L, LG, or LNG	Storage . . . . .STG
Circuit Breaker . . . . .CIR BKR	Library . . . . .LIB	Switch . . . . .SW or S
Cleanout . . . . .CO	Light . . . . .LT	Telephone . . . . .TEL
Cleanout Door. . . . . .COD	Limestone . . . . .LS	Terra Cotta . . . . .TC
Clear Glass. . . . . .CL GL	Linen Closet . . . . .L CL	Terrazzo . . . . .TER
Closet. . . . . .C, CL, or CLO	Lining . . . . .LN	Thermostat . . . . .THERMO
Cold Air . . . . .CA	Living Room. . . . . .LR	Threshold . . . . .TH
Cold Water . . . . .CW	Louver LV	Toilet. . . . . .T
Collar Beam . . . . .COL B	Main . . . . .MN	Tongue and Groove. . . . . . . .T & G
Concrete . . . . .CONC	Marble . . . . .MR	Tread . . . . .TR or T
Concrete Block . . . . .CONC B	Masonry Opening . . . . .MO	Typical . . . . .TYP
Concrete Floor . . . . .CONC FL	Material . . . . .MATL	Unfinished . . . . .UNF
Conduit . . . . .CND	Maximum . . . . .MAX	Unexcavated. . . . . .UNEXC
Construction. . . . . .CONST	Medicine Cabinet . . . . .MC	Utility Room . . . . .URM
Contract . . . . .CONT	Minimum. . . . . .MIN	Vent V
Copper . . . . .COP	Miscellaneous . . . . .MISC	Vent Stack. . . . . .VS
Counter . . . . .CTR	Mixture. . . . . .MIX	Vinyl Tile . . . . .V TILE
Cubic Feet . . . . .CU FT	Modular . . . . .MOD	Warm Air. . . . . .WA
Cut Out . . . . .CO	Mortar . . . . .MOR	Washing Machine . . . . .WM
Detail. . . . . .DET	Moulding. . . . . .MLDG	Water. . . . . .W
Diagram . . . . .DIAG	Nosing . . . . .NOS	Water Closet . . . . .WC
Dimension . . . . .DIM	Obscure Glass . . . . .OBSC sL	Water Heater. . . . . .WH
Dining Room . . . . .DR	On Center . . . . .OC	Waterproof. . . . . .WP
Dishwasher . . . . .DW	Opening . . . . .OPNG	Weather Stripping. . . . . .WS
Ditto . . . . .DO	Outlet . . . . .OUT	Weephole. . . . . .WH
Double-Acting. . . . . .DA	Overall . . . . .OA	White Pine . . . . .WP
Double Strength Glass . . . . . . .DSG	Overhead. . . . . .OVHD	Wide Flange . . . . .WF
Down. . . . . .DN	Pantry . . . . .PAN	Wood . . . . .WD
Downspout . . . . .DS	Partition . . . . .PTN	Wood Frame. . . . . .WF
Drain . . . . .D or DR	Plaster . . . . .PL or PLAS	Yellow Pine . . . . .YP
Drawing . . . . .DWG	Plastered Opening. . . . . .PO	

Fig. 7-20 Commonly used abbreviations *(Courtesy of Pinellas Voc-Tech Institute)*

walls are dimensioned to the outside face of the wood frame with an added dimension for the veneer.

The location of windows and doors is usually shown by a dimension to their centerline. The direction of swing of casement windows and doors is shown. The type of window or door is shown by an appropriate symbol. A letter or number near the symbol is a reference to the door and window schedule for more complete information.

Each room on the floor plan is identified. Partitions are usually, but not always, dimensioned to their centerline. In case the dimension is to the faces of the partition, the thickness of the partition is usually given, Figure 7-21.

Openings, such as for doors, in interior partitions may not always be dimensioned. In case no dimensions for openings are given, their location must be scaled or estimated. For instance, it can be clearly seen if a closet door is centered on the closet. If a door is to be located near a corner and no dimension is given, it is the responsibility of the carpenter either to place the door opening near enough to the corner so that any door casing later applied can be scribed (fitted) to the corner, or far enough away to allow convenient finishing (painting, wallpapering) between the casing and the corner. A small space (1 inch or less) between the outside edge of the side casing and a corner is very difficult to finish.

The floor plan also shows the location of staircases. Also shown is the direction of travel and the number of risers (steps) in the staircase. Details of the staircase are shown in another drawing, if necessary, Figure 7-22.

A floor plan shows the location, style and size of chimneys, fireplaces and hearths. The dimensions of the fireplace opening, size of the flue, size of the hearth, and kind of material to be used may be given.

The location of bathroom and kitchen fixtures is shown by appropriate symbols, abbreviations, and notations. Kitchen and bathroom details are usually made clearer on other drawings.

For simple structures, the electrical wiring may be shown on the floor plan. Appropriate symbols are used to show the location of switches, outlets, and fixtures. Curved lines may be drawn from switches to the fixtures they control. The location of the circuit breaker panel may also be shown.

Section reference lines may be found in various places on the floor plan, Figure 7-23. Each line is clearly numbered for reference to the section drawing. The direction of view is indicated by arrowheads.

Many notations are given on floor plans. These cover such items as the size, spacing, and direction of floor or ceiling framing members above the floor being viewed; the finish floor material; and others, depending on the complexity of the structure, Figure 7-24.

THE PLOT PLAN

The plot plan (refer to Figure 7-9) shows the shape and dimensions of the lot. The dimensions of boundary lines are usually given in decimal feet. Sometimes the compass direction of each boundary line is also given. Streets on which the lot borders are sometimes shown.

The shape and location of the building is given. Dimensions, usually in feet and inches, show the distance from the boundary lines to the building (*setback*). In some plot plans, the shape of the roof is given.

The shape, size, location and kind of materials used for walks and driveways are also found on plot plans. Dimensions for these objects are usually given in feet and inches. Details of their construction are found in another drawing.

The elevation of the finished first floor is usually shown clearly. This elevation and all others are generally given in decimal feet. Because the foundation is constructed first, a calculation must be made to find the elevation of the top of the foundation wall. The total thickness of the material from the finish floor to the top of the foundation wall must be known and then subtracted from the finish floor elevation to determine the elevation of the top of the foundation wall. This thickness can be determined from an exterior wall section.

A reference point for determining elevations of various parts of the construction is shown on some part of the plot plan and is clearly marked. This reference point may be the

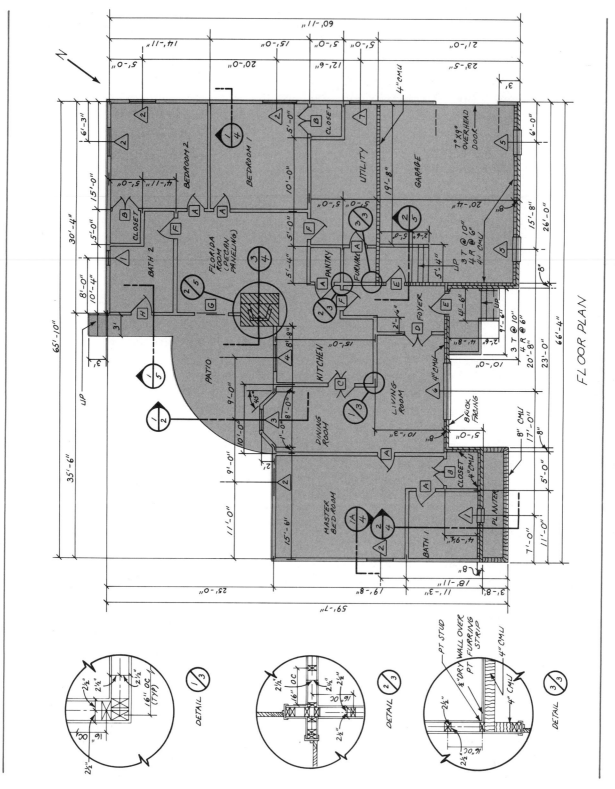

Fig. 7-21 Much information is found in a floor plan. *(Courtesy of Pinellas Voc-Tech Institute)*

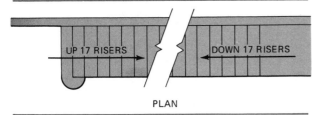

Fig. 7-22 Detail of a staircase as shown in a floor plan

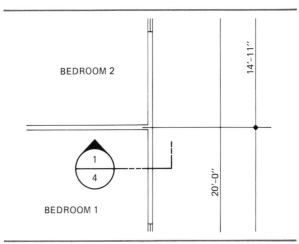

Fig. 7-23 A section reference line *(Courtesy of Pinellas Voc-Tech Institute)*

2 X 6 CEILING JOISTS 16" O.C. ABOVE

Fig. 7-24 Example of notation found on a floor plan for ceiling joists

actual elevation taken from an established elevation in relation to sea level and is called a *bench mark*. This bench mark is located on some permanent object, on or near the lot, which will not be moved or destroyed, at least until the construction is complete. In many cases, the bench mark is given in arbitrary elevation of 100.00 feet.

Contour lines show the elevation of the ground. These are irregular, curved lines connecting points of the same elevation, which is shown in decimal feet by a number placed near the line. Contour lines may show a difference in elevation from one to five feet depending on how specifically the contour of the land needs to be shown. The closer the lines are together, the steeper the slope that is indicated. Widely spaced contour lines indicate a gradual slope. On some plans, dotted contour lines are drawn to indicate the existing grade and solid lines are drawn to indicate the new grade.

A North compass heading is shown on the plot plan. The location and kind of existing and new trees and shrubs are indicated. Utility poles and drains may sometimes be shown.

FOUNDATION PLAN

The foundation plan (refer to Figure 7-10) is something like a floor plan. In addition to showing the size and shape of walls and openings, a dotted line inside and outside of supporting walls is drawn to indicate the footing. The location, type, and size of columns and their footings and the direction and size of girders supported by columns are shown. Found also on the foundation plan are specifications for the floor, and information on chimneys, fireplaces,

stairways, heating units, and areaways. *Areaways* are retaining walls of metal or concrete to keep backfill away from foundation openings.

ELEVATIONS

From the exterior elevations, the general shape and design of the building can be determined; and the location of any steps, porches, dormers, and chimneys can be seen, Figure 7-25. *Footings* and *foundation walls* below the grade level may be shown with hidden lines.

Some important dimensions are floor to floor heights, distance from grade level to finished floor, height of window openings from the finished floor, and distance from the ridge to the top of the chimney.

The kind and size of exterior siding is shown by the use of symbols and notations as is decorative trim, such as columns, balustrades, entrances, and special treatment around doors and windows.

The elevations show the windows and doors in their exact location. Their *style* and *size* are identified by appropriate symbols and

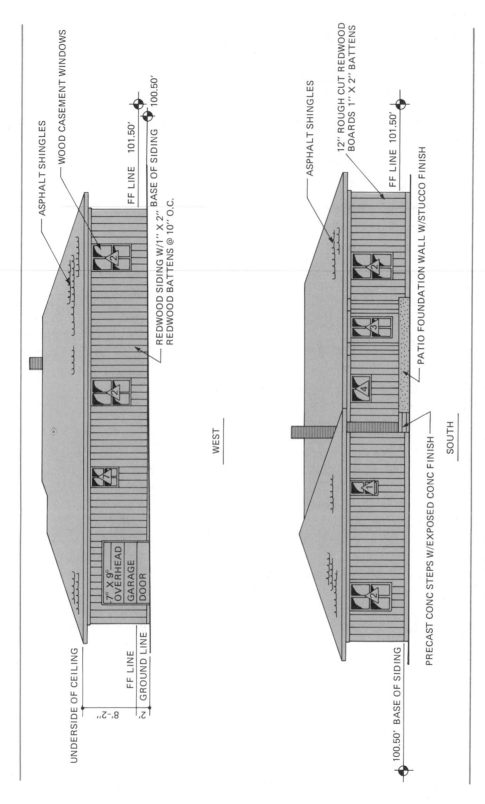

Fig. 7-25 Exterior elevations *(Courtesy of Pinellas Voc-Tech Institute)*

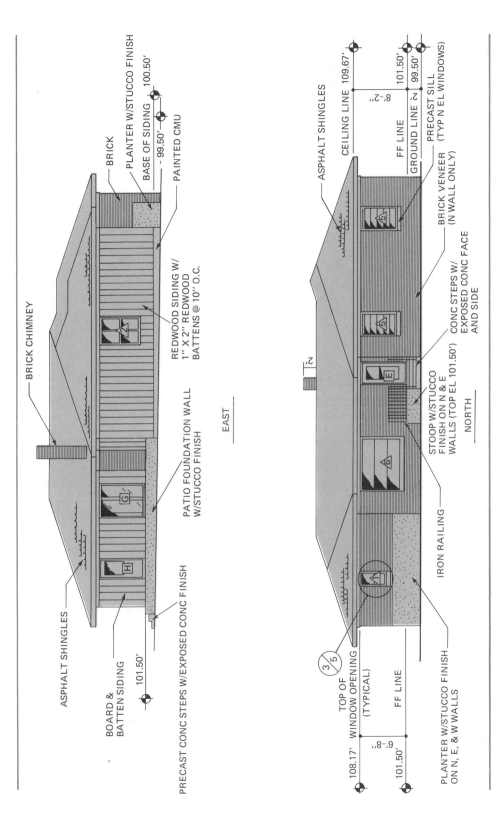

Fig. 7-25 (Continued)

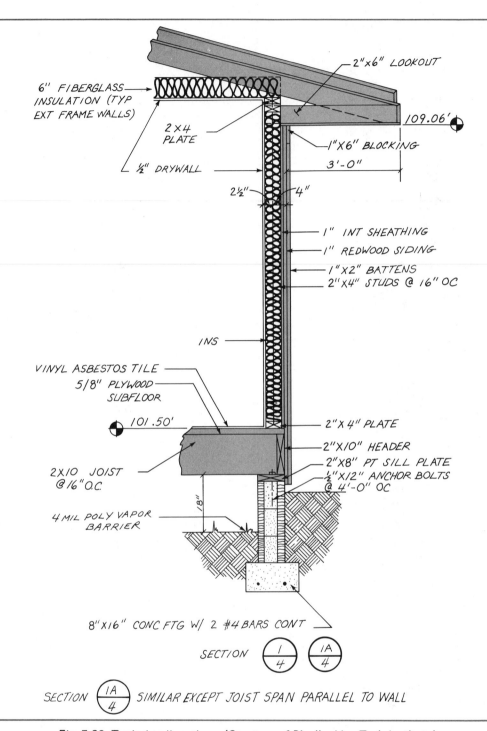

6" FIBERGLASS INSULATION (TYP EXT FRAME WALLS)

2"x6" LOOKOUT

2 X 4 PLATE

109.06'

1"X6" BLOCKING

3'-0"

½" DRYWALL

2½"   4"

1" INT SHEATHING

1" REDWOOD SIDING

1"X2" BATTENS

2"X4" STUDS @ 16" OC

INS

VINYL ASBESTOS TILE

5/8" PLYWOOD SUBFLOOR

101.50'

2"X 4" PLATE

2"X10" HEADER

2"X8" PT SILL PLATE

½"X12" ANCHOR BOLTS @ 4'-0" OC

2X10 JOIST @ 16"O.C.

8"

4 MIL POLY VAPOR BARRIER

8"X16" CONC FTG W/ 2 #4 BARS CONT

SECTION $\frac{1}{4}$ $\frac{1A}{4}$

SECTION $\frac{1A}{4}$ SIMILAR EXCEPT JOIST SPAN PARALLEL TO WALL

Fig. 7-26 Typical wall sections *(Courtesy of Pinellas Voc-Tech Institute)*

notations. Other openings, such as vents and louvers, are shown in place and designated.

The type of roofing material, the roof pitch, and the cornice style may also be determined from the exterior elevations. A number of other things may be shown on exterior elevations depending on the complexity of the structure.

SECTIONS

A number of section drawings are found in a set of prints. The number depends on what is needed for a complete understanding of the construction. One of the most important sections which is almost always found in a set of prints, is the *exterior wall section*, Figure 7-26. This view gives the builder much information:

- The size of the footing, foundation wall, basement slab (if any), and the material used in the construction.

- Sill details, such as the size of the sill plate, size and spacing of anchor bolts, size and spacing of floor joists, type of subfloor and finish floor, and the grade height at that point.

- Exterior wall information such as sizes of plates and studs, stud spacing, type and size of insulation, placement and kind of vapor barrier, and kind of interior and exterior wall finish and trim.

- Cornice and roof details such as amount of cornice overhang; cornice framing; size, location, and kind of material used for cornice trim, pitch of roof; size and spacing of ceiling joists; kind and size of ceiling insulation; kind of ceiling finish; and kind and size of roof sheathing and roofing.

It must be emphasized that the information given in one wall section does not necessarily apply for all walls, or for all parts of the same wall. The wall section or any other sections being viewed apply only to that part of the construction located by the section reference lines.

Some sectional views show only a small part of the construction. Usually drawn at a larger scale, they may also be called details.

DETAILS

It is necessary to show some features of construction at a larger scale because enough information cannot be given in the space of a smaller scale drawing, Figure 7-27. Usually the types of details needed for light construction consist of the following:

- Sectional views through parts of the building such as through a girder, foundation, fireplace, or cornice

- Sectional, plan views or elevations of special window, door, or entrance arrangements

- Elevations of kitchens and bathroom walls

- Elevations, plans, and sections of walls in other rooms that contain some special features

- Details of exterior and interior trim — a notation "See Detail" is used on the small scale drawings to call attention to the fact that more information about that particular part can be found in a detail drawing. When several details are drawn, they are designated as Detail 1 and Detail 2, for example.

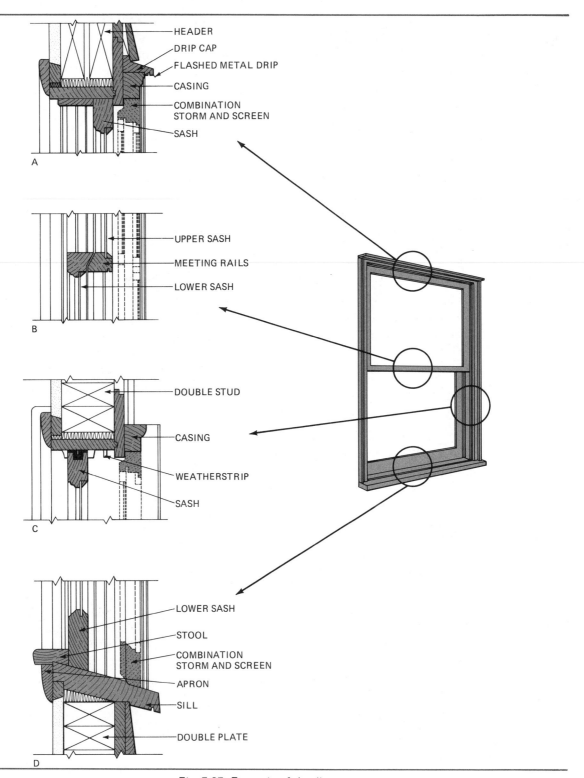

HEADER
DRIP CAP
FLASHED METAL DRIP
CASING
COMBINATION STORM AND SCREEN
SASH

A

UPPER SASH
MEETING RAILS
LOWER SASH

B

DOUBLE STUD
CASING
WEATHERSTRIP
SASH

C

LOWER SASH
STOOL
COMBINATION STORM AND SCREEN
APRON
SILL
DOUBLE PLATE

D

Fig. 7-27 Example of details

REVIEW QUESTIONS

Select the most appropriate answer.

1. A view looking from the top downward is called
   a. an elevation.
   b. a perspective.
   c. a plan.
   d. a section.

2. A view showing a vertical cut through the construction is called
   a. an elevation.
   b. a detail.
   c. a plan.
   d. a section.

3. The most commonly used scale for floor plans is
   a. 1/4″ = 1′-0″.
   b. 3/4″ = 1′-0″.
   c. 1 1/2″ = 1′-0″.
   d. 3″ = 1′-0″.

4. To determine a missing dimension,
   a. use the scale rule.
   b. calculate it.
   c. read the specifications.
   d. find a typical wall section.

5. Centerlines are indicated by a
   a. series of short, uniform dashes.
   b. series of long and short dashes.
   c. long dash followed by two short dashes.
   d. a solid, broad, dark line.

6. Which of the dimensions below would be found on a blueprint?
   a. 3′
   b. 3 ft
   c. 3′-0″
   d. 36″

7. On what drawing would the setback of the building from the property lines be found?
   a. Floor plan
   b. Plot plan
   c. Elevation
   d. Foundation plan

8. To find out which edge of doors are to be hinged, look at the
   a. elevations.
   b. framing plan.
   c. floor plan.
   d. wall section.

9. The elevation of the finished first floor is usually found on the
   a. plot plan.
   b. floor plan.
   c. foundation plan.
   d. framing plan.

10. An exterior wall stud height can best be determined from the
    a. floor plan.
    b. framing elevation.
    c. wall section.
    d. specifications.

SECTION

2

Rough Carpentry

BUILDING LAYOUT 8

OBJECTIVES

After completing this unit, the student will be able to

- *accurately set up and use the builder's level and the transit-level for leveling, determining elevations, and laying out angles.*

- *read a plot plan, build batter boards, and accurately establish layout lines for a building using the builder's level and transit-level.*

- *use the 6-8-10 method for squaring corners and check the layout for accuracy.*

Before construction begins, lines must be laid out to show the location and elevation (height) of the building foundation. Accuracy in laying out these lines is essential to comply with local zoning ordinances. Zoning laws govern, among other things, the location of the building on the lot, the size of the building in relation to the lot, and the use of building in regard to the setback of the building from property lines. In addition, accurate layout lines provide for a foundation that is level and to specified dimensions. Accuracy in the beginning makes the work of the carpenter and other construction workers easier at a later time. For example, a level floor frame of the correct dimension is much easier to construct on a foundation that is accurately laid out than on one that is not.

It is generally the carpenter's responsibility to lay out building lines. The carpenter must determine the location of the building from the plot plan and accurately use the necessary tools.

BUILDER'S LEVEL

One of the specialized tools used in building layout is the builder's level, Figure 8-1. This tool is used for establishing or determining elevations, leveling, and laying out angles of any number of degrees.

The builder's level consists of a *telescope* to which a *spirit level* is mounted. The telescope is adjusted to a level position by means of four *leveling screws* which rest on a *base leveling*

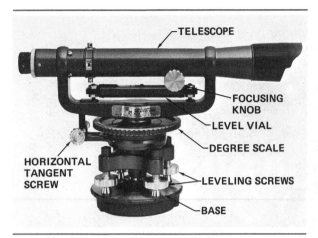

Fig. 8-1 The builder's level *(Courtesy of David White)*

plate. In higher quality levels, the base plate is part of the instrument. In less expensive models, the base plate is part of the *tripod.* Cross hairs are located within the telescope to aim directly at a point. The instrument revolves horizontally on a circle which is graduated in degrees. A pointer, or *index,* inside the circle indicates the number of degrees the telescope is turned. A hook, centered below the instrument, is provided for suspending a plumb bob. The plumb bob is used to place the level directly over a particular point.

SETTING UP AND ADJUSTING THE BUILDER'S LEVEL

Open and adjust the legs of the tripod to a convenient height for the operator. Spread the legs of the tripod well apart and firmly place its feet into the ground. If on a smooth surface, make sure the points on the feet hold without slipping. When set up, the top of the tripod should be as level as possible by sighting by eye. With the top of the tripod close to level, adjustment of the instrument is made easier.

Lift the instrument from its case by the frame, not by the telescope. Make sure the clamp screw is loose so the telescope revolves freely. While holding onto the frame, secure the instrument to the tripod by the means provided. The attaching mechanism varies according to the manufacturer and the quality of the instrument.

Accurate leveling of the builder's level is important. Care must be taken not to damage the instrument. Never use force on any parts of the instrument. All moving parts turn freely and easily by hand. Excessive pressure on the leveling screws may damage the threads or the base plate. Unequal tension on the screws will cause the instrument to wobble slightly on the base plate.

Line up the telescope directly over two opposite leveling screws. Turn the screws in opposite directions with forefingers and thumbs by moving the thumbs toward or away from each other, as the case may be, to center the bubble in the spirit level, Figure 8-2.

Rotate the telescope 90° over the other two opposite leveling screws and repeat the procedure. Feel that each of the screws have the

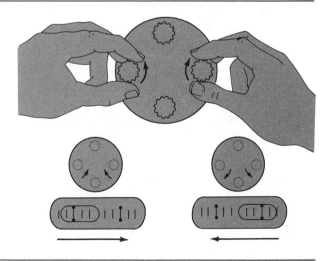

Fig. 8-2 Level the instrument by moving thumbs toward or away from each other.

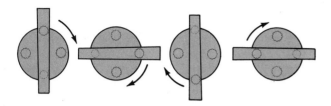

Fig. 8-3 The instrument is level when the bubble remains centered as the telescope is revolved in a complete circle.

same, but not too much, tension. Return to the original position; check and make minor adjustments. Continue adjustments until the bubble remains centered when the instrument is revolved in a complete circle, Figure 8-3.

SIGHTING THE LEVEL

To sight an object, rotate the telescope and sight over its top, aiming it at the object. Look through the telescope and focus it until the object becomes clear. Keep both eyes open. This eliminates squinting, does not tire the eyes, and gives the best view through the telescope, Figure 8-4. Center the cross hairs on the object by moving the telescope slightly left or right. The horizontal cross hair is used for reading elevations. The vertical cross hair is used when laying out angles.

LEVELING

When the instrument is leveled, any point on the line of sight is exactly level with any other point. To level one point with another, have a helper hold the end of a rule or leveling rod on the starting point. Read the vertical distance from the point, Figure 8-5. The rule or rod must be held vertical and facing the instrument. Rotate the instrument to the other point. Have the rod or rule moved vertically until the same reading is sighted. Mark the point at the end of the rule.

The folding rule is used often when sighting short distances. For longer sightings, the leveling rod is used because of its clearer graduations. In construction work, the rods used are marked in feet, inches, and eighths of an inch. The graduations are 1/8 inch wide and 1/8 inch apart, Figure 8-6.

Fig. 8-4 When sighting through the instrument, keep both eyes open. *(Courtesy of David White)*

Fig. 8-5 Leveling the top of a concrete form by sighting at a folding rule. *(Courtesy of Portland Cement Association)*

Fig. 8-6 The leveling rod is marked in feet, inches, and eighths of an inch. *(Courtesy of David White)*

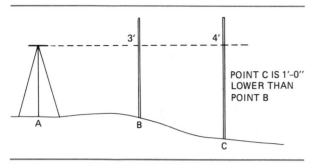

POINT C IS 1'-0''
LOWER THAN
POINT B

Fig. 8-7 Determining the difference in
elevation of two points

DETERMINING AND ESTABLISHING ELEVATIONS

The difference in elevation of two or more points is easily determined with the use of the builder's level. The difference in the rod readings at various points is the difference in elevation of those points. The drawing in Figure 8-7 shows how the difference in elevation is determined. If the rod reading at B is 3 feet and the reading at C is 4 feet, then point B is 1 foot higher than point C.

Establishing differences in elevations is done for such things as setting stakes to grade the site and to establish the height of the foundation of the building. There must be a starting point for establishing these differences.

Usually an arbitrary point is located somewhere on the job site and is called a bench mark (BM). The bench mark is part of, or marked on, an object that will not be disturbed during construction operations. The bench mark is usually designated as having an elevation of 100'-0''. Therefore, a finish floor specified to have an elevation of 105'-0'' would be located 5 feet higher than the bench mark. The location and elevation of the bench mark is clearly shown on the plot plan.

LAYING OUT ANGLES

To lay out angles, the leveling instrument must be set up over a point. The plumb bob is suspended from the instrument, Figure 8-8. Move the tripod and instrument over the approximate point. Shift the instrument on the tripod head until the plumb bob is directly over the point. Level the instrument as described previously.

Sight to the first point. If the point is above or below the line of sight, have a straightedge held plumb from the point using a hand level, Figure 8-9. Sight the edge of the straightedge and lock the telescope in position. Without disturbing the instrument, turn the circle scale until the index points at zero. Unlock the

HANGING THE
PLUMB BOB

TO HANG THE
PLUMB BOB,
ATTACH CORD TO
THE PLUMB BOB
HOOK ON THE
TRIPOD AND KNOT
THE CORD AS
ILLUSTRATED.

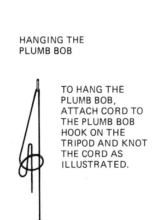

Fig. 8-8 To locate the instrument directly over a point, a plumb bob is suspended from the level as shown.

Fig. 8-9 Locating a point below the line of sight
(a) using a hand level and a straightedge or (b) using a plumb bob

telescope and turn it the desired number of degrees. For a square corner, this is 90°. Lock the telescope in position. Have the helper move the straightedge in a plumb position until the edge lines up with the vertical cross hair. Mark the point when lined up.

READING THE CIRCLE SCALE AND VERNIER

The circle scale is divided into degrees. In a full circle there are 360 degrees. Each degree is divided into 60 minutes. Each minute is divided into 60 seconds.

Usually the circle scale (outside ring) is divided into four segments of 90° each. To obtain degree readings it is only necessary to read the exact degree on the circle where it lines up with zero on the index, Figure 8-10.

For more precise readings, the *vernier scale* is used. The vernier scale is used to read minutes of a degree. If the vernier zero does not line up exactly with a degree mark on the circle, notice the last degree mark passed. On the vernier scale, locate a vernier mark that coincides with a circle mark. This indicates the reading in degrees and minutes, Figure 8-11. The carpenter is not usually concerned with any finer readings.

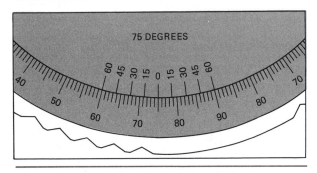

Fig. 8-10 Reading the circle for whole degrees

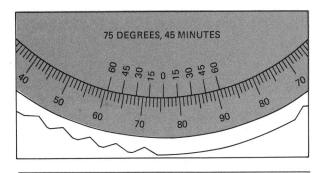

Fig. 8-11 Use the vernier scale to read degrees and minutes; read up the vernier scale either left or right to the marks that coincide.

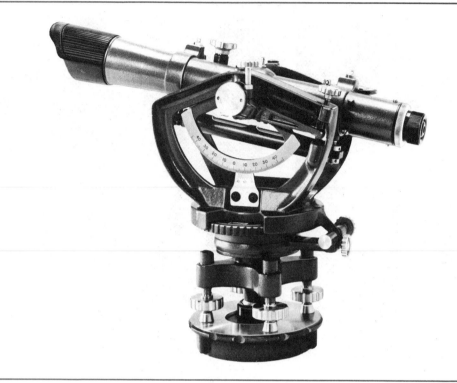

Fig. 8-12 The transit-level may be moved up and down 45° each way *(Courtesy of David White)*

TRANSIT-LEVEL

The transit-level, Figure 8-12, is similar to the builder's level except that its telescope can be moved up and down 45 degrees in each direction. To lay out angles, it can be used without the necessity of using a straightedge and hand level or plumb bob. It is also used effectively for laying out straight lines and plumbing walls and uprights, Figure 8-13.

There are many models of leveling instruments available. To become familiar with more sophisticated levels, study the manufacturer's literature. No matter what type of level is used, the basic procedures are the same.

LAYING OUT BUILDING LINES

The carpenter needs a guide for the dimensions and location of a building. This guide is in the form of a drawing, called a plot plan. See Unit 7 Blueprint Reading.

STAKING THE BUILDING

Find the survey rods that mark the corners of the property. Do not guess at the property lines. On the front property line, drive stakes just outside each survey rod. Drive a nail partway home in the top of each stake exactly on

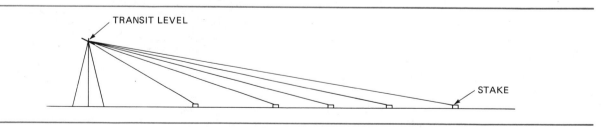

Fig. 8-13 Laying out a straight line using the transit-level

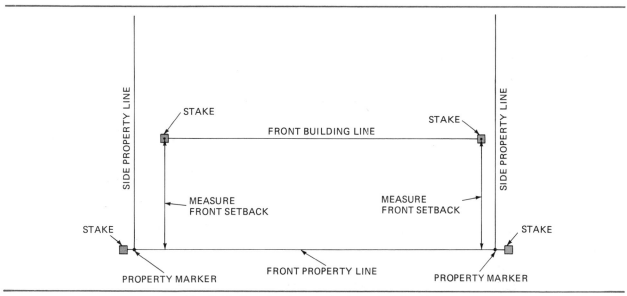

Fig. 8-14 Step 1 — Locate the front building line.

the property line. Stretch and secure a line between each stake marking the front property line exactly.

Measure in on each side from the front property line the specified front setback. Drive stakes on each end. Stretch a line between the stakes to mark the front line of the building, Figure 8-14. Measure back from the front building line the width of the building, drive

stakes, and lay out the rear building line in a manner similar to that used when laying out the front building line, Figure 8-15.

Note: All measurements must be made on the level. If the land slopes, the tape is held level with a plumb bob suspended from it, Figure 8-16.

Along the front building line, measure in from the side property line, the specified side

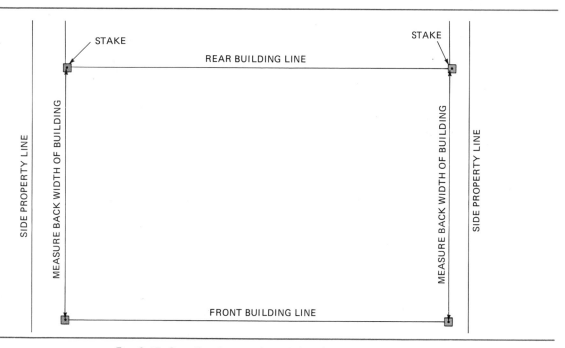

Fig. 8-15 Step 2 — Locate the rear building line.

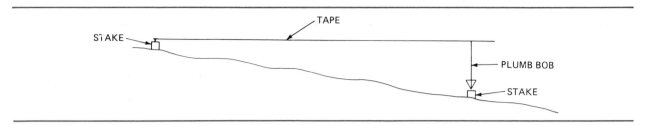

Fig. 8-16 For building layouts, measurements must be taken on the level.

setback of the building and drive a stake. Drive a nail in the top of the stake to the exact side setback. From this nail, measure along the front building line and drive another stake. Drive a nail in the top of the stake marking the exact length of the building, Figure 8-17.

Set up the level over one of the stakes marking the front corner of the building. Move the level so the plumb bob is directly over the

nail driven in the top of the stake. Sight along the front property line to the nailhead in the stake of the opposite front corner. Swing the telescope 90°; locate and drive the stake for the rear corner of the building. Drive a nail in the top of the stake marking the rear corner exactly, Figure 8-18. Locate the stake for the other corner of the building by measuring from the established front and side building line stakes.

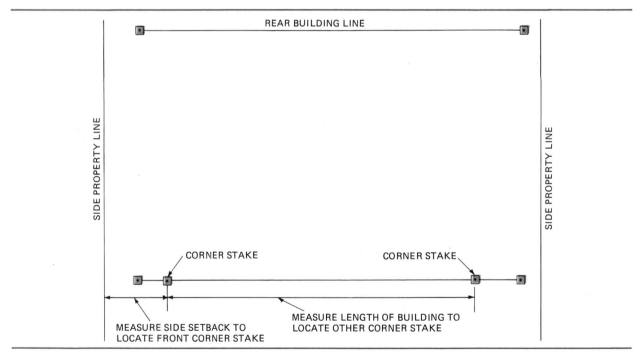

Fig. 8-17 Step 3 — Locate the front corners of the building.

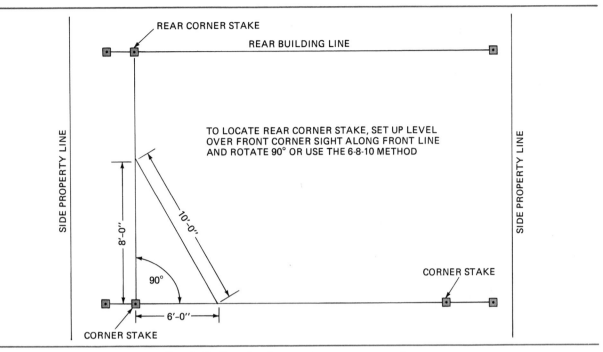

REAR CORNER STAKE

REAR BUILDING LINE

TO LOCATE REAR CORNER STAKE, SET UP LEVEL OVER FRONT CORNER SIGHT ALONG FRONT LINE AND ROTATE 90° OR USE THE 6-8-10 METHOD

SIDE PROPERTY LINE

SIDE PROPERTY LINE

8'-0''

10'-0''

90°

CORNER STAKE

6'-0''

CORNER STAKE

Fig. 8-18 Step 4 — Locate the rear corner stake.

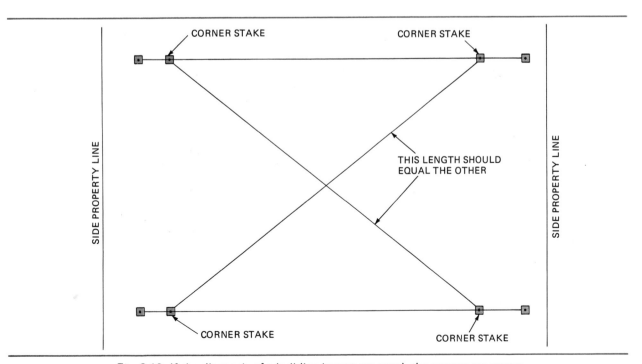

CORNER STAKE

CORNER STAKE

SIDE PROPERTY LINE

SIDE PROPERTY LINE

THIS LENGTH SHOULD EQUAL THE OTHER

CORNER STAKE

CORNER STAKE

Fig. 8-19 If the diagonals of a building layout are equal, the corners are square.

Because stakes, other than for the front corners, may have to be moved slightly for accuracy, some carpenters mark these corners with a spike pushed into the ground through a small square of thin cardboard.

Check the accuracy of the work by measuring diagonally from corner to corner. The diagonal measurements should be the same. Make adjustments if the measurements differ, Figure 8-19.

Irregular-shaped buildings are laid out in the same manner. All corner stakes are located by measuring from the established front and side building lines, Figure 8-20.

LAYING OUT A SQUARE CORNER BY THE 6-8-10 METHOD

A right angle corner may be laid out by the 6-8-10 method in the absence of a builder's level or transit-level. Measure 6 feet from the corner stake along the front building line. Mark the line by pushing a small brad through it. Measure and mark 8 feet on the side building line. Move the rear end of the side line until the distance between the marks is 10 feet. The corner is then square, Figure 8-21.

This method is based on the fact that any triangle with sides of 3, 4, and 5, or multiples thereof, is a right triangle. For example, a triangle with sides of 12, 16, and 20 is also a right triangle. Carpenters use measurements of 6, 8, and 10 feet because they find them more convenient.

ERECTING BATTER BOARDS

Batter boards are wood frames to which building layout lines are secured. Batter boards consist of horizontal members called *ledgers* attached to stakes driven in the ground. The ledgers are fastened in a level position to the stakes, usually at the same height as the foundation wall, Figure 8-22.

Batter boards are erected in such a manner that they will not be disturbed during excavation. The purpose of corner stakes is to help in locating batter boards and building lines. The corner stakes are removed when excavation begins.

Drive 2''x4'' stakes into the ground about 4 feet outside the building lines at each corner. Set up the builder's level about center on the building location. Sight to the bench mark and record the sighting. Determine the difference between the bench mark sighting and the height of the ledgers. Sight and mark each corner

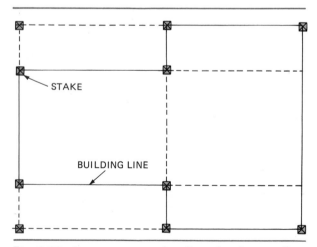

Fig. 8-20 Locate corner stakes first, then intermediate stakes, when staking out an irregular building.

Fig. 8-21 Laying out a square corner by the 6-8-10 method

Fig. 8-22 Batter boards are erected for a large commercial building.

stake. Attach ledgers to the stakes so that the top edge of each ledger is on the mark. Brace the batter boards for strength if necessary, Figure 8-23.

Stretch lines between batter boards directly over the nailheads in the original corner stakes. Locate the position of the lines by suspending a plumb bob directly over the nailheads. When the lines are accurately located, make a saw cut on the top edge of the ledger to permanently mark the location, Figure 8-24.

Check the accuracy by measuring the diagonals to see if they are equal. Saw cuts are also often made on batter boards to mark the location of the foundation footing. The footing extends outside of the foundation wall.

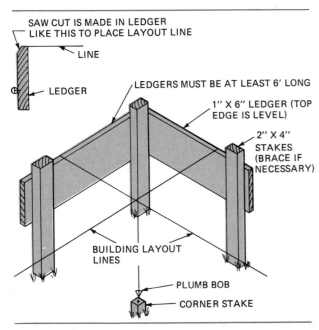

Fig. 8-23 Typical batter board construction

Fig. 8-24 Layout lines are located on batter boards by suspending a plumb bob directly over corner stakes.

REVIEW QUESTIONS

Select the most appropriate answer.

1. The minimum setback of a building from property lines is governed by the
   a. building code.
   b. board of appeals.
   c. location of the lot.
   d. zoning ordinance.

2. The location of the building on the lot is determined from the
   a. foundation plan.
   b. floor plan.
   c. architect.
   d. plot plan.

3. The builder's level is used for
   a. laying out straight lines.
   b. reading elevations.
   c. plumbing walls.
   d. all of these.

4. When sighting objects through the telescope of the builder's level,
   a. use a sun shade.
   b. keep both eyes open.
   c. close one eye.
   d. have the cross hairs centered on the object.

5. The horizontal cross hair is used for
   a. laying out angles.
   b. laying out straight lines.
   c. plumbing walls and posts.
   d. reading elevations.

6. The graduations of a leveling rod are
   a. 1/8 inch wide and 1/8 inch apart.
   b. 1/4 inch wide and 1/4 inch apart.
   c. 1/16 inch wide and 1/16 inch apart.
   d. 1/8 inch wide and 1/4 inch apart.

7. If the rod reading at point A is 6'-0'' and the reading at point B is 3'-6'' then
   a. the elevation at point B is 9'-6''.
   b. point A is higher than point B.
   c. point A has an elevation of 2'-4''.
   d. point B is higher than point A.

8. The reference for establishing elevations on a construction site is called
   a. starting point.
   b. reference point.
   c. bench mark.
   d. sight mark.

9. To read minutes of a degree, when laying out angles with the builder's level, use the
   a. base plate.
   b. index.
   c. outside circle.
   d. vernier scale.

10. A right-angle triangle has sides of
    a. 2, 4, and 6 feet.
    b. 3, 5, and 7 feet.
    c. 6, 8, and 12 feet.
    d. 12, 16, and 20 feet.

CONCRETE FORM CONSTRUCTION 9

OBJECTIVES

After completing this unit, the student will be able to

- *construct footing forms and forms for slabs, walks, and driveways.*
- *construct concrete forms for foundation walls.*
- *construct forms for columns and piers.*
- *lay out and build forms for stairs.*

The construction of concrete forms is the responsibility of the carpenter. The forms must be constructed to shape concrete to specified dimensions and be strong enough to withstand the tremendous pressure exerted against them. If a form fails to hold, costly labor and materials are lost. Time must be spent in dismantling and cleanup. The formwork must be erected all over again. Cleanup is difficult because the concrete has set up by that time. If reinforcing rods are used in the concrete, the cleanup and rebuilding is even more costly. The carpenter must know that the forms will hold before the concrete is placed.

CONCRETE

An understanding of the characteristics and handling of concrete is helpful in the construction of concrete forms.

Concrete is a mixture of Portland cement, fine and coarse aggregate, and water. The fine aggregate is usually sand. The coarse aggregate is usually crushed stone or gravel. Other aggregates are used for special purposes.

The mix may vary according to the strength desired. For example, a strong mix has 1 part Portland cement, 2 parts sand, and 3 parts stone. This is called a 1-2-3 mix. A 1-3-5 mix produces a concrete that is not as strong. The engineer or architect determines the mix needed for a particular job.

Most concrete is delivered to the job site in ready-mix trucks. Sawdust, nails and other debris should be removed from the forms. The

131

inside surfaces are brushed with oil to make form removal easier. Also, before concrete is placed, the forms and sub-grade are moistened with water. This is done to prevent rapid absorbing of water from the concrete.

Concrete is *placed*, not poured. Excessive water should never be added so that concrete flows into forms without working it. Too much water results in poor quality concrete. The mixture should be as stiff as possible and yet be workable. *Slump tests* are made by supervisors on the job to determine how wet or dry the concrete is, Figure 9-1. In the slump test, concrete is placed in a cone-shaped metal container which is open on both ends. When the container is removed, the concrete slumps down. The amount of slump determines its wetness.

Concrete should be placed where needed. It should not be pushed or dragged any more than is necessary. It should not be dropped more than 3 or 4 feet. Drop chutes are used in high, narrow wall forms. To eliminate voids or honeycombs in the concrete, it should be thoroughly spaded or vibrated as it goes into the form. The sides of the form may be tapped with a hammer to assure a smooth surface to the concrete face. Concrete must not be placed at a rapid rate, especially in high forms. Rapid placing leaves the concrete in the bottom still in a fluid state and exerting great lateral pressure on the forms at the bottom.

FOOTING FORMS

The *footing* for a foundation wall is a base to spread the load over a wider area to prevent settling. The footing is formed separately from the foundation wall in most cases. In residential construction, often the footing width is twice the wall thickness and the depth is the same as the wall thickness, Figure 9-2. However, to be certain, consult local building codes. For larger buildings, architects or engineers design the footings to carry the load imposed upon them. Usually these footings are strengthened by placing reinforcing rods of specified size and spacing in them. The placing and tying of these rods is not the job of the carpenter. However, the carpenter must work closely with those who install them.

Footings must be located below the frost line (the point below the surface to which the ground freezes) because water expands when frozen. Foundations above the frost line will heave and buckle when the ground freezes. In extreme northern climates, footings must be placed up to 7 feet below the surface, Figure 9-3. In tropical climates, footings are located just below the grade level.

In areas where the soil is stable, no formwork is necessary. A trench is dug to the width and depth of the footing and the concrete carefully placed into the trench, Figure 9-4.

Fig. 9-1 A slump test shows the wetness or dryness of a concrete mix.

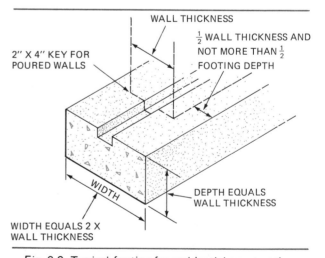

WALL THICKNESS

½ WALL THICKNESS AND NOT MORE THAN ½ FOOTING DEPTH

2" X 4" KEY FOR POURED WALLS

WIDTH

DEPTH EQUALS WALL THICKNESS

WIDTH EQUALS 2 X WALL THICKNESS

Fig. 9-2 Typical footing for residential construction

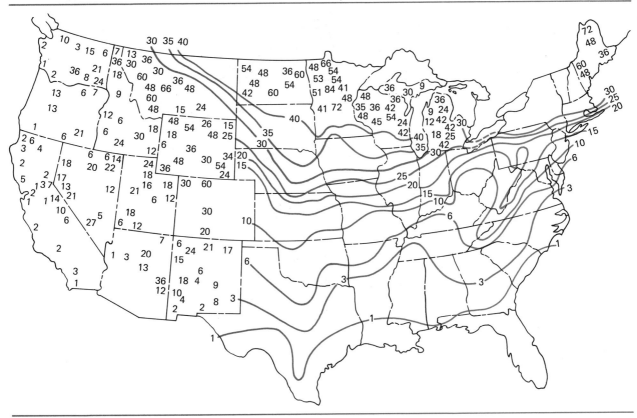

Fig. 9-3 Frost line penetration in the United States
(Courtesy of U.S. Department of Commerce)

Fig. 9-4 In stable soils, no footing formwork is necessary.

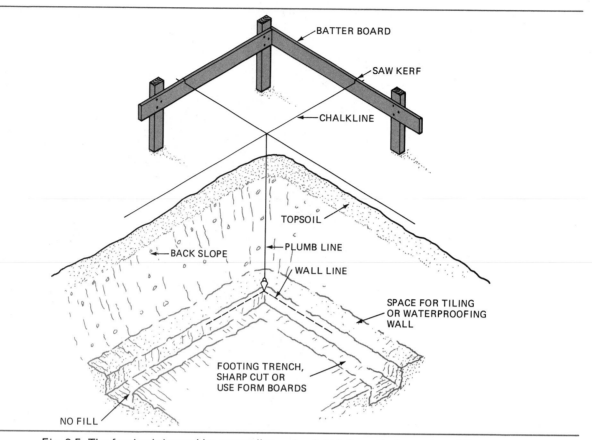

Fig. 9-5 The footing is located by suspending a plumb bob from the batter board lines.

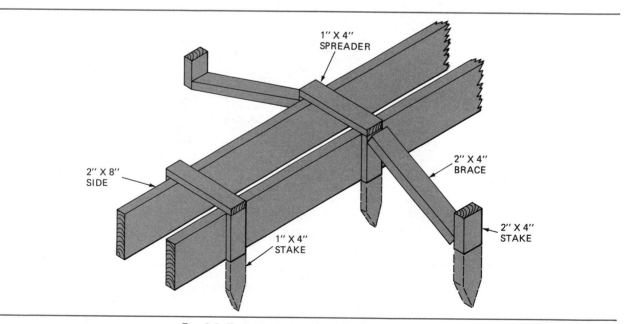

Fig. 9-6 Typical construction of a footing form

LOCATING FOOTINGS

To locate the footings, stretch lines on the batter boards in line with the outside of the footing. Suspend a plumb bob from the batter board lines at each corner and drive stakes. Stretch lines between the stakes to the correct elevation of the top of the footing, Figure 9-5.

BUILDING WALL FOOTING FORMS

When building wall footing forms, erect the outside form first. Usually 2x8 lumber is used for the side of the form and 1x4 stakes are used to hold the sides in position. Fasten the sides by driving nails through the stakes. Use duplex nails for easy removal. Keep the top inside corner of the form as close as possible to the line **without touching it.** Be sure the form does not touch the line. If the form touches the line at any point, then the line is no longer stretched straight. Form the outside of the footing in this manner all around. Space stakes 4 to 6 feet apart or as necessary to hold the form straight.

Before erecting the inside forms, cut a number of 1-inch by 3-inch *spreaders*. These are nailed to the top edges of the form to tie the two sides together at the correct distance apart. Erect the inside forms in a manner similar to that used in erecting the outside forms. Place stakes for the inside forms opposite those holding the outside form. Level across from the outside form to determine the height of the inside form. Fasten the spreaders across the form to the stakes.

Brace the stakes where necessary to hold the forms straight. In many cases, no wood bracing is necessary. Footing forms are sometimes braced by shoveling earth or placing large stones against the outside of the forms. Figure 9-6 shows the construction of a typical wall footing form.

KEYWAYS

A keyway is formed in the footing by pressing 2x4 lumber into the fresh concrete, Figure 9-7. The keyway form is beveled on both edges for easy removal after the concrete has set. The purpose of a keyway is to provide a lock between the footing and the foundation wall. This joint helps the foundation wall resist the pressure of the back-filled earth against it. It also helps to prevent seepage of water into the basement. In some cases, where the design of the keyway is not important, 2x4 pieces are not beveled on the edges, but are pressed into the fresh concrete at an angle.

STEPPED WALL FOOTINGS

When the foundation is to be built on sloped land, it is sometimes necessary to *step* the footing (form it at different levels). In building stepped footing forms, the thickness of the footing must be maintained. The vertical part of each step must not exceed the footing thickness. The horizontal part of the step must be at least twice the vertical, Figure 9-8. Vertical boards are placed between the forms to retain the concrete at each step.

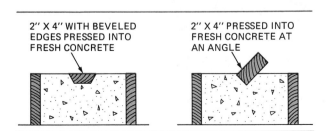

Fig. 9-7 Methods of forming keyways in the footing

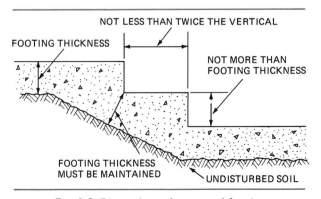

Fig. 9-8 Dimensions of a stepped footing

COLUMN FOOTINGS

Concrete for interior footings for columns, piers, posts, and chimneys is usually placed at the same time as the wall footings. The size of the footing varies according to what it has to support. The size of the footing is determined from the foundation plan.

The forms for this type of footing are usually built by spiking 2x8 pieces together in a square or rectangular shape to the specified size.

Measurements are laid out on the wall footing forms to locate the interior footings. Lines are stretched from opposite sides of the

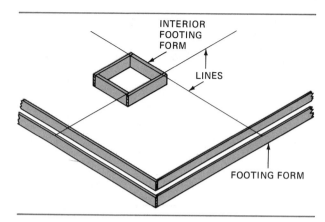

Fig. 9-9 Locating interior footings

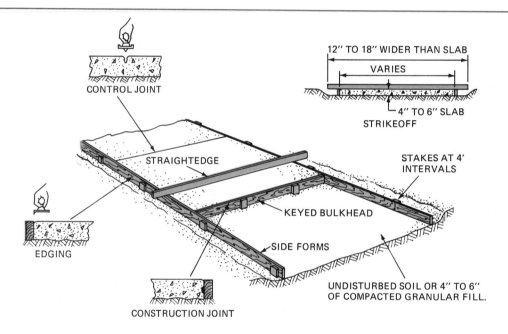

Fig. 9-10 Forms for walks are built so water drains from the surface.
(Courtesy of Portland Cement Association)

wall footing forms to locate the position of the interior forms. The interior forms are laid in position corresponding to the stretched lines, Figure 9-9. Stakes are driven and forms fastened in a position so that the top edges are level with the wall footing forms.

SLAB, WALK, AND DRIVEWAY FORMS

Building forms for slabs, walks, and driveways is similar to building continuous footing forms. Forms for floor slabs are built level, and usually 2x4 lumber is used for the sides of the form.

Forms for walks and driveways are usually built so water will drain from the surface of the concrete. In these cases, grade stakes must be established and grade lines carefully followed, Figure 9-10.

In many instances, walks and driveways are curved. To form the curve, 1/4-inch plywood or hardboard is used for small radius curves. For curves of a long radius, 1-inch lumber is used, or saw kerfs (cuts) are made close together in 2-inch lumber which is then bent until the saw kerfs close, Figure 9-11.

WALL FORMS

Foundation walls are usually formed by using panels. Panel construction simplifies the erection and stripping of formwork. Standard panel size is 36 inches wide by 8 feet high. Any larger size panel cannot be handled easily by one person. Narrower panels are used when necessary as fillers. Panels may be made of boards or special form plywood backed by 2x4 or 2x3 studs (vertical members). The edges of the panels are notched, and holes are bored in the intermediate studs to receive snap ties.

SNAP TIES

Snap ties are available in many styles and sizes. They hold the wall forms the desired distance apart and support both sides against the

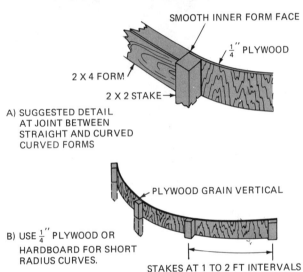

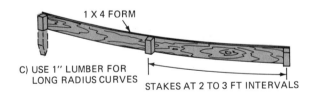

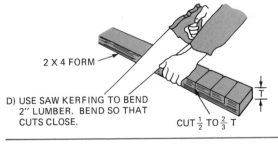

A) SUGGESTED DETAIL AT JOINT BETWEEN STRAIGHT AND CURVED CURVED FORMS

SMOOTH INNER FORM FACE
$\frac{1}{4}$" PLYWOOD
2 X 4 FORM
2 X 2 STAKE

B) USE $\frac{1}{4}$" PLYWOOD OR HARDBOARD FOR SHORT RADIUS CURVES.

PLYWOOD GRAIN VERTICAL
STAKES AT 1 TO 2 FT INTERVALS

C) USE 1" LUMBER FOR LONG RADIUS CURVES

1 X 4 FORM
STAKES AT 2 TO 3 FT INTERVALS

D) USE SAW KERFING TO BEND 2" LUMBER. BEND SO THAT CUTS CLOSE.

2 X 4 FORM
CUT $\frac{1}{2}$ TO $\frac{2}{3}$ T

Fig. 9-11 Forming for curved walks and driveways

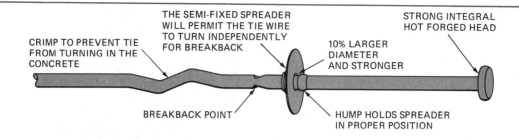

Fig. 9-12 Snap tie *(Courtesy of Dayton Sure-Grip and Shore Company)*

lateral pressure of the concrete, Figure 9-12. These ties reduce the need for external bracing and greatly simplify the erection of wall forms. They are nicked or weakened so that they may be broken off within the concrete after removal of the forms. The small holes remaining are easily patched.

WALERS

The snap ties are wedged against horizontal members called *walers*. Walers may be 4x4 lumber with holes bored for the snap ties or doubled 2x4 pieces with spaces between them. Figure 9-13 shows the construction of a typical wall form.

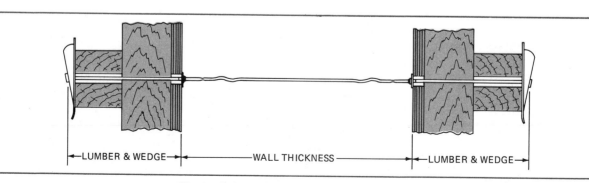

Fig. 9-13 Cross section of a typical wall form

The vertical spacing of the snap ties and walers depends on the thickness and height of the wall. The vertical spacing is closer together near the bottom because there is more lateral pressure from the concrete there than at the top, Figure 9-14. Ties are spaced 18 inches apart horizontally under ordinary circumstances.

For low wall forms, less than 4 feet in height, the panels are laid horizontal with vertical walers spaced every 3 or 4 feet, Figure 9-15.

ERECTING WALL FORMS

After the concrete placed in the footing has hardened sufficiently (usually at least 3 days), the forms are removed and cleaned. The salvaged forms are reused.

Lines are stretched on the batter boards in line with the outside of the foundation wall. A plumb bob is suspended from the layout lines to the footing. Marks that are plumb with the layout lines are placed on the footing at each corner. A chalk line is snapped on the top of the footing between the corner marks outlining the outside of the foundation wall.

Stack the number of panels necessary to form the inside of the wall in the center of the excavation. Place the panels needed for the outside of the wall so they are laid around the walls of the excavation.

Erect the outside wall forms first. Set all corner panels in place by nailing through the shoe (bottom horizontal 2x4 piece) into the green concrete. (Concrete is called green during the first few days and before it hardens.) Use duplex nails through the shoe (bottom) and into the corners. Make sure the corners are plumb by testing with a hand level.

Fill in between the corners with panels, nailing into the green concrete where necessary to keep the inside surface of the panels on the chalk line. Place snap ties in the grooves between panels as work progresses. Tie panels together by driving U-shaped clamps over the edge 2x4s or by spiking them together with duplex nails. Use filler panels as necessary to complete each wall section.

After the panels for the outside of the wall have been erected, place snap ties in the intermediate holes. Erect the panels for the inside of the wall, keeping joints between panels opposite to those for the outside of the wall. Insert the other end of the snap ties between panels and in center holes as panels are erected.

CAUTION!

EXCESSIVE POUR RATES ARE DANGEROUS TO MEN AND MATERIALS.

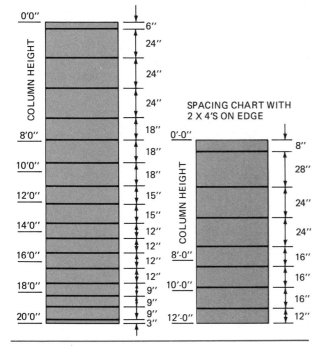

Fig. 9-14 Horizontal stiffeners are spaced closer together near the bottom than they are at the top.

Fig. 9-15 When constructing low wall forms, the panels are laid horizontal.

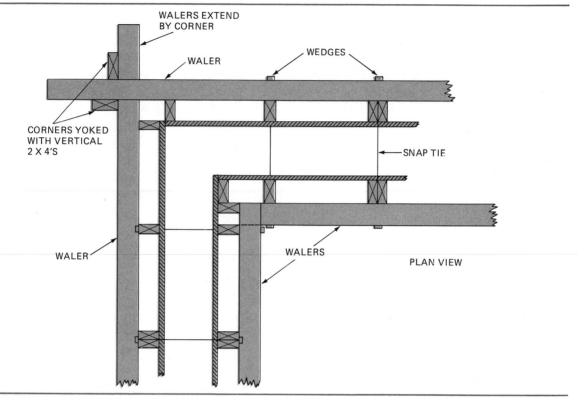

Fig. 9-16 The outside corners of wall forms are yoked with vertical 2x4s.

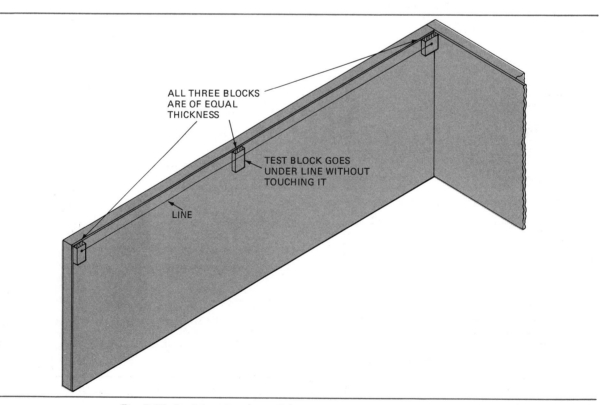

Fig. 9-17 Straightening the wall form with a line and test blocks

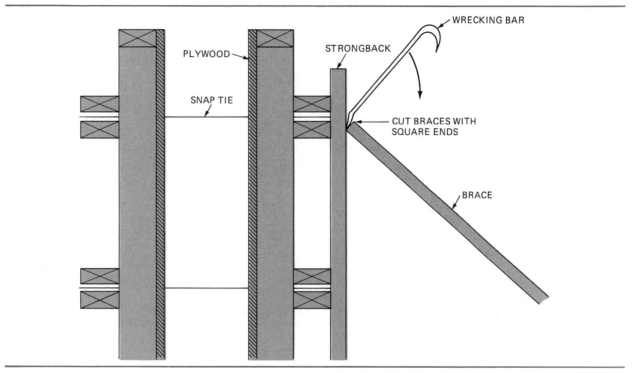

Fig. 9-18 Cut braces with square ends. This allows for
easy prying when straightening the formwork.

When all panels are in place, install the walers, letting the snap ties come through them and wedge into place. Let the ends of the walers extend by the corners of the formwork. Reinforce the corners with vertical 2x4s as shown in Figure 9-16. This is called *yoking* the corners. Brace the walls inside and out as necessary to straighten them. Wall forms are straightened by sighting from corner to corner by eye. Another method of straightening is by stretching a line from corner to corner at the top of the form over two blocks of the same thickness. Move the forms until a test block of equal thickness passes just under the line, Figure 9-17. Braces are cut with square ends and spiked into place. The sharp corner of the square ends helps hold the braces in place and allows easy prying with a bar to tighten the braces and move the forms, Figure 9-18. There is no need to make a bevel cut on the ends of braces.

After the wall forms have been straightened and braced, chalk lines are snapped on the inside of the form for the height of the foundation wall. Nails are driven partway in at intervals along the chalk line as a guide for screeding

(leveling) the top of the wall, Figure 9-19. As soon as the wall is screeded, anchor bolts are set in the fresh concrete. Care must be taken to set the anchor bolts at the correct height and at specified locations.

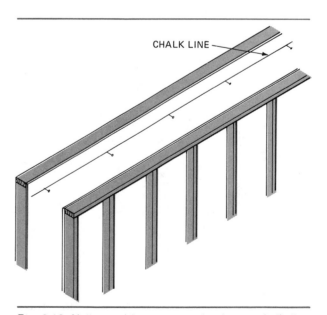

Fig. 9-19 Nails are driven partway in along a chalk line as a guide to screed the top of the wall.

OPENINGS IN CONCRETE WALLS

In many cases, openings must be formed in concrete foundation walls for such things as windows, doors, ducts, and beams. The forms used for providing the larger openings are called *bucks.* The buck is usually made of 2-inch stock with its width the same as the thickness of the foundation wall, Figure 9-20. They are made to the specified dimension and

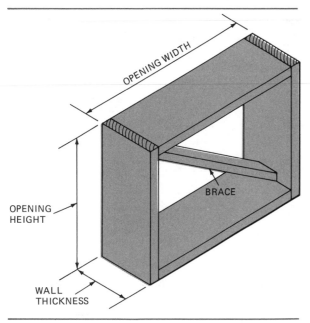

Fig. 9-20 Construction of a typical window buck

installed against the inside face of the outside panels. Duplex nails through the outside panels hold the bucks in their proper place. The inside wall panels are then installed against the other side of the bucks. Nails are driven through the inside wall panels to secure the bucks on the inside. Nailing blocks or strips are often fastened to the outside of the bucks. These are beveled on both edges to lock them into the concrete when the form is stripped. They provide for nailing window and door frames in the openings.

Recesses in the top of the foundation wall sometimes need to be made to receive the ends of girders (beams). A box of the size needed is made and fastened to the inside wall panel, Figure 9-21.

COLUMN FORMS

Column forms are built in a manner similar to that used for wall forms, except that snap ties are not used. Use panels to simplify erection and stripping. Build and erect two panels to the exact thickness of the column. Build panels for the opposite sides of the column to overlap the previously built panels. Plumb and nail the corners together with duplex nails. Install

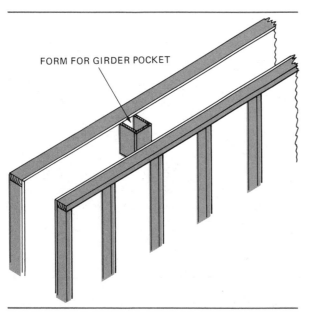

Fig. 9-21 Box form to provide pocket in foundation wall for girder

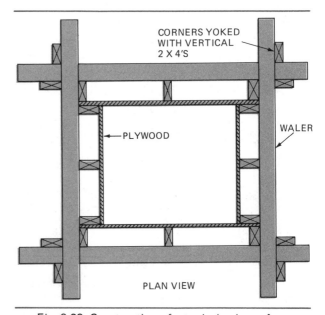

Fig. 9-22 Construction of a typical column form

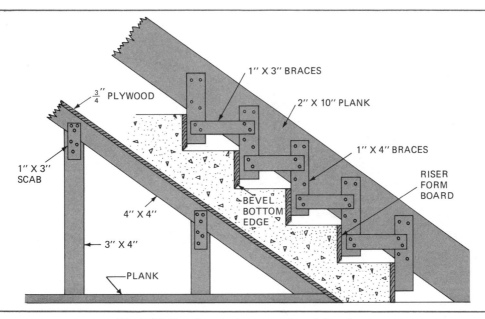

Fig. 9-23 Beveling the bottom edge of riser form
permits troweling of the entire tread.

walers around the column forms letting their ends extend beyond the corners. Spike them together where they overlap. Install walers closer together at the bottom as in wall forming. Yoke the corners, Figure 9-22. Brace the formwork securely to hold it plumb. Triangular strips placed in the corners of the form are sometimes used to bevel the column corners.

STAIR FORMS

Detailed information on stair terms and layout is found later in Unit 16 Stair Framing. Emphasis here is on building forms for concrete stairs.

When stairs are formed between two existing walls, the rise and run of each step is laid out on the walls. Planks are ripped to width to correspond to the height of each riser. The planks are then beveled on their bottom edge and wedged securely in position to the layout lines. Beveling the bottom edge of the plank permits the mason to trowel the entire surface of the tread, Figure 9-23. After the riser planks are secured in position, they are braced between

the ends to keep them from bowing outward due to the pressure of the concrete, Figure 9-24.

In cases where the end of the stairs is to be open, panels are erected on each end. The stairs are laid out on the inside surfaces of the panels. *Cleats* (small blocks of wood) are nailed at each riser location. The formwork for the risers is

Fig. 9-24 Stairs formed between two existing walls
(Courtesy of Portland Cement Association)

Fig. 9-25 Typical form construction for concrete stairs having open ends
(Courtesy of Portland Cement Association)

Fig. 9-26 Only a hammer is needed to tie panels or corners together when metal wedges are used.
(Courtesy of Symons Corporation)

then nailed to the cleats and braced between the two ends, Figure 9-25.

ECONOMY AND CONSERVATION IN FORM BUILDING

Economical concrete construction depends a great deal on the reuse of forms. Forms should be designed and built to facilitate stripping and reuse. Use panels to build forms whenever possible. Use only as many nails as necessary. Nails are used mostly to hold the forms in position until they are braced.

Care must be taken when stripping forms to prevent damage to the panels so they can be reused. Stripped forms should be cleaned of all adhering concrete and stacked neatly.

Long lengths of lumber can often be used without trimming. Random length walers can

be allowed to extend beyond the forms. Studs need not be cut off at the top of a wall form. There is no need to spend a lot of time in cutting lumber to exact lengths. The important thing is to form the concrete to specified dimensions without spending too much time in unnecessary fitting.

CONCRETE FORMING SYSTEMS

A concrete forming system consists of manufactured items for concrete form construction. Among these are specially designed panels of steel or a combination of steel and wood. The panels are tied together with metal wedges, Figure 9-26. A variety of panel sizes is available.

Small column forms can be erected quickly with less material because no walers are necessary. The corners are wedged eliminating the need for yoking them, Figure 9-27. Large

Fig. 9-27 Walers or yoking is not necessary to form small columns when forming systems are used. *(Courtesy of Symons Corporation)*

Fig. 9-28 Large rectangular column form with walers and strongbacks attached — the forming system eliminates the need for yoking the corners. *(Courtesy of Symons Corporation)*

Fig. 9-29 Walers are easily installed on forming systems when special hardware is used. *(Courtesy of Symons Corporation)*

Fig. 9-30 Adjustable braces are available with forming systems. *(Courtesy of Symons Corporation)*

Fig. 9-31 Steel inside corners of a forming system make fast erection for formwork possible. *(Courtesy of Symons Corporation)*

column forms may require walers but no external corner bracing is necessary, Figure 9-28.

Walers are easily installed on wall forms when the forming system hardware is used, Figure 9-29. Adjustable braces are available; few are required to hold the wall straight, Figure 9-30. Steel inside corners are used with wall panels for fast and easier forming of the inside corner, Figure 9-31. Other accessories are available, such as molded beam forms of reinforced fiberglass, Figure 9-32, and Domeforms® for forming beams and slabs simultaneously, Figure 9-33.

Forming systems require a considerable initial outlay of funds, but the investment is saved many times over because of low labor costs and reusable forming system components. A number of forming systems are available for various types of work. Systems are made for light, residential work and for heavy-duty concrete formwork. With forming systems, fewer ties and materials are used. For special jobs, manufacturers also make custom-made formwork specially designed for a particular job. The time saved by using custom-made formwork on a job is well worth the cost of the formwork, Figure 9-34.

ESTIMATING CONCRETE

Sometimes it is the carpenter's responsibility to order concrete for the job. Ready-mix concrete is sold by the cubic yard. To determine the number of cubic yards of concrete needed for a job, find the number of cubic feet and divide by 27. For example, a wall 8 inches thick, 8 feet high, and 36 feet long requires 7.1 cubic yards of concrete ($2/3 \times 8 \times 36 \div 27 = 7.1$). In this formula, always change all dimensions to feet or fractions of a foot. In the above example, the 8-inch thickness is changed to 2/3 of a foot.

Fig. 9-32 Slab and beam formed with a molded fiberglass beam form *(Courtesy of Symons Corporation)*

Fig. 9-33 Domeforms® of fiberglass or steel form beams and floors easily.

Fig. 9-34 Custom-made formwork was reused for all 70 floors of this high-rise building. *(Courtesy of Symons Corporation)*

REVIEW QUESTIONS

Select the most appropriate answer.

1. The inside surfaces of forms are oiled to
   a. protect the forms from moisture.
   b. prevent the loss of moisture from concrete.
   c. strip the forms easier.
   d. prevent honeycombs in the concrete.

2. Rapid placing of concrete
   a. prevents adequate vibrating.
   b. may burst the forms.
   c. separates the aggregate.
   d. causes voids and honeycombs.

3. Unless footings are placed below the frost line,
   a. the foundation will settle.
   b. the foundation will heave and crack.
   c. excavation is difficult in winter.
   d. problems with water and moisture will result.

4. Spreaders for footing forms are used
   a. to allow easy placement of the concrete.
   b. to prevent forcing the form out of alignment.
   c. because they are easier to fasten.
   d. because they maintain the footing width.

5. The sides of a residential footing form are usually
   a. 2x4s.   c. 2x8s.
   b. 2x6s.   d. 2x10s.

6. The horizontal surface of a stepped footing must be at least
   a. 4 feet.   c. the vertical distance.
   b. twice the vertical distance.   d. the thickness of the footing.

7. The usual size of a standard wall form panel is
   a. 4 feet by 8 feet.   c. 32 inches by 8 feet.
   b. 3 feet by 8 feet.   d. 2 feet by 8 feet.

8. Reinforcing the walers at the corners with vertical 2x4s is called
   a. double locking the corners.   c. yoking the corners.
   b. interlocking the corners.   d. tying the corners.

9. To guide the screeding of the top of a foundation wall
   a. space nails along a chalk line.
   b. use the top of the form as a guide.
   c. use a chalk line only.
   d. add enough water to the concrete so it will flow level.

10. When erecting concrete formwork
    a. drive extra nails for added strength.   c. use as few nails as possible.
    b. drive all nails home.   d. use duplex nails throughout.

FLOOR FRAMING 10

OBJECTIVES

After completing this unit, the student will be able to

- *describe platform, balloon, and plank and beam framing; and identify framing members.*
- *describe several energy and material conservation framing methods.*
- *build and install girders, erect columns, and lay sills.*
- *lay out and install floor joists.*
- *frame openings in floors.*
- *lay out, cut, and install bridging.*
- *apply subflooring.*

Wood frame construction is used for most homes for several important reasons, such as economy, durability, and variety. The cost for wood frame construction is generally less than for other types of construction. Fuel and air-conditioning expenses are reduced because wood frame construction provides for better insulation. Wood frame homes are very durable. Many existing wood frame homes are hundreds of years old. Many different architectural styles are possible because wood can be shaped easily, producing many different patterns.

PLATFORM FRAME CONSTRUCTION

The platform frame, sometimes called the western frame, is the type of frame mostly used in residential construction, Figure 10-1. In this type of construction, the floor is built and the walls erected on top of it. When more than one story is built, the second floor platform is erected on top of the walls of the first story.

A platform frame is easy to erect because at each floor level a flat surface is provided on which to work. A common practice is to assemble wall framing units on the floor and then tilt the units up in place.

Because lumber shrinks mostly across its width and thickness, a disadvantage of the platform frame is the relatively large amount of settling caused by the shrinkage of the horizontal frame members. However, because of the equal amount of horizontal lumber, the shrinkage is equal throughout the building.

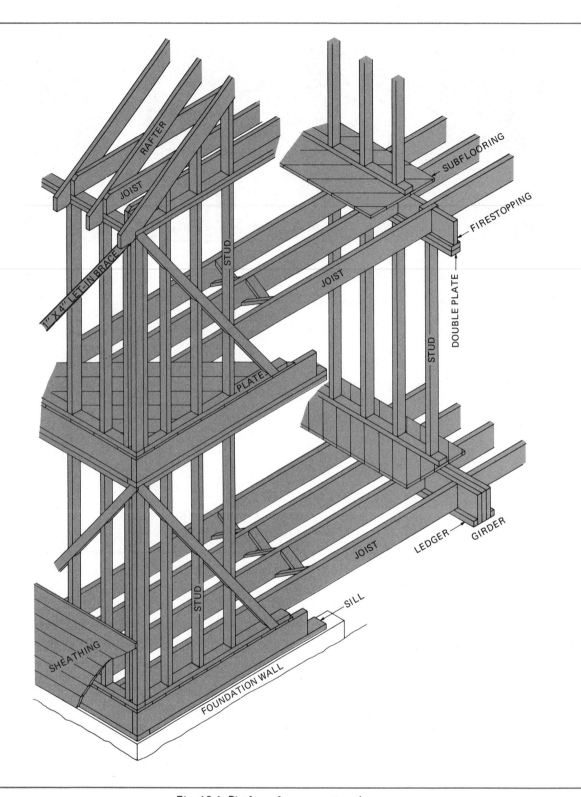

Fig. 10-1 Platform frame construction

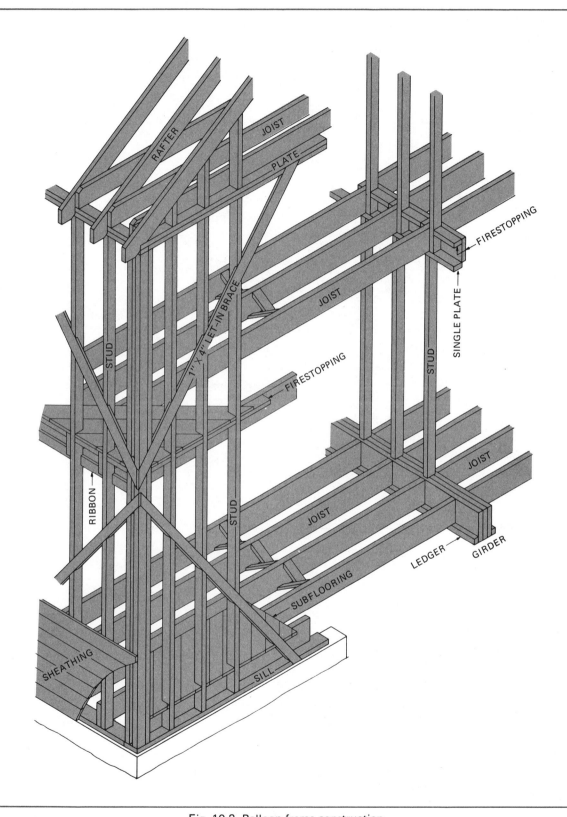

Fig. 10-2 Balloon frame construction

BALLOON FRAME CONSTRUCTION

In the balloon frame, the studs and first floor joists rest on the sill. The second floor joists rest on a 1"x4" ribbon which is cut in flush with the inside edges of the studs, Figure 10-2.

Because the studs run from sill to plate, settling caused by shrinkage of lumber is held to a minimum in the exterior walls. This is a preferred frame when the exterior wall finish is to be brick or stone veneer or stucco because there is little movement between the wood frame and the masonry veneer, Figure 10-3.

To prevent unequal settling of the frame due to shrinkage, the studs of the bearing partitions rest directly on the girder. Sometimes a balloon frame is used in combination with a platform frame, especially in the construction of multilevel homes, to avoid uneven settling due to lumber shrinkage.

In a balloon frame, the walls need to be firestopped; in the platform frame, the wall plates act as firestops. *Firestops* are a form of blocking of air passages in strategic locations in the frame of a structure to prevent the spread of fires through the building. Firestops must be installed in the following locations:

- In all stud walls, partitions and furred spaces at ceiling and floor levels
- Between stair stringers at the top and bottom
- Between chimneys, fireplaces, and wood framing
- At the ends of floor joists and over the supports
- All other locations as required by building codes

PLANK AND BEAM FRAME CONSTRUCTION

The plank and beam frame uses fewer but larger pieces. Timbers are used for joists, posts, and rafters; matched planks (tongue and grooved) are used for floors and roof sheathing, Figure 10-4. Usually the roof plank is left exposed and serves as the finished ceiling.

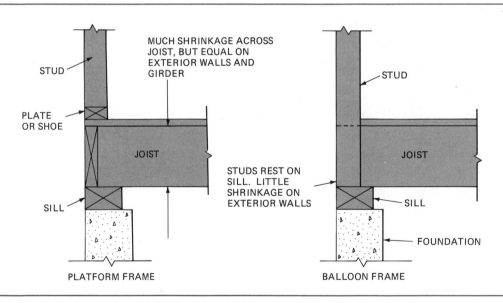

Fig. 10-3 Comparison of shrinkage in exterior walls of platform frame and balloon frame

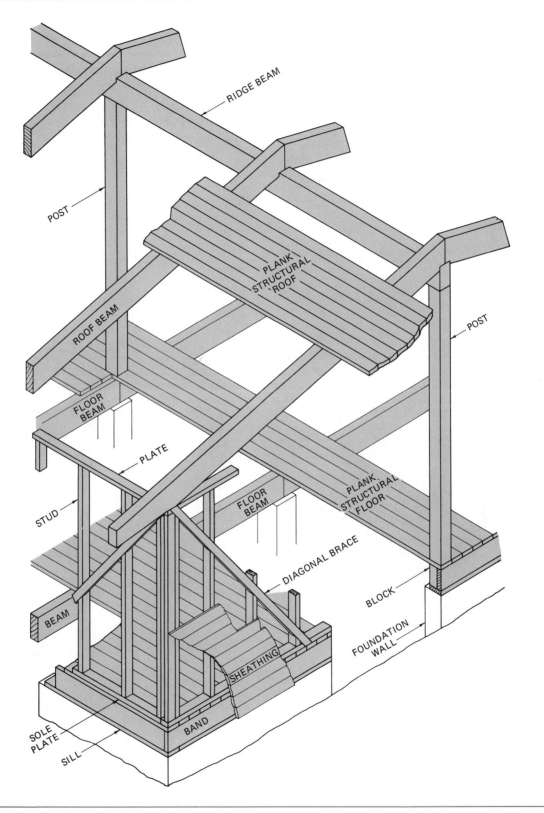

RIDGE BEAM

POST

PLANK STRUCTURAL ROOF

ROOF BEAM

POST

FLOOR BEAM

PLATE

STUD

FLOOR BEAM

PLANK STRUCTURAL FLOOR

DIAGONAL BRACE

BLOCK

BEAM

FOUNDATION WALL

SHEATHING

SOLE PLATE

BAND

SILL

Fig. 10-4 Plank and beam framing

Beams are spaced up to 8 feet on centers permitting opportunity for large glass areas in walls. For the attachment of exterior and interior finish, additional framing is necessary between wall posts.

Because of the fewer number of pieces used, well planned plank and beam framing saves labor costs. Cross-bridging is eliminated and fewer fasteners are used.

ENERGY AND MATERIAL CONSERVATION FRAMING METHODS

There has been much concern and thought about conserving energy and materials in building construction. Several systems have been devised which differ from conventional framing methods and use less material and labor.

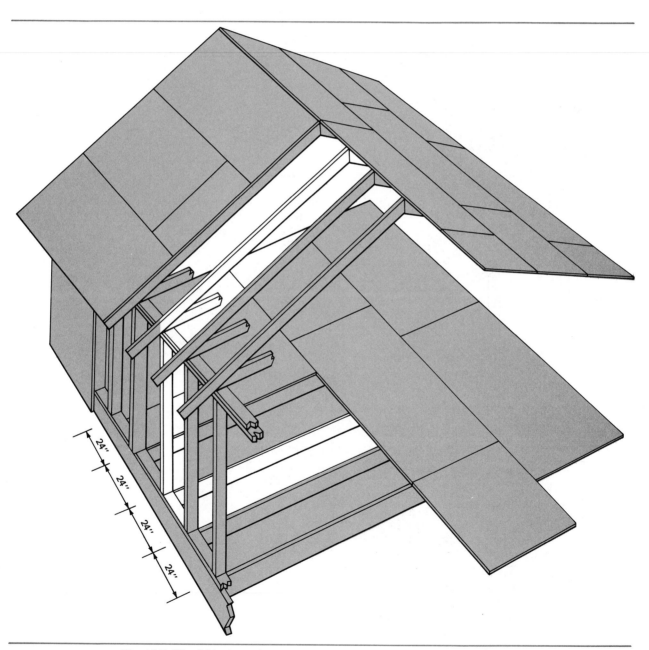

Fig. 10-5 The 24-inch module system of framing uses less material and labor.

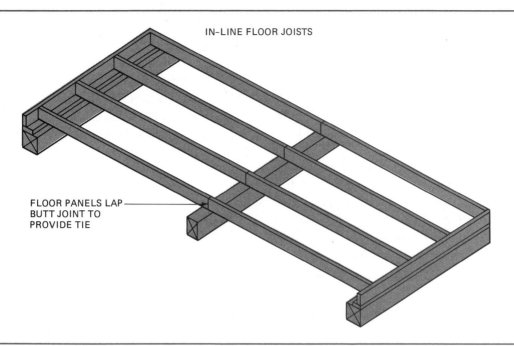

Fig. 10-6 In-line floor joists make installation of plywood subflooring simpler.

THE 24-INCH MODULE METHOD

One of the new methods uses plywood over lumber framing spaced on a 24-inch module. All framing — floors, walls, and roofs — is spaced 24 inches on center (O.C.). Joists, studs, and rafters line up with each other, Figure 10-5.

Floors. For maximum savings, a single layer of 3/4-inch tongue and grooved plywood is used over joists spaced 24 inches (O.C.). In-line floor joists are used to make installation of the plywood floor easier, Figure 10-6. The use of adhesive when fastening the plywood floor is recommended. Gluing increases stiffness and prevents squeaky floors, Figure 10-7.

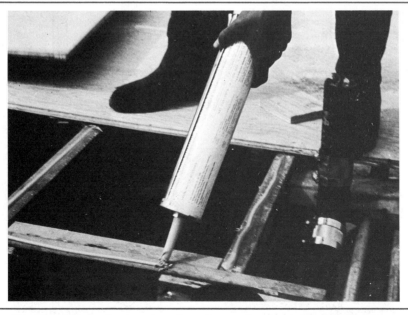

Fig. 10-7 Using adhesive when fastening subflooring makes the floor frame stiffer and stronger. *(Courtesy of American Plywood Association)*

Walls. Studs up to 10 feet long with 24-inch spacing can be used in single-story buildings. Stud height for two-story buildings should be limited to 8 feet.

A single layer of plywood acts as both sheathing and exterior siding. In this case, the plywood must be least 1/2-inch thick, Figure 10-8. If two-layer construction is used, 3/8-inch plywood is acceptable.

Wall openings are planned to be located so that at least one side of the opening falls on the module. Whenever possible, window and door sizes are selected so that the rough opening width is a multiple of the module, Figure 10-9. Also, locate partitions at normal wall stud positions if possible.

Roofs. Rafters or trusses are spaced 24 inches on center in line with the studs under 3/8-inch plywood sheathing. Increasing the spacing of the roof framing members and using a thinner roof sheathing conserves material.

HOUSE DEPTHS

House depths that are not evenly divisible by four waste floor framing and sheathing. Lumber for floor joists is produced in increments of 2 feet. A 25-foot house depth uses the same total linear footage of standard floor joist lengths as a house depth of 28 feet. Full 48-inch-width plywood panels can be used without ripping for 24-, 28-, and 32-foot house depths.

REDUCING THE CLEAR SPAN OF FLOOR JOISTS

The need to select large-size floor joists can be avoided by reducing the clear span. This can be done by using wider sill plates and center bearing plates, Figure 10-10. When the use of wider plates makes it possible to use smaller joists, the added cost of the plates is little when compared to the savings achieved.

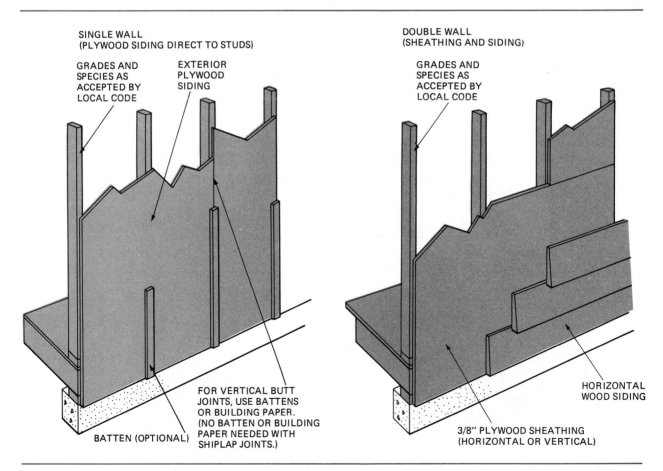

Fig. 10-8 Single-layer and double-layer exterior wall covering

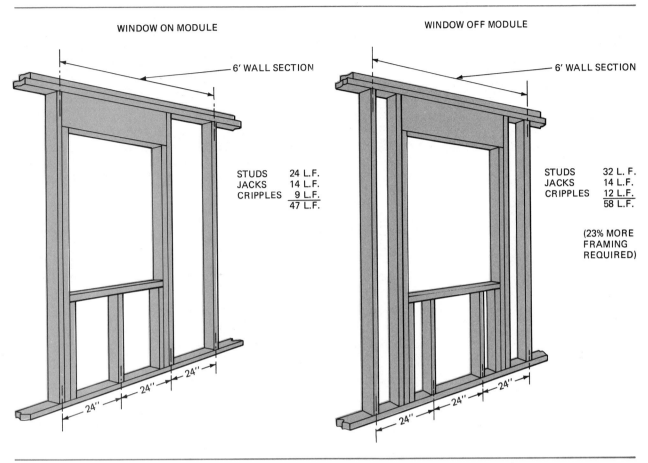

Fig. 10-9 To conserve materials, locate wall openings so they fall on the spacing module.

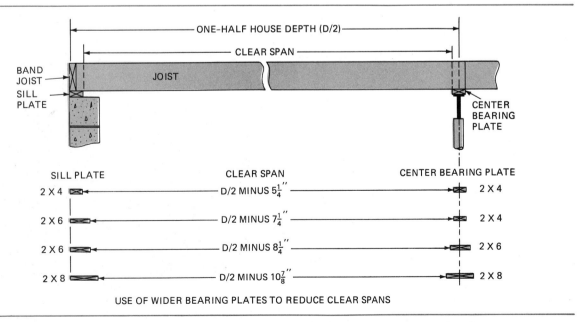

Fig. 10-10 Smaller floor joists can sometimes be used if the clear span is reduced.
(Courtesy of Southern Forest Products Association)

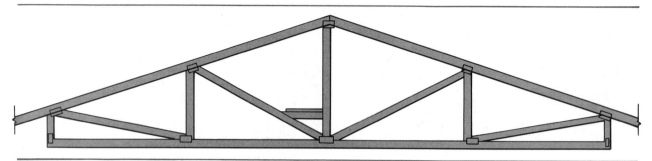

Fig. 10-11 Modified truss design accommodates 12″ ceiling insulation without compressing at eaves.

THE ARKANSAS SYSTEM

An energy saving construction system developed by the Arkansas Power and Light Company uses 2x6 wall studs spaced 24 inches O.C. This permits using a full 6-inch insulation in the exterior walls. A modified truss accommodates 12 inches of insulation in the ceiling without compressing it at the eaves, Figure 10-11.

With the increased thickness of insulation in floors and ceilings, smaller heating and cooling units are needed and substantial savings in energy costs are achieved.

FLOOR FRAMING

A floor frame consists of girders, columns, sills, joists, bridging, and subfloor fastened together in such a manner as to support the loads expected on the floor. The floor frame is started after the foundation is complete and set (hardened).

GIRDERS

Girders are heavy beams which support the inner ends of the floor joists. They may be made of solid wood or built up of two or more 2-inch planks. Steel S or W (I-shaped) beams with a 2-inch thick wood plate are sometimes used. In addition, MICRO-LAM® (registered trademark of Trus Joist Corporation, Boise, Idaho) beams are available for use as girders. Each beam is composed of many layers of specially graded veneers, glue-laminated under heat and pressure, with all grain in the same

direction to form a strong, stable product, Figure 10-12. These beams are available in three depths: 9 1/2 inches, 11 7/8 inches, and 14 1/4 inches. They can be used separately or in multiples to fit almost any application.

The size of wood or steel girders is best determined by reference to blueprints and building codes. To select the size of a MICRO-LAM® girder, use the table in Figure 10-13. Caution must be taken that conditions do not exceed those on which the table is based.

If built-up girders are used, the members are fastened together with spikes spaced about 32 inches apart. The end joints are planned so that they are located over supports, Figure 10-14.

Fig. 10-12 MICRO-LAM® beams are made up of many layers of veneer with all grain parallel. *(Courtesy of Trus Joist Corporation)*

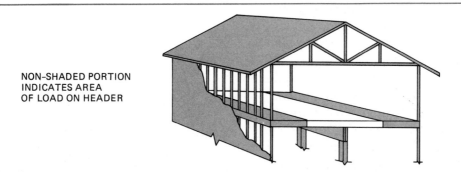

NON-SHADED PORTION
INDICATES AREA
OF LOAD ON HEADER

COLUMN SPACING	FLOOR JOIST SPAN (Use 1/2 the sum of the joist spans on both sides of the beam)								
	10'	11'	12'	13'	14'	15'	16'	17'	18'
8'	2 - 9 1/2"	2 - 9 1/2"	2 - 9 1/2"	2 - 9 1/2"	2 - 9 1/2"	2 - 9 1/2"	2 - 9 1/2"	2 - 9 1/2"	2 - 9 1/2"
10'	2 - 9 1/2"	2 - 9 1/2"	2 - 9 1/2"	2 - 9 1/2"	2 - 9 1/2"	2 - 9 1/2"	3 - 9 1/2" or 2 - 11 7/8"	3 - 9 1/2" or 2 - 11 7/8"	3 - 9 1/2" or 2 - 11 7/8"
12'	3 - 9 1/2" or 2 - 11 7/8"	3 - 9 1/2" or 2 - 11 7/8"	3 - 9 1/2" or 2 - 11 7/8"	3 - 9 1/2" or 2 - 11 7/8"	2 - 11 7/8"	2 - 11 7/8"	2 - 11 7/8"	3 - 11 7/8" or 2 - 14 1/4"	3 - 11 7/8" or 2 - 14 1/4"
14'	2 - 11 7/8"	2 - 11 7/8"	3 - 11 7/8" or 2 - 14 1/4"	3 - 11 7/8" or 2 - 14 1/4"	3 - 11 7/8" or 2 - 14 1/4"	3 - 11 7/8" or 2 - 14 1/4"	3 - 11 7/8" or 2 - 14 1/4"	2 - 14 1/4"	3 - 14 1/4"
16'	3 - 11 7/8" or 2 - 14 1/4"	3 - 11 7/8" or 2 - 14 1/4"	2 - 14 1/4"	2 - 14 1/4"	3 - 14 1/4"	3 - 14 1/4"	3 - 14 1/4"	3 - 14 1/4"	3 - 14 1/4"

• Based on residential loading of 40 PSF live load and 10 PSF dead load.
• Deflection limited to $L/360$ at live load.
• Calculations are based on continuous floor joist span and simple or continuous beam span conditions.
• For applications other than shown, special engineering may be required.

Fig. 10-13 Table for selecting the size of a MICRO-LAM® girder *(Courtesy of Trus Joist Corporation)*

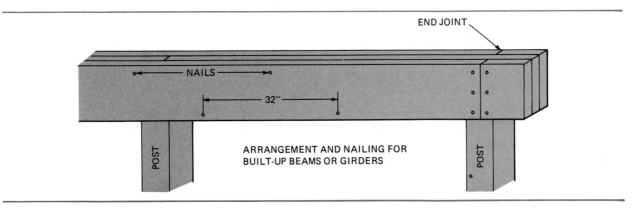

END JOINT

NAILS

32"

POST

POST

ARRANGEMENT AND NAILING FOR
BUILT-UP BEAMS OR GIRDERS

Fig. 10-14 Spacing of fasteners and joints of a built-up girder

The ends of the girder are supported by a pocket formed in the foundation wall if its top is to be the same height as the sill. The pocket should provide at least a 4-inch bearing for the girder and should be wide enough to provide 1/2-inch clearance on both sides and the end, Figure 10-15. This allows any moisture in this area to be evaporated by a circulation of air, thus preventing absorption of moisture into the girder and the resulting decay of the timber. The pocket should be deep enough to provide for an iron bearing plate under the girder. Wood shims are not suitable for use under girders because the weight imposed on them compresses the wood.

If girders are to be raised to provide more headroom, *ledgers* may be fastened to the sides or joist hangers used to provide support for the floor joists. The girder ends rest either on the foundation wall or on the sill.

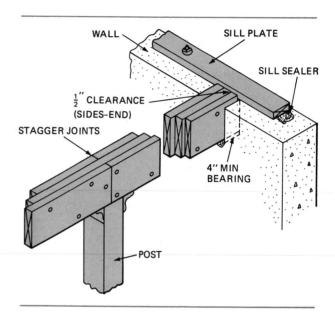

Fig. 10-15 The girder pocket in the foundation wall should be large enough to provide air space around the end of the girder.

INSTALLING GIRDERS

When installing the girder, cut it to length and shape the ends, if necessary, to fit on the bearing seats in the end walls. Slide the assembled girder along the top of the foundation wall to its bearing seat. Support the girder in the center and keep it from turning over. Tip the girder in position with its crowned edge up.

Install temporary supports under the girder and brace them so the girder is at the desired height and is straight. A line stretched from sill to sill can be used as a guide to straighten the girder.

COLUMNS

Girders may be supported by framed walls, wood posts, or steel columns. Usually, columns of steel pipe filled and reinforced with concrete are used. These are commonly called *lally columns*. Metal plates are used at the top and bottom of the columns to distribute the load over a wider area. The plates have predrilled holes so that they may be fastened to the girder. Notched sections prevent the columns from slipping off the plates. Column size should be determined from the blueprints.

INSTALLING COLUMNS

When installing columns, measure accurately from the column footing to the bottom of the girder. An accurate way of measuring is to hold a strip of lumber on the column footing and mark it at the bottom of the girder. Transfer this mark to the column. Deduct the thickness of the top and bottom plate. To mark around the column so it has a square end, wrap a sheet of paper around it and mark along the edge of the paper.

Cut through the metal along the line, using a hacksaw. Snap the column waste off by hitting the end with a hammer.

Install the columns in a plumb position. Fasten the top plates to the girder with lag screws. If the girder is steel then holes must be drilled and the plates bolted to the girder, or the plates may be welded to the girder. The bottom of the columns are held in place when the finish concrete basement floor is placed around them.

SILLS

Sills are horizontal members of a frame that lie directly on the foundation wall. Sills

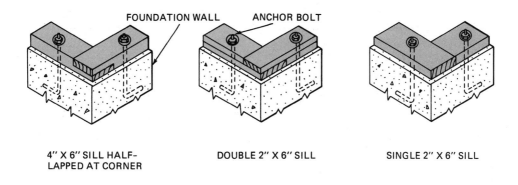

FOUNDATION WALL   ANCHOR BOLT

4" X 6" SILL HALF-
LAPPED AT CORNER

DOUBLE 2" X 6" SILL

SINGLE 2" X 6" SILL

Fig. 10-16 Sill details at corner

may consist of single 2x6 lumber or, in many cases, a doubled 2x6. Sometimes a solid 4x6 is used for sills. Usually 6-inch widths are used, but the width may vary depending on the type of construction, Figure 10-16.

The sill is usually, but not always, attached to the foundation wall with anchor bolts spaced a maximum of 8 feet on center. To take up irregularities between the foundation wall and the sill, a sill sealer is used. The sill sealer is an insulating material used to seal against drafts, dirt, and insects. It comes in rolls of 50 feet. Its nominal thickness is 1 inch, but compresses to as little as 1/32 inch.

INSTALLING SILLS

Sills must be installed so they are straight, level, and to the specified dimension. The level of all other framing members depends on the care taken with the installation of the sill.

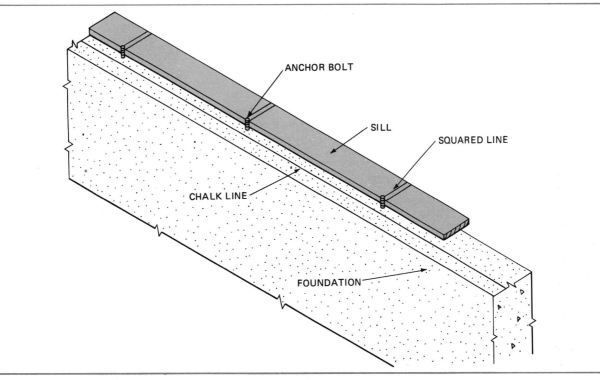

ANCHOR BOLT
SILL
SQUARED LINE
CHALK LINE
FOUNDATION

Fig. 10-17 Square lines across the sill to locate the center of anchor bolts.

Remove washers and nuts from the anchor bolts. Snap a chalk line on the top of the foundation wall in line with the inside edge of the sill. Sometimes the outside edge of the sill is flush with the outside of the foundation wall. Other times it is set in the thickness of the wall sheathing depending on the type of construction. In the case of brick veneered exterior walls, the sill plate may be set back even further.

Cut the sill sections to length and hold in place against the anchor bolts. Square lines across the sill on each side of the bolts, Figure 10-17. Measure the distance from the center of each bolt to the chalk line. Transfer this distance at each bolt location to the sill measuring from the inside edge.

Bore holes in the sill for each anchor bolt. Bore the holes at least 1/8-inch oversize to allow for adjustments. Place the sill sections in position over the anchor bolts after installing the sill sealer, if used. The inside edges of the sill sections should be on the chalk line. Replace the nuts and washers. Be careful not to tighten the nuts too much, especially if the concrete wall is green, because this may result in cracking the wall.

Shim all low spots, if any, with wood shingles so that the sill is level. Grout spaces between the sill and foundation wall with mortar. After the mortar has dried remove the shim shingles and grout the spaces left by their removal.

PROTECTION AGAINST TERMITES AND DECAY

The subterranean termite is an insect which eats wood and other similar substance for its nourishment. The termite attacks wood frame structures in contact with the ground or structures above the ground by means of shelter tubes attached to foundation walls, piers, or other members in contact with the ground. However, the termite colony can only survive when it maintains contact with soil moisture. A barrier separating wood from earth is an effective method of preventing damage by termites.

In crawl spaces and other concealed areas, clearance between the bottom of floor joists and the ground should be at least 18 inches. Wood sills should be at least 8 inches above exposed earth on the exterior of the building, Figure 10-18.

Termite shields should be of not less than 26-gauge galvanized iron or other suitable metal installed in an approved manner on the top of all foundation walls and piers and around all pipes leading from the ground, Figure 10-19. Joints should be locked and soldered.

Use pressure treated or naturally durable wood when control of termites may not be prevented by construction practices alone. Framing members on a concrete or masonry slab in direct contact with the earth should be pressure treated or naturally durable wood. The heartwood or redwood, bald cypress, and eastern red cedar are particularly resistant to termite attack.

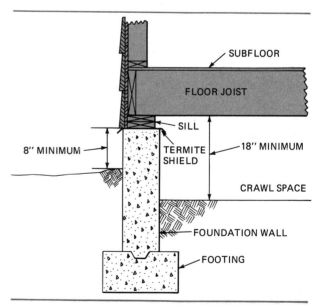

Fig. 10-18 Recommended distance for wood above ground for termite protection

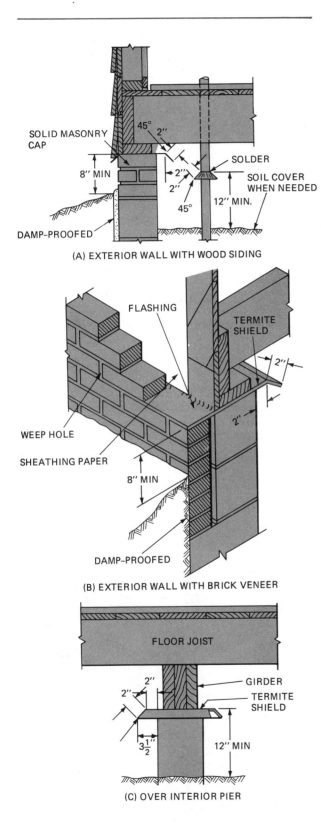

(A) EXTERIOR WALL WITH WOOD SIDING

SOLID MASONRY CAP

45°
2"

SOLDER

8" MIN

2"
2"

45°

SOIL COVER WHEN NEEDED

12" MIN.

DAMP-PROOFED

FLASHING

TERMITE SHIELD

2"

2"

WEEP HOLE

SHEATHING PAPER

8" MIN

DAMP-PROOFED

(B) EXTERIOR WALL WITH BRICK VENEER

FLOOR JOIST

2"
2"

GIRDER

TERMITE SHIELD

3 1/2"

12" MIN

(C) OVER INTERIOR PIER

Fig. 10-19 Installation of termite shields

FLOOR JOISTS

Floor joists are horizontal members of a frame that rest on and transfer the load to sills and girders. In residential construction, they are usually made of 2-inch lumber placed on edge. They are generally spaced 16 inches O.C. in conventional framing, but may be spaced 12 or 24 inches O.C. depending on the type of construction and the load.

Steel joists are available for use in floor framing. The size of solid lumber or steel joists should be determined from the construction drawings.

The Trus Joist Corporation of Boise, Idaho makes a residential TJI® joist that has a plywood web, pressure-fitted between flanges of MICRO-LAM® laminated veneer lumber or specially selected, southern pine, solid lumber, Figure 10-20. The joists are available in 9 1/2-inch and 11 7/8-inch depths and in lengths up to 60 feet with 1 1/2-inch holes in the web 13 inches O.C. for easy installation of wiring. Additional holes may be cut on the job in

Fig. 10-20 The residential TJI® joist has a plywood web pressure-fitted between flanges. *(Courtesy of Trus Joist Corporation)*

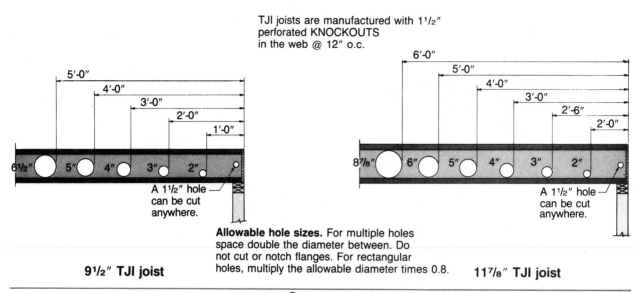

TJI joists are manufactured with 1¹/₂″ perforated KNOCKOUTS in the web @ 12″ o.c.

9¹/₂″ TJI joist

Allowable hole sizes. For multiple holes space double the diameter between. Do not cut or notch flanges. For rectangular holes, multiply the allowable diameter times 0.8.

A 1¹/₂″ hole can be cut anywhere.

11⁷/₈″ TJI joist

A 1¹/₂″ hole can be cut anywhere.

Fig. 10-21 Allowable holes in TJI® joists *(Courtesy of Trus Joist Corporation)*

accordance with the chart shown in Figure 10-21. A residential floor span chart and construction details are shown in Figure 10-22.

Joists should rest on at least 1 1/2 inches of bearing on wood and 3 inches on masonry. In platform construction, floor joists butt against a *header joist* which is fastened to the sill, Figure 10-23. In a balloon frame, joists are cut flush with the outside edge of the sill, Figure 10-24.

Joists may rest directly on the top or may be framed into the side of the girder in a number of ways, Figure 10-25-A and B. If joists are lapped over the girder, the minimum amount of lap is 4 inches and the maximum overhang is 12 inches.

A bevel cut called a fire cut is made on the ends of floor joists in masonry walls. The fire cut prevents the masonry wall from collapsing if

	JOIST DEPTH	
O.C. SPACING	9 1/2″	11 7/8″
12″	18′-2″	21′-9″
16″	16′-6″	19′-9″
19.2″	15′-6″	18′-6″
24″	14′-4″	17′-1″
32″	11′-4″	14′-4″

1. Based on floor load of 40# live load and 10# dead load.
2. Based on deflection of ᴸ/360 at live load.
3. Assumes composite action with single layer of nailed plywood decking for deflection only.

NOTE: For other load conditions, refer to allowable uniform load chart.

Fig. 10-22 TJI® residential floor joist span chart *(Courtesy of Trus Joist Corporation)*

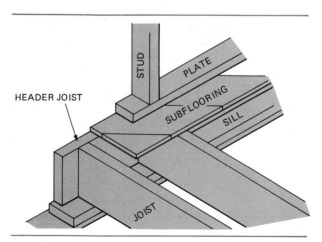

Fig. 10-23 First-floor platform framing at sill

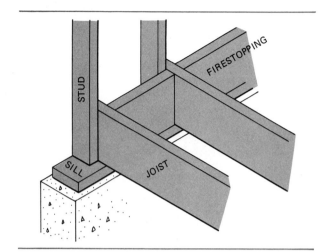

Fig. 10-24 First-floor balloon framing at sill

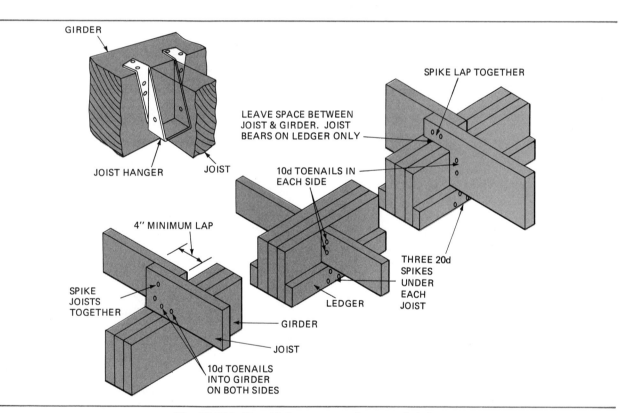

Fig. 10-25A Wood joists supported on wood girder

165

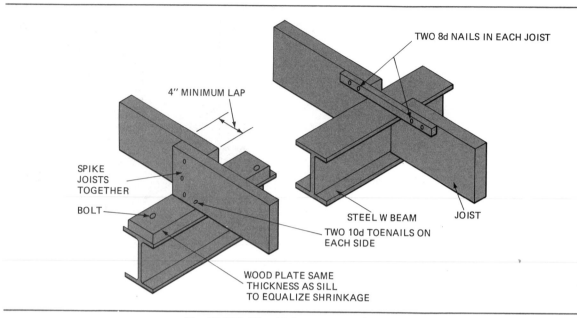

Fig. 10-25B Wood joists supported on steel girder

the joist is burned through and falls in a fire, Figure 10-26.

NOTCHING AND BORING OF JOISTS

Notches in the bottom or top of floor joists should not exceed 1/6 of the joist depth and should not be located in the middle third of the joist span. Notches on the ends should not exceed 1/4 of the joist depth.

Holes bored in joists for piping or wiring should not be larger than 1/3 of the joist depth and should not be closer than 2 inches to the top or bottom of the joist, Figure 10-27.

LAYING OUT FLOOR JOIST LOCATIONS

Floor joist locations are laid out on the inside of the header joist in platform construction. In balloon construction, the layout is

Fig. 10-26 A fire cut is made on the end of floor joists in masonry walls.

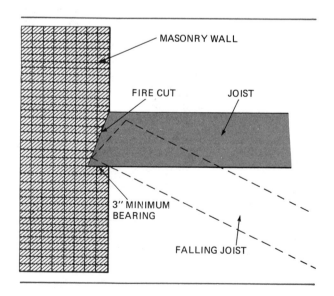

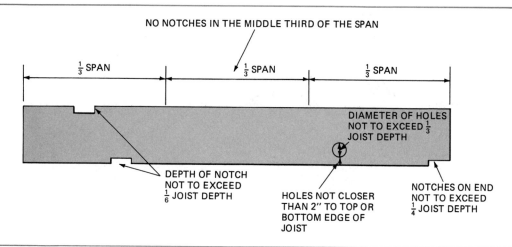

Fig. 10-27 Allowable notches and holes in floor joists

made directly on the sill. A squared line marks the side of the joist and an *X* to one side of the line indicates on which side of the line the joist is to be placed, Figure 10-28.

When plywood subflooring is to be used, floor joists must be laid out so that the ends of the plywood sheets fall directly on the center of floor joists. Start the joist layout by measuring from the end of the sill the on center spacing.

Measure back 1/2 the thickness of the joist and square a line across the sill or header joist. This line indicates the side of the joist. Place an *X* on the side of the line on which the joist is to be placed, Figure 10-29.

From the squared line, measure and mark the spacing of the joists along the length of the building. Place an *X* on the same side of each line as for the first joist location.

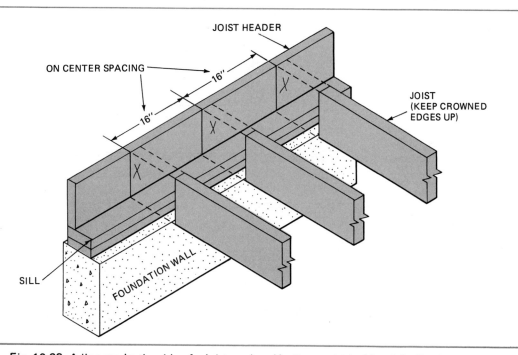

Fig. 10-28 A line marks the side of a joist, and an *X* tells on which side of the line it is placed.

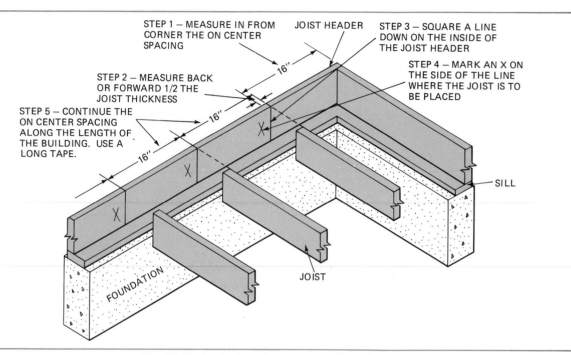

STEP 1 — MEASURE IN FROM CORNER THE ON CENTER SPACING

JOIST HEADER

STEP 3 — SQUARE A LINE DOWN ON THE INSIDE OF THE JOIST HEADER

STEP 4 — MARK AN X ON THE SIDE OF THE LINE WHERE THE JOIST IS TO BE PLACED

STEP 2 — MEASURE BACK OR FORWARD 1/2 THE JOIST THICKNESS

STEP 5 — CONTINUE THE ON CENTER SPACING ALONG THE LENGTH OF THE BUILDING. USE A LONG TAPE.

16"

16"

16"

SILL

FOUNDATION

JOIST

Fig. 10-29 Laying out the location of floor joists

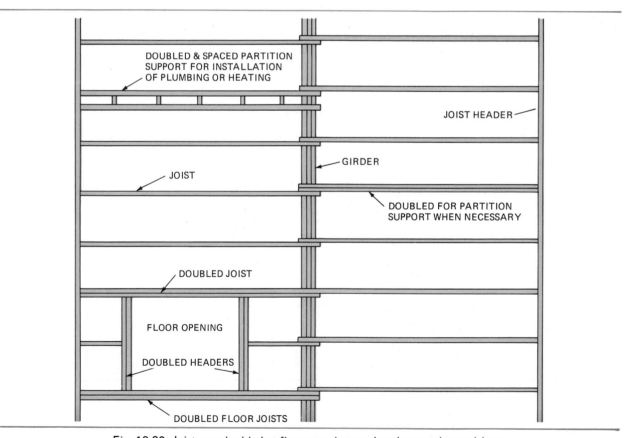

DOUBLED & SPACED PARTITION SUPPORT FOR INSTALLATION OF PLUMBING OR HEATING

JOIST HEADER

GIRDER

JOIST

DOUBLED FOR PARTITION SUPPORT WHEN NECESSARY

DOUBLED JOIST

FLOOR OPENING

DOUBLED HEADERS

DOUBLED FLOOR JOISTS

Fig. 10-30 Joists are doubled at floor openings and under certain partitions.

When measuring for the spacing of the joists, use a tape stretched along the length of the building. Using a tape in this manner is more accurate. Measuring and marking each spacing individually generally causes a gain in the spacing.

Study the plans for the location of floor openings and identify the marks that are not for full length floor joists. Shortened floor joists for floor openings are called *tail joists* and are usually identified by changing the *X* to a *T*. Indicate, also, where joists are to be doubled for floor openings or for partition supports or wherever additional floor joists are required, Figure 10-30.

Transfer the layout to a long strip of wood. Nail pieces together, if necessary. This is called a layout rod. Use the rod to make the layout on the girder and on the opposite wall. If the joists are in-line, *Xs* are made on the same side of the mark. If the joists are lapped, an *X* is placed on both sides of the mark at the girder and on the opposite side of the mark on the other wall, Figure 10-31.

INSTALLING FLOOR JOISTS

After the header joist is fastened in place and joist locations marked, all full length joists are fastened in position. Each joist is carefully sighted along its length by eye and installed with its crowned edge up. Any joist with a severe crook or other warp is rejected, Figure 10-32.

Toenail the joists to the sill and girder with 10d common nails. Drive three 16d or 20d spikes through the header joists into the end of each floor joist. Spike the joists together if they lap at the girder, Figure 10-33. When all floor joists are in position, they are sighted by eye from end to end and straightened. They are held straight by a strip of 1x3s tacked to the top of the joists about in the middle of the span.

TJI® joists are installed using standard tools and can be easily cut to any required

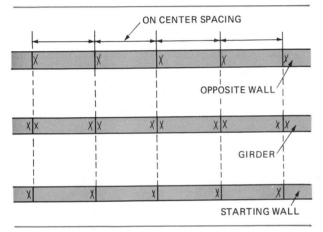

Fig. 10-31 Layout lines for the floor joists on the girder and the opposite wall (joists lapped over the girder)

Fig. 10-32 Installing floor joists — Sight each joist and install with crowned edge up. *(Courtesy of Georgia-Pacific)*

Fig. 10-33 A floor frame with joists lapped over the girder *(Courtesy of Georgia-Pacific)*

Fig. 10-34 Installing TJI® floor joists *(Courtesy of Trus Joist Corporation)*

length on the job site. The wide, straight wood flanges on the joist make nailing easier, Figure 10-34. Long lengths, up to 60 feet, eliminate lapping over girders or walls. TJI® joists must be installed in accordance with installation specifications shown in Figure 10-35.

DOUBLING FLOOR JOISTS

For added strength, doubled floor joists must be securely fastened together, with their top edges even. In most cases, the top edges do not lie flush. To bring them flush with each other, toenail down through the top edge of the higher one while squeezing both together tightly by hand. Use as many toenails as necessary to

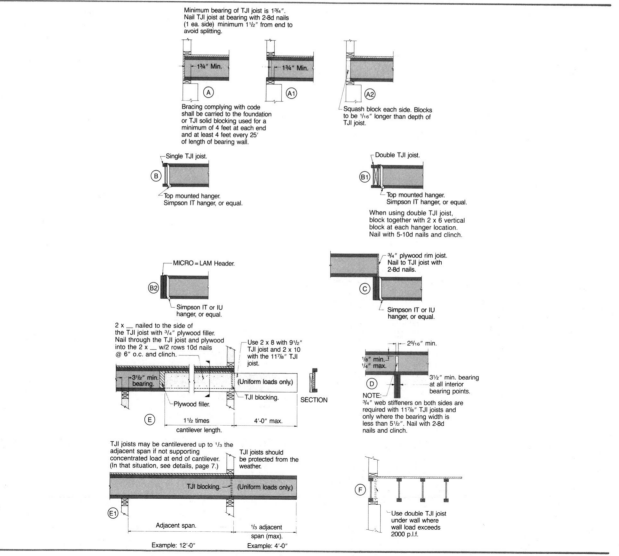

Fig. 10-35 Installation details for TJI® joists *(Courtest of Trus Joist Corporation)*

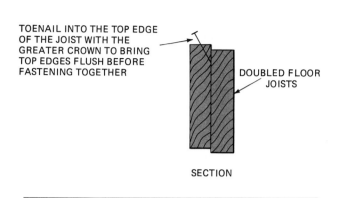

TOENAIL INTO THE TOP EDGE OF THE JOIST WITH THE GREATER CROWN TO BRING TOP EDGES FLUSH BEFORE FASTENING TOGETHER

DOUBLED FLOOR JOISTS

SECTION

Fig. 10-36 Bring the tops of doubled floor joists flush by toenailing them.

bring the top edges flush with each other, Figure 10-36. Then fasten the two pieces securely together by driving 16d spikes at a slight angle through the sides, staggered from top to bottom, about 2 feet apart.

FRAMING FLOOR OPENINGS

To frame an opening in a floor, first set the trimmer joists in place. *Trimmer joists* are full length joists that run alongside the opening. Cut headers to length by taking the measurement at the sill. *Headers* are members of the opening that run at right angles to the floor joists. Taking the measurement at the sill where the trimmers are fastened, rather than at the opening, is re-required because of any bow in the trimmer joists. Mark the location of the tail joists on the headers at the sill. Fasten the first header in position by driving spikes through the side of the trimmer into the ends of the headers. Fasten the tail joists in position. Double up the headers and finally double up the trimmer joist. Figure 10-37 shows the sequence of operations used to frame a floor opening.

BRIDGING

Bridging is installed between floor joists at intervals not exceeding 8 feet. Its purpose is to distribute a concentrated load on the floor over a wider area. Although under certain conditions, bridging may not be required, many builders install it because of customary practice.

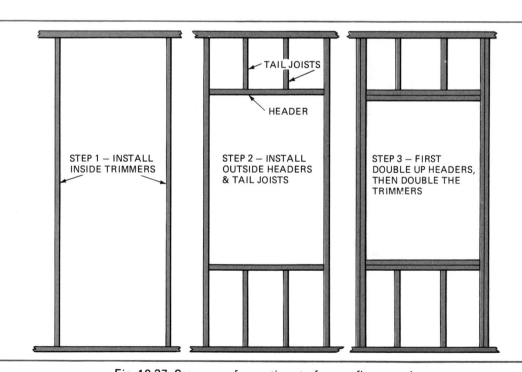

TAIL JOISTS

HEADER

STEP 1 – INSTALL INSIDE TRIMMERS

STEP 2 – INSTALL OUTSIDE HEADERS & TAIL JOISTS

STEP 3 – FIRST DOUBLE UP HEADERS, THEN DOUBLE THE TRIMMERS

Fig. 10-37 Sequence of operations to frame a floor opening

Bridging may be solid wood, wood cross bridging, or metal cross bridging, Figure 10-38. Usually solid wood bridging is the same size as the floor joists and installed in an offset fashion to permit end nailing.

Wood cross bridging should be at least nominal 1x3 lumber with two 6d nails at each end. It is placed in double rows that cross each other in the joist space.

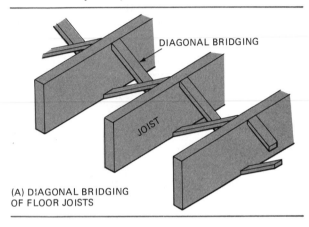

(A) DIAGONAL BRIDGING OF FLOOR JOISTS

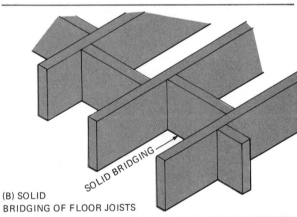

(B) SOLID BRIDGING OF FLOOR JOISTS

(C) COMPRESSION BRIDGING INSTALLS VERY QUICKLY AFTER SUB-FLOORING IS IN PLACE.

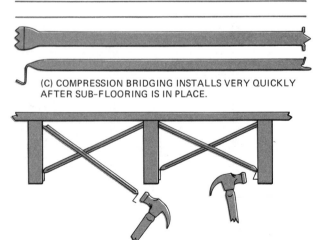

Fig. 10-38 Types of bridging

Metal cross bridging is available in different lengths for particular joist size and spacing. Usually of 18-gauge steel, 3/4-inch wide, it comes in a variety of styles. It is applied in a way similar to that used for wood cross bridging.

LAYING OUT AND CUTTING WOOD CROSS BRIDGING

Wood cross bridging may be laid out using a framing square. Determine the actual distance between floor joists and the actual depth of the joist. For example, 2x10 floor joists 24 inches O.C. measure 22 1/2 inches between them. The actual depth of the joist is 9 1/4 inches.

Hold the framing square on the edge of a piece of bridging stock so that the 9 1/4-inch mark of the tongue lines up with the lower edge of the stock and the 22 1/2-inch mark of the blade lines up with the upper edge of the stock. Mark lines along the tongue and blade across the stock.

Turn the square, keeping the same face up and mark along the tongue, Figure 10-39. The bridging may then be cut using a radial arm saw with the blade tilted and a stop set for duplicate lengths.

Another method of cutting bridging to length is by having a helper hold two strips of bridging material together at an angle between the floor joists. Using a handsaw, cut both pieces at the same time using the side of the joist

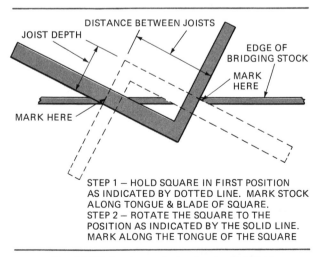

STEP 1 – HOLD SQUARE IN FIRST POSITION AS INDICATED BY DOTTED LINE. MARK STOCK ALONG TONGUE & BLADE OF SQUARE.
STEP 2 – ROTATE THE SQUARE TO THE POSITION AS INDICATED BY THE SOLID LINE. MARK ALONG THE TONGUE OF THE SQUARE

Fig. 10-39 Laying out wood cross-bridging with a framing square

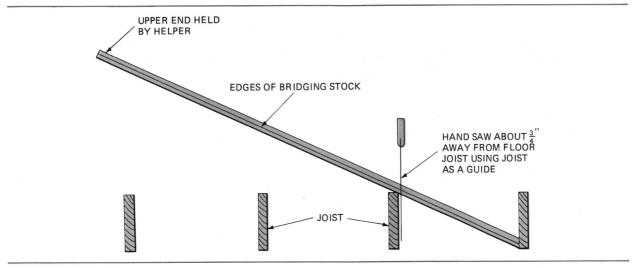

Fig. 10-40 Cutting cross-bridging with a handsaw

as a guide for cutting the angle, Figure 10-40. With this method, bridging is cut to the correct length no matter what the joist spacing may be.

INSTALLING BRIDGING

Determine the location of the bridging and snap a chalk line across the tops of the floor joists. Square down from the chalk line to the bottom edge of the floor joists on both sides.

Solid Wood Bridging. To install solid wood bridging, cut the pieces to length and install pieces in every other joist space on one side of the chalk line. Fasten the pieces by spiking through the joists into their ends. Keep the top edges flush with the floor joists.

Install pieces in the remaining spaces on the opposite side of the line in a similar manner. Installing pieces in every other space keeps from bowing the floor joists.

Wood Cross Bridging. To install wood cross bridging, start two 6d nails in one end of the bridging. Fasten it flush with the top of the joist on one side of the line. Nail only the top end. The bottom ends are not nailed until the subfloor is laid.

Fasten another piece of bridging in the same space to the other joist flush with the top and on the opposite side of the line. Continue installing bridging in the other spaces but al-

ternating the location so that the top ends of the bridging in each space are opposite each other.

Metal Cross Bridging. Metal cross bridging is fastened in a manner similar to that used for wood cross bridging. The method of fastening may differ according to the style of the bridging. Usually the bridging is fastened to the top of the joists through predrilled holes in the bridging. Because the metal is thin, nailing to the top of the joists does not interfere with the subfloor.

Some types of metal cross bridging have steel prongs which are driven into the side of the floor joists. This type can be installed from below after the subfloor is laid.

SUBFLOORING

Plywood is generally used for subflooring, although square-edge or tongue-and-grooved boards are also used. Plywood can be applied faster. Also, the finish floor can be laid in any direction, Figure 10-41.

Boards may be applied at right angles to the floor joists or diagonally. Boards laid diagonally permit the finish floor to be laid either parallel or at right angles to the floor joists. Diagonal subflooring also provides stiffness to the entire building. End joints in subflooring should be made over the joists unless

Fig. 10-41 Applying plywood subflooring
(Courtesy of Georgia-Pacific)

end-matched boards are used. Boards should not exceed 8 inches in width.

Plank decking should be edge and end matched. Each plank should rest on at least one support and be fastened with two 16d spikes. The planks should be continuous over more than one span, if possible, for greater strength and stiffness.

APPLYING PLYWOOD SUBFLOORING

The most popular spacing of joists for plywood floors is 16 inches O.C. using 1/2-inch or 5/8-inch plywood. With 24 inch spacing, plywood at least 3/4 inch thick must be used.

Plywood panels are fastened with 8d common nails spaced 6 inches apart around panel edges and 10 inches apart on intermediate joists. For stapling, use staples with a 1 5/8-inch leg spaced 4 inches apart on panel edges and 7 inches apart on intermediate supports for 1/2-inch plywood. For 5/8-inch plywood, fasten 2 1/2 inches apart at panel edges and 4 inches apart in the center.

Start the plywood subflooring by laying the first row to a chalk line. Leave a 1/16-inch space at all panel end joints to allow for expansion.

Fasten the second row and stagger the end joints. Leave a 1/8-inch space between panel edges. All end joints are made over joists.

Continue laying and fastening plywood sheets in this manner until the entire floor is covered. Leave out sheets where there are to be openings in the floor.

Snap chalk lines across the edges and ends of the building. Trim overhanging plywood with a portable electric saw.

APPLYING BOARD SUBFLOORING

When applying board subflooring start the first row of boards flush with the edge of the floor. End joints should come over floor joists. Lay the second row, staggering the ends joints from those in the first row. Drive the edges up tight to the first row by toenailing into the edge of the boards. Facenail 6-inch wide boards with two 8d common nails in each floor joist. Use three nails in 8-inch boards. Continue applying boards in this manner until the floor is covered. Snap lines and trim all overhanging ends.

Diagonal flooring is started by snapping a chalk line at a 45° angle to the joists. To lay out a 45° angle, measure out from one corner along the side and end the same distance. Snap a line between the points.

ESTIMATING MATERIALS

- To determine the amount of floor joists needed in a floor frame, divide the length of the building by the spacing and add one. Multiply by the number of rows of floor joists. Add the number needed for doubling and for header joists.

- To determine the amount of wood cross bridging needed, multiply the length of the building by three for each row of bridging.

- To determine the amount of plywood subflooring required, divide the floor area by 32 to find the number of sheets of plywood needed.

- For 6-inch board subflooring, add 20 percent to the floor area. Add 15 percent to the floor area when using 8-inch boards. Add 5 percent more if the boards are laid diagonally.

REVIEW QUESTIONS

Select the most appropriate answer.

1. A platform or western frame is easy to erect because
   a. only one-story buildings are constructed with this type of frame.
   b. each platform may be constructed on the ground.
   c. at each level a flat surface is provided on which to work.
   d. less framing members are required.

2. One of the advantages of the balloon frame is that
   a. the bottom plates act as firestops.
   b. there is little vertical shrinkage in the frame.
   c. the second floor joists rest on a ribbon instead of a plate.
   d. it is stronger, stiffer and better able to resist lateral pressures.

3. A heavy beam that supports the inner ends of floor joists is called a
   a. pier.   c. stud.
   b. girder.   d. sill.

4. That member of a floor frame that rests directly on the foundation wall is called a
   a. pier.   c. stud.
   b. girder.   d. sill.

5. To mark a square end on a round column
   a. use a square.
   b. measure down from the existing end.
   c. use a pair of dividers.
   d. wrap a piece of paper around it.

6. Anchor bolts are spaced a maximum of
   a. 6 ft O.C.   c. 10 ft O.C.
   b. 8 ft O.C.   d. 12 ft O.C.

7. To protect against termites keep wood in crawl spaces and other concealed areas at least
   a. 8 inches above the ground.
   b. 12 inches above the ground.
   c. 18 inches above the ground.
   d. 24 inches above the ground.

8. When floor joists rest on wood they should have at least
   a. 4 inches of bearing.
   b. 3 1/2 inches of bearing.
   c. 2 1/2 inches of bearing.
   d. 1 1/2 inches of bearing.

9. Floor joists should have a minimum lap of
   a. 2 inches.   c. 6 inches.
   b. 4 inches.   d. 8 inches.

10. It is extremely important when installing floor joists to
    a. fasten them securely with enough nails.
    b. fasten the girder end first.
    c. have the crowned edges up.
    d. notch and size the ends for any differences in width.

EXTERIOR WALL FRAMING

11

OBJECTIVES

After completing this unit, the student will be able to

- *identify and describe the function of each part of the wall frame.*

- *determine the length of exterior wall studs and the size of rough openings and lay out a story pole.*

- *build corner posts and partition intersections and describe several methods of forming them.*

- *lay out the wall plates.*

- *construct and erect wall sections to form the exterior wall frame.*

- *plumb, brace, and straighten the exterior wall frame.*

- *apply wall sheathing.*

PARTS OF A WALL FRAME

The wall frame consists of a number of different parts. The student should understand the function of each and know the location and usual sizes.

PLATES

The top and bottom horizontal members of a wall frame are called *plates*. The bottom member is called a *soleplate*, but is also referred to as the *bottom plate* or *shoe*. The top members are called *top plates* and consist usually of doubled 2-inch stock, Figure 11-1. In a balloon frame, the soleplate is not used because the studs rest directly on the sill.

STUDS

Studs are vertical members of the wall frame that run full length between plates. Shortened studs that do not extend full length because of an opening in the wall are called *cripple studs* or *jack studs*. Studs are usually 2x4s, but 2x6s are used when 6-inch insulation is desired in exterior walls.

HEADERS

Headers run at right angles to studs and form the top of window, door, and other wall openings, such as fireplaces. At the bottom of a window opening, they are called *rough sills* and are usually a single 2x4 or 2x6 because they

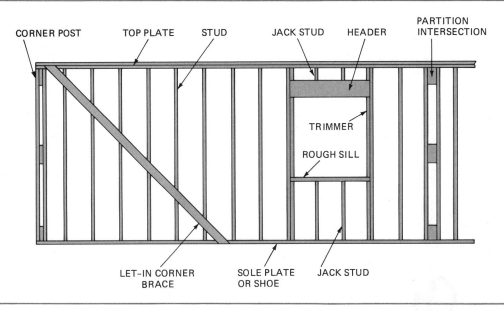

Fig. 11-1 Parts of an exterior wall frame

carry little load. However, many carpenters prefer to use a double rough sill to provide more surface on which to fasten window trim.

Headers are made of solid 4-inch or 6-inch lumber or two pieces of 2-inch lumber with 1/2-inch or 2 1/2-inch spacers between them. The spacers are required to give the header thickness the full 3 1/2-inch or 5 1/2-inch width of the stud. In many buildings when the opening must be supported without increasing the header size, the top of a wall opening may be trussed to provide support, Figure 11-2.

Headers must be strong enough to support the load across the opening. The width or depth of the header depends on the width of the opening. Check drawings, specifications, or codes for header size.

However, when the opening is close to the top plate, the header completely fills the space between the plate and the top of the opening.

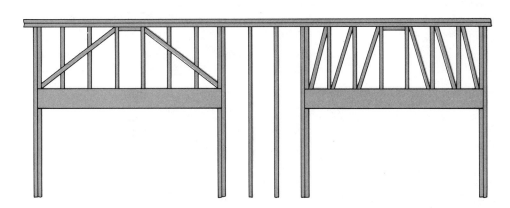

Fig. 11-2 Sometimes the top of a wall opening is trussed instead of increasing the header depth. Two methods of trussing an opening are shown.

177

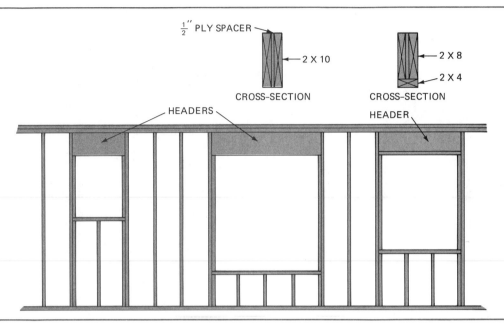

Fig. 11-3 It is common practice to use the same header width and completely fill the space between the opening and the top plate for all openings regardless of the opening width.

In this case, the same size header is used for all wall openings, regardless of the width of the opening, Figure 11-3. This eliminates the need of installing short jack studs above the header.

MICRO-LAM® beams, previously described in Unit 10 Floor Framing, may also be used as headers, Figure 11-4. Because of their strength, they can be used for wider openings, or to carry greater loads than solid lumber of the same size.

For multistory structures where a greater load is imposed on the header, the size is designated by the architect or specified in local building codes. The size of MICRO-LAM® headers for various applications may be determined using the tables shown in Figure 11-5.

Fig. 11-4 MICRO-LAM® beams used as headers *(Courtesy of Trus Joist Corporation)*

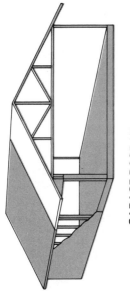

NON-SHADED PORTION
INDICATES AREA
OF LOAD ON HEADER

GARAGE DOOR HEADER TABLE
For Single Story Applications

ROOF LOAD	125% NO - SNOW 20# L.L. + 10# D.L.			125% NON SNOW 20# L.L. + 20# D.L.			115% SNOW 25# L.L. + 10# D.L.			115% SNOW 30# L.L. + 10# D.L.			115% SNOW 40# L.L. + 10# D.L.		
DOOR OPENING SIZE \ ROOF TRUSS SPAN IN FEET WITH 24" SOFFIT ASSUMED	9'-3"	16'-3"	18'-3"	9'-3"	16'-3"	18'-3"	9'-3"	16'-3"	18'-3"	9'-3"	16'-3"	18'-3"	9'-3"	16'-3"	18'-3"
22'	1-9 1/2"	2-11 7/8" or 3-9 1/2"	2-11 7/8"	1-9 1/2"	2-11 7/8"	2-14 1/4"	1-9 1/2"	2-11 7/8"	2-14 1/4" or 3-11 7/8"	1-9 1/2"	2-11 7/8"	2-14 1/4" or 3-11 7/8"	1-11 7/8" or 2-9 1/2"	2-14 1/4" or 3-11 7/8"	3-14 1/4"
24'	1-9 1/2"	2-11 7/8" or 3-9 1/2"	2-11 7/8"	1-9 1/2"	2-11 7/8"	2-14 1/4"	1-9 1/2"	2-11 7/8"	2-14 1/4" or 3-11 7/8"	1-9 1/2"	2-11 7/8"	2-14 1/4" or 3-11 7/8"	1-11 7/8" or 2-9 1/2"	2-14 1/4" or 3-11 7/8"	3-14 1/4"
26'	1-9 1/2"	2-11 7/8" or 3-9 1/2"	2-11 7/8"	1-9 1/2"	2-11 7/8"	2-14 1/4"	1-9 1/2"	2-11 7/8"	2-14 1/4" or 3-11 7/8"	1-9 1/2"	2-11 7/8"	2-14 1/4" or 3-11 7/8"	1-11 7/8" or 2-9 1/2"	2-14 1/4" or 3-11 7/8"	3-14 1/4"
28'	1-9 1/2"	2-11 7/8" or 3-9 1/2"	2-11 7/8"	1-9 1/2"	2-11 7/8"	2-14 1/4"	1-9 1/2"	2-11 7/8"	2-14 1/4" or 3-11 7/8"	1-11 7/8" or 2-9 1/2"	2-14 1/4" or 3-11 7/8"	2-14 1/4" or 3-11 7/8"	1-11 7/8" or 2-9 1/2"	2-14 1/4" or 3-11 7/8"	3-14 1/4"
30'	1-9 1/2"	2-11 7/8"	2-11 7/8"	2-9 1/2"	2-14 1/4"	2-14 1/4"	1-9 1/2"	2-11 7/8"	2-14 1/4" or 3-11 7/8"	1-11 7/8" or 2-9 1/2"	2-14 1/4" or 3-11 7/8"	2-14 1/4" or 3-11 7/8"	1-11 7/8" or 2-9 1/2"	3-11 7/8"	3-14 1/4"
32'	1-9 1/2"	2-11 7/8"	2-14 1/4" or 3-11 7/8"	2-9 1/2"	2-14 1/4"	2-14 1/4"	1-11 7/8" or 2-9 1/2"	2-11 7/8"	2-14 1/4" or 3-11 7/8"	1-11 7/8" or 2-9 1/2"	2-14 1/4" or 3-11 7/8"	3-14 1/4"	1-11 7/8" or 2-9 1/2"	3-14 1/4"	3-14 1/4"
34'	1-9 1/2"	2-11 7/8"	2-14 1/4" or 3-11 7/8"	2-9 1/2"	2-14 1/4"	3-14 1/4"	1-11 7/8" or 2-9 1/2"	2-14 1/4" or 3-11 7/8"	2-14 1/4" or 3-11 7/8"	1-11 7/8" or 2-9 1/2"	2-14 1/4" or 3-11 7/8"	3-14 1/4"	2-9 1/2"	3-14 1/4"	3-14 1/4"
36'	1-9 1/2"	2-11 7/8"	2-14 1/4" or 3-11 7/8"	2-9 1/2"	2-14 1/4"	3-14 1/4"	1-11 7/8" or 2-9 1/2"	2-14 1/4" or 3-11 7/8"	2-14 1/4" or 3-11 7/8"	1-11 7/8" or 2-9 1/2"	2-14 1/4" or 3-11 7/8"	3-14 1/4"	2-9 1/2"	3-14 1/4"	3-14 1/4"

NOTE:
- This table is for headers carrying roof load only. For compound loads (i.e. 2nd floor **and** roof) refer to table below.
- For loads in excess of those shown, refer to MICRO = LAM® ALLOWABLE LOAD TABLE.
- Deflection limited to L/240 at live load or L/180 at total load.
- Reduction in live load for 125% load condition only, per UBC section 2306.

Fig. 11-5 MICRO-LAM® beam header span tables *(Courtesy of Trus Joist Corporation)*

GARAGE DOOR HEADER TABLE
For Two Story Applications

	DOOR OPENING SIZE	ALLOWABLE LOAD IN LBS. PER LINEAL FOOT (PLF)			
		9'-3"	16'-3"	18'-3"	
MEMBER QTY. & SIZE	2 pcs 9 1/2"	989	245	173	
	3 pcs 9 1/2"	1484	367	259	
	2 pcs 11 7/8"	1505	479	341	
	3 pcs 11 7/8"	2258	718	508	
	2 pcs 14 1/4"	1995	712	562	
	3 pcs 14 1/4"	2989	1061	843	

Instructions for sizing:

1. Calculate roof load (plf) by multiplying 1/2 truss span + 2' soffit times total load per sq. ft.
2. Calculate floor load (plf) by multiplying 1/2 span of floor joists times total load per sq. ft.
3. Add weight of wall (60 plf in most cases.)
4. Add roof load (plf), floor load (plf), and wall load (plf) to obtain total load (plf).
5. From chart at right, select a header that will carry the total load (plf).

WINDOW HEADER MAXIMUM SPAN TABLE
For 1 Story Application

Header Size	Roof Total Load	30 PSF (125%)	40 PSF (115%)	50 PSF (115%)
	1- 9 1/2"	8'-6"	7'-9"	7'-4"
	2- 9 1/2"	11'-0"	10'-4"	9'-8"
	1-11 7/8"	9'-11"	9'-2"	8'-8"
	2-11 7/8"	12'-6"	11'-7"	10'-11"

Table shows maximum span capacity of indicated members assuming 10# dead load with 36' roof truss with 2' soffit.

WINDOW HEADER MAXIMUM SPAN TABLE
For 2 Story Application

Header Size	Roof Total Load	30 PSF (125%)	40 PSF (115%)	50 PSF (115%)
	1- 9 1/2"	6'-9"	5'-6"	4'-9"
	2- 9 1/2"	9'-9"	8'-10"	8'-3"
	1-11 7/8"	8'-10"	8'-3"	7'-9"
	2-11 7/8"	11'-2"	10'-7"	10'-2"

Table shows maximum span capacity of indicated members assuming 10# dead load with 36' roof truss with 2' soffit.

Fig. 11-5 (Continued)

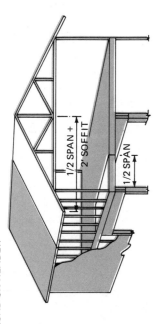

NON-SHADED PORTION INDICATES AREA OF LOAD ON HEADER

1/2 SPAN + 2' SOFFIT

1/2 SPAN

NON-SHADED PORTION INDICATES AREA OF LOAD ON HEADER

NON-SHADED PORTION INDICATES AREA OF LOAD ON HEADER

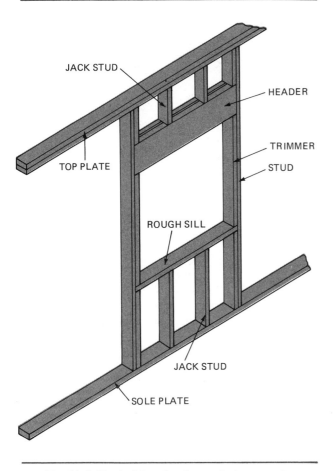

Fig. 11-6 Typical framing for a window opening

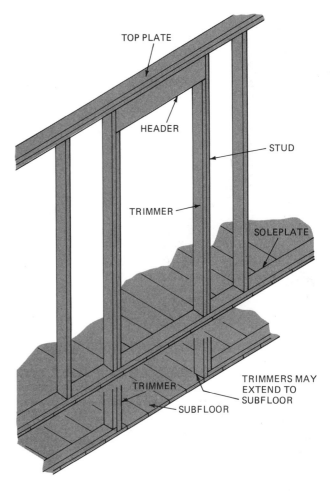

Fig. 11-7 Typical framing for a door opening

TRIMMERS

Trimmers are shortened studs that support the headers and are fastened to the studs on each side of the opening. In window openings they should fit snugly between the header and rough sill, Figure 11-6. In door openings, the trimmers, sometimes called door jacks, fit between the header and the soleplate. Some carpenters prefer to cut out the soleplate between the studs of a door opening and let the bottom end of the trimmers rest on the subfloor, Figure 11-7. However, cutting out the soleplate may make it awkward when building wall sections on the subfloor and erecting them into place. In some cases, door trimmers are left out until the wall section is erected. The soleplate is then cut out and trimmers installed.

CORNER POSTS

Corner posts are the same length as studs. They are constructed in a manner that provides an outside and an inside corner on which to fasten the exterior and interior wall coverings. Corner posts are built in a number of ways. Most of the time, three 2x4s are used. Sometimes a solid 4x6 is used along with a single 2x4, Figure 11-8. With special gypsum board clips, a corner post may be made of only two 2x4s or 2x6s, Figure 11-9.

PARTITION INTERSECTIONS

Wherever interior partitions meet an exterior wall, extra studs need to be put in the exterior wall to provide wood for fastening the

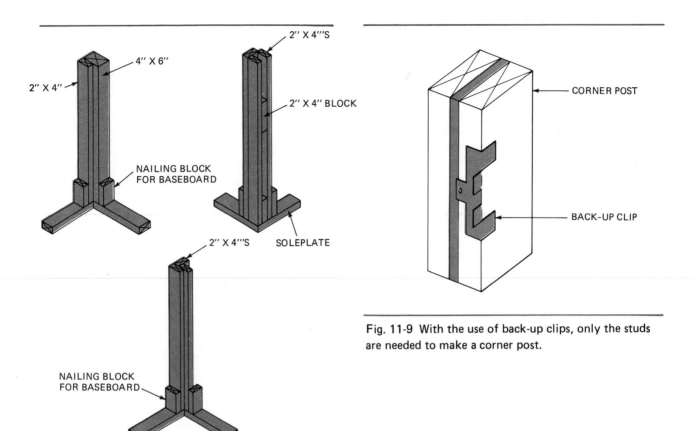

Fig. 11-8 Several methods of making corner posts are used.

Fig. 11-9 With the use of back-up clips, only the studs are needed to make a corner post.

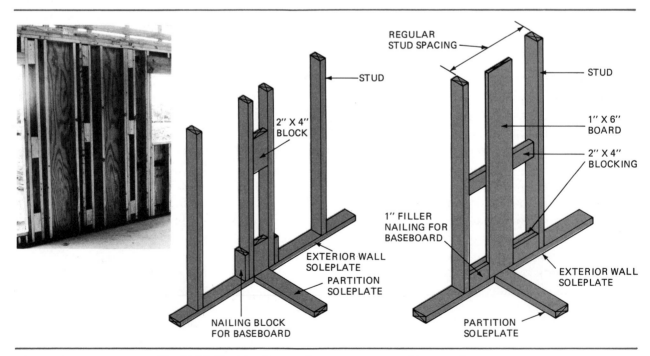

Fig. 11-10 Partition intersections are constructed in several ways.

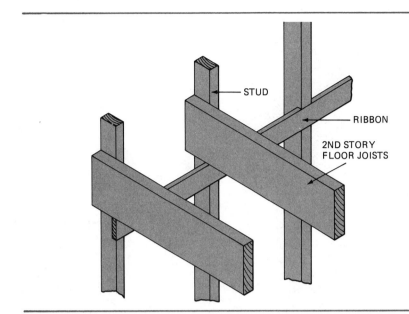

Fig. 11-11 Ribbons are used to support floor joists in a balloon frame.

interior wall covering. In most cases, the partition intersection is made of two studs spiked to the edge of 2x4 blocks about a foot long, with one block placed at the bottom, one at the top, and another about center on the studs.

Another method is to maintain the regular spacing of the studs and install blocking between them wherever partitions occur. The block is set back from the inside edge of the stud the thickness of a board. A 1x6 board is then fastened vertically on the inside of the wall so that it is centered on the partition, Figure 11-10.

RIBBONS

Ribbons are horizontal members of the exterior wall frame in balloon construction and are used to support the second floor joists. The inside edge of the wall studs are notched so that the ribbon lays flush with the edge, Figure 11-11. Ribbons are usually made of 1x4 stock. Notches in the stud should be made carefully so the ribbon fits snugly in the notch to prevent the floor joists from settling.

BRACES

Generally, no wall bracing is required if boards are applied diagonally, if plywood is used for wall sheathing, or if fiberboard or gypsum board is applied vertically. In other cases, such as when insulating board sheathing is used, walls are braced by using nominal 1x4 continuous diagonal strips set into the face of the studs, top plate, and soleplate at each corner of the building, Figure 11-12, or by 2x4s cut in between the studs.

EXTERIOR WALL FRAMING

Plans or blueprints usually indicate the height from finished floor to finished ceiling. This dimension is found in a wall *section* which is a drawing of a vertical cut through the building.

DETERMINING THE LENGTH OF STUDS

From this dimension, the carpenter must calculate the length of the stud so that the

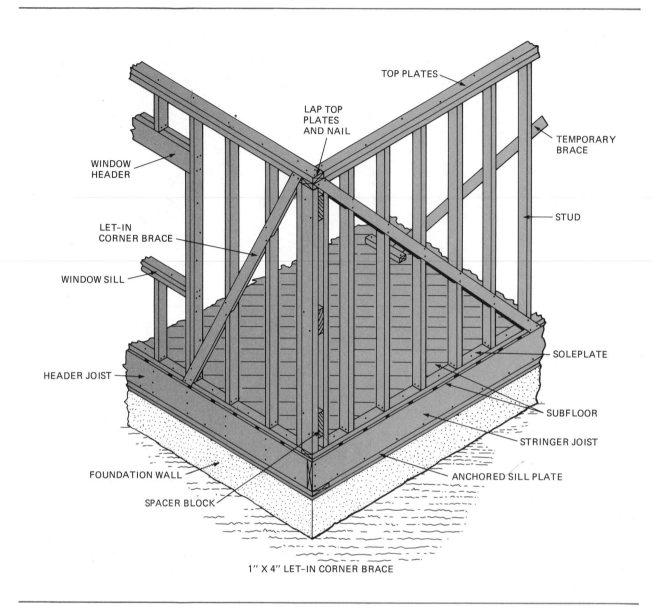

TOP PLATES

LAP TOP
PLATES
AND NAIL

TEMPORARY
BRACE

WINDOW
HEADER

STUD

LET-IN
CORNER BRACE

WINDOW SILL

SOLEPLATE

HEADER JOIST

SUBFLOOR

STRINGER JOIST

FOUNDATION WALL

ANCHORED SILL PLATE

SPACER BLOCK

1" X 4" LET-IN CORNER BRACE

Fig. 11-12 Wall braces

distance from finish floor to ceiling will be as specified. To determine the stud length, the carpenter must know the thickness of the finish floor and the ceiling thickness below the joist.

From the specified floor to ceiling height, add on the thickness of the finish floor and the thickness of the ceiling below the ceiling joist. Deduct the combined thickness of the top plates and the soleplates to find the length of the stud, Figure 11-13. This applies only to platform frame construction. For example, the

finish floor to ceiling height is shown to be 7'-9''. Add 3/4 inch for the finish floor and 1 1/4 inches for the ceiling thickness below the joist (3/4'' for furring and 1/2'' for gypsum board). The total is 7'-11''. Deduct the combined thickness of the top plates and soleplates (4 1/2 inches), leaving a stud length of 7'-6 1/2''.

For balloon frame construction, find the height from the sill to the top of the top plates by adding to the combined finish floor to ceiling heights, the thickness of all ceilings from the

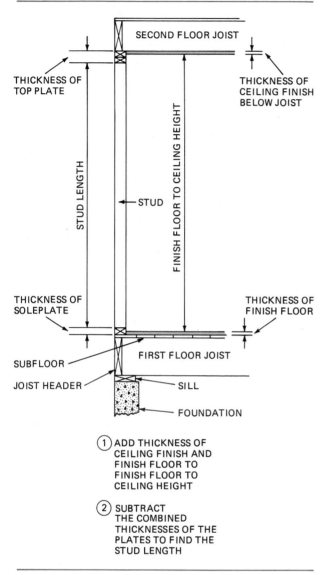

Fig. 11-13 Finding the length of a stud in an exterior wall frame of platform construction

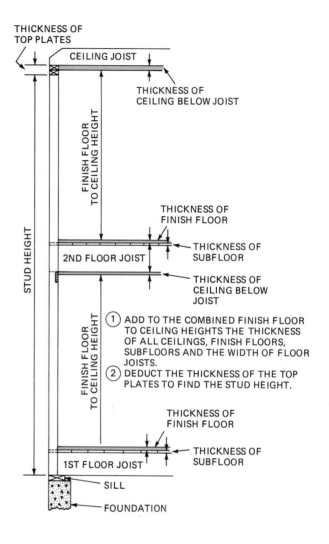

Fig. 11-14 Finding the length of a stud in a balloon frame

finish to the joists, the thickness of all finish floors and subfloors, and the combined width of all floor joists. Then deduct the total thickness of the top plates, Figure 11-14.

The location of the ribbon that supports the second floor joists in a balloon frame is found by adding to the specified first floor to ceiling height, the thickness of the first floor ceiling, finish floor, subfloor, and the width of the first floor joist. Measure this distance from the bottom end of the stud to the top edge of the ribbon.

STORY POLE

The *story pole* is usually a straight strip of 1x2 or 1x3 lumber. Its length is the distance from the subfloor to the bottom of the joist above. Lines are squared across its width indicating the location of the soleplates and top plates, headers, and rough sills for all openings. From the story pole, the length of studs, jack studs, and trimmers can be determined, even if opening heights differ. The thickness of rough sills and the depth, or width, of top headers are

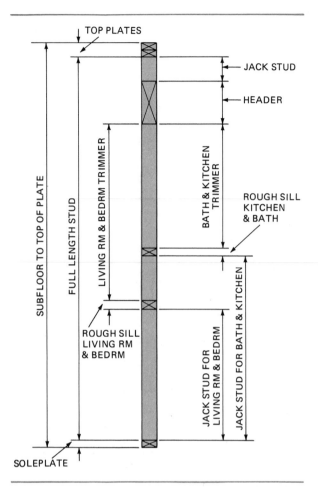

Fig. 11-15 Layout of a typical story pole

also indicated and identified on the story pole, Figure 11-15. A different story pole is made for each floor of the building. As framing progresses, the story pole is used to test the accurate location of framing members in the wall.

DETERMINING THE SIZE OF WALL OPENINGS

The width and height of rough openings are not usually indicated in the plans. It is the carpenter's responsibility to determine the rough opening size from the door and window schedule.

A door and window schedule in a set of plans identifies each different door or window by a number or a letter, Figure 11-16. The location of these units is usually shown on the floor plan by placing the identifying number or letter close to the door or window symbol. Dimensions or measurements found on plans are usually to the centerline of doors and windows.

The door and window schedule contains the kind, style, manufacturer's model number, and size of each unit. It is from this information that the rough opening sizes are determined.

DOOR SCHEDULE			
MARK	SIZE	TYPE	REMARKS
1	1 3/4"x3'-0"x7'-0"	Paneled	Morgan 3819
2	1 3/4"x2'-8"x6'-8"	Paneled	Morgan 119
3	1 3/8"x2'-8"x6'-8"	Flush	Morgan — Birch — Hollow Core
4	1 3/8"x2'-6"x6'-8"	Louver	Morgan 611
5	1 3/8"x2'-0"x6'-8"	Flush	Morgan — Birch — Hollow Core

WINDOW SCHEDULE			
MARK	SIZE	TYPE	REMARKS
A	5'-7 1/2"x12'-3	W6N5	Andersen Casement Bow
B	5'-7 1/2"x6'-10 1/2"	4N5	Andersen Casement
C	3'-7 1/8"x3'-7"	2N3	Andersen Casement
D	3'-7 1/8"x6'-10 1/2"	4N3	Andersen Casement
E	1'-7 3/8"x1'-7 3/8"	2817	Andersen Basement Utility
F	5'-0"x8'-0"	G805	Andersen Gliding

Fig. 11-16 A typical door and window schedule found in building plans

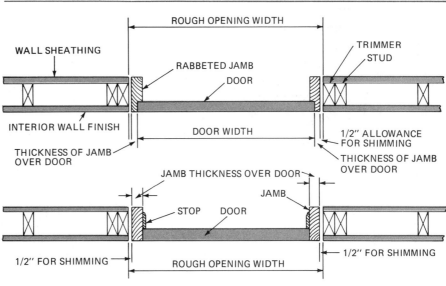

Fig. 11-17 Determining the rough opening width of a door opening

NOTE: ROUGH OPENING WIDTHS FOR THE SAME SIZE DOOR MAY VARY ACCORDING TO THE THICKNESS OF THE JAMB STOCK.

ROUGH OPENING SIZES FOR EXTERIOR DOORS

The rough opening for an exterior door must be larger than the size of the door to allow for the thickness of the door frame plus an allowance for shimming the frame to a level and plumb position. Usually 1/2 inch is allowed for shimming at the top and both sides. The amount allowed for the door frame itself depends on the thickness of the door frame material over the door.

The sides and top of a door frame are called *jambs* and may vary in thickness. Sometimes rabbeted jambs are used. At other times nominal 1- or 2-inch lumber is used with a separate stop applied, Figure 11-17.

The bottom member of the door frame is called a *sill*. In some sections it is called a *stool*. Sills may be hardwood, metal, or a combination of wood and metal. The type of sill and its thickness must be known in order to figure the rough opening height, Figure 11-18.

ROUGH OPENING SIZES FOR WINDOWS

Many different kinds of windows are manufactured by a number of firms. Because of this variety and because construction methods differ, it is best to consult the manufacturer's catalog to obtain the rough opening sizes. These

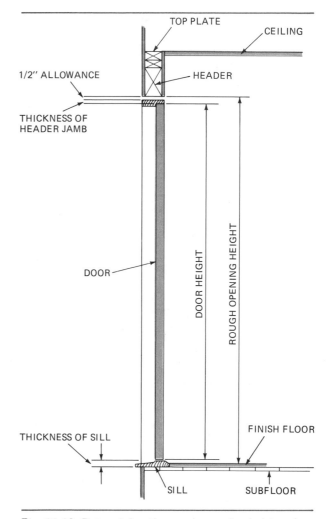

Fig. 11-18 Determining the rough opening height of an exterior door opening

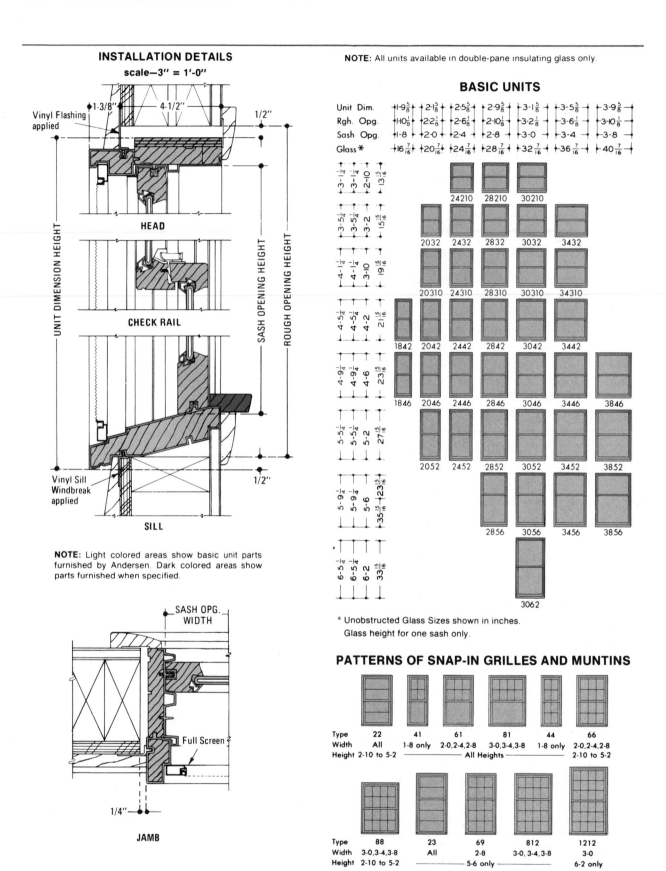

INSTALLATION DETAILS
scale—3" = 1'-0"

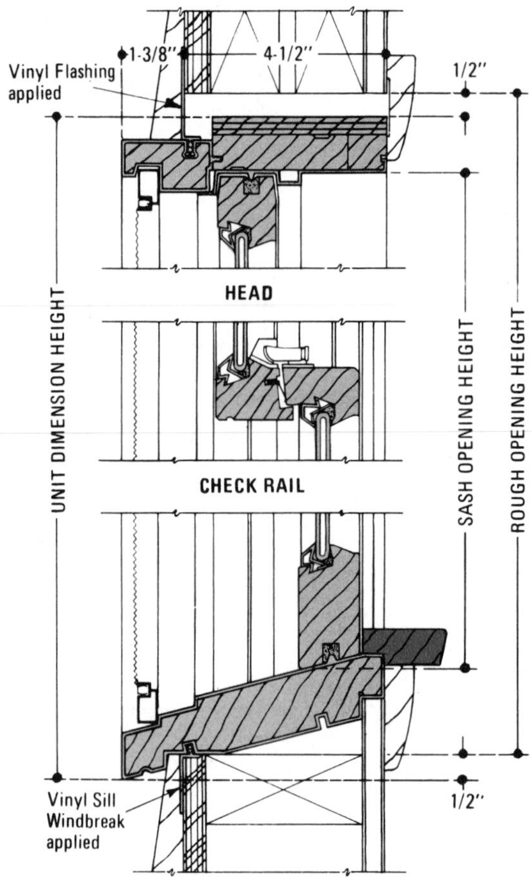
1-3/8"
4-1/2"
Vinyl Flashing applied
1/2"
HEAD
CHECK RAIL
UNIT DIMENSION HEIGHT
SASH OPENING HEIGHT
ROUGH OPENING HEIGHT
Vinyl Sill Windbreak applied
1/2"
SILL

NOTE: Light colored areas show basic unit parts furnished by Andersen. Dark colored areas show parts furnished when specified.

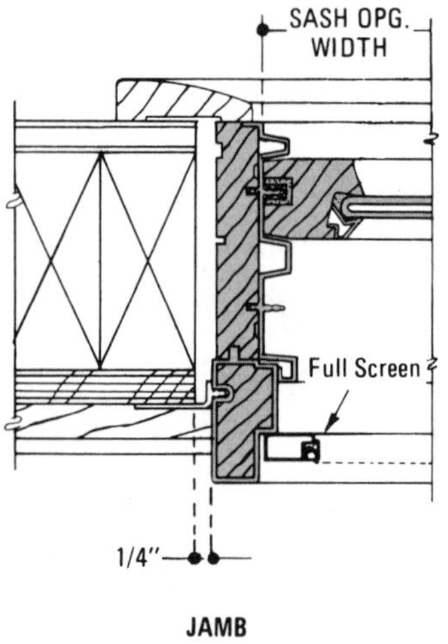
SASH OPG. WIDTH
Full Screen
1/4"
JAMB

NOTE: All units available in double-pane insulating glass only.

BASIC UNITS

Unit Dim.	1-9 5/8	2-1 5/8	2-5 5/8	2-9 5/8	3-1 5/8	3-5 5/8	3-9 5/8
Rgh. Opg.	1-10 1/8	2-2 1/8	2-6 1/8	2-10 1/8	3-2 1/8	3-6 1/8	3-10 1/8
Sash Opg.	1-8	2-0	2-4	2-8	3-0	3-4	3-8
Glass *	16 7/16	20 7/16	24 7/16	28 7/16	32 7/16	36 7/16	40 7/16

3-1 1/4	3-1 1/4	2-10	13 15/16
3-5 1/4	3-5 1/4	3-2	15 15/16
4-1 1/4	4-1 1/4	3-10	19 15/16
4-5 1/4	4-5 1/4	4-2	21 15/16
4-9 1/4	4-9 1/4	4-6	23 15/16
5-5 1/4	5-5 1/4	5-2	27 15/16
5-9 1/4	5-9 1/4	5-6	35 15/16 / 23 15/16
6-5 1/4	6-5 1/4	6-2	33 15/16

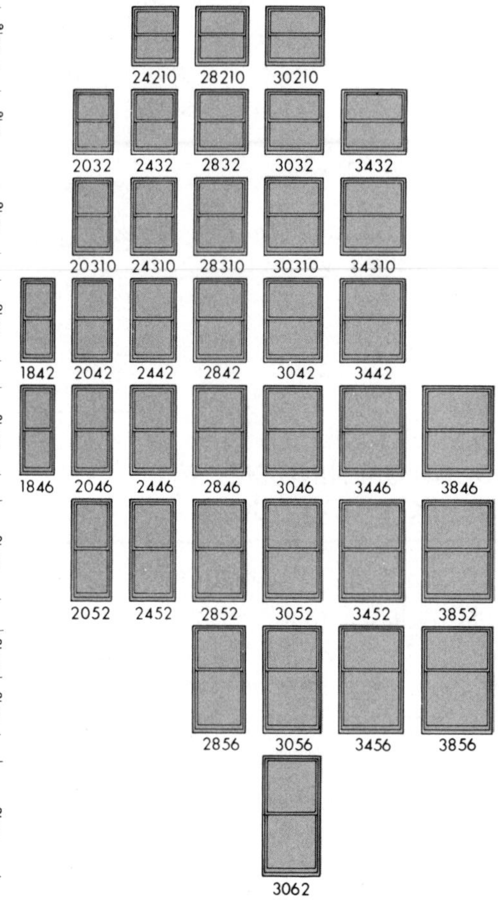
24210 28210 30210
2032 2432 2832 3032 3432
20310 24310 28310 30310 34310
1842 2042 2442 2842 3042 3442
1846 2046 2446 2846 3046 3446 3846
2052 2452 2852 3052 3452 3852
2856 3056 3456 3856
3062

* Unobstructed Glass Sizes shown in inches.
Glass height for one sash only.

PATTERNS OF SNAP-IN GRILLES AND MUNTINS

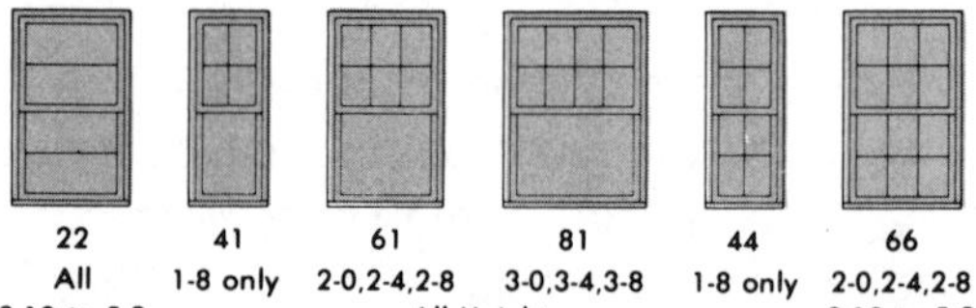

Type	22	41	61	81	44	66
Width	All	1-8 only	2-0,2-4,2-8	3-0,3-4,3-8	1-8 only	2-0,2-4,2-8
Height	2-10 to 5-2	All Heights				2-10 to 5-2

Type	88	23	69	812	1212
Width	3-0,3-4,3-8	All	2-8	3-0, 3-4, 3-8	3-0
Height	2-10 to 5-2	5-6 only			6-2 only

Fig. 11-19 Sample of a manufacturer's catalog showing rough opening sizes for window units *(Courtesy of Anderson Corporation)*

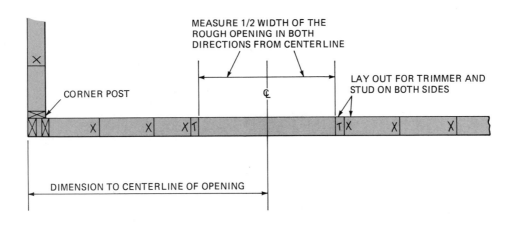

MEASURE 1/2 WIDTH OF THE
ROUGH OPENING IN BOTH
DIRECTIONS FROM CENTERLINE

LAY OUT FOR TRIMMER AND
STUD ON BOTH SIDES

CORNER POST

DIMENSION TO CENTERLINE OF OPENING

Fig. 11-20 Laying out a rough opening width on the soleplate

catalogs show the style and size of the window unit and give the rough opening (R.O.) for each unit, Figure 11-19.

LAYING OUT THE PLATES

To lay out the plates, measure in at the corners the thickness of the wall, and snap lines on the subfloor between the marks. Fasten the soleplate so that the inside edge is to the chalk line. Just *tack* the soleplate in position if it is later to be removed to assemble wall sections. (Nails are not driven home, and only as many as are needed to hold the piece in place are used.)

In some cases, the soleplate is fastened securely. The wall sections are built without the soleplate. They are erected, and then the studs are toenailed to the soleplate. With this method only full length studs are used to make up wall sections. Openings are framed after the sections are erected.

WALL OPENINGS

From the blueprints, determine the centerline dimension of all the openings in the wall and lay these out on the soleplate. Measure in each direction from the centerline one-half the width of the rough opening. Square lines at these points across the soleplate and mark a *T* for the trimmers.

From the squared lines, measure back the stud thickness, square lines across, and mark *X*s for the full length studs on each side of the openings, Figure 11-20. For convenient erection of wall sections, all joints in the soleplate should fall on the center of a full length stud.

PARTITION INTERSECTIONS

On blueprints, interior partitions are usually dimensioned to their centerline. Mark on the soleplate the centerline of all partitions intersecting the wall. From the centerlines measure in each direction one-half the partition stud thickness. Square lines across the soleplate and mark *X*s outside the lines for the location of partition intersection studs, Figure 11-21.

STUDS AND JACK STUDS

After all openings and partitions have been laid out, start laying out all full length studs and jack studs. Proceed in the same manner as laying out floor joists.

Measure in from the outside corner the regular stud spacing. From this mark, measure in one direction or the other, one-half the stud thickness, and square a line across the soleplate. Place an *X* on the side of the line where the stud will be located.

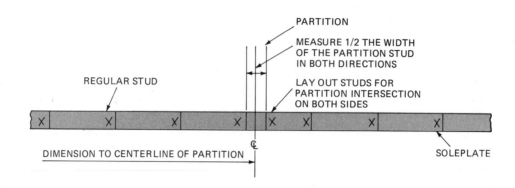

Fig. 11-21 Laying out a partition intersection on the soleplate

Stretch a tape along the length of the sole-plate from this first stud location. Square lines across the soleplate at each specified stud spacing. Place *X*s on the same side of the line as the first line.

Where openings occur, mark a *J* on the side of the line instead of an *X* to indicate the location of jack studs, Figure 11-22. All regular and jack studs should line up with the floor joists below in platform construction.

LAYING OUT THE TOP PLATE

Use straight lengths of lumber for the top plate. Place a length alongside the soleplate. With one end flush with the end of the soleplate, mark and cut the other end of the top plate so it is centered on a stud. Cut enough top plates

for the length of the wall. All joints in the top plate should fall on the center of a stud, Figure 11-23. For easy erection of wall sections, joints in the top plate should coincide with those in the soleplate. Tack the top plate in position alongside the soleplate and transfer the marks on the soleplate to the top plate, Figure 11-24.

ASSEMBLING AND ERECTING WALL SECTIONS

To assemble a wall section, make up the necessary number of corner posts and partition intersections. Cut the full length studs to length. Make up and cut to length all headers and rough sills. From the story pole, determine the length of and cut all trimmers and jack

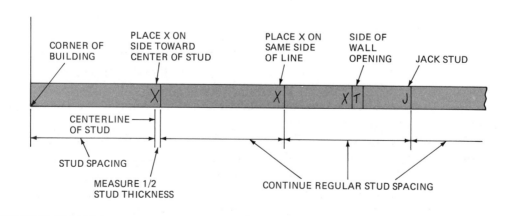

Fig. 11-22 Laying out the soleplate for regular and jack studs

Fig. 11-23 Joints in the top plate should fall on the center of a stud.

studs. Identify headers, jacks, and trimmers when necessary, if rough openings are different.

Separate the top plate from the soleplate. Stand them on edge and place all full length studs, corner posts, and partition intersections in between them. Make sure not to turn the plates around to be certain the layout lines coincide.

Carefully sight the full length studs and place them on the subfloor with their crowned edges down. It will be difficult in later stages for those who apply the interior finish, if no attention is paid to the manner in which studs

are installed in the wall. A stud that is fastened in position with its crowned edge out, next to one with its crowned edge in, will certainly present problems later on.

It is important to lay all studs on the sub-floor with the crowned edges down so that when the wall is erected, each stud in the wall will be bowed in. This is much better than having them alternately bowed in and out, which would present problems when the interior finish is later applied.

Fasten each stud, corner post, and partition intersection in the proper position by driving 16d spikes through the plates into the ends of the members. From the story pole, mark the position of all headers and rough sills on the full length studs on each side of each opening. Fasten the headers and rough sills in position by driving 16d spikes through the studs into their ends.

Fasten all jack studs in position between top plates and headers and between soleplates and rough sills. Fasten all trimmers in position between headers and rough sills for window openings and between the header and soleplate for door openings. It may be necessary to toe-nail the ends of some members with 8d nails if spikes cannot be driven in their ends. Remember, when doubling up studs, to toenail their edges flush before spiking them together.

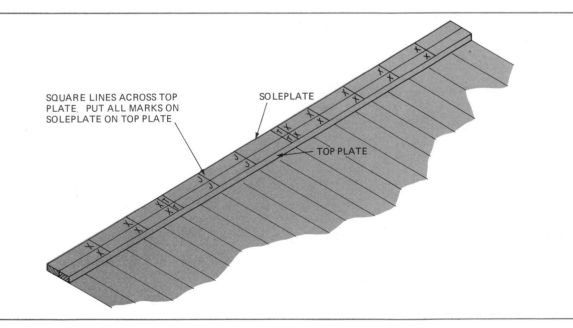

Fig. 11-24 Transferring the layout of a soleplate to the top plate

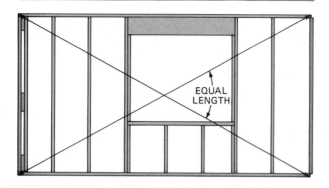

Fig. 11-25 When the diagonal measurements of a wall section are equal, the section is square.

BRACING END SECTIONS

If let-in braces are to be used, it is much easier to notch out the studs and plates and install the braces while the wall section is lying on the subfloor. To make sure the wall section is square before laying out for the brace, measure each section from corner to corner both ways. If the measurements are the same, the section is square. If the measurements are not the same, rack the section by pushing on one corner until the corner-to-corner measurements are equal, Figure 11-25.

When the section is square, tack the brace in position against the studs and plates. If the brace is to be at a 45° angle, measure in from the corner the same distance as the height of the wall. Mark the studs and plates on each edge

of the brace. Remove the brace and mark the depth of cut on each member. Cut out the notches, using a portable electric circular saw and wood chisel, and fasten the brace in the notches using two 8d common nails in each framing member, Figure 11-26.

To erect the wall, raise one section at a time into position and spike the soleplate to the chalk line with 16d spikes spaced about 16 inches on center into the floor joists or joist header. Sections are fastened together by nailing the top plates to the projecting end of the studs of the adjoining wall sections.

Brace each section temporarily as erected by fastening one end of a 1x6 board to a 2x4 block which has been nailed to the subfloor and the other end to the side of a stud, Figure 11-27. Do not drive nails in the braces all the way in or use duplex head nails.

PLUMBING AND BRACING THE CORNERS

If the walls have not been previously braced, or if no permanent bracing is required, all corners must be plumbed and temporarily braced. Cut 1x6 braces for both sides of each corner with their ends cut at a 45° angle. Fasten the top end of the brace to the top plate near the corner post on the inside of the wall. Temporary braces are fastened to the inside of the wall so they can remain in position until the

Fig. 11-26 A let-in corner brace

Fig. 11-27 Raising an exterior wall section — the sheathing has already been applied.

exterior wall sheathing is applied and even longer — until it is necessary to remove them for the application of the interior wall finish. Care must be taken not to let the ends of the braces extend beyond the corner post or top plate because this might interfere with the application of wall sheathing or the ceiling or roof framing.

Plumb the corner post, and fasten the bottom end of the brace. Use duplex headed nails or drive nails only partway in for easy removal. For accurate plumbing of the corner posts, use a 6-foot level with aluminum blocks attached to each end. The blocks prevent the level from resting against the surface of the corner post; any bow or irregularity in the surface affects the accurate reading of the level.

An alternative method is to use a shorter than 6-foot level in combination with a long straightedge. On the ends of the straightedge, two blocks of equal thickness are fastened to keep the edge of the straightedge away from the surface of the corner post.

DOUBLING THE TOP PLATES

Usually, the next step in the framing of exterior walls is doubling the top plate. Plates are doubled by lapping the joints at the corners, Figure 11-28. Toenail into the edges to bring them flush, and then spike together over every stud with 16d spikes staggered from side to side. Stagger the joints in the top member from the bottom as far as possible.

Where bearing partitions intersect the exterior wall, a space is left for the top member of the bearing partition to lap the plate of the exterior wall, Figure 11-29.

STRAIGHTENING THE WALLS

After the corner posts have been plumbed and the top plates doubled, the tops of the walls must be straightened and braced. To straighten each wall, use three blocks of 1-inch lumber of exactly the same thickness. To assure equal thickness of the blocks, cut them from the same piece. Fasten two blocks to the top plate — one at each corner.

Stretch a line tightly over and between the two corner blocks. Use the third block as a test block. Move the wall in or out at about 8- or 10-foot intervals until the test block just clears

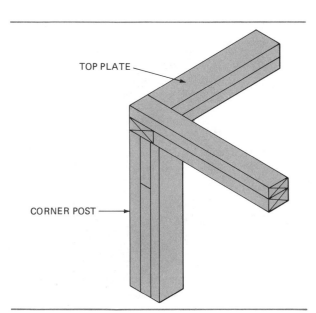

Fig. 11-28 Top plate is lapped at the corners.

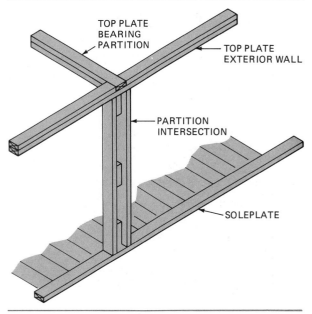

Fig. 11-29 The top plate of bearing partitions laps the exterior wall.

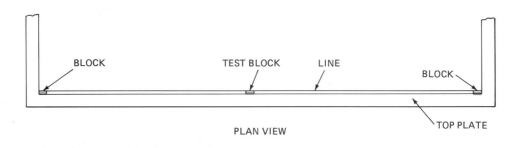

TOP PLATE

PLAN VIEW

Fig. 11-30 Straightening the wall with a stretched line and test block

the line when held against the top plate, Figure 11-30. Brace the walls securely at these points in the manner described previously for temporary braces.

Another method of bracing the exterior wall is by using a *spring brace.* This method is used in some sections of the country. Use a 1x8 board about 12 or 14 feet long. Nail the top end securely with four or five 8d common nails driven home to the underside of the top plate. Nail the bottom end to the subfloor in the same manner so the top of the wall has to be brought in. Spring the board upward by installing a piece of the same material in a vertical position between it and the subfloor about midway on the brace, Figure 11-31. The more the brace is sprung upward, the more the wall is brought inward. Bring the wall inward in this manner so the test block just fits between the stretched line

and the top plate without touching the line. When the wall is straight, fasten the uprights by nailing through the brace and toenailing to the subfloor.

Use as many braces as necessary to straighten the wall along its entire length. Fasten short 2x4 blocks against both sides of the corner posts and partition intersections and down on the soleplate to provide fastening for the ends of the baseboard, Figure 11-32.

WALL SHEATHING

Wall sheathing covers the exterior walls and may consist of boards, plywood, fiberboard, gypsum board, or rigid foam board. Boards may be applied diagonally or horizontally. Diagonal

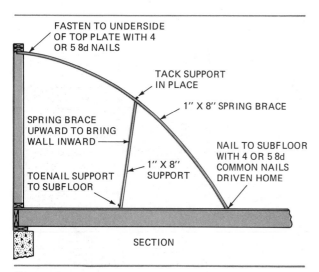

Fig. 11-31 Straightening a spring brace to hold the wall in a straight line

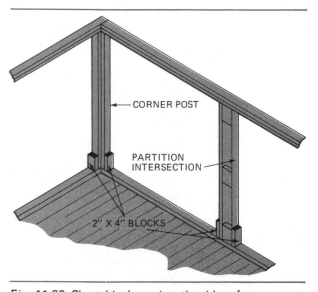

Fig. 11-32 Short blocks against the sides of corner posts and partition intersections provide fastening for the ends of the baseboard.

Fig. 11-33 Plywood is used at the corners in combination with fiberboard sheathing to eliminate corner bracing.

Fig. 11-34 Rigid foam insulation being used for exterior wall sheathing *(Courtesy of Dow Chemical Corporation)*

sheathed walls require no other bracing and make the frame stiffer and stronger than boards applied horizontally. Use two 8d common nails at each stud for 6- and 8-inch boards. Use three nails at each stud for 10- and 12-inch boards. End joints must occur over the center of studs.

Plywood sheathing may be applied horizontally or vertically. A minimum 3/8-inch thickness is recommended when the sheathing is to be covered by a type of exterior siding. Greater thicknesses are recommended when the sheathing also acts as the exterior finish. Use 6d nails spaced 6 inches apart on the edges and 12 inches apart on intermediate studs.

Plywood is often used in combination with fiberboard, rigid foam, or gypsum board. Plywood is applied vertically on both sides of the corner to eliminate any other corner bracing, Figure 11-33.

Fiberboard sheathing, commonly called insulation board, is a soft, synthetic sheet material. Common sheet sizes are 2'x8' and 4'x8'. Usually the edges and ends are matched (tongue and grooved). Standard thicknesses are 1/2 inch and 25/32 inch. Most sheets are coated with asphalt for water resistance.

All softboard sheathing, Figure 11-34, is fastened with roofing nails spaced 3 inches apart around the edges and 6 inches apart in the center. Fasteners must be kept a minimum of 3/8 inch from the edges. It may be applied

vertically or horizontally. When applied vertically, no additional wall bracing is necessary.

Gypsum sheathing consists of a treated gypsum filler between sheets of water-resistant paper. Usually 1/2 inch thick, 2 feet wide, and 8 feet long, the sheets have matched edges to provide a tighter wall. Because of the soft material, roofing nails must be used to fasten gypsum sheathing. Space the nails about 4 inches around the edges and 8 inches in the center.

Gypsum board sheathing is used when a more fire-resistant sheathing is required. Rigid foam sheathing is used when greater insulation is required, Figure 11-35. A disadvantage of using

Fig. 11-35 Applying insulation board sheathing before the walls are erected

195

fiberboard, gypsum board, or rigid foam sheathing is that the exterior siding cannot be nailed to it. Either the siding must be fastened to the studs, or special fasteners must be used.

ESTIMATING MATERIALS FOR EXTERIOR WALLS

To estimate the amount of material needed for exterior walls,

- first determine the total linear feet. Then, figure one stud for every linear foot of wall. This allows for the extras needed for corner posts, for partition intersections, and for doubling at openings.

- For plates, multiply the total linear feet of wall by 3 (1 soleplate and 2 top plates) and add 10 percent for waste.

- For headers and rough sills, calculate the size for each opening and add together the material needed for different sizes.

- For wall sheathing, multiply the total linear feet of wall by the wall height to find the total number of square feet. For board sheathing, add a percentage for waste to find the total number of board feet. Add 20 percent for 6-inch boards, 15 percent for 8-inch boards and 10 percent for 10-inch boards.

- To find the number of sheets of plywood, fiberboard, or gypsum sheathing divide the total wall area by the number of square feet in each sheet. Add about 5 percent for waste.

- For all sheathing, deduct the area of any large openings from the total wall area. Small wall openings are disregarded.

SUMMARY OF WAYS TO ASSEMBLE AND ERECT EXTERIOR WALLS

You must remember that there are a number of ways to assemble and erect exterior walls. The method described in this unit is probably the most commonly used. As pre-viously mentioned, another method is to fasten the soleplate securely, nail all full length studs to the top plate, then raise the sections, and toe-nail the studs to the soleplate. In this method, the rough openings are framed after the wall is erected and then the sheathing is applied. This method is used when the framing crew is small.

Still another method consists of framing and sheathing the entire wall on the subfloor. Sometimes the windows and doors are installed and even the wall siding applied. A disadvantage of this method is that the wall is very heavy to raise into position and may require special raising equipment. Sometimes complete wall sections with doors, windows, and siding installed are prefabricated in the shop, transported to the job site, and erected in position.

ALL-WEATHER WOOD FOUNDATION

A wood foundation built of pressure-preservative-treated lumber and plywood is called the *ALL-WEATHER WOOD FOUNDATION (AWWF)*, Figure 11-36. It can be used to support light-frame buildings, such as houses, apartments, schools, and office buildings. It is accepted by major model building codes and by federal agencies such as the VA and FHA. For complete information on the All-Weather Wood Foundation system contact the American Plywood Association or the National Forest Products Association.

LUMBER AND PLYWOOD

All lumber and plywood must be pressure treated in accordance with strict requirements. To assure compliance with pressure-treated requirements, all material should be stamped AWPB-FDN (American Wood Preservers Bureau-Foundation). Members more than 8 inches above the ground need not be pressure treated.

The most commonly used plywood for the AWWF is APA Rated Sheathing, Exposure 1. APA Rated Sheathing, Exterior can also be used. If appearance is a factor, use A-C Ext., B-C Ext., C-C Plugged or MDO (Medium Density Overlay). All should be Group 1 species. See

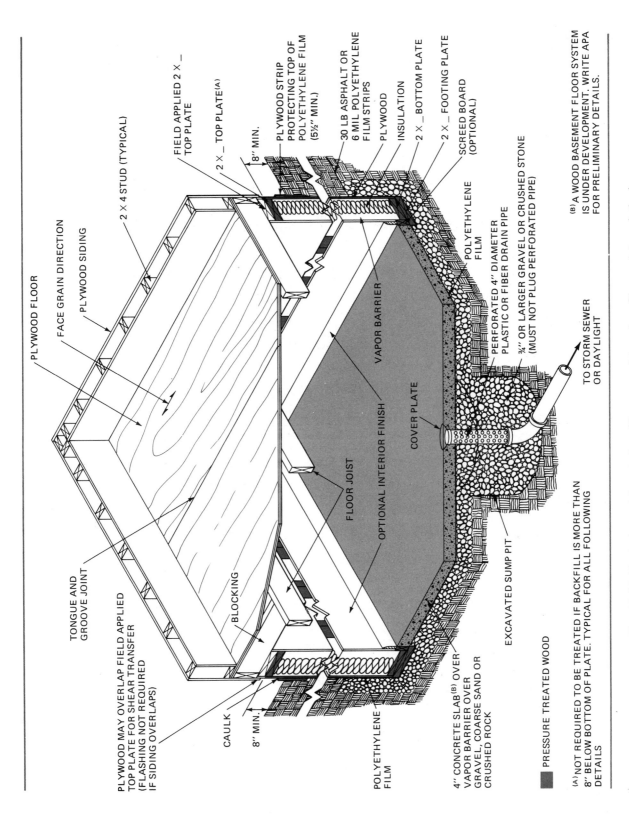

TONGUE AND GROOVE JOINT

PLYWOOD MAY OVERLAP FIELD APPLIED TOP PLATE FOR SHEAR TRANSFER (FLASHING NOT REQUIRED IF SIDING OVERLAPS)

PLYWOOD FLOOR

FACE GRAIN DIRECTION

PLYWOOD SIDING

2 × 4 STUD (TYPICAL)

FIELD APPLIED 2 × _ TOP PLATE

2 × _ TOP PLATE(A)

8″ MIN.

PLYWOOD STRIP PROTECTING TOP OF POLYETHYLENE FILM (5½″ MIN.)

30 LB ASPHALT OR 6 MIL POLYETHYLENE FILM STRIPS

PLYWOOD

INSULATION

2 × _ BOTTOM PLATE

2 × _ FOOTING PLATE

SCREED BOARD (OPTIONAL)

POLYETHYLENE FILM

PERFORATED 4″ DIAMETER PLASTIC OR FIBER DRAIN PIPE

¾″ OR LARGER GRAVEL OR CRUSHED STONE (MUST NOT PLUG PERFORATED PIPE)

TO STORM SEWER OR DAYLIGHT

EXCAVATED SUMP PIT

COVER PLATE

VAPOR BARRIER

OPTIONAL INTERIOR FINISH

FLOOR JOIST

BLOCKING

CAULK

8″ MIN.

POLYETHYLENE FILM

4″ CONCRETE SLAB(B) OVER VAPOR BARRIER OVER GRAVEL, COARSE SAND OR CRUSHED ROCK

▬ PRESSURE TREATED WOOD

(A) NOT REQUIRED TO BE TREATED IF BACKFILL IS MORE THAN 8″ BELOW BOTTOM OF PLATE. TYPICAL FOR ALL FOLLOWING DETAILS

(B) A WOOD BASEMENT FLOOR SYSTEM IS UNDER DEVELOPMENT. WRITE APA FOR PRELIMINARY DETAILS.

Fig. 11-36 Typical wood foundation (Courtesy of American Plywood Association)

Figure 11-37 for minimum grade and thicknesses of plywood.

The size of framing members for AWWF depends upon the load, species and grade of lumber, stud height, and backfill height. Minimum sizes for crawl space and basement wall members are shown in Figure 11-38.

A footing plate is the bottom member of the foundation. Minimum footing plate sizes are shown in Figure 11-39.

MINIMUM PLYWOOD GRADE AND THICKNESS FOR BASEMENT CONSTRUCTION
30 pcf equivalent-fluid density soil pressure

Height of fill (inches)	Stud spacing (inches)	Face grain across studs[2]			Face grain parallel to studs		
		Grade[3]	Minimum thickness[1]	Span Rating	Grade[3]	Minimum thickness[1,5]	Span Rating
24	12	B	1/2	32/16	B	1/2	32/16
	16	B	1/2	32/16	B	1/2 (4 ply or 5 layer)	32/16
48	12	B	1/2	32/16	B A	1/2[6] (5 layer) 1/2	32/16 32/16
	16	B	1/2	32/16	A B	5/8 3/4	42/20 48/24
72	12	B	1/2	32/16	A	5/8[6] (5 layer)	42/20
	16	A[4]	1/2[6]	32/16	B	3/4[6]	48/24
86	12	B	1/2[6]	32/16	A B	5/8[6] 3/4[6]	42/20 48/24
	16	A B	5/8 3/4	42/20 48/24	A	5/8[7]	42/20

[1] Crawl space sheathing may be 3/8 inch for face grain across studs and maximum 3 foot depth of unequal fill.

[2] Minimum 2 inch blocking between studs required at all horizontal panel joints more than 4 feet below adjacent ground level (also where noted in details).

[3] Minimum plywood grades conforming to "U.S. Product Standard PS 1, Construction and Industrial Plywood" are:
   A. Structural I Rated Sheathing Exposure 1
   B. Rated Sheathing Exposure 1
If a major portion of the wall is exposed above ground, a better appearance may be desired. The following Exterior grades would be suitable:
   A. Structural I A-C, Structural I B-C or Structural I C-C (Plugged)
   B. A-C Exterior Group 1, B-C Exterior Group 1, C-C (Plugged) Exterior Group 1 or MDO Exterior Group 1 or Ungrooved Group 1 303 plywood siding.

[4] For this combination of fill height and panel grade, only Structural I A-C or Structural I Rated Sheathing Exterior may be substituted for improved appearance.

[5] When face grain is parallel to studs, plywood panels of the required thickness, grade and Span Rating may be any construction permitted except as noted in the table for minimum number of plies required.

[6] For this fill height, thickness and grade combination, panels which are continuous over less than three spans (across less than three stud spacings) require blocking 2 feet above bottom plate. Offset adjacent blocks and fasten through studs with two 16d corrosion resistant nails at each end.

[7] For this fill height, thickness and grade combination, panels require blocking 16 inches above bottom plate. Offset adjacent blocks and fasten through studs with two 16d corrosion resistant nails at each end.

Fig. 11-37 Minimum plywood grade and thickness for basement construction
(Courtesy of American Plywood Association)

MINIMUM STRUCTURAL REQUIREMENTS FOR BASEMENT WALLS

Wall height — 8 feet. Roof supported on exterior walls. Floors supported on interior and exterior bearing walls.[1]
30 lbs. per cu. ft. equivalent-fluid density soil pressure — 2000 lbs. per sq. ft. allowable soil bearing pressure.

Construction	House width (feet)	Height of fill (inches)	Lumber species and grade[2]	Stud and plate size (nominal)	Stud spacing (inches)	Lumber species and grade[2]	Stud and plate size (nominal)	Stud spacing (inches)
			Roof — 40 psf live; 10 psf dead / Ceiling — 10 psf / 1st floor — 50 psf live and dead / 2nd floor — 50 psf live and dead			Roof — 30 psf live; 10 psf dead / Ceiling — 10 psf / 1st floor — 50 psf live and dead / 2nd floor — 50 psf live and dead		
2 Stories	32 or less	24	D	2x6	16	D	2x6	16
		48	D	2x6	16	D	2x6	16
		72	A	2x6	16	A	2x6	16
			B	2x6	12	B	2x6	12
			C	2x8	16	D	2x8	16
			D	2x8	12			
		86	A*	2x6	12	A*	2x6	12
			B	2x8	16	B	2x8	16
			C	2x8	12	C	2x8	12
	24 or less	24	D	2x6	16	D	2x6	16
		48	D	2x6	16	D	2x6	16
		72	C	2x6	12			
			D	2x8	16	C	2x6	12
		86	A*	2x6	12			
			B	2x8	16			
			C	2x8	12	D	2x8	12
1 Story	32 or less	24	B	2x4	16	C	2x4	16
			D	2x4	12	D	2x4	12
			D	2x6	16	D	2x6	16
		48	D	2x6	16	D	2x6	16
		72	A	2x6	16	A	2x6	16
			B	2x6	12	C	2x6	12
			D	2x8	16	D	2x8	16
		86	A*	2x6	12	A*	2x6	12
			B	2x8	16	B	2x8	16
			C	2x8	12	D	2x8	12
	28 or less	24	B	2x4	16			
			D	2x4	12	D	2x4	16
			D	2x6	16			
		48	D	2x6	16	B	2x4	12
		72	C	2x6	12	C	2x6	12
		86				A*	2x6	12
						B	2x8	16
			D	2x8	12	D	2x8	12
	24 or less	24	D	2x4	16	D	2x4	16
		48	B	2x4	12	B	2x4	12
		72	C	2x6	12	C	2x6	12
		86				A*	2x6	12
						B	2x8	16
			D	2x8	12	D	2x8	12

[1] Studs and plates in interior bearing walls supporting floor loads only must be of lumber species and grade "D" or higher. Studs shall be 2 inches by 4 inches at 16 inches on center where supporting one floor and 2 inches by 6 inches at 16 inches on center where supporting 2 floors. Footing plate shall be 2 inches wider than studs.

[2] Species, species groups and grades having the following minimum (surfaced dry or surfaced green) properties as provided in the National Design Specification.

	A	B	C	D
F_b (repetitive member) psi 2x6, 2x8	1,750	1,450	1,150	975
2x4	2,050	1,650	1,350	1,150
F_c psi 2x6, 2x8	1,250	1,050	875	700
2x4	1,250	1,000	825	675
F_{c1} psi	385	385	245	235
F_v psi	95	95	75	70
E psi	1,800,000	1,600,000	1,400,000	1,100,000
Typical lumber grades	Douglas fir No. 1 / Southern Pine No. 1 KD	Douglas fir No. 2 / Southern Pine No. 2 KD	Hem fir No. 2	Lodgepole Pine No. 2 / Northern Pine No. 2 / Ponderosa Pine No. 2

Where indicated (*), length of end splits or checks at lower end of studs not to exceed width of piece.

Fig. 11-38 Minimum structural requirements for foundation walls *(Courtesy of American Plywood Association)*

MINIMUM STRUCTURAL REQUIREMENTS FOR CRAWL-SPACE WALL FRAMING

Apply to installations with outside fill height not exceeding 4 feet and wall height not exceeding 6 feet. Roof supported on exterior walls. Floors supported on interior and exterior bearing walls.[1]   30 lbs. per cu. ft. equivalent fluid density soil pressure — 2000 lbs. per sq. ft. allowable soil bearing pressure.

Construction	House width (feet)	Uniform load conditions					
		Roof — 40 psf live; 10 psf dead Ceiling — 10 psf 1st floor — 50 psf live and dead 2nd floor — 50 psf live and dead			Roof — 30 psf live; 10 psf dead Ceiling — 10 psf 1st floor — 50 psf live and dead 2nd floor — 50 psf live and dead		
		Lumber species and grade[2]	Stud and plate size (nominal)	Stud spacing (inches)	Lumber species and grade[2]	Stud and plate size (nominal)	Stud spacing (inches)
2 Stories	32 or less	B	2x6	16	B	2x6	16
					D	2x6	12
	28 or less	D	2x6	12	D	2x6	12
	24 or less	D	2x6	12	C	2x6	16
1 Story	32 or less	B	2x4	12	A	2x4	16
		B	2x6	16	B	2x4	12
		D	2x6	12	D	2x6	16
	28 or less	A	2x4	16	B	2x4	12
		D	2x6	16	D	2x6	16
	24 or less	D	2x6	16	C	2x4	12

[1] Studs and plates in interior bearing walls supporting floor loads only must be of lumber species and grade "D" or higher. Studs shall be 2 inches by 4 inches at 16 inches on center where supporting one floor and 2 inches by 6 inches at 16 inches on center where supporting 2 floors. Footing plate shall be 2 inches wider than studs.

[2] Species, species groups and grades having the following minimum (surfaced dry or surfaced green) properties as provided in the National Design Specification.

		A	B	C	D
F_b (repetitive member) psi	2x6	1,750	1,450	1,150	975
	2x4	2,050	1,650	1,350	1,150
F_c psi	2x6	1,250	1,050	875	700
	2x4	1,250	1,000	825	675
F_{c1} psi		385	385	245	235
F_v psi		95	95	75	70
E psi		1,800,000	1,600,000	1,400,000	1,100,000
Typical lumber grades		Douglas fir No. 1 Southern Pine No. 1 KD	Douglas fir No. 2 Southern Pine No. 2 KD	Hem fir No. 2	Lodgepole Pine No. 2 Northern Pine No. 2 Ponderosa Pine No. 2

Where indicated (*), length of end splits or checks at lower end of studs not to exceed width of piece.

Fig. 11-38 (Continued)

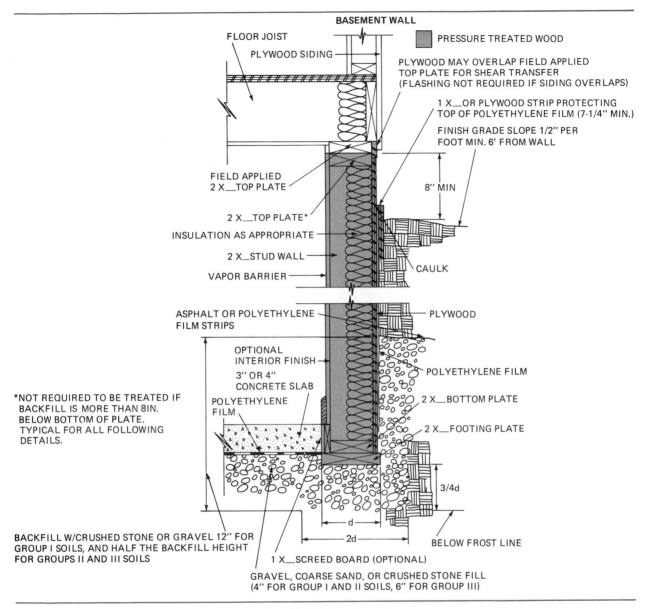

BASEMENT WALL

FLOOR JOIST

PLYWOOD SIDING

PRESSURE TREATED WOOD

PLYWOOD MAY OVERLAP FIELD APPLIED
TOP PLATE FOR SHEAR TRANSFER
(FLASHING NOT REQUIRED IF SIDING OVERLAPS)

1 X__OR PLYWOOD STRIP PROTECTING
TOP OF POLYETHYLENE FILM (7-1/4" MIN.)

FINISH GRADE SLOPE 1/2" PER
FOOT MIN. 6' FROM WALL

FIELD APPLIED
2 X__TOP PLATE

8" MIN

2 X__TOP PLATE*

INSULATION AS APPROPRIATE

2 X__STUD WALL

VAPOR BARRIER

CAULK

ASPHALT OR POLYETHYLENE
FILM STRIPS

PLYWOOD

OPTIONAL
INTERIOR FINISH

3" OR 4"
CONCRETE SLAB

POLYETHYLENE FILM

POLYETHYLENE
FILM

2 X__BOTTOM PLATE

2 X__FOOTING PLATE

*NOT REQUIRED TO BE TREATED IF
BACKFILL IS MORE THAN 8IN.
BELOW BOTTOM OF PLATE.
TYPICAL FOR ALL FOLLOWING
DETAILS.

3/4d

d

2d

BELOW FROST LINE

BACKFILL W/CRUSHED STONE OR GRAVEL 12" FOR
GROUP I SOILS, AND HALF THE BACKFILL HEIGHT
FOR GROUPS II AND III SOILS

1 X__SCREED BOARD (OPTIONAL)

GRAVEL, COARSE SAND, OR CRUSHED STONE FILL
(4" FOR GROUP I AND II SOILS, 6" FOR GROUP III)

Fig. 11-38 (Continued)

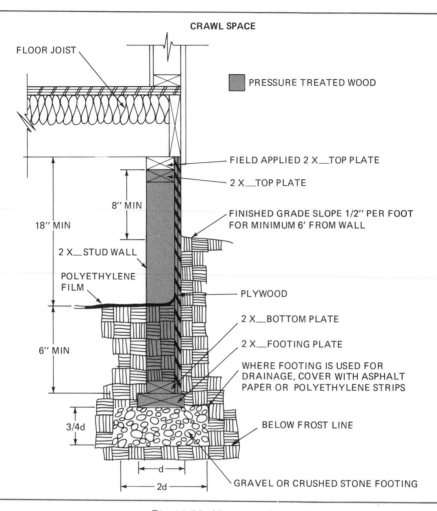

CRAWL SPACE

FLOOR JOIST

PRESSURE TREATED WOOD

FIELD APPLIED 2 X__TOP PLATE

2 X__TOP PLATE

8" MIN

18" MIN

FINISHED GRADE SLOPE 1/2" PER FOOT FOR MINIMUM 6' FROM WALL

2 X__STUD WALL

POLYETHYLENE FILM

PLYWOOD

2 X__BOTTOM PLATE

2 X__FOOTING PLATE

6" MIN

WHERE FOOTING IS USED FOR DRAINAGE, COVER WITH ASPHALT PAPER OR POLYETHYLENE STRIPS

3/4d

BELOW FROST LINE

d

2d

GRAVEL OR CRUSHED STONE FOOTING

Fig. 11-38 (Continued)

MINIMUM FOOTING PLATE SIZE[1, 2]
For house constructions and sizes in Figure 11-38

House width (feet)	Roof — 40 psf live; 10 psf dead Ceiling — 10 psf 1st floor — 50 psf live and dead 2nd floor — 50 psf live and dead		Roof — 30 psf live; 10 psf dead Ceiling — 10 psf 1st floor — 50 psf live and dead 2nd floor — 50 psf live and dead	
	2 stories	1 story	2 stories	1 story
32	2x10	2x8	2x10[3]	2x8
28	2x10	2x8	2x8	2x6
24	2x8	2x6	2x8	2x6

[1] Footing plate shall be not less than species and grade combination "D" on page 200.

[2] Where width of footing plate is 4 inches (nominal) or more wider than that of stud and bottom plate, use 3/4 inch thick continuous treated plywood strips with face grain perpendicular to footing; minimum grade APA Rated Sheathing 48/24 EXP 1 marked PS 1. Use plywood of same width as footing and fasten to footing with two 6d galvanized nails spaced 16 inches.

[3] This combination of house width and height may have 2x8 footing plate when second floor design load is 40 psf live and dead load.

Fig. 11-39 Minimum footing plate size *(Courtesy of American Plywood Association)*

A. General Nailing Schedule		
JOINT DESCRIPTION	MINIMUM[1] NAIL SIZE	NUMBER OR SPACING
Bottom plate to footing plate — Face nail	10d	12" o.c.
Bottom plate to stud — End nail — 2" plate	10d	2
— 1" plate	8d	2
Top plate to stud — End nail minimum (See Table B)	16d	2
Upper top plate to top plate — Face nail minimum (See Table B) (No overlap of plywood)	10d	8" o.c.
Header joist to upper top plate — Toe nail minimum (See Table C)	8d	16" o.c.
	8d	3
Joist to upper top plate — Toe nail minimum (See Table C)	10d	2
End joist to plate (joists parallel to wall) — Toe nail minimum (See Table C)	8d	4" o.c.
Plywood flooring to blocking at end walls (See Table D)	—	—
Window header support studs to window sill — End nail minimum	16d	2
Window sill to studs under — End nail minimum (See Table B)	16d	2
Window header to stud — End nail	16d	4
Knee wall top plate to studs — End nail	16d[2]	2
Knee wall bottom plate to studs — End nail	8d[2]	2
Knee wall top plate to foundation wall — Toe nail	16d[2]	1 per stud
Knee wall stud over 5' long to foundation wall stud — Toe nail at mid-height of stud	16d[2]	2 per stud
Knee wall bottom plate to footing plate — Face nail	8d[2]	2 per stud space
Window, door or beam pocket header support stud to stud — Face nail	10d	24" o.c.
Corner posts — stud to stud — Face nail	16d	16" o.c.

[1] Heavy loads may require more or larger fasteners or framing anchors. All lumber-to-lumber fasteners below grade may be hot-dipped or hot-tumbled galvanized, except as noted.

[2] Stainless-steel nails required.

B. Minimum Nailing Schedules: Top Plate to Stud and Plate to Plate Connections[1]							
Height of fill (inch)	Treated lumber species[5]	End-nail treated top plate to treated studs		Face-nail untreated top plate to treated top plate			
				No overlap of plywood		3/4" plywood overlap	
		Nail size[2]	Number per joint	Nail size[2]	Spacing (inch)	Nail size[2]	Spacing (inch)
24	All	16d	2	10d	8	10d	16
48	All	16d	2	10d	8	10d	16
72	A, B	16d[3]	3	10d	6	10d	8
	C, D	16d[3]	4	10d	4[4]	10d	4[4]
86	A, B	20d[3]	3	10d	3[4]	10d	4[4]
	C, D	20d[3]	4	10d	2[4]	10d	3[4]

[1] Based on 30 pcf equivalent-fluid density soil pressure and dry lumber.

[2] Hot-dipped, hot-tumbled or stainless steel common wire nails.

[3] Alternatively, may use "U" type framing anchor or hanger with nails and steel plate meeting requirements of DF12.4 and having a minimum load capacity (live plus dead load, normal duration) of 340 pounds in species combination "B" (see page 199)

[4] Alternatively, two nails 2 1/2 inches apart across the grain at twice the spacing indicated may be used.

[5] See page 199 for required properties of lumber species.

Fig. 11-40 Nailing schedules *(Courtesy of American Plywood Association)*

C. Minimum Nailing Schedules: Floor Joists to Wall Connections[1]

Joists perpendicular to wall:

Height of fill (inch)	Joist spacing (inch)	Toe-nail[2] header joist to plate — Nail size[5]	Toe-nail[2] header joist to plate — Spacing (inch)	Toe-nail[2] each joist to plate — Nail size[5]	Toe-nail[2] each joist to plate — No. per joist	Framing anchor[4] each joist to plate
48 or less	16	8d	16	8d	3	none
		10d	16	10d	2	none
	24	8d	8	8d	3	none
		10d	8	10d	2	none
72	16	8d	8	8d	3	none
		10d	8	10d	2	none
		8d	16	none	none	1
	24	10d	8	10d	3	none
		8d	16	none	none	1
86	16	8d	8	none	none	1
	24	8d	4	none	none	1

Joists parallel to wall:

Height of fill (inch)	Blocking[3] between joists, spacing (inch)	Toe-nail[2] end joist to plate — Nail size[5]	Toe-nail[2] end joist to plate — Spacing (inch)	Toe-nail[2] blocking to plate — Nail size[5]	Toe-nail[2] blocking to plate — No. per block	Framing anchor[4] each block to plate
48 or less	No blocking	8d	4	none	none	none
72	48	8d	4	8d	3	none
		10d	4	10d	2	none
		10d	6	10d	4	none
		8d	6	none	none	1
86	24	8d	4	none	none	1

[1] Based on 30 pcf equivalent-fluid density soil pressure and dry lumber. Untreated top plate not less than species combination "D" from page 199, or species Group III from National Design Specification, Table 8-1A.

[2] Toe-nails driven at angle of approximately 30° with the piece and stated approximately one-third the length of the nail from the end or edge of the piece.

[3] See Table D for additional spacing requirements for blocking, and for subfloor to blocking nailing schedule.

[4] Framing anchors shall have a minimum load capacity (live load plus dead load, normal duration) of 320 pounds per joist. If plate or joist is species combination "C" or "D", then rated load capacity of anchors when installed in species combination "B" shall be not less than 395 pounds per joist.

[5] Common wire steel nails.

D. Minimum Nailing Schedules: Subfloor to End Wall Blocking[2,3]

Height of fill (inch)	"A" or "B" — Block spacing (inch)	"A" or "B" — Nails per block[1] 6d	"A" or "B" — Nails per block[1] 8d	"C" or "D" — Block spacing (inch)	"C" or "D" — Nails per block[1] 6d	"C" or "D" — Nails per block[1] 8d
60	48	4	3	48	6	4
72	48	8	6	48	11	8
	24	3	2	24	5	3
86	24	7	5	24	9	7

[1] Common wire nails. Nails shall be spaced 2 inches on centers or more; where block length requires, nails may be in two rows.

[2] See Table C for additional requirements for block spacing and nailing.

[3] Based on 30 pcf equivalent-fluid density soil pressure and dry lumber.

[4] See page 199 for minimum properties of lumber species combinations. Other species in Group III, National Design Specification, Table 8-1A, require the same nailing as species grade combination "D".

Fig. 11-40 (Continued)

FASTENERS

Stainless steel fasteners are recommended for below grade attachment of plywood to lumber. Galvanized fasteners are recommended for framing above and below grade, and for plywood above grade. A recommended nail size and spacing is shown in the nailing schedules in Figure 11-40.

FOOTINGS

Crushed stone, gravel, or coarse sand are used for footings under foundation walls and for fill under the basement slab. A minimum four-inch thickness is laid as a base for the floor slab. The width and thickness under foundation walls depend on the width of the footing plate. In most cases, the footing is twice the width of the footing plate and its depth is three-quarters of the footing plate width. A concrete footing may be used, but it should rest on gravel, or drains through the concrete must be provided, Figure 11-41.

MOISTURE CONTROL AND DRAINAGE

A good drainage system is important in keeping any type of foundation (masonry or wood) dry and trouble-free. A perimeter drain may be used if conditions permit, Figure 11-42, but the use of a *sump* is recommended for wood foundations, Figure 11-43.

It is very important to slope the ground away from the foundation and to use down-spouts and splash blocks to direct water away from the building. In full basement construction, panel joints are caulked, and a 6-mil polyethylene sheet is used to cover the below-grade portions of the foundation wall.

CONSTRUCTING THE FOUNDATION WALL

Typical wall panels are built and the footing plate is attached to panel sections. Joints in the footing plate are offset from those of the wall section. Corners are framed in the same manner as in exterior wall panelized construction. Wall sections are constructed so all vertical joints between plywood sheathing are backed by a stud, Figure 11-44. Specifications for framing beam pockets and other openings in panel sections should be followed closely.

The first two panels are installed at a corner. Once the first corner is in place and leveled, remaining sections are erected. When all wall sections are up, the untreated top plate is applied and the entire structure is squared and straightened. Backfilling takes place only after the basement floor is poured and the first-floor framing and plywood subfloor are installed.

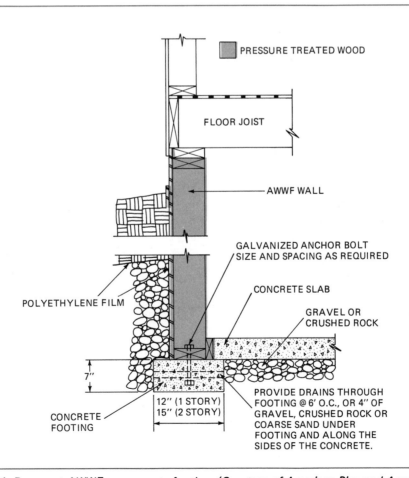

□ PRESSURE TREATED WOOD

FLOOR JOIST

AWWF WALL

GALVANIZED ANCHOR BOLT SIZE AND SPACING AS REQUIRED

POLYETHYLENE FILM

CONCRETE SLAB

GRAVEL OR CRUSHED ROCK

7"

12" (1 STORY)
15" (2 STORY)

CONCRETE FOOTING

PROVIDE DRAINS THROUGH FOOTING @ 6' O.C., OR 4" OF GRAVEL, CRUSHED ROCK OR COARSE SAND UNDER FOOTING AND ALONG THE SIDES OF THE CONCRETE.

Fig. 11-41 Basement AWWF on concrete footing *(Courtesy of American Plywood Association)*

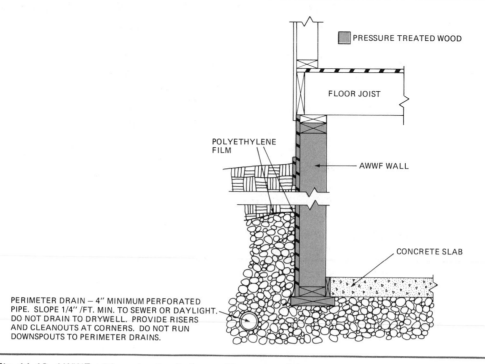

PRESSURE TREATED WOOD

FLOOR JOIST

POLYETHYLENE FILM

AWWF WALL

CONCRETE SLAB

PERIMETER DRAIN – 4" MINIMUM PERFORATED PIPE. SLOPE 1/4"/FT. MIN. TO SEWER OR DAYLIGHT. DO NOT DRAIN TO DRYWELL. PROVIDE RISERS AND CLEANOUTS AT CORNERS. DO NOT RUN DOWNSPOUTS TO PERIMETER DRAINS.

Fig. 11-42 AWWF with perimeter drainage *(Courtesy of American Plywood Association)*

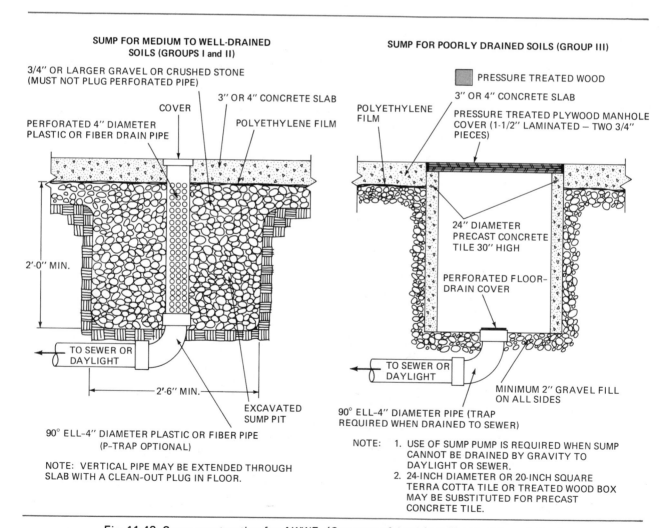

SUMP FOR MEDIUM TO WELL-DRAINED SOILS (GROUPS I and II)

3/4" OR LARGER GRAVEL OR CRUSHED STONE (MUST NOT PLUG PERFORATED PIPE)

COVER

3" OR 4" CONCRETE SLAB

PERFORATED 4" DIAMETER PLASTIC OR FIBER DRAIN PIPE

POLYETHYLENE FILM

2'-0" MIN.

TO SEWER OR DAYLIGHT

2'-6" MIN.

EXCAVATED SUMP PIT

90° ELL-4" DIAMETER PLASTIC OR FIBER PIPE (P-TRAP OPTIONAL)

NOTE: VERTICAL PIPE MAY BE EXTENDED THROUGH SLAB WITH A CLEAN-OUT PLUG IN FLOOR.

SUMP FOR POORLY DRAINED SOILS (GROUP III)

PRESSURE TREATED WOOD

3" OR 4" CONCRETE SLAB

POLYETHYLENE FILM

PRESSURE TREATED PLYWOOD MANHOLE COVER (1-1/2" LAMINATED – TWO 3/4" PIECES)

24" DIAMETER PRECAST CONCRETE TILE 30" HIGH

PERFORATED FLOOR-DRAIN COVER

TO SEWER OR DAYLIGHT

MINIMUM 2" GRAVEL FILL ON ALL SIDES

90° ELL-4" DIAMETER PIPE (TRAP REQUIRED WHEN DRAINED TO SEWER)

NOTE: 1. USE OF SUMP PUMP IS REQUIRED WHEN SUMP CANNOT BE DRAINED BY GRAVITY TO DAYLIGHT OR SEWER.
2. 24-INCH DIAMETER OR 20-INCH SQUARE TERRA COTTA TILE OR TREATED WOOD BOX MAY BE SUBSTITUTED FOR PRECAST CONCRETE TILE.

Fig. 11-43 Sump construction for AWWF *(Courtesy of American Plywood Association)*

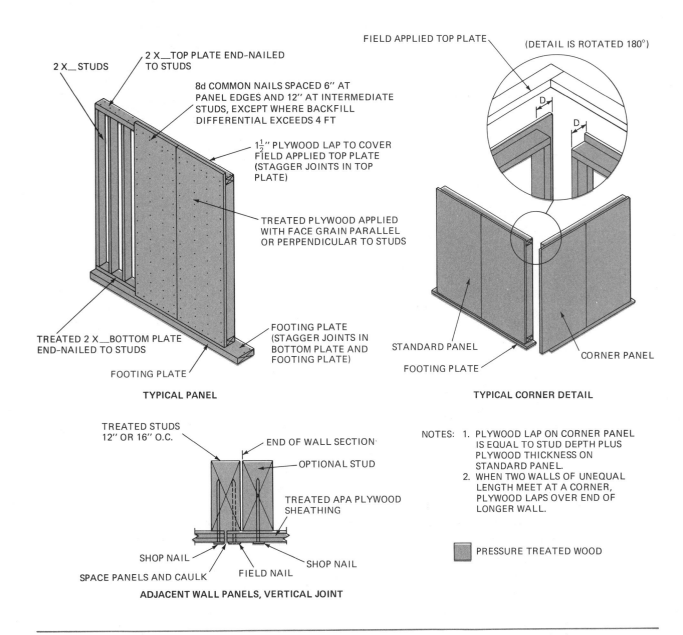

2 X __ STUDS

2 X __ TOP PLATE END-NAILED
TO STUDS

8d COMMON NAILS SPACED 6'' AT
PANEL EDGES AND 12'' AT INTERMEDIATE
STUDS, EXCEPT WHERE BACKFILL
DIFFERENTIAL EXCEEDS 4 FT

$1\frac{1}{2}$'' PLYWOOD LAP TO COVER
FIELD APPLIED TOP PLATE
(STAGGER JOINTS IN TOP
PLATE)

TREATED PLYWOOD APPLIED
WITH FACE GRAIN PARALLEL
OR PERPENDICULAR TO STUDS

TREATED 2 X __ BOTTOM PLATE
END-NAILED TO STUDS

FOOTING PLATE
(STAGGER JOINTS IN
BOTTOM PLATE AND
FOOTING PLATE)

FOOTING PLATE

TYPICAL PANEL

FIELD APPLIED TOP PLATE

(DETAIL IS ROTATED 180°)

D

D

STANDARD PANEL

FOOTING PLATE

CORNER PANEL

TYPICAL CORNER DETAIL

TREATED STUDS
12'' OR 16'' O.C.

END OF WALL SECTION

OPTIONAL STUD

TREATED APA PLYWOOD
SHEATHING

SHOP NAIL

SHOP NAIL

SPACE PANELS AND CAULK

FIELD NAIL

ADJACENT WALL PANELS, VERTICAL JOINT

NOTES: 1. PLYWOOD LAP ON CORNER PANEL
IS EQUAL TO STUD DEPTH PLUS
PLYWOOD THICKNESS ON
STANDARD PANEL.
2. WHEN TWO WALLS OF UNEQUAL
LENGTH MEET AT A CORNER,
PLYWOOD LAPS OVER END OF
LONGER WALL.

PRESSURE TREATED WOOD

Fig. 11-44 AWWF panel construction *(Courtesy of American Plywood Association)*

REVIEW QUESTIONS

Select the most appropriate answer.

1. The top and bottom horizontal members of a wall frame are called
   a. headers.   c. trimmers.
   b. plates.   d. sills.

2. The horizontal wall member supporting the load over an opening is called a
   a. header.   c. plate.
   b. rough sill.   d. truss.

3. Shortened studs above and below openings are called
   a. shorts.   c. jacks.
   b. lame.   d. stubs.

4. The finish floor to ceiling height in a platform frame is specified to be 7'-10''. The finish floor is 3/4'' thick and the ceiling material below the joist is 1 1/2'' thick. A single soleplate and a double top plate are used, each of which has an actual thickness of 1 1/2''. What is the stud height?
   a. 7'-6 3/4''   c. 7'-9 1/4''
   b. 7'-7 3/4''   d. 7'-11 1/2''

5. A rabbeted jamb is 3/4'' thick. Allowing 1/2'' on each side for shimming the frame, what is the rough opening width for a door that is 2'-8'' wide?
   a. 2'-9 1/2''   c. 2'-11 1/2''
   b. 2'-10 1/2''   d. 3'-0 1/2''

6. A story pole shows
   a. the length of headers.
   b. the length of rough sills.
   c. the length of trimmers.
   d. the width of the rough opening.

7. Short blocks fastened to the sides of corner posts and partition intersections and against the soleplate provide
   a. added strength to the corner posts and partition intersections.
   b. more secure fastening of the corner posts and partition intersections to the soleplate.
   c. a surface to attach lines when straightening walls.
   d. more surface on which to fasten the ends of baseboard.

8. When assembling wall sections, studs are laid on the floor with the crowned edge
   a. up.   c. in either direction.
   b. down.   d. alternated.

9. Exterior walls are straightened by
   a. sighting by eye.
   b. using a line stretched between two blocks and testing with another block.
   c. using a plumb bob dropped to the soleplate at intervals along the wall.
   d. by sighting along the length of the wall using a builder's level.

INTERIOR ROUGH WORK

12

After completing this unit, the student will be able to

- *assemble, erect, brace, and straighten bearing partitions.*
- *determine and make rough openings for interior doors.*
- *lay out, cut, and install ceiling joists.*
- *apply furring strips to attach ceiling material.*
- *lay out and erect nonbearing partitions and install backing in walls for fixtures.*
- *apply plaster grounds and metal and gypsum lath.*

Interior rough work is that which is constructed in the inside of structure and later covered by some type of finish work. The interior rough work described in this unit includes partitions, ceiling joists, furring strips, wall backing, and plaster grounds. As stated previously, as much care must be taken with rough work as with finish work.

BEARING PARTITIONS

Bearing partitions usually run the length of the building and are placed in or near the center directly over the girder. When two bearing partitions are used, they are each placed directly above a girder. Bearing partitions support the inner ends of the joists above and are erected in the same manner as are exterior walls. Bearing partitions are not required when trusses are used for the roof frame. A doubled top plate is required. Headers across openings must be strong enough to support the load imposed on them.

Common practice is to erect only bearing partitions before the room is framed. Nonbearing partitions are usually framed after the roof is in place and made tight.

LAYOUT AND FRAMING OF THE BEARING PARTITION

From the floor plan, determine the location of the bearing partitions. As for all other partitions, dimensions are usually given to the centerline. Measure one-half of the thickness of the wall from the centerline and snap a chalk line on the subfloor. Lay the edge of the soleplate to the chalk line, tacking it in position.

Lay out the openings, partition intersections, and all regular and jack studs. Lay the top plate next to the soleplate and transfer the layout of the soleplate to the top plate. All joints in the plates should come in the center of a stud.

Frame, erect, straighten, and temporarily brace the bearing partition in the same manner as for exterior walls. Double the top plate, letting it lap the plate of the exterior wall, Figure 12-1.

ROUGH OPENINGS SIZES FOR INTERIOR DOORS

The rough opening width for an interior door opening is found by adding to the door width twice the thickness of the jamb stock and one inch for shimming (1/2" for each side). The rough opening height is found by adding the sum of the thickness of the finish floor, one-half inch for clearance between the finish floor and the bottom of the door, the thickness of the head jamb, and one-half inch clearance between the head jamb and the rough header, to the height of the door.

Usually no threshold is used under interior doors. However, in case a threshold is used,

substitute its thickness for the one-half inch clearance allowed under the door, Figure 12-2.

CEILING JOISTS

Ceiling joists generally run from the exterior walls to the bearing partition across the width of the building. Construction practices vary according to geographical location, traditional practices, and, at times, the size and style of the building. The size of ceiling joists is based on the span, spacing, load, and the kind and grade of lumber used. Determine the size and spacing from the plans or from local building codes.

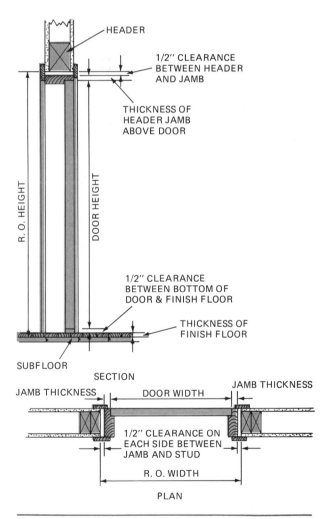

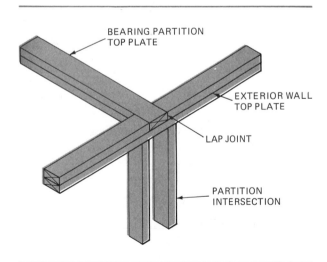

Fig. 12-1 The top plate of the bearing partition laps the plate of the exterior wall.

Fig. 12-2 Figuring the rough opening size for an interior door

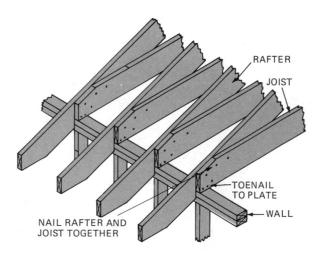

Fig. 12-3 Ceiling joists are located so they can be fastened to the side of rafters.

METHODS OF INSTALLING JOISTS

At the exterior walls, the ceiling joists are placed so that the roof rafters can be fastened to their sides. In a conventionally framed roof the rafters must be fastened to the sides of the joists to keep the walls from spreading due to the weight of the roof exerting an outward thrust on the exterior walls. The opposing rafters should be in line and directly over the studs in the exterior wall, Figure 12-3.

Sometimes, the ceiling joists are installed in-line. In this case, their ends butt each other at the centerline of the bearing partition. The joint must be *scabbed* to tie the joists together, Figure 12-4. (*Scabs* are short boards fastened to the side of the joist and centered on the joint. They should be a minimum of 24 inches long.)

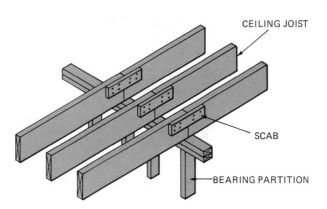

Fig. 12-4 The joint of in-line ceiling joists must be scabbed at the bearing partition.

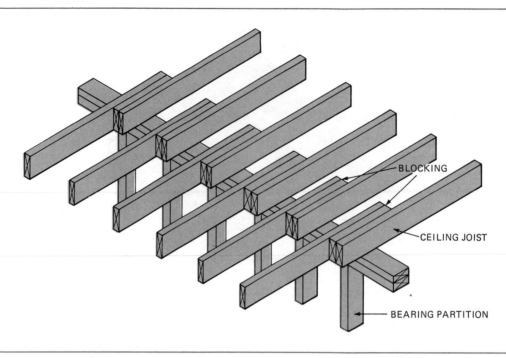

Fig. 12-5 Short blocks are sometimes used between ceiling joists at the bearing partition.

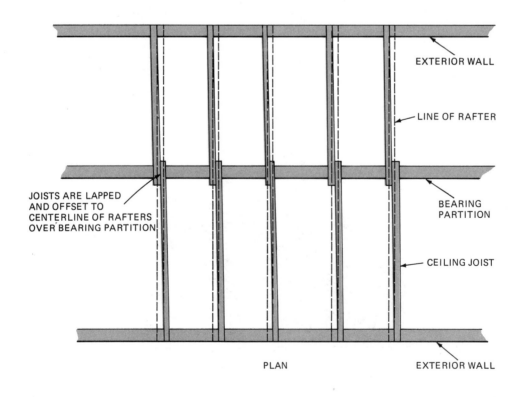

Fig. 12-6 Ceiling joists may be offset at the bearing partition.

The location for the ceiling joists on each exterior wall is on the same side of the rafter location when joists are in-line.

Another method of installing ceiling joists is to place the joists on opposite sides of the rafter on the exterior walls. On the bearing partition, lap the joists with a short block of joist stock between them, Figure 12-5.

Some carpenters install ceiling joists by using the same layout on the exterior walls just mentioned, but offset the joists on the bearing partition with the centerline of the lap falling on the centerline of the rafters, Figure 12-6. This method is recommended only when *furring strips* are applied to the bottom edge of ceiling joists for attachment of the ceiling finish. If the ceiling finish is to be attached directly to the joists, difficulty may be encountered when fastening the ceiling finish because of the offset of the joists.

CUTTING THE ENDS OF CEILING JOISTS

The ends of ceiling joists on the exterior walls usually project above the top of the rafter, especially when the roof is low-pitched. These ends must be cut to the pitch of the roof, flush with or slightly below the top edge of the rafter. Lay out the cut, using a framing square, in the manner shown in Figure 12-7, and cut one joist for a pattern. Use the pattern to mark the rest. Make sure when laying out the joists that you sight each for a crown. Make the cut on the crowned edge so that the crowned edge is

up when the joists are installed. Cut the ends of all joists before installing them.

STUB JOISTS

Usually ceiling joists run parallel to the end walls. However, in the case of low-pitched hip roofs, *stub joists* that run at right angles to the end wall must be installed. The use of stub joists allows clearance for the bottom edge of the rafters that run at right angles to the end wall. Figure 12-8 shows how stub joists are framed.

FRAMING CEILING JOISTS TO A BEAM

In many cases, the bearing partition does not run the length of the building because of large room areas. Some type of beam is needed to support the ceiling joists between the end of the bearing partition and the exterior wall.

If the beam is to project below the ceiling, it is installed in the same manner as a header for an opening. The joists are then installed over the beam in the same manner as over the bearing partition.

If the beam is to be raised in order to make a flush and continuous ceiling below, then the ends of the beam are set on the top of the bearing partition and the exterior end wall. The joists are butted to the beam and may be

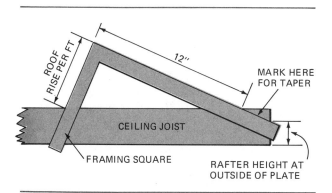

Fig. 12-7 Laying out the taper on the ends of ceiling joists with a framing square.

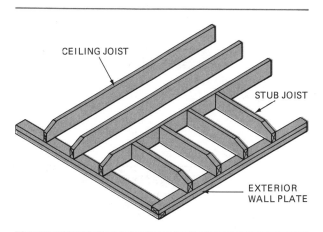

Fig. 12-8 Stub joists are used for low-pitched hip roofs.

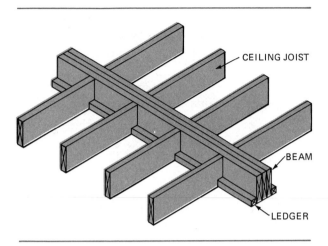

Fig. 12-9 Framing ceiling joists into a beam for a flush ceiling

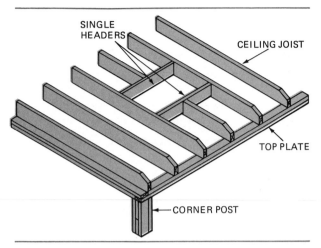

Fig. 12-10 Joists and headers need not be doubled for small ceiling openings.

supported by a ledger strip or by metal joist hangers, Figure 12-9.

OPENINGS

Openings in ceiling joists may need to be made for such things as chimneys and attic access (*scuttle*). Large openings are framed in the same manner as for floor joists. For small openings, there is no need to double the joists or headers, Figure 12-10.

LAYOUT AND SPACING OF CEILING JOISTS

Ceiling joists are installed before the rafters. Spacing of the joists and rafters should be the same. If the spacing of the joists and rafters is the same as that of the studs in the exterior wall, use the same layout rod as used for the studs; or, just square up from either side of each stud and across the top plate. Mark an *X* on the side of the line over the stud for the

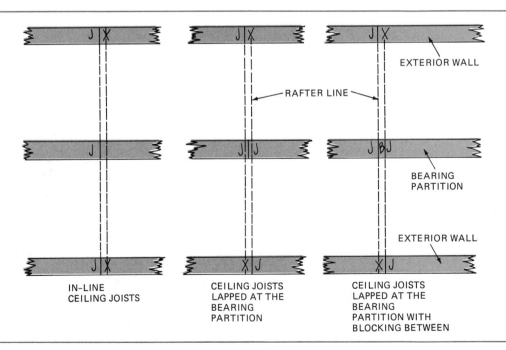

Fig. 12-11 Methods of laying out ceiling joists

rafter. Place a *J* on the other side of the line for the location of the ceiling joist.

On the bearing partition and the opposite exterior wall, use the same layout if the ceiling joists are to be installed in-line. If the joists are to be lapped on the bearing partition, the joist layout is made on one side of the rafter for one exterior wall and on the opposite side of the rafter on the other exterior wall.

The bearing partition is laid out for short blocks placed in line with the rafters and joists placed on each side of the block. If no blocks are used, then the bearing partition is laid out with lines coinciding with the centerline of the rafters. Joists are fastened to each side of the line. Figure 12-11 shows the layouts for various methods of installing ceiling joists.

INSTALLING CEILING JOISTS

The joists on the end walls are placed so the outside face is flush with the inside of the wall, Figure 12-12. This provides fastening for the ends of furring strips or the ceiling finish. All other joists are fastened in position with their sides to the layout lines. If the joist ends have not been precut, sight each for a crown and install each joist with the crowned edge up. Re-

ject any badly warped joists. Toenail joists with at least two 10d nails into the plates. Spike lapped joists together with at least four 10d nails.

FURRING STRIPS

Furring strips for ceilings, sometimes called *strapping* or *stripping,* are usually 1x3 wood strips. They are fastened to the underside of ceiling joists for the attachment of the ceiling finish. They are fastened at right angles to the ceiling joists in rows of 12, 16, or 24 inches on center depending on the type of ceiling finish to be used, Figure 12-13.

Advantages of Using Furring Strips. Often, furring strips are omitted and the ceiling finish is attached directly to the ceiling joists. However, many builders prefer to install furring strips for a number of reasons. One important reason is that electricians can run wiring at right angles to the joists without boring holes in each one. This is especially important when there is a finished living space on the floor above.

Another reason is that furring strips can be spaced to accommodate any size ceiling material. Ceiling tile, for instance, cannot be fastened directly to the joists because the joist spacing may be too great. It is recommended that metal lath be fastened 12 inches on center.

With the use of furring strips, the base for the ceiling can be straightened wherever necessary

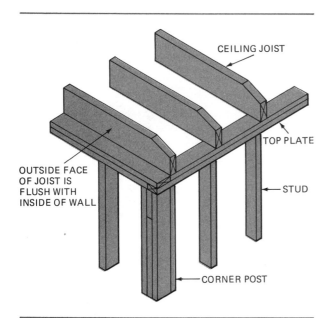

Fig. 12-12 The end ceiling joist is located with its outside face flush with the inside of the wall.

Fig. 12-13 Furring strips are used for the attachment of ceiling finish.

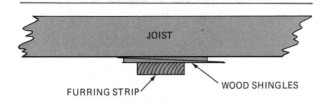

Fig. 12-14 Straightening the furring strips by shimming with wood shingles

by shimming the strips, Figure 12-14. Also, insulation can be laid between the joists without fastening because the furring strips hold it up.

LAYING OUT FURRING STRIPS

Usually the furring strips are installed before any of the nonbearing partitions are erected if plaster or gypsum board is to be used. If ceiling tiles are to be used, then the nonbearing partitions are erected, and each room is individually laid out for the furring strips. The layout of furring strips for ceiling tile is discussed later in Unit 25 Ceiling Finish.

Lay out the spacing of furring strips on the bottom edge of the ceiling joist against each end wall. Starting from the side wall, measure in the desired spacing of the furring strips. Measure back one-half the width of the furring strip and square a line across the bottom edge of the joist. Place an *X* on the side of the line on which the strip is to be fastened. From the squared line, continue the spacing across the end wall, placing *X*s on the same side of the line. Make a similar layout on the opposite end wall, Figure 12-15. Snap chalk lines between the marks on the bottom edges of the joists for the location of the strips.

INSTALLING FURRING STRIPS

Start the installation by constructing a working platform of convenient height. Usually short sawhorses with planks laid on top are used. This platform can be moved where convenient as work progresses.

Cut the first strip so that when one end butts against the end wall, the other end falls on

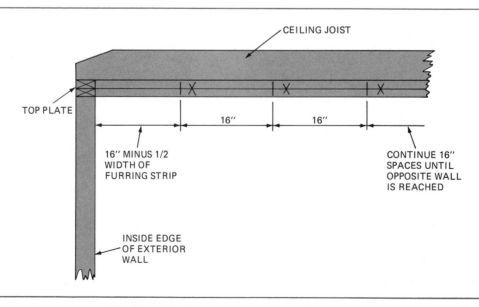

Fig. 12-15 Laying out the location of furring strips

the center of a joist. Cut ends using a handsaw while up on the platform. Steady the strip on your knee while cutting. There is no need to square a line across the piece or to make tight joints on the ends of furring strips. Just sight by eye for squareness when cutting. This eliminates getting up and down from the platform and saves much time and energy.

Apply the first strip with its end against the end wall and its edge against the side wall. Drive two 8d common nails into each joist. Continue applying strips until the opposite end wall is reached. Fasten the other strips to the chalk line in a similar manner, Figure 12-16. Toenail the edge of the strip, if necessary, to straighten it and bring it to the chalk line. The last strip is placed with its edge against the bearing partition. If no bearing partition is used, the last strip is placed against the opposite side wall.

Some nonbearing partitions run parallel to the furring strips. Additional furring is required to provide fastening for the ceiling material on each side of the partitions. Install 1x6 furring between regular furring strips so its centerline falls on the centerline of the partitions, Figure 12-17.

Sight each row of furring by eye and shim down any places necessary to straighten the rows. Use wood shingles split into about 2-inch widths for shims.

NONBEARING PARTITIONS

Nonbearing partitions carry no load and just divide the floor area into rooms. Most building codes state that the studs may be spaced up to 28 inches on center and may be set with the long dimension parallel to the wall. Openings may be framed with single studs and headers.

INSTALLING SOLEPLATES

The location of the partitions is found on the floor plans. Dimensions are usually given to their centerline. Locate each partition on the subfloor and snap chalk lines to the partition side. Mark where each partition ends and all openings.

Fasten the soleplate to the subfloor keeping the edge to the chalk line. Straighten the soleplate as necessary by toenailing 8d common nails into its edges. When possible, spike into

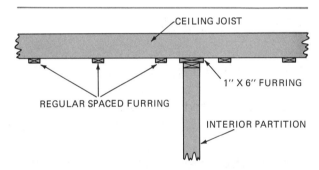

Fig. 12-16 Installing furring strips

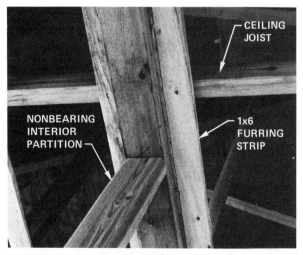

Fig. 12-17 Install 1″x6″ furring for partitions that run between, and parallel to, regular furring

Fig. 12-18 Soleplates for nonbearing interior partitions are fastened to the subfloor.

the floor joists below with one spike into each floor joist staggered along the length of the soleplate. In case they run between and parallel to the joists, toenail into both edges about 16 inches apart. Cut all plates to length and leave spaces for openings, Figure 12-18.

LOCATING TOP PLATES

The position of the top plate must be marked on the ceiling. Hang a plumb bob from the ceiling so it is centered on the edge of the soleplate on one end of the partition. Mark the ceiling at this point, Figure 12-19. Do the same

at the other end of the partition. Snap a chalk line between the marks. Locate the top plates for all other partitions in the same manner.

LAYING OUT THE PLATES

Cut the top plate the same length as the soleplate. Although the soleplate has been cut out for openings, the top plate runs uncut for the total length of the partition, in most cases.

Just tack the top plate on top of the soleplate so that it can be removed later. Lay out the plates in the same manner as for exterior walls using the specified stud spacing, Figure 12-20. Build corner posts and partition intersections as necessary. Find the stud length and cut all full length studs.

CONSTRUCTING THE NONBEARING PARTITION

Separate the top plate from the soleplate. Place all full length studs, corner posts, and partition intersections on the floor with their crowned edges in the same direction and their ends against the fastened soleplate. Spike the top plate to these members to the layout lines.

Raise the section into position and toenail the studs to the soleplate. Use four 8d common nails in each stud, two on each side. Fasten the

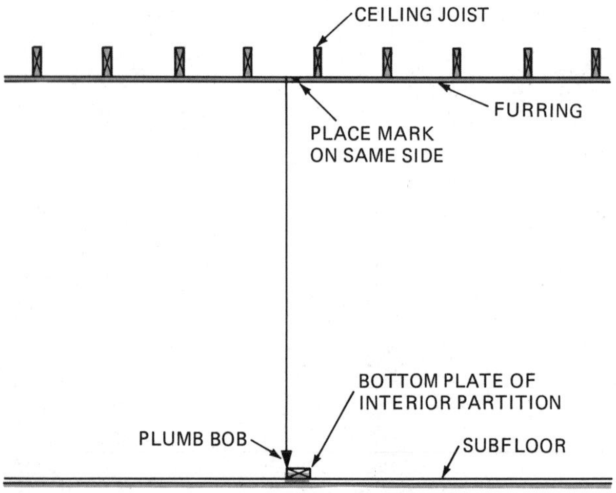

Fig. 12-19 Hang a plumb bob from the ceiling to locate the top plate of interior partitions.

Fig. 12-20 Laying out the plates for interior partitions

Fig. 12-21 Straighten the end stud of interior partitions by toenailing into the center block of the partition intersection. Test with a straightedge.

Fig. 12-22 The wall thickness may be increased by fastening furring strips to the edges of the studs.

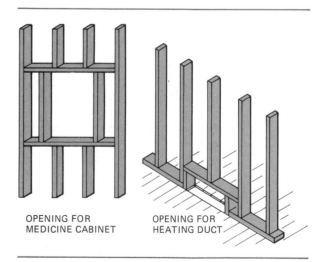

OPENING FOR MEDICINE CABINET   OPENING FOR HEATING DUCT

Fig. 12-23 Miscellaneous openings in interior partitions

top plate into position to the chalk line by toenailing 8d common nails into the furring strips.

IMPORTANT: Straighten the end stud that butts a wall by toenailing through its edge into the center block of the partition intersection. Test for straightness by using a long straightedge against its edge, Figure 12-21.

Construct all nonbearing partitions in a similar manner. For ease of erecting, erect the longer partititons first and then the shorter cross partitions, such as for closets.

Bathroom and kitchen walls containing plumbing are made thicker. Sometimes 2x6 plates and studs are used or a double 2x4 partition is erected to accommodate the pipes in the wall. If the wall thickness needs to be increased only slightly, furring strips may be added to the edges of the studs and plates, Figure 12-22.

OPENINGS

Besides door openings, the carpenter must frame openings for heating and air-conditioning ducts, medicine cabinets, electrical panels, and similar items. If the items do not come in a stud space, the stud must be cut and a header installed, Figure 12-23. When ducts run through the floor, the soleplate and subfloor must be cut out. The reciprocating saw is a useful tool for making these cuts.

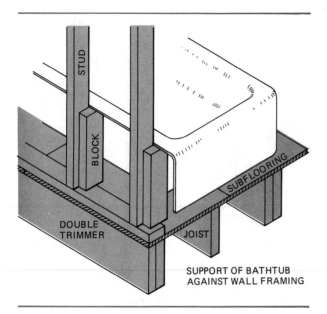

SUPPORT OF BATHTUB
AGAINST WALL FRAMING

Fig. 12-24 Supporting the inner edge of a
bathtub with blocking

BLOCKING AND BACKING

Blocking is installed to support bathtubs as shown in Figure 12-24. Backing is installed for lavatories and shower curtain and drapery rods. The backing may be installed by notching the studs or by cutting the backing between the studs, Figure 12-25.

Plumbing rough-in work varies with the make and style of plumbing fixtures. The experienced carpenter obtains the rough-in schedule from the plumber and installs backing

in the proper location for such things as bathtub faucets, showerheads, lavatories, laundry basins, and toilet tanks, Figures 12-26 and 12-27.

PLASTER

Plaster has been used for many years for wall and ceiling finish. It is made from a white mineral called *gypsum,* which is mined from the earth in rock form; the rock is crushed and ground to a powder form. It is then mixed with sand and a binder to make plaster. The plaster powder is mixed with water to form a paste. When spread thinly on a flat surface, it dries to form a hard and smooth surface. The surface on which the plaster is spread is called a *plaster base.*

PLASTER BASES

In the past, thin wood strips called *wood lath* were used as a plaster base. They are still found in many of the older buildings. They were applied with spaces between them which served to form a *key* or lock for the plaster. Wood lath is no longer used, except perhaps in remodeling to match the existing construction. At the present time, *gypsum lath,* commonly called *rock lath,* and *metal lath* are used.

Gypsum Lath. Gypsum lath is a thin board consisting of a plaster core sandwiched between

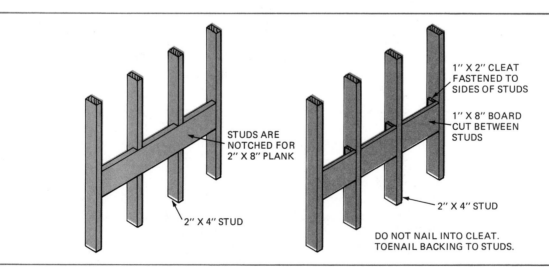

STUDS ARE
NOTCHED FOR
2" X 8" PLANK

2" X 4" STUD

1" X 2" CLEAT
FASTENED TO
SIDES OF STUDS

1" X 8" BOARD
CUT BETWEEN
STUDS

2" X 4" STUD

DO NOT NAIL INTO CLEAT.
TOENAIL BACKING TO STUDS.

Fig. 12-25 Methods of installing backing for plumbing fixtures

a special paper covering. On one side the paper is porous to form a tight bond with the plaster. Some gypsum lath is perforated with 2/4-inch holes spaced 4 inches apart which improves the bond. This material is also available with an aluminum foil back which acts as a vapor barrier and reflective insulation.

Gypsum lath is made in several sizes. The standard size is 16x48 inches, although other sizes are available. For studs spaced 16 inches on center, 3/8-inch thickness is used. Use 1/2-inch thickness for studs spaced 24 inches on center.

Metal Lath. Another type of lath is made of sheet metal. It is available in widths of 24 and 27 inches and in lengths of 8, 10, and 12 feet. The sheet is slit and expanded creating numerous

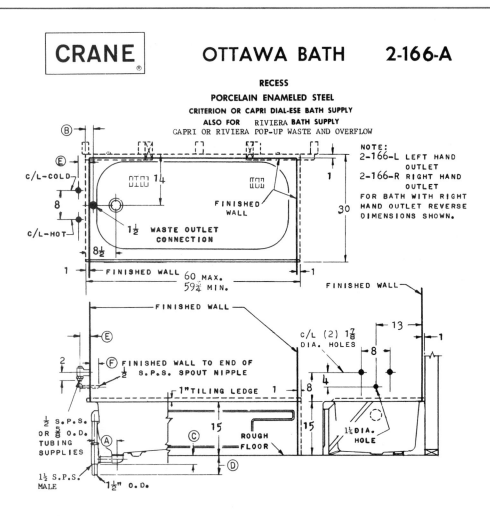

Fig. 12-26 Typical roughing-in schedule for a bathtub

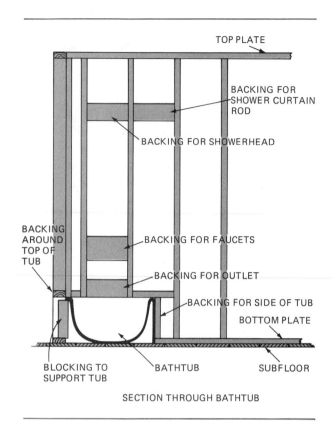

Fig. 12-27 Backing and blocking required for bathtub

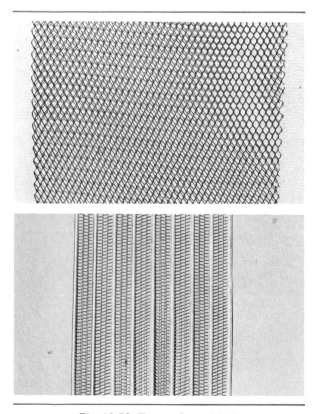

Fig. 12-28 Types of metal lath

openings to key the plaster, Figure 12-28. The metal is expanded in various forms such as diamond mesh, ribbed, or flat ribbed. Metal lath is used mostly in commercial construction and is commonly called *wire lath.* In residential construction, it may be used on ceilings or in shower and bathtub areas. Metal lath requires three coats of plaster, one scratch coat, one brown coat, and one finish coat. Gypsum lath requires only two coats, the brown coat and the finish coat. Total thickness of the lath and plaster combined is 7/8 inch.

APPLYING GYPSUM LATH

Gypsum lath is applied with its long dimension across the studs or joists. The end joints are staggered with none occurring in line with the sides of openings, Figure 12-29. Use blued lathing nails 1 1/8 inches long for 3/8-inch lath and 1 1/4 inches long for 1/2-inch lath. Drive 5 nails into each framing member across the width of the sheet, not less than 3/8 inch from ends and edges. For 24-inch wide, use six nails in each framing member. Be careful not to let the nail heads cut into the face paper. Gypsum lath may be cut to length by scoring one side with a utility knife, bending and breaking the sheet along the scored line. Then score the back side on the bend and snap off the waste, Figure 12-30. A compass saw or wallboard saw is used to make cutouts for pipes or electrical outlets.

Fig. 12-29 Gypsum lath is applied with its end joints staggered.

Fig. 12-30 Cutting gypsum board lath

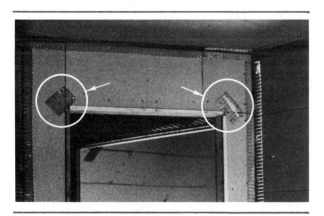

Fig. 12-31 Reinforce the tops of windows and door openings with small strips of metal lath.

Fig. 12-32 Interior corners are reinforced with wire lath commonly called cornerite.

REINFORCING

To prevent plaster cracks, certain sections are reinforced with expanded metal. Fasten small strips diagonally across the upper corners of all window and door openings, Figure 12-31. Inside corners at the intersection of walls and ceilings are reinforced with metal lath called *cornerite.* These are strips bent at a 90° angle to fit in the corners and project about 2 inches out on each side, Figure 12-32. *Corner beads* of expanded metal are applied on all exterior corners. Use a straightedge and install them level and plumb. Plumb corner beads from the *plaster grounds* (strips of wood used as guides to control thickness of plaster) installed at the base. Where the corner beads meet at the corner of an opening, make sure the bead of each is exactly flush, Figure 12-33.

APPLYING METAL LATH

Metal lath is applied so that its edges and ends lap. No extra reinforcing is necessary except using corner beads at the exterior corners. Fasten the expanded metal with large head blued lathing nails spaced about 6 inches on center into each framing member. Cut expanded metal where necessary using tin snips.

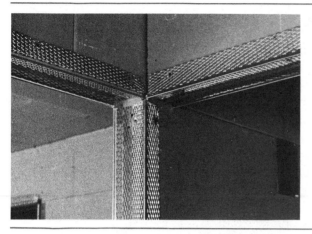

Fig. 12-33 Exterior corners are reinforced with corner bead.

CAUTION: Handle the cut sheets carefully. The sharp cut ends of the expanded metal can cut deeply into the skin.

Plaster grounds are strips of wood used as guides by the plasterer to control the thickness of the plaster. In large commercial construction, metal bent to various shapes is sometimes used for grounds. The thickness of the ground is the total thickness of the lath and plaster combined. Recommended thickness is 7/8 inch. They are located at the base of walls and around door and window openings, Figure 12-34. They are placed in a

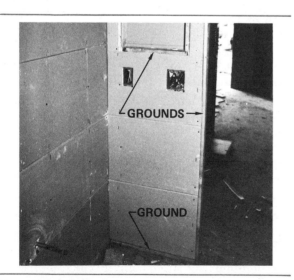

Fig. 12-34 Plaster grounds serve as a guide to control the thickness of plaster.

position so the joint between the plaster and the ground will later be covered by the interior trim. They are installed wherever it is imperative that a straight and even plaster line be provided for high quality application of interior trim. Plaster grounds must be installed in a straight and true line.

Usually 7/8 x 2 1/2-inch strips are used at the base. Wider strips may be used if the baseboard will cover them. Around door and window openings and at other locations usually 7/8 x 1-inch strips are used.

In some cases, the inside edges of the window frame act as grounds when the jamb is the proper width. In most cases, the grounds are fastened permanently in position and left in place after the plaster is applied. However, they may be installed temporarily, especially around door openings, and removed when the plastering is completed, Figure 12-35.

The carpenter is responsible for the installation of plaster bases and grounds unless a specialist installs them. This is also the case for the installation of partitions framed with steel studs.

NONCOMBUSTIBLE PARTITIONS

Steel studs are used when noncombustible partitions are required, usually in commercial buildings, Figure 12-36. Steel partitions are framed in a manner similar to that used for wood partitions. Self-drilling screws are used to

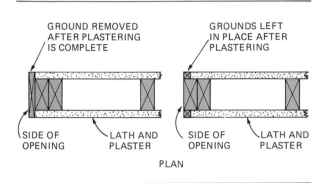

Fig. 12-35 Grounds may be installed temporarily (and later removed) or be installed permanently around openings.

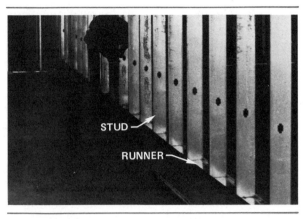

Fig. 12-36 Steel studs are used in commercial buildings where noncombustible walls are required.

fasten the frame members together. These screws have a cutting edge on their point and are driven with a portable power screwgun.

The noncombustible wall components consist of channel-shaped studs and runners. *Runners* are set at floor and ceiling to receive studs which are then attached with screws. Stud webs have prepunched holes for easier installation of pipe and conduit.

Sections come factory-cut to length, but occasionally some cutting is done on the job. A portable electric circular saw or a radial arm saw with a metal cutting blade or a reinforced abrasive cut-off wheel is used for job-site cutting needs.

A full spectrum of steel studs is available that ranges from light-duty drywall studs to heavy-gauge structural studs. Figure 12-37 shows a steel exterior wall section being erected.

Fig. 12-37 A steel exterior wall section is being erected.

REVIEW QUESTIONS

Select the most appropriate answer.

1. Bearing partitions
   a. have a single top plate.
   b. carry no load.
   c. are constructed similar to exterior walls.
   d. are erected after the roof is tight.

2. The top plate of the bearing partition
   a. laps the plate of the exterior wall.
   b. is a single member.
   c. butts the top plate of the exterior wall.
   d. is applied after the ceiling joists are installed.

3. What is the rough opening height of a door opening for a 6'-8'' door if the finish floor is 3/4-inch thick, 1/2-inch clearance is allowed between the door and the finish floor, and the jamb thickness is 3/4 inch?
   a. 6'-9''   c. 6'-10''
   b. 6'-9 1/2''   d. 6'-10 1/2''

4. If ceiling joists are installed in-line,
   a. their end joints lap at the bearing partition.
   b. the joint over the bearing partition must be scabbed.
   c. they are fastened to opposite sides of the rafter.
   d. blocks are placed between them at the bearing partition.

5. The ends of ceiling joists are cut to the pitch of the roof
   a. for easy application of the roof sheathing.
   b. so they will not project above the rafters.
   c. so their crowned edges will be up.
   d. after they are fastened in position.

6. Stub joists
   a. run at right angles to regular ceiling joists.
   b. are used on low pitched common roofs.
   c. are short pieces cut between regular joists.
   d. span the bearing partition and exterior wall.

7. Strapping and stripping are other names given to furring strips fastened to the underside of ceiling joists
   a. to hold the ceiling joists straight from end to end.
   b. for the attachment of the ceiling material.
   c. to make the erection of nonbearing partitions easier.
   d. to provide an air space between the joists and the ceiling.

8. The top plate of nonbearing partitions is located by
   a. measuring from the exterior walls the same distance as the soleplate.
   b. by plumbing with a transit from the soleplate.
   c. by hanging a plumb bob from the ceiling to the soleplate.
   d. any of the above methods.

9. The end stud of partitions that butt against another wall
   a. must be straightened.
   b. must be fastened securely to the wall.
   c. is left out until the wall is erected.
   d. must not be fastened in its center.

10. Gypsum board lath, commonly called rock lath, must be fastened to each framing member with at least
    a. 3 nails.   c. 5 nails.
    b. 4 nails.   d. 6 nails.

SCAFFOLDS 13

OBJECTIVES

After completing this unit, the student will be able to

- *name the parts of and erect safe single pole and double pole wood scaffolds.*
- *erect metal scaffolding according to recommended safe procedures.*
- *build a staging with wall brackets.*
- *install roof brackets and build a safe roof staging.*
- *set up scaffolding with ladder, pump, trestle jacks, and horses.*
- *describe the safety rules for using ladders.*

Scaffolds, also called staging, are temporary working platforms. They are constructed at convenient heights above the floor or ground so that the work can be performed with speed and safety. They must be built strong enough to support workers, tools, and materials and to provide an extra safety margin, Figure 13-1. All scaffolds must be capable of supporting without failure at least four times the maximum intended load.

Carpenters usually erect scaffolding and must be familiar with the different types and methods of construction to provide a safe working platform. The type of scaffold depends on its location, the kind of work being performed, the distance above ground, and the load it is required to support.

Fig. 13-1 Scaffolds are temporary working platforms, used so that work can be performed easily at various heights.

Fig. 13-2 A light-duty single pole scaffold

Fig. 13-3 A light-duty double pole scaffold

WOOD SCAFFOLDS

Wood scaffolds are of the single pole or double pole type. These are used when working on walls. The single pole is used when it can be attached to the wall with no interference to the work, Figure 13-2. The double pole type is used when the staging cannot be fastened to the wall or when it must be kept clear of the wall for the application of materials, Figure 13-3. Wood scaffolds are designated as light, medium, or heavy duty according to the loads they are required to support.

SCAFFOLDING TERMS

The vertical members of a scaffold are called *poles.* All poles should be set plumb and bear on a footing of sufficient size and strength to spread the load to prevent the pole from settling, Figure 13-4. If wood poles need to be spliced for additional height, the ends are squared so the upper pole rests squarely on the lower pole. The joint is scabbed on at least two adjacent sides with scabs at least four feet long. The scabs are fastened to the poles so they overlap the butted ends equally, Figure 13-5.

Bearers or *putlogs* are horizontal members that run from building to pole in a single pole staging or from pole to pole at right angles to the building in a double pole staging. They are set with their greatest dimension vertical and must be long enough to project a few inches outside of the staging pole. When placed against

Fig. 13-4 Scaffold poles are set on footings or pads to prevent settling.

Fig. 13-5 Splicing a scaffold pole for additional height

Fig. 13-6 Bearers must be notched into a short block to be placed against the side of a building.

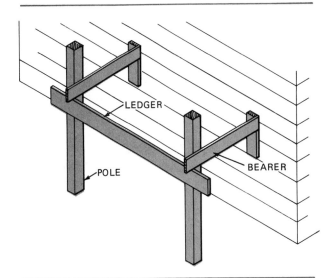

Fig. 13-7 Ledgers run horizontally from pole to pole and support the bearers.

the side of a building, they must be fastened to a notched block, Figure 13-6.

Ledgers run horizontally from pole to pole, parallel with the building, and support the bearers. Ledgers must be long enough to extend over two pole spaces. They must be overlapped at the pole and not spliced between them, Figure 13-7.

Braces are diagonal members to stiffen the scaffolding and prevent the poles from moving or buckling. Full diagonal face bracing is applied across the entire face of the scaffold in both directions. On medium- and heavy-duty double pole scaffolds the inner row of poles is braced in the same manner, Figure 13-8. Cross bracing is also provided between the inner and outer

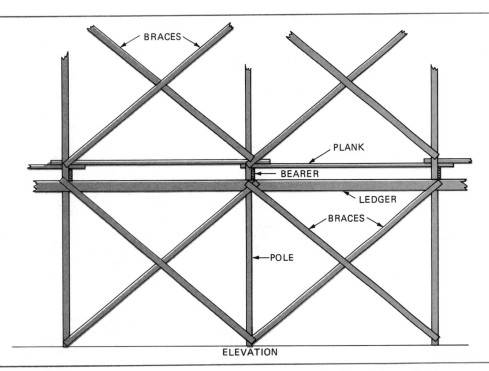

Fig. 13-8 Diagonal bracing is applied across the entire face of the scaffold.

229

sets of poles on all double pole scaffolds. All braces are spliced on the poles.

Staging planks rest on the bearers and are laid with the edges close together so the platform is tight with no spaces through which tools or materials can fall. All planking should be Scaffold Grade or its equivalent and have the ends banded with steel to prevent excessive checking. Overlapped planks should extend at least six inches beyond the bearer. Where the end of planks butt each other to form a flush floor, the butt joint is placed at the centerline of the pole. Each end rests on separate bearers and is secured to prevent movement. End planks should not overhang the bearer by more than 12 inches, Figure 13-9.

Guardrails are installed on all open sides and ends of scaffolds that are more than 10 feet in height. The top rail is usually of 2x4 lumber and is fastened to the poles about 42 inches above the working platform. A middle rail of 1x6 lumber and a toeboard with a minimum height of 4 inches are also installed, Figure 13-10.

SIZING AND SPACING OF MEMBERS

Carpenters must know the minimum sizes of lumber to use and the maximum spacing of the members in order to build a safe scaffold. See Figure 13-11 for these recommendations.

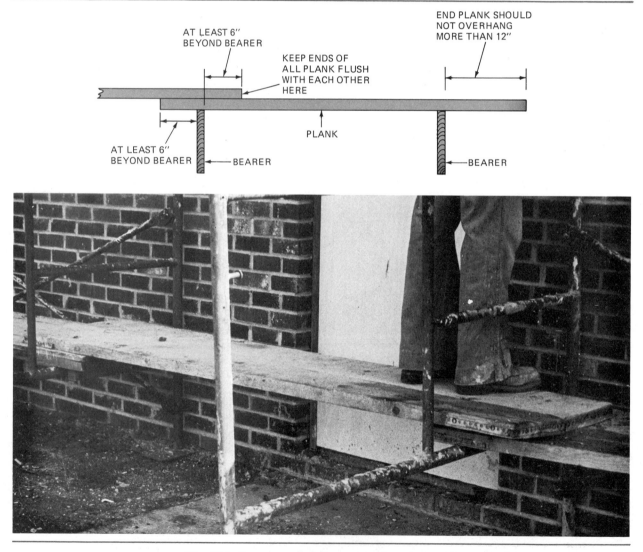

Fig. 13-9 Recommended placement of staging plank

230

ERECTING LIGHT-DUTY SINGLE POLE SCAFFOLDS

Determine the height of the first stage and mark it on both ends of the building. Snap a chalk line between the marks. Fasten a 1x8 bearer on each end so that it projects out from the side wall the desired distance and its top edge is on the chalk line. The maximum distance from wall to pole for this size bearer is 3'-0". Allow enough extra length for fastening to the wall and pole and to extend beyond the pole a few inches. Use at least five 8d duplex nails through the bearer into the wall.

Fasten intermediate bearers to the side wall with their top edges to the chalk line. Use at least five 8d duplex nails into each notched block. Use 16d duplex nails if the blocks are of 2-inch lumber. Space the bearers not more than 10 feet apart. Measure out from the wall and mark each bearer the same distance from the wall for the location of the poles.

Lean the necessary number of poles against the wall alongside each bearer. Fasten the end bearer to the pole at the mark so that the pole is plumb and the bearer is level. Again use at least five 8d duplex nails. Sighting each member by eye is usually sufficient. Place a footing under the pole if there is danger of the pole sinking into the ground.

Fasten the rest of the bearers to the poles in a similar manner. Sight each bearer by eye with the first one so that the top edges are in line. Sight each pole with the first one so that all poles line up.

Install ledgers under the bearers by fastening them to the outside of each pole. Use at

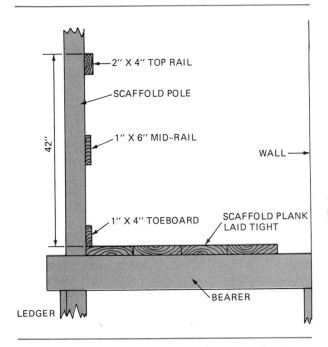

Fig. 13-10 Construction of a guardrail

	Light Duty		Medium Duty	Heavy Duty
WOOD STAGING				
MINIMUM SIZES AND MAXIMUM SPACING OF MEMBERS				
	20 Ft	60 Ft	60 Ft	60 Ft
Poles	2"x4"	4"x4"	4"x4"	4"x6"
Pole Spacing	10'-0"	10'-0"	8'-0"	6'-0"
Scaffold Width	5'-0"	5'-0"	5'-0"	5'-0"
Bearers to 3'-0"	2"x4" or 1"x8"	2"x4" or 1"x8"		
Bearers to 5'-0"	2"x6"	2"x6"	2"x10"	2"x10"
Ledgers	1"x4"	1 1/4"x9"	2"x10"	2"x10"
Plank	2"x10"	2"x10"	2"x10"	2"x10"
Braces	1"x4"	1"x4"	1"x4"	2"x4"
Vertical Distance Between Stages	7'-0"	9'-0"	7'-0"	6'-6"

Fig. 13-11 Minimum sizes and maximum spacing of wood staging members

least two 8d duplex nails into each pole. Remember that ledgers should not be spliced between poles.

Install 1x4 diagonal cross braces across the face of the staging. Nail one end of the brace to the pole just above the bearer and the other end near the bottom of the post. Crisscross the braces between each set of poles. Use two 8d duplex nails in each end of the brace. Drive one nail where the braces cross to tie them together. If the end of the nail protrudes, bend it over.

Lay two staging planks side by side on the bearers in every other section. Let the plank overhang the bearer by the recommended distance. Usually two planks are sufficient for a light-duty scaffold. Plank the remaining sections by laying the plank on those previously laid. Make sure these planks also overhang the bearer at least six inches. These planks should be of equal length so that the ends are flush with each other, Figure 13-12. Unequal lengths create a danger because the worker may not expect two drops in level in such a short distance. In the center of the span fasten a short piece of 1x6 lumber across the planks to tie them together and strengthen them. Use five 8d common nails driven home into each plank, Figure 13-13.

Upper stages of a single pole scaffold are built in the same manner except that the poles are already erected. Always raise scaffolds to the desired height. Do not use ladders, boxes, or other makeshift devices to increase their height. If staging poles need to be spliced, do this in the recommended manner.

ERECTING A DOUBLE POLE SCAFFOLD

Double pole scaffolds should be set as near to the wall of a building as possible without hampering the work being done. They are constructed in a way similar to that used for single pole scaffolds except that the inner end of the bearers are fastened to an inner set of poles.

If the ground is not level, snap a line across the wall of the building at the height of the first stage. Stand the first inner pole upright against the wall and mark it at the same height as the chalk line. Fasten a bearer to it so its top edge is on the mark. Stand the outside post up and fasten the bearer to it so the top edge of the bearer is level. Temporary braces may be required to hold the posts upright. At least two workers are needed. Fasten diagonal cross braces between the bearer and the bottom of the posts.

Continue making sets of inner and outer posts until the desired number is built. Tie the sets together with ledgers on the inside and outside faces. Cross brace the outside face on

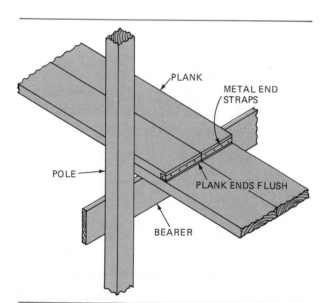

Fig. 13-12 The ends of overlapped staging plank should be flush with each other.

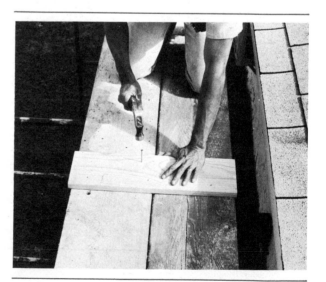

Fig. 13-13 Strengthen staging planks with a short piece of 1x6 across their corners.

light-duty staging. Cross brace both the outside and inside faces on medium- and heavy-duty staging.

Install the planking in the same manner as for single pole staging. Upper stages are built in the same manner as the first except that the height can be determined by measuring up from the first stage.

If the height or length of a double pole scaffold exceeds 25 feet, it must be secured to the building at intervals not greater than 25 feet vertically and horizontally. Ties may be made of 1x4 lumber or reinforcing rod wire.

METAL SCAFFOLDING

Metal scaffolding consists of manufactured tubular welded end frames with accessories such as folding braces, screw legs, and ladders. They are easily erected in sections. Each section consists of two end frames and two folding braces, Figure 13-14. The braces are inserted into pins at the top and bottom of the end frame and locked into position. These hold the end frames securely so the erected scaffold is always plumb, square, and rigid. The end frames have built-in ladders for climbing to the platform.

The scaffolding is extended lengthwise by adding braces and end frames until the desired length is reached. Adjustable screw legs fit in the bottom of the legs of the end frames to level the staging. However, on extremely sloping ground, metal scaffolding may be difficult to level because of the limit of the screw legs. In

any case, the bottom legs must have adequate footing to keep the staging from settling.

The height of the scaffold is increased simply by stacking end frames on top of each other. The bottoms of the legs of one end frame slide into the tops of the legs of another. To prevent movement, the scaffold is secured to the structure at intervals not to exceed 26 feet vertically and 30 feet horizontally. Staging planks are installed on the cross members of the end frames in the same manner as on wood scaffolds. Guardrails of the same type as on wood scaffolds must be installed on metal staging that is more than 10 feet above the ground.

MOBILE SCAFFOLDS

Freestanding mobile scaffold towers may be made with metal scaffolding simply by inserting casters in the bottom of the legs. Casters must be of approved type to support four times the maximum intended load. They must also be provided with a positive locking device to hold the scaffold in position. The height of mobile towers must not exceed four times the smaller base dimension, Figure 13-15. Platforms must be tightly planked to the full width of the scaffold except for necessary entrance openings.

Move scaffolds only on level floors free of obstructions and openings. To avoid the possibility of tipping the mobile scaffold over,

Fig. 13-14 Erecting metal wall scaffolding

Fig. 13-15 The height of a mobile scaffold must not exceed four times its smaller base dimension.

Fig. 13-16 Wall brackets are used frequently on wood frame residential construction.

apply the force necessary to move as close to the base as possible.

WALL BRACKETS

Scaffold plank may be supported by folding metal wall brackets. Made of angle iron, when opened they form a triangular shape, Figure 13-16. A bolt attached to the top member passes through a hole in the wall sheathing. A short block of 2x4 lumber with a hole bored for the bolt is placed against at least two studs on the inside. The bracket is then tightened with a nut against the block, Figure 13-17. Accessories are available that may be attached to the brackets for installing guardrails.

INSTALLING WALL BRACKETS

Snap a chalk line across the wall at the desired height of the staging. Bore 5/8-inch holes through the wall sheathing along the chalk

Fig. 13-17 Wall brackets are tightened against a 2x4 that crosses at least two studs.

line and in stud spaces not more than 8 feet apart. Insert the bolt of one bracket in each of the holes. On the inside insert a prebored 2x4 over the bolt. The 2x4 must cross over at least two studs. Tighten the nut on the bolt. Lay plank on the brackets in the same manner as for other scaffolds. If the outside wall sheathing is thin, insert a short piece of at least 1x6 stock between the bottom of the bracket and the sheathing, crossing over at least two studs. This prevents the bottom of the bracket from puncturing the sheathing when weight is applied. Install guardrails if the staging is over 10 feet in height.

ROOFING BRACKETS

Roofing brackets are used when the pitch of the roof is too steep for carpenters to work without slipping, Figure 13-18. Usually any roof with more than a 5-inch rise per- foot requires roof brackets. Roofing brackets are made of wood or metal and are adjustable for roofs of different pitches.

A metal plate at the top of the bracket has three slots in which to drive nails to fasten the bracket to the roof. The bottom of the slot is made round and large enough to slip over the nailhead. This is necessary to remove the bracket from the fasteners without pulling the nails. The bracket is simply tapped upward

from the bottom and then lifted over the nail-heads. The nails that remain are then driven home.

APPLYING ROOF BRACKETS

Roof brackets are usually used when the roof is being shingled. When roof sheathing or paper is being applied, 2x4s are fastened to the roof to keep the worker from slipping. On steep pitched roofs, brackets are necessary when shingling, not only to keep the worker from slipping, but to hold the roofing materials. Apply roof brackets in rows a vertical distance apart that can be reached without climbing off the roof bracket staging below.

On asphalt shingled roofs, place the brackets at about 8-foot horizontal intervals so the top end of the bracket is just below the next course of shingles. Nail the bracket over a joint or cutout in the tab of the shingle course below. This is necessary so that no joint or cutout in the course above will fall in line with the nails holding the bracket and cause a leak in the roof. Use three 8d common nails driven home. Try to get at least one nail into a rafter.

Open the brackets so the top member is approximately level or slightly leaning toward the roof. Place staging plank on the top of the brackets. Overlap them as in wall scaffolds. Keep the inner edges against the roof for greater

Fig. 13-18 Roofing brackets are used when the roof pitch is too steep for carpenters to work without slipping.

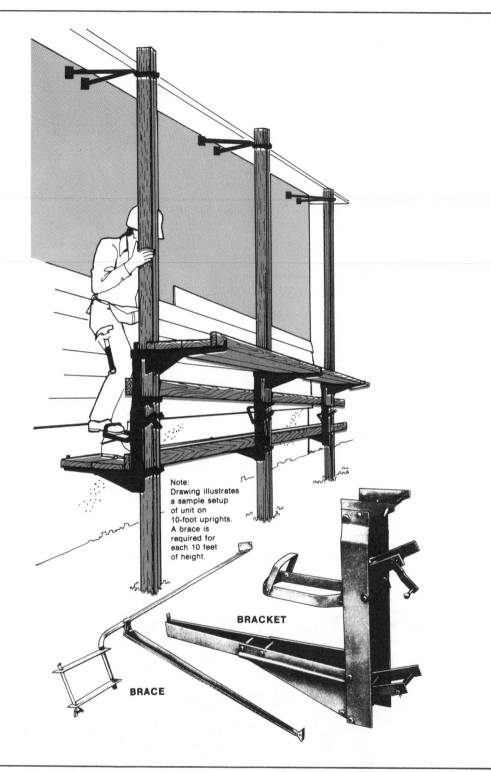

Note:
Drawing illustrates
a sample setup
of unit on
10-foot uprights.
A brace is
required for
each 10 feet
of height.

BRACKET

BRACE

Fig. 13-19 A pump jack scaffold *(Courtesy of Jim and Slim's Tool Supply)*

support. A toeboard made of 1x6 or 1x8 lumber is usually placed flat on the roof with its bottom edge on top of the brackets to protect the new roofing from wear and tear of the workers toes when the roofing has progressed that far.

PUMP JACK SCAFFOLDS

Pump jack scaffolds consist of 4x4 poles, a pump jack mechanism, and a metal bracket, Figure 13-19. The metal bracket is attached at intervals and near the top of the pole. The arms of the bracket, extending from both sides of the pole at a 45° angle, are attached to the sidewall or roof to hold the pole steady.

The pump jack mechanism is foot operated and is pumped up the pole. The mechanism has brackets on which to place the staging plank and others in which to fit a guardrail or platform to place materials. Reversing a lever allows the staging to be pumped downward. Pump jack scaffolds are used widely for wood shingling of sidewalls, where staging must be kept away from the walls, and when a continuous comfortable working height is desired.

However, pump jack scaffolds have their limitations. They should not be used when the working load exceeds 500 pounds. Poles must not exceed 30 feet in height. Braces must be installed at a maximum vertical spacing of not more than 10 feet. In order to pump the bracket past a brace location, temporary braces are used. The temporary bracing is installed about 4 feet above the original bracing. Once the bracket is past the location of the original brace, the original brace can be reinstalled and the temporary brace removed. No more than two persons are permitted at one time between any two supports.

Usually pump jack poles are constructed of two 2x4s spiked together. They must be fastened with the joint parallel to the bracket with 10 d common nails not less than 12 inches apart staggered uniformly from opposite outside edges.

LADDER JACKS

Ladder jacks are metal brackets installed on ladders to hold staging plank. At least two

ladders and two jacks are necessary for a section. Ladders should be heavy-duty ladders, free from defects, and placed not more than 8 feet apart. They should have devices to keep them from slipping. The ladder jack should bear on the side rails in addition to the ladder rungs. If bearing on the rungs only, the bearing area should be at least 10 inches on each rung. No more than two persons should occupy any 8 feet of ladder jack scaffold at any one time. The platform width must not be less than 18 inches and planks must overlap the bearing surface by at least 10 inches.

TRESTLES AND HORSES

Trestles and horses are used when a low scaffold on a level surface is desired. This type of scaffold is used frequently in the building interior by plasterers and dry-wall applicators working on ceilings.

Trestles are manufactured metal horses that are adjustable in height at about 3-inch intervals. They are clamped to a timber on which the staging planks are placed. Metal braces hold the trestle rigid. The size of the timber depends on how far apart the end jacks are placed.

Low scaffolds are also supported by wood horses. For light-duty work, horses and trestles should not be spaced more than 8 feet apart. Do not use horses that have become weak or defective. The horizontal bearing member of the horse should be at least 3x4 inches. The horse legs should be made of 1 1/4x4 1/2 stock.

If a horse scaffold is arranged in tiers, no more than two tiers should be used. The legs of the horses in the upper tier should be nailed to the planks and each tier cross braced.

LADDERS

Carpenters must often use ladders to work from or to reach working platforms above the ground. Most commonly used ladders are the stepladder and the extension ladder. They are usually made of wood, aluminum, or fiberglass. Make sure all ladders are in good condition before using them.

237

CAUTION: When using a stepladder, open the legs fully so the brackets are straight and locked. Make sure the ladder does not wobble. If necessary, place a shim under the leg to steady the ladder. Never work above the second step from the top. Do not use the ledge in back of the ladder as a step. The ledge is used to hold tools and materials only. Move the ladder as necessary to avoid overreaching. Make sure all materials and tools are removed from the ladder before moving it.

Fig. 13-20 Raising an extension ladder

Extension Ladders. To raise an extension ladder, place its feet against a solid object. Pick up the other end, walk forward under the ladder, pushing upward on each rung until the ladder is upright, Figure 13-20. With the ladder upright and close to a wall, extend the ladder by pulling on the rope with one hand while holding the ladder upright with the other. Raise the ladder to the desired height and make sure the locks are securely hooked over the rungs on both sides.

Lean the top of the ladder against the wall and move the base out until the distance from the wall is about one-fourth the vertical height. If the ladder is used to reach a roof or working platform, it must extend above the top support at least three feet.

CAUTION: Be careful of overhead power lines when using metal ladders or wet wood ladders. Metal and wet wood ladders conduct electricity. Contact with power lines could result in electrocution.

When the ladder is in position, shim one leg, if necessary, to prevent wobbling. Face the ladder when climbing and grasp the side rails or rungs with both hands.

JOB-MADE LADDERS

At times it is necessary to build a ladder on the job. These are usually short, straight ladders not more than 24 feet in length. The side rails are made of clear, straight-grained 2x4 stock spaced 15 to 20 inches apart. *Cleats,* or rungs, are cut from 1x4 stock and inset into the edges of the side rails not more than 1/2 inch; or, filler blocks are sometimes used on the rails between the cleats. Cleats are uniformly spaced at 12 inches top to top.

SAFETY

It must be emphasized again that the safety of those working at height depends on properly constructed scaffolds. Those who have the responsibility of constructing scaffolds must be thoroughly familiar with the sizes, spacing, and fastening of scaffold members and other scaffold construction techniques.

REVIEW QUESTIONS

Select the most appropriate answer.

1. The vertical members of a scaffold are called
   a. columns.   c. poles.
   b. piers.   d. uprights.

2. Bearers support
   a. ledgers.   c. rails.
   b. plank.   d. braces.

3. Staging plank should be at least
   a. 2"x6".   c. 2"x10".
   b. 2"x8".   d. 2"x12".

4. Overlapped staging planks should extend beyond the bearer at least
   a. 3 inches.   c. 8 inches.
   b. 6 inches.   d. 12 inches.

5. End planks should not overhang the bearer by more than
   a. 6 inches.   c. 10 inches.
   b. 8 inches.   d. 12 inches.

6. On a light-duty wood scaffold, bearers should not be spaced horizontally more than
   a. 8 feet.   c. 12 feet.
   b. 10 feet.   d. 14 feet.

7. The minimum size of cross braces in a wood scaffold is
   a. 1"x2".   c. 1"x4".
   b. 1"x3".   d. 2"x4".

8. A double pole wood scaffold must be secured to the building if its length or height exceeds
   a. 20 feet.   c. 30 feet.
   b. 25 feet.   d. 35 feet.

9. The height of a mobile scaffold must not exceed the minimum base dimension by
   a. three times.   c. five times.
   b. four times.   d. six times.

10. Guardrails must be installed on all scaffolds more than
    a. 10 feet in height.   c. 20 feet in height.
    b. 16 feet in height.   d. 24 feet in height.

ROOF FRAMING: GABLE AND SHED ROOFS

14

OBJECTIVES

After completing this unit, the student will be able to

- *describe several roof types and define roof framing terms.*
- *describe the members of gable and shed roofs.*
- *lay out a common rafter and erect an equal-pitched gable roof.*
- *lay out and install gable studs.*
- *frame and erect a shed roof.*
- *erect a trussed roof.*

ROOF TYPES

Several roof styles are in common use. These roofs are described in the following material and are illustrated in Figure 14-1.

- The most common roof style is the *gable roof*. Two sloping roof surfaces meet at the top, forming triangular shapes at each end of the building called *gables*.

- The *shed roof* slopes in one direction only. It is sometimes referred to as a lean-to and is commonly used on additions to existing structures. It is also used extensively on contemporary homes.

- The *hip roof* slopes upward from all walls of the building to the top. This style is used when the same overhang is desired all around the building. The hip roof diminishes the height of a building and eliminates maintenance of the sidewalls above the eaves.

- An *intersecting roof* is required on buildings that have wings. Where two roofs intersect, valleys are formed, requiring several types of rafters.

- The *gambrel roof* is a variation of the gable roof. It has two slopes on each side instead of one. The lower slope is made much steeper than the upper slope.

- The *mansard roof* is a variation of the hip roof and has two slopes on each of the four sides. It is framed somewhat like two separate hip roofs.

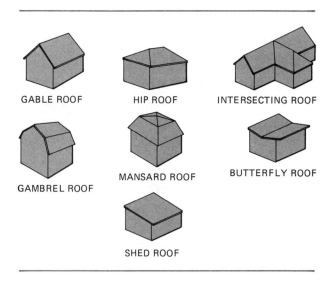

GABLE ROOF   HIP ROOF   INTERSECTING ROOF

GAMBREL ROOF   MANSARD ROOF   BUTTERFLY ROOF

SHED ROOF

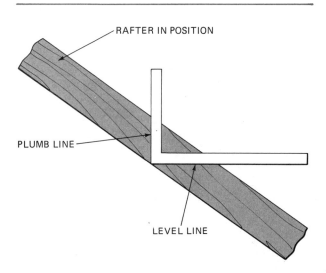

RAFTER IN POSITION

PLUMB LINE

LEVEL LINE

- The *butterfly roof* is an inverted gable roof and is used on many modern homes.
- Other roof styles are used which are a combination of the styles just mentioned.

ROOF FRAMING TERMS

It is important for the carpenter who wants to become proficient in roof framing to be familiar with roof framing terms. These terms are defined in the following material and illustrated in Figure 14-2.

- The *span* of a roof is the horizontal distance covered by the roof. This is usually the width of the building measured from the outer faces of the frame.
- A *rafter* is one of the sloping members of a roof frame and supports the roof covering.
- The *total run* of a rafter is the horizontal distance over which the rafter rises. In most cases, this is one-half of the span.
- The *ridge* is the uppermost horizontal line of the roof.
- The *total rise* is the total vertical distance that the roof rises.
- The *line length* of a rafter is the hypotenuse (longest side) of a right triangle formed with its base as the total run and its altitude as the total rise. The line length gives no consideration to the thickness or width of the stock.

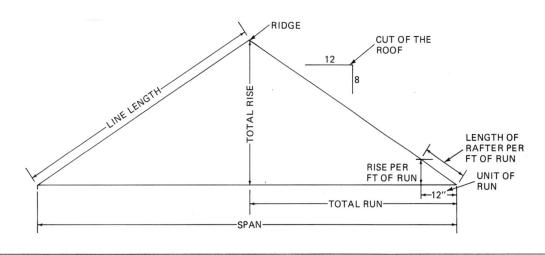

Fig. 14-2 Roof framing terms

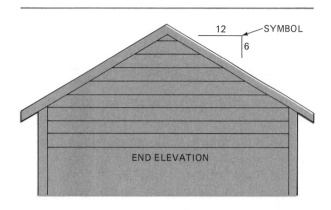

Fig. 14-3 Architectural symbol shows the slope of the roof.

- The *pitch* is the amount of slope of a roof. This is usually expressed as a fraction. The pitch is found by dividing the total rise by the span. For example, if the span of the building is 32 feet and total rise is 8 feet, then 8 divided by 32 equals 1/4. The roof is said to be 1/4 pitch.

- In most cases the slope of a roof is also expressed in the number of inches of vertical rise per foot of horizontal run. This is called the *rise in inches*. Every foot of horizontal run is called the *unit of run*. In most drawings, the slope, also called the *cut* of a roof, is indicated by a symbol showing the rise in inches per foot of run (4 and 12, 6 and 12, or 8 and 12, for example) as shown in Figure 14-3.

- A *plumb cut* is any cut on the rafter which is vertical when the rafter is in position.

- A *level cut* is any cut on the rafter which is horizontal when the rafter is in position.

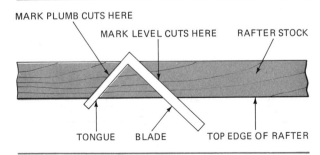

Fig. 14-4 Plumb cuts are laid out on the tongue of the square, and level cuts are laid out on the blade.

When laying out rafters, always use the blade of the framing square (wide side) for marking level cuts. Use the tongue of the square for laying out plumb cuts. This is important because in a complicated cut, it is sometimes difficult to determine which part of the square to mark against, Figure 14-4.

CALCULATING ROUGH RAFTER LENGTHS

The rough length of a rafter from plate to ridge can be calculated using the line length method. Let the blade of the square represent the total run and the tongue of the square represent the total rise. Using a scale of one inch to the foot, measure the distance from blade to tongue from a given total rise and total run to find the length of the rafter, Figure 14-5. Add for any overhang of the rafter beyond the plate.

Example: Assume the total rise of a rafter is 9 feet and the total run is 12 feet. Locate 9 inches on the tongue of the square and 12 inches on the blade and measure the diagonal. This will be found to be 15 inches. Using a scale of 1 inch equals 1 foot, the rough rafter length from plate to ridge is 15 feet. Standard stock length of 16 feet would have to be used for the rafter.

Fig. 14-5 Scaling the rough length of a common rafter

LAYING OUT A COMMON RAFTER

An equal-pitched gable roof is a roof that has an equal slope on both sides intersecting the ridge in the center of the span. This type roof is the simplest to frame because it is composed of only two parts, the common rafter and the ridge, Figure 14-6.

The common rafter extends at right angles from the plate to the ridge. It is called a *common rafter* because it is common to all types of roofs and is used as a basis for laying out other kinds of rafters.

The *ridge board,* although not absolutely necessary, is used because it simplifies the erection of the roof and provides a means of tying the rafters together before the roof is sheathed. Erecting a roof frame is called "raising" the roof.

Laying out a common rafter consists of making two cuts. The cut at the top is called the *plumb cut* or *ridge cut* and fits against the ridge board. The cut at the bottom is called the *bird's mouth* or *seat cut* and fits against the top plate of the wall. In addition to laying out the two cuts, the distance between the cuts must be determined.

Start by laying a straight piece of rafter stock across two sawhorses. Sight the stock for a crook. If there is any slight crook, the carpenter stands on the side which has the crown. This is the top edge of the rafter. Select the straightest piece possible, because this is used as a pattern to mark the rest of the rafters.

LAYING OUT THE RIDGE CUT

To lay out the ridge cut, lay the square down on the side of the stock at its extreme left end. Hold the tongue of the square with the left hand and the blade with the right hand. Move the square until the outside edge of the tongue and the edge of the stock nearest you line up with the specified rise in inches, and the blade of the square and the edge of the stock line up with 12 inches. Mark along the outside edge of the tongue for the plumb cut at the ridge, Figure 14-7.

Example: Assume the cut of the roof is 6 and 12. Hold the square so the 6-inch mark on the tongue and the 12-inch mark on the blade line up with the top edge of the rafter stock. Mark along the outside edge of the tongue of the square.

FINDING THE LENGTH OF A COMMON RAFTER

The distance between the plumb cut at the ridge and the seat cut may be found by the step-off method or by using the rafter tables. The *step-off method* is based on the unit of run (12 inches). The rafter stock is stepped-off for each unit of run until the desired number of units or parts of units are stepped-off.

Fig. 14-6 The common rafters and the ridge form the gable roof.

Fig. 14-7 Laying out the plumb cut of the common rafter at the ridge

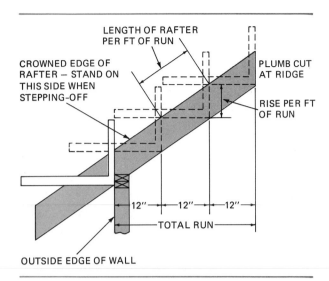

Fig. 14-8 Stepping-off the common rafter

To use the step-off method, first lay out the ridge cut and mark where the blade intersects with the top edge of the rafter. Hold the square at the same angle and shift it until the tongue lines up with this mark. Move the square in a similar manner until the total run of the rafter is laid out, Figure 14-8. Mark a plumb line at the last step. This line will be parallel to the ridge cut.

Example: If the roof has a total run of 16 feet, the square is moved 16 times.

If the total run contains a fractional part of a foot, step off for the whole number of feet. Shift the square and mark along the outside of the blade the fractional part. Move the square so the outside of the tongue lines up with the mark and mark a plumb line along the tongue of the square, Figure 14-9.

Example: If the roof has a total run of 16 feet, 8 inches, step off 16 times. Hold the

square for the 17th step, but mark along the blade at the 8-inch mark. With the square in the same position move it to the mark and lay out the plumb cut of the bird's mouth by marking along the tongue.

Several aids may be used to keep the square in the same position for each step during the step-off process; framing square gauges may be attached to the square to be used as stops against the top edge of the rafter. These gauges are attached to the tongue for the desired rise per foot of run and to the blade at the unit of run, Figure 14-10.

A small straightedge clamped across the tongue and blade of the square at the desired location will also insure that the square is held in the same position for every step. The square is moved by sliding the straightedge against the top edge of the rafter stock, Figure 14-11.

A *pitch board* may also be made and used in place of the framing square, Figure 14-12. It is made by laying out the rise and run on a scrap piece of board. A short piece of 1"x2" stock is fastened to the long edge as a guide.

Rafter tables are available that give information for finding the length and cuts on various kinds of rafters. They come in booklet form and are also stamped on the side of most framing squares, Figure 14-13. On the square, the inch marks indicate the rise of the roof per foot of run. Directly below the inch mark is the length of the rafter in inches per foot of run. To find the length of a common rafter, multiply the

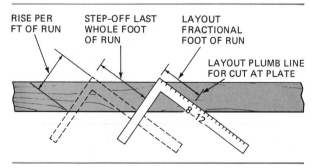

Fig. 14-9 Laying out a fractional part of the total run of a common rafter

Fig. 14-10 Framing square gauges are attached to the square to hold it in the same position for every step-off.

Fig. 14-11 A straightedge attached to the square aids the step-off procedure.

Fig. 14-12 A pitch board may be used instead of the framing square for stepping-off.

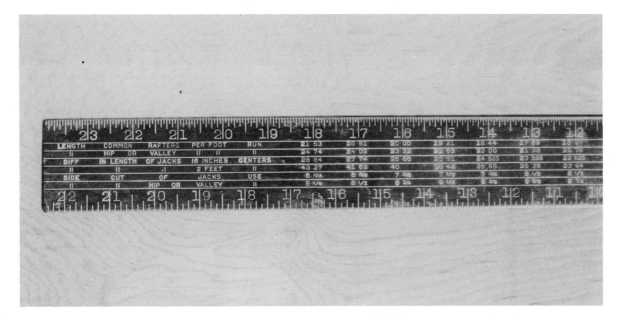

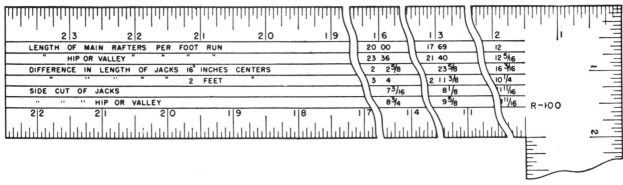

Fig. 14-13 Rafter tables are found on the framing square.

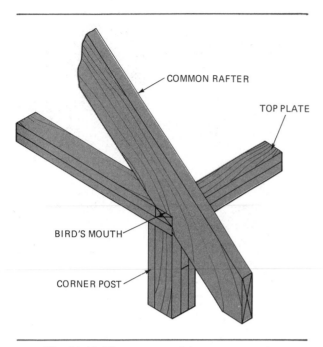

COMMON RAFTER

TOP PLATE

BIRD'S MOUTH

CORNER POST

Fig. 14-14 The seat cut of the common rafter

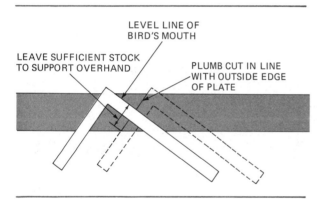

LEVEL LINE OF
BIRD'S MOUTH

LEAVE SUFFICIENT STOCK
TO SUPPORT OVERHAND

PLUMB CUT IN LINE
WITH OUTSIDE EDGE
OF PLATE

Fig. 14-15 Laying out the level line
of the bird's mouth

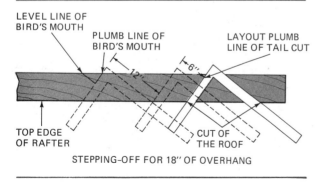

LEVEL LINE OF
BIRD'S MOUTH

PLUMB LINE OF
BIRD'S MOUTH

LAYOUT PLUMB
LINE OF TAIL CUT

12″

6″

TOP EDGE
OF RAFTER

CUT OF
THE ROOF

STEPPING-OFF FOR 18″ OF OVERHANG

Fig. 14-16 Laying out the overhang
of the common rafter

length found in the table by the total run of the rafter.

Example: Find the length of a common rafter with a rise of 8 inches per foot of run for a building 28 feet wide.

Reading below the 8-inch mark on the square, the length of the common rafter per foot of run is found to be 14.42 inches. Multiplying 14.42 inches by 14, which is the total run of the rafter, gives 201.88 inches. Dividing by 12 to change this total to feet gives 16.82 feet. Changing 0.82 foot to inches by multiplying by 12 gives 9.84 inches. Changing 0.84 inch to sixteenths of an inch by multiplying by 16, gives slightly over 13/16 inch, making the total length of the rafter 16′-9 13/16″ long.

Mark the plumb cut at the ridge. Measure along the top of the rafter from the ridge cut for the length of the rafter as found from the rafter tables. Mark at this point and make another plumb cut at the seat. The bird's mouth, overhang, tail cut, and shortening are laid out in the same manner as in the step-off method.

THE BIRD'S MOUTH

The *bird's mouth* of the rafter is a combination of a level cut and a plumb cut. The level cut rests on top of the wall plate and the plumb cut fits snugly against the outside edge, Figure 14-14.

On the plumb line, which has already been laid out with the last step, measure down from the top edge of the rafter a distance that will leave sufficient stock to insure enough strength to support any overhang, and mark the plumb line. This is usually a minimum of two-thirds the width of the stock. Holding the square in the same position as for stepping-off, slide it along the stock until the blade lines up with the mark. Mark along the blade for the level cut, Figure 14-15.

OVERHANG OF THE RAFTER

In most plans, the rafter overhang is given in terms of a level measurement. To lay out the overhang, simply step off the necessary amount of run from the plumb line of the bird's mouth, Figure 14-16. The *tail cut* is the cut at the end

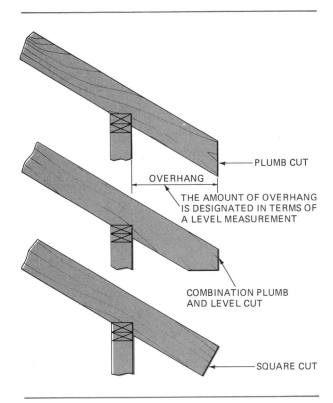

Fig. 14-17 Various tail cuts of the common rafter

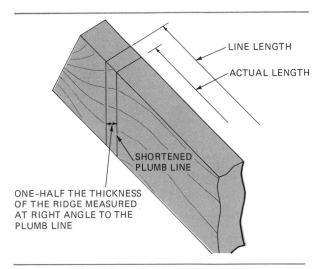

Fig. 14-18 Shortening the common rafter

of the rafter overhang and may be a plumb cut, level cut, a combination of cuts, or a square cut, Figure 14-17. It is common practice, however, to let the rafter tails run *wild,* that is, longer than needed. They are cut off in a straight line after the roof frame is erected. In some cases, where the rafter tails are left exposed, decorative curved cuts may be made.

SHORTENING THE COMMON RAFTER

Because the length of the rafter at this point is to the centerline of the ridge, the rafter must be shortened to allow for the thickness of a ridge board if one is used, Figure 14-18. If no ridge board is used, no shortening is necessary.

To shorten the rafter, measure back at a right angle from the plumb line at the ridge a distance equal to one-half the thickness of the ridge board. Lay out another plumb line at this point for the actual length of the rafter. **Note: Shortening is always measured at right angles to the ridge cut, regardless of the pitch of the roof.**

CUTTING THE REQUIRED NUMBER OF RAFTERS

To find the number of rafters required for a gable roof, divide the length of the building by the spacing of the rafters, add one, and multiply the total by two. **Example:** A building is 42 feet long and the rafter spacing is 16 inches O.C. Divide 42 by 1 1/3 (16 inches divided by 12 equals 1 1/3 feet) to get 31 1/2 spaces. Change 31 1/2 to the next whole number to get 32. Add 1 to make 33. Multiply by 2 to show that 66 rafters are needed.

Pile the required number of rafter stock with the crowned edges all facing in the same direction. Lay the pattern on each piece so the crowned edge of the stock will be the top edge of the rafter. Mark and make the necessary cuts. Usually a portable electric saw is used to make these cuts.

CAUTION: On some cuts that are at a sharp angle with the edge of the stock (i.e. the level cut of the seat) the guard may not retract. In this case, retract the guard by hand until the cut is made a few inches into the stock. Never wedge the guard in an open position to overcome this difficulty.

In addition to solid lumber, TJI® joists may be used for rafters. Construction details and span tables are shown in Figure 14-19.

247

LAYING OUT THE RIDGE BOARD

Lay a sufficient number of boards alongside the wall plate. The width of the boards should be equal to the length of the plumb cut of the rafter.

Transfer the rafter layout at the plate to the ridge board. Joints in the ridge board are centered on a rafter. The total length of the ridge board should be the same as the length of the building, unless there is an over-

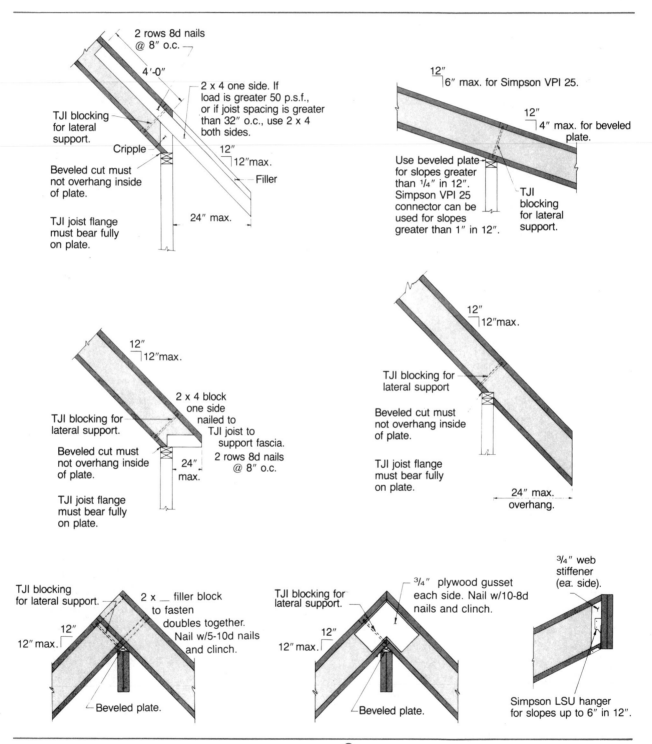

Fig. 14-19 Construction details and span table for TJI® rafters *(Courtesy of Trus Joist Corporation)*

RESIDENTIAL ROOF SPAN CHART

O.C. Spacing	Joist Depth	Non-Snow (125%) 20 PSF Live Load 10 PSF Dead Load	Non-Snow (125%) 20 PSF Live Load 20 PSF Dead Load	Snow (115%) 25 PSF Live Load 10 PSF Dead Load	Snow (115%) 30 PSF Live Load 10 PSF Dead Load	Snow (115%) 40 PSF Live Load 10 PSF Dead Load	Snow (115%) 50 PSF Live Load 10 PSF Dead Load
12"	9 1/2"	25'-0"	22'-10"	23'-9"	22'-9"	21'-0"	19'-9"
	11 7/8"	30'-0"	27'-0"	28'-0"	27'-0"	25'-0"	23'-6"
16"	9 1/2"	22'-9"	20'-6"	21'-6"	20'-7"	19'-0"	17'-10"
	11 7/8"	27'-0"	24'-4"	25'-7"	24'-6"	22'-7"	21'-1"
19.2"	9 1/2"	21'-4"	19'-4"	20'-3"	19'-4"	17'-10"	16'-8"
	11 7/8"	25'-4"	23'-0"	24'-0"	23'-0"	21'-1"	19'-8"
24"	9 1/2"	19'-9"	17'-10"	18'-8"	17'-10"	16'-4"	14'-6"
	11 7/8"	23'-6"	21'-2"	22'-3"	21'-1"	19'-5"	16'-8"
32"	9 1/2"	17'-10"	16'-0"	16'-10"	16'-0"	13'-2"	10'-10"
	11 7/8"	21'-2"	19'-0"	19'-10"	19'-1"	15'-1"	12'-7"
48"	9 1/2"	15'-4"	11'-10"	12'-6"	10'-10"	8'-8"	7'-3"
	11 7/8"	18'-0"	13'-8"	14'-4"	12'-6"	9'-11"	8'-4"

1. Roof joists to be sloped 1/8" in 12" minimum. No camber provided.
2. Maximum deflection is limited to L/180 at total load.
NOTE: For loads not shown, refer to allowable uniform load chart.

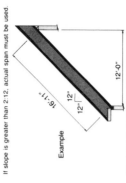

If slope is greater than 2:12, actual span must be used.

16'-11"   12"   12"   12'-0"

Example

Actual span is 16'-11". Use 16'-11" as joist span when selecting joist and spacing from above chart.

Fig. 14-19 (Continued)

hang at the gable ends. If there is an overhang add the necessary amount on both ends.

Erect a scaffold, if necessary, in the center of the building at a convenient height from the ridge, usually about 5 feet below the ridge. Lay the ridge board on top of the scaffold in the same direction it was laid out. Be careful not to turn the pieces around.

Lean the cut rafters against the building, spacing them along each side, with the ridge cut up. Select four straight pieces for the end rafters, commonly called the *rake* rafters. At least three carpenters are needed to raise the roof: one on each side of the building and one at the ridge. Raise one end rafter into position and fasten it at the seat, using 12d or 16d nails toenailed into the plate. Raise the opposite rafter and fasten it at the seat. Let the ridge cut of the two opposing rafters butt each other without fastening.

Near the end of the first section of ridge board, raise a pair of opposing rafters in a similar manner. Slip the ridge board up from the bottom and through the ridge cuts of the two pairs of opposing rafters. Fasten one rafter by nailing through the ridge board into the plumb cut. (Use three 8d common nails if the ridge board is 1 inch. Use 16d nails if the ridge board is 2 inches.) The opposing rafter will have to be moved slightly to the side. Fasten the opposing rafter in position by nailing through the ridge

board at an angle. Drive the nail home with a nail set if necessary. Keep the top edges of the rafters and ridge board flush with each other.

Plumb and brace this section. Plumb the section by dropping a plumb bob from the end of the ridge board to the outside edge of the plate at the end of the building, Figure 14-20. Brace temporarily from attic floor to ridge.

Fasten the rest of the rafters in this section in place. Erect first one rafter and then the opposing one. Do not erect a number of rafters on one side only. In addition to fastening rafters to the wall plate, fasten them to the sides of ceiling joists.

Raise all other sections in a similar manner. Check the roof for plumb on both ends. Brace the roof frame securely with braces fastened to the under edges of the rafters running from plate to ridge. Brace both sides of the roof in this manner.

Note: Applying the brace to the underside of the roof frame allows the roof to be sheathed without disturbing the braces, Figure 14-21.

If the rafter tails have not previously been cut, measure and mark the end rafters for the amount of overhang. This is usually a level measurement. Plumb the marks up to the top edge of the rafters. Snap a line between these two marks across the top edges of all the rafters. Using a level, plumb down on the side of each rafter from the chalk line. Using a portable

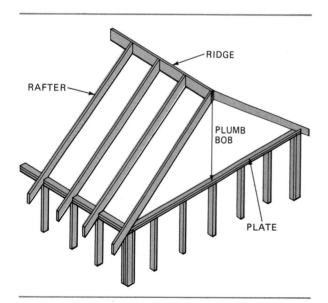

Fig. 14-20 Plumbing the end of a gable roof

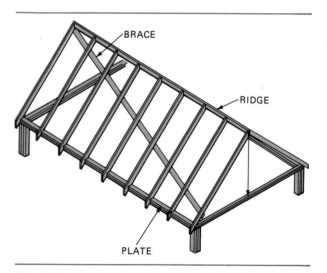

Fig. 14-21 The gable roof is braced on the underside of the rafters. This allows the roof to be sheathed without disturbing the braces.

Fig. 14-22 Lookouts support the rake overhang.

electric saw, cut each rafter starting the cut from the top edge and following the line carefully.

RAKE OVERHANG

If an overhang is specified at the rakes, *lookouts* (horizontal structural members) must be installed to support the rake rafter. These are shown in Figure 14-22.

INSTALLING GABLE STUDS

The triangular areas formed by the rake rafters and the wall plate at the ends of the building are called gables and must be framed with studs. These studs are called *gable studs* The bottom ends are cut square and fit against the top of the wall plate. The top ends fit snugly against the bottom edge and inside face of the end rafter. They are cut off flush with the top edge of the rafter, Figure 14-23.

The end wall plate is laid out for the location of gable studs. Plumb down from the center of the ridge board to the wall plate and square a line across the plate. Measure one-half a stud space on each side of the centerline for the studs on each side of center. From these marks, lay out the stud spacing to the outside of the building. With a straightedge and level, plumb these marks up to the inside face of the end rafters to lay out the top end of the gable studs.

One method of finding the length and cut of the stud is to stand the stud plumb on the

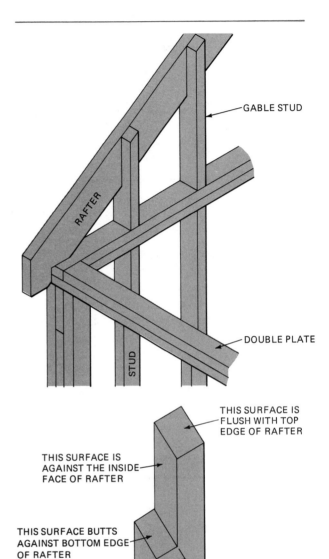

Fig. 14-23 The cut at the top end of the gable stud

mark at the plate. Hold the stud against the inside face of the rafter and mark it along the bottom and top edge of the rafter, Figure 14-24. Remove the stud and use a scrap piece of rafter stock to mark the depth of cut on the stud. Mark and cut all studs in a similar manner. The studs are fastened by toenailing to the plate and by nailing through the rafter into the edge of the stud. Care must be taken when

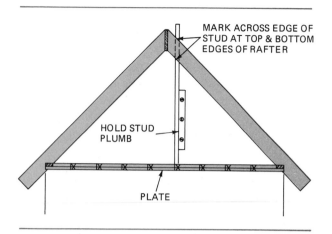

**Fig. 14-24 Finding the length and
cut of the gable stud**

installing gable studs not to force the end rafters and create a crown in them. Sight the top edge of the end rafters for straightness as gable studs are installed. After all gable studs are installed, the end ceiling joist is spiked to the inside edges of the studs. Any crook is taken out of the ceiling joist as it is fastened.

LAYING OUT GABLE STUDS BY A COMMON DIFFERENCE

Gable studs that are spaced equally have a common difference in length. Each stud is shorter by the same amount, Figure 14-25. Once the length of the first stud and the common difference is known, gable studs can be easily laid out and cut on the ground.

To find the common difference in the length of gable studs, multiply the spacing in feet by the rise per foot of the roof.

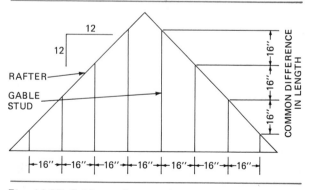

Fig. 14-25 Gable studs spaced equally have a common difference in length.

Example: Gable studs are spaced 16 inches O.C. The roof rises 6 inches per foot of run. Change 16 inches to 1 1/3 feet. Multiply 1 1/3 by 6 to get 8. The common difference is 8 inches.

The quickest method of finding the length of the first stud is to place it in position and mark it as described above. The first stud should be the longest one. All other stud lengths can be found by marking each 8 inches shorter than the previous one. Studs for both sides may be cut using one pattern, if the layout is made from the centerline. Two length patterns are required if the layout is off center.

LAYING OUT A SHED ROOF RAFTER

A *shed roof rafter* is different than common rafter because it has two bird's mouths instead of one, Figure 14-26. The layout is the same as the common rafter. However, the total run of the rafter is the width of the building minus the thickness of one of the walls.

To lay out the rafter, deduct the thickness of one of the walls from the width of the building. Step-off the rafter using the remainder as the run, and lay out two bird's mouths exactly the same.

Shed roofs are raised by toenailing the rafters into the plate at the designated spacing. The bird's mouths must be kept snug against the walls.

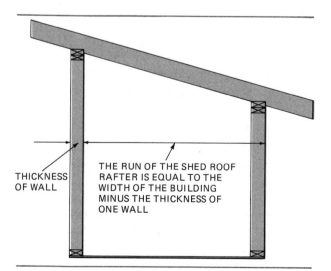

Fig. 14-26 The shed roof rafter has two bird's mouths.

ROOF TRUSSES

Fig. 14-27 Trusses are used extensively for roof framing.

Roof trusses are extensively used in residential and commercial construction, Figure 14-27. Because of their design, they can support the roof over a wide span. This eliminates the need of load bearing partitions below. With trusses, the roof is framed in much less time. Because of their design, much usable attic space is lost, however.

The truss consists of upper and lower *chords* and diagonals called *web members*. The upper chords act as rafters and the lower chords as ceiling joists. Joints are fastened securely with metal or wood gusset plates, Figure 14-28.

TRUSS DESIGN

Fig. 14-28 The members of roof trusses are securely fastened with metal gussets.

Most trusses are made in truss fabricating plants and transported to the job site. Trusses are designed by engineers to support the prescribed loads. Trusses may also be built on the job, but approved designs must be used. The carpenter should not attempt to design a truss. Approved and tested designs for job-built trusses are available from the American Plywood Association.

The most common truss design for residential construction is the *Fink* truss, Figure 14-29. Other truss shapes are designed to meet special requirements, Figure 14-30.

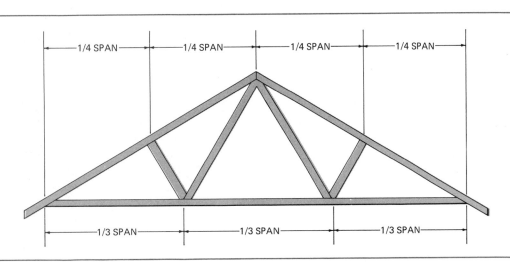

Fig. 14-29 The Fink truss is widely used in residential construction.

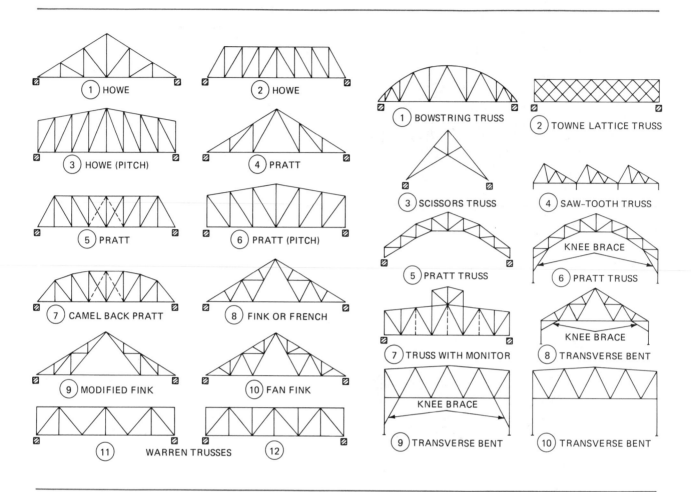

Fig. 14-30 Truss shapes are designed for special requirements.

ERECTING A TRUSS ROOF

The trusses are placed upside down, hanging on the wall plates, toward one end of the building. The gable end truss is installed first by swinging it up into place and bracing it securely in a plumb position. It is fastened to the plate by toenailing. Framing anchors are used to fasten trusses to the wall plate, Figure 14-31.

The other trusses are erected in a similar manner at the specified spacing. Spacing strips are fastened to the trusses at the top and bottom. Braces are installed to keep the frame from collapsing.

Trusses for spans of 40 feet and over require the use of a crane to lift them in position. One at a time is lifted, fastened, and braced in place.

Fig. 14-31 Metal brackets are used to fasten trusses to the wall plate.

REVIEW QUESTIONS

Select the most appropriate answer.

1. Using a scale of 1'' equals 1'-0'', what is the rough length of a rafter (to the next whole number of feet) that has a total rise of 8 feet and a total run of 16 feet?
   a. 17
   b. 18
   c. 19
   d. 20

2. When laying out a common rafter, stand on
   a. the crowned edge side.
   b. the hollowed edge side.
   c. either side.
   d. the side that will be the bottom edge.

3. The unit of run of the common rafter is
   a. 8 inches.
   b. 12 inches.
   c. 16 inches.
   d. 24 inches.

4. If a fractional part of a step must be laid out, mark along the
   a. inside of the tongue of the square.
   b. outside of the tongue of the square.
   c. inside of the blade of the square.
   d. outside of the blade of the square.

5. The minimum amount of stock left above the bird's mouth of the common rafter to insure enough strength to support the overhang is usually
   a. one-quarter of the rafter stock width.
   b. one-half of the rafter stock width.
   c. two-thirds of the rafter stock width.
   d. three-quarters of the rafter stock width.

6. If a ridge board is used, the common rafter
   a. is shortened by the thickness of the ridge board.
   b. is shortened by one-half the thickness of the ridge board.
   c. is shortened by one-half the 45° thickness of the ridge board.
   d. is not shortened.

7. What is the line length of a common rafter from the centerline of the ridge to the plate with a rise of 5 inches per foot of run, if the building is 28'-0'' wide?
   a. 14'-8 1/2''
   b. 15'-2''
   c. 15'-6''
   d. 16'-2''

8. The required number of common rafters, excluding any overhang at the rakes, for a building 48'-0'' long, is
   a. 37.
   b. 49.
   c. 74.
   d. 98.

9. The common difference in the length of gable studs spaced 24 inches O.C. for a roof with a pitch of 8 inches rise per foot of run is
   a. 8 inches.
   b. 12 inches.
   c. 16 inches.
   d. 20 inches.

10. The total run of a shed roof rafter is
   a. one-half the width of the building.
   b. the span of the roof.
   c. the span minus the thickness of one of the walls.
   d. the span minus the thickness of both walls.

ROOF FRAMING: HIP AND INTERSECTING ROOFS

15

OBJECTIVES

After completing this unit, the student will be able to

- *describe the members of hip and intersecting roofs.*
- *lay out the members of and frame a hip roof.*
- *lay out the members of and frame an intersecting roof.*
- *apply roof sheathing.*

THE HIP ROOF

The hip roof contains common rafters, a ridge, and also hip rafters and hip jack rafters, Figure 15-1. Because the roof slopes upward from all walls of the building, a special rafter is required where the slopes meet. This rafter is called a *hip rafter*. The *hip jack rafter* is somewhat like a common rafter except its top end butts against the hip rafter instead of the ridge. It could be called a shortened common rafter.

RUN OF THE HIP RAFTER

Because the hip rafter runs at a 45° angle to the plate, the amount of horizontal distance it covers is greater than the common rafter. Because the unit of run of the common rafter is 12 inches and because the hip rafter must rise the same amount as the common rafter with the

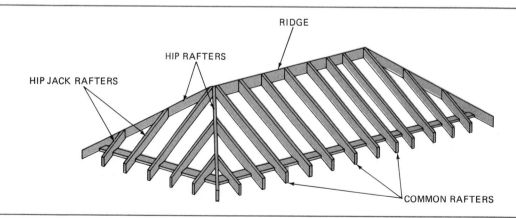

Fig. 15-1 Members of a hip roof frame

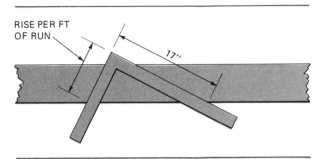

Fig. 15-2 The unit of run of the hip
rafter is 17 inches.

same number of steps (units of run), the unit of run for the hip rafter is increased. Therefore, the diagonal of a 12-inch square is used as the unit of run for the hip rafter; this is 16.97 inches or, for all practical purposes, 17 inches, Figure 15-2.

For example, if the pitch of a hip roof is an 8-inch rise per foot of run, the common rafter is laid out by holding the square at 8 and 12, but the hip rafter would be laid out by holding the square at 8 and 17 and taking the same number of steps for each rafter. The number of steps taken is equal to half the span of the building in feet.

ESTIMATING THE ROUGH LENGTH OF THE HIP RAFTER

Let the tongue of the square represent the common rafter length and the blade of the square represent half the span (total run of the common rafter). Measure across the square at these two points using a scale of 1 inch equals 1 foot, Figure 15-3.

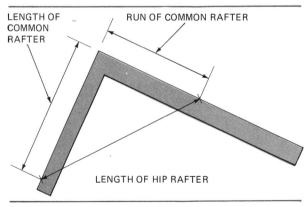

LENGTH OF COMMON RAFTER

RUN OF COMMON RAFTER

LENGTH OF HIP RAFTER

Fig. 15-3 Scaling across the framing square to determine the rough length of the hip rafter

Example: The common rafter length is 15 feet, and the run is 12 feet. A measurement across these points scales off to 19'-2 1/2". It will be necessary to order 20-foot lengths for the hip rafters unless extra is needed for overhang.

LAYING OUT THE HIP RAFTER

The ridge cut of a hip rafter is a compound angle; that is, the cut is made at an angle through its thickness and also across its side. These cuts on rafters are called *cheek cuts*. The hip jack rafter also is cut in this manner, Figure 15-4. A single cheek cut or a double cheek cut may be made on the hip rafter at the ridge according to the way it is framed.

To make the ridge cut, lay the rafter on its side so the crowned edge will be its top edge. Hold the square for the rise of the roof per foot of run on the tongue and 17 inches on the blade and mark along the tongue. Measure back at right angles from the plumb line, one-half the thickness of the rafter stock, and lay out another plumb line. Square both of these lines across the top edge of the stock. Draw a centerline on the top edge. Connect a diagonal line through the center. The direction of the diagonal line depends on the position of the rafter in the roof.

THE DOUBLE CHEEK CUT

The double cheek cut is laid out in a similar manner to that used for single cheek cuts. Measure at right angles to the plumb line one-half

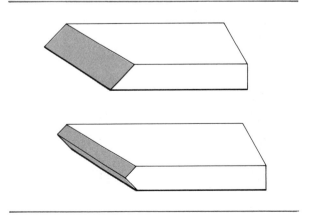

Fig. 15-4 Single and double cheek cuts

the thickness of the stock. Square both lines across the top edge of the rafter. Draw a centerline on the top edge. Connect the diagonals from centerline to plumb line, Figure 15-5. This method of laying out cheek cuts gives accurate results regardless of the pitch of the roof.

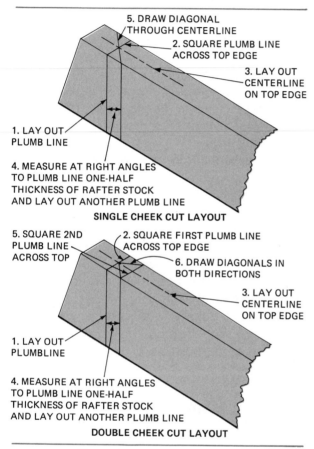

Fig. 15-5 Laying out single and double cheek cuts

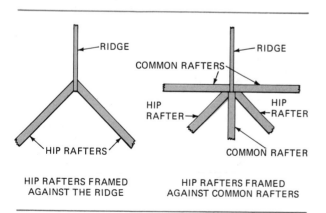

Fig. 15-6 Methods of framing the hip rafter at the ridge

SHORTENING THE HIP RAFTER

The amount the hip rafter is shortened depends on how it is framed at the ridge. Two methods are used. In one method, the hip rafter is framed against the ridge. In another method, the hip rafter is framed against the common rafters, Figure 15-6.

When framed against the ridge, the hip rafter is shortened by one-half the 45° thickness of the ridge board. When framed against common rafters, the hip is shortened one-half the 45° thickness of the common rafter.

Remember that all measurements for shortening must be taken at right angles to the plumb line. See Figure 15-7 on how to shorten the hip rafter for both methods of framing.

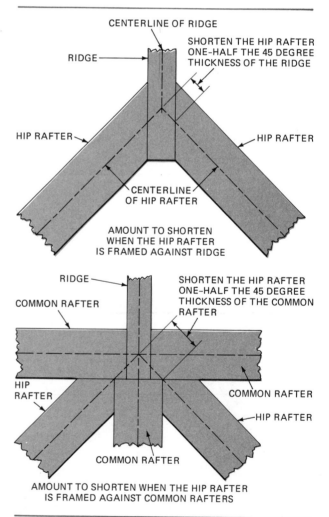

Fig. 15-7 Amount to shorten the hip rafter for both methods of framing

LAYING OUT
THE HIP RAFTER LENGTH

The length of the hip rafter is laid out in a manner similar to that used for the common rafter and may be found either by the step-off method or by the rafter tables. Remember to always start any layout for length from the original plumb line before any shortening is done.

In the step-off method, the number of steps for the hip rafter is the same as for the common rafter in the same roof. The same rise is used, but the unit of run for the hip is 17. For example, for a roof with a rise of 6 inches per foot of run, the square is held at 6 and 12 for the common rafter and 6 and 17 for the hip rafter. If the total run contains a fractional part of a foot, the same fractional part of 17 must be used to determine the extra length of the hip rafter.

Example: If the total run of the roof is 15'-8", step off 15 times using 17 as the unit of run. The 8 inches is 2/3 foot. Two-thirds of 17 is 11 4/12 inches. Measure this dimension along the blade of the square when held at the last step. One set of graduations on the framing square is divided into twelfths of inches. Mark it on the stock and move the square until the tongue lines up with the mark. Mark along the tongue for the plumb cut in line with the outside edge of the plate, Figure 15-8.

USING RAFTER TABLES TO
FIND HIP RAFTER LENGTH

Finding the length of the hip using the tables found on the framing square is similar to finding the length of the common rafter. However, the figures from the second line are used. These figures give the length of the hip rafter in inches for every foot of run. It is only necessary to multiply this figure by the number of feet of run.

Example: Find the length of a hip rafter for a roof with a rise of 8 inches per foot of run with a total run of 14 feet. On the square below the 8-inch mark, it is found that the length of the hip rafter per foot of run is 18.76 inches.

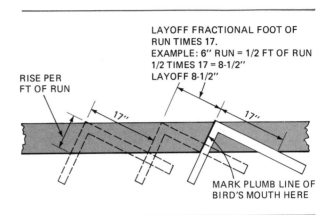

Fig. 15-8 Stepping-off a fractional part of a foot of run on the hip rafter

This multiplied by 14 equals 262.64 inches. Dividing by 12 gives 21.87 feet. Multiplying 0.87 foot by 12 gives 10.44 inches. Change 0.44 inch to 7/16 inch by multiplying by 16. The total length of the hip rafter, then, is 21'-10 7/16".

If the total run contains a fractional part of a foot, multiply the figure in the tables by the whole number of feet plus the fractional part. Mark off the length obtained on the top edge of the hip rafter stock from the original plumb line at the ridge before any shortening is done.

THE HIP RAFTER BIRD'S MOUTH

Like the common rafter, the bird's mouth of the hip rafter is a combination of plumb and level cuts. The last plumb line made when laying out the rafter length is the plumb cut for the seat of the rafter. When making the level cut of the seat, consideration must be given to dropping the hip.

DROPPING THE HIP RAFTER

Because the hip rafter is at the intersection of the side and end slopes of the roof, the top outside corners project above the plane of the roof. These corners may be beveled off (called *backing the hip*), Figure 15-9. Or, the level part of the bird's mouth may be cut higher.

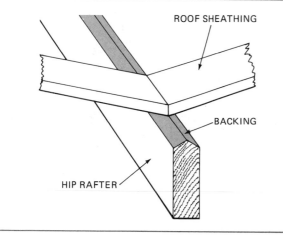

Fig. 15-9 "Backing" the hip

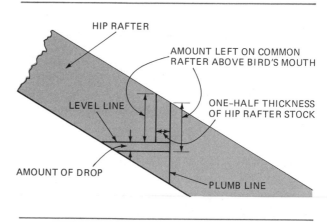

Fig. 15-10 Layout to determine the amount to drop the hip rafter

Dropping the hip is much easier and is the most frequently used process. Measure down from the top edge of the rafter along the plumb line of the bird's mouth a distance equal to the amount of stock left on the common rafter above the plate. Draw a level line at this point. From the plumb line, measure — at right angles toward the ridge end of the rafter — one-half the thickness of the rafter stock. Draw another plumb line. Measure down from the top edge along this line the same distance as measured down previously. Draw another level line at this point. The right angle distance between the two level lines is the amount of drop. The level cut of the bird's mouth is made at the top level line, Figure 15-10.

If it is preferred to back the hip, the level cut at the bird's mouth is made at the bottom level line. The amount of backing is the same as the amount of drop.

OVERHANG OF THE HIP RAFTER

If it is desired to lay out the overhang of the hip rafter, use the same number of steps as used when laying out the overhang of the common rafter. Use 17 as a unit of run. The tail cut is usually a double cheek cut.

THE HIP JACK RAFTER

The *hip jack rafter* runs at right angles from the plate to the hip rafter. Its bird's mouth and

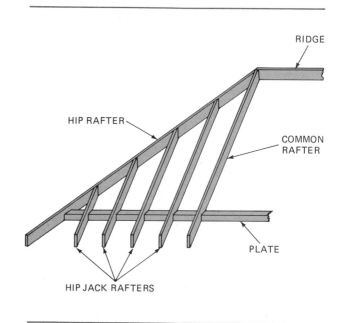

Fig. 15-11 Hip jack rafters are shortened common rafters.

overhang are the same as for the common rafter. It is actually a shortened common rafter and has the same unit of run, Figure 15-11.

FINDING THE LENGTH OF HIP JACK RAFTERS

Hip jack rafters are framed in pairs against the side of the hip rafter. Each set is shorter than the next set by the same amount. This is

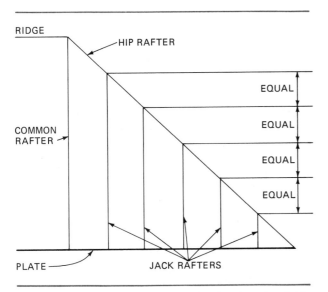

Fig. 15-12 Common difference in the length of hip jack rafters

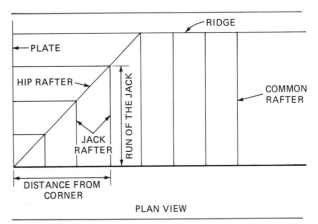

Fig. 15-13 The total run of any hip jack rafter is equal to its distance from the corner of the building.

called the common difference, Figure 15-12. This difference is found in the rafter tables on the framing square for jacks 16 and 24 inches O.C.

Once the length of the longest jack is found, the length of all others can be found by making each set shorter by the common difference. In order to find the length of the longest jack, its total run must be known.

FINDING THE TOTAL RUN OF THE LONGEST JACK

The total run of any pair of jack rafters and the outside edge of the plate form a square. Since all sides of a square are equal, the total run of any hip jack rafter is equal to its distance from the corner of the building. At the plate, measure from the corner of the building to the centerline of the longest jack to find its total run, Figure 15-13.

Find the length of the jack by stepping-off or by using the rafter tables in the same way as for common rafters except that the total run of the jack as found above is used to determine the number of steps. Lay out the bird's mouth and overhang exactly the same as you did the common rafter.

SHORTENING THE HIP JACK RAFTER

Since the hip jack rafter meets the hip rafter at a 45° angle, it must be shortened

one-half the 45° angle thickness of the hip rafter stock. Measure the distance at right angles to the original plumb line and draw another plumb line.

MAKING THE CHEEK CUT

The hip jack rafter has a single cheek cut. Square the shortened plumb line across the top of the rafter. Draw a centerline along the top edge. Measure back at right angles from the last plumb line a distance equal to one-half the jack rafter stock and lay out another plumb line. On the top edge, draw a diagonal from the last plumb line through the intersection of the center and the second plumb line, Figure 15-14. The direction on the diagonal depends on which side of the hip rafter it is framed.

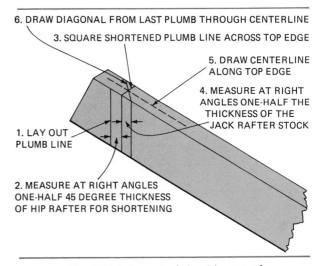

6. DRAW DIAGONAL FROM LAST PLUMB THROUGH CENTERLINE

3. SQUARE SHORTENED PLUMB LINE ACROSS TOP EDGE

5. DRAW CENTERLINE ALONG TOP EDGE

4. MEASURE AT RIGHT ANGLES ONE-HALF THE THICKNESS OF THE JACK RAFTER STOCK

1. LAY OUT PLUMB LINE

2. MEASURE AT RIGHT ANGLES ONE-HALF 45 DEGREE THICKNESS OF HIP RAFTER FOR SHORTENING

Fig. 15-14 Layout of the ridge cut of the hip jack rafter

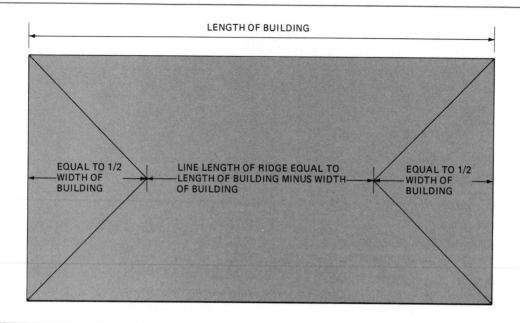

Fig. 15-15 Determining the line length of the hip roof ridge

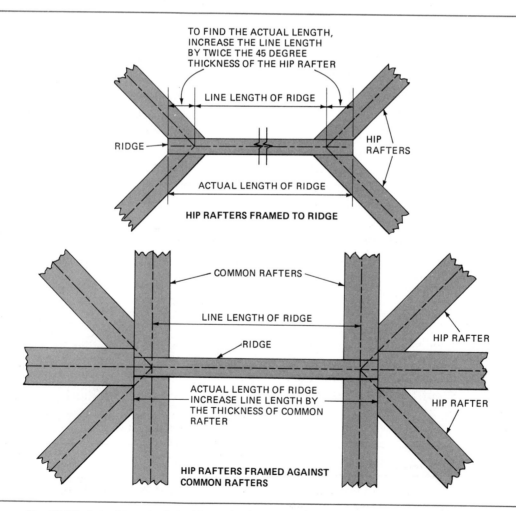

Fig. 15-16 Actual length of the hip roof ridge for both methods of framing the hip rafter

Lay out a pattern for each different hip jack length. Make each pattern shorter than the first by the common difference found in the rafter tables. The patterns are then used to cut all the jack rafters necessary to frame the roof.

FINDING THE LENGTH OF THE HIP ROOF RIDGE

The total run of the hip rafter is the diagonal of a square whose sides are equal to the run of the common rafter. Since the run of the common rafter is equal to one-half the width of the building, the line length of the ridge is found by subtracting the width of the building from the length of the building, Figure 15-15.

The actual length of the hip roof ridge must be cut longer than the line length, however. The amount of increase depends on the construction. If the hip rafters are framed against the ridge, increase the length at *each* end by one-half the 45° thickness of the hip rafter. If the hip rafters are framed against the common rafters, the line length of the ridge is increased at *each* end one-half the thickness of the common rafter stock, Figure 15-16.

RAISING THE HIP ROOF

Lay out the ridge board in a manner similar to the common roof. Erect the common roof section in a manner described previously. Brace this section temporarily. Next, install the hip rafters by placing them diagonally to the first. When the hip rafters are in position, they also act as permanent braces for the roof frame.

Fasten jack rafters to the plate and to hip rafters in pairs. As each pair of jacks is fastened, sight the hip rafter along its length and keep it straight. Any bow in the hip is straightened by driving the jack a little tighter against the bow side of the hip as the roof is framed.

THE INTERSECTING ROOF

Buildings of irregular shape require a roof for each section. These roofs meet at an intersection and are called *intersecting roofs*. The roof may be a gable, a hip, or a combination of both. Because of the intersection, several special kinds of rafters are required. Figure 15-17 illustrates these rafters which are also described in the following material.

- *Valley rafters* form the intersection of the slopes of two roofs. If the heights of both roofs are different, two kinds of valley rafters are required. The *supporting valley rafter* runs from the plate to the ridge of the main roof. The *shortened valley rafter* runs from the plate to the supporting valley rafter.

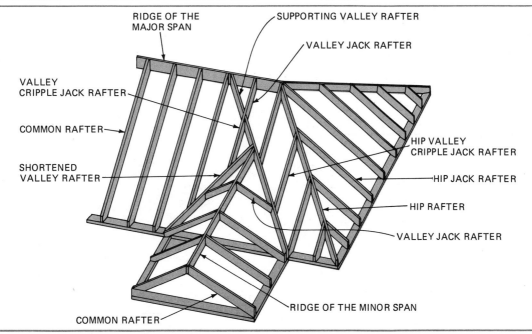

Fig. 15-17 Members of the intersecting roof frame

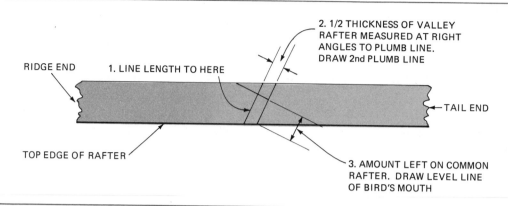

Fig. 15-18 Bird's mouth of the valley rafter

- *Valley jack rafters* run from the ridge to the valley rafter.

- The *valley cripple jack rafter* runs between the supporting and shortened valley rafter.

- The *hip-valley cripple jack rafter* runs between a hip rafter and a valley rafter.

- Hip and valley rafters run at a 45° angle to the plate. All jack rafters run at a 90° angle to the plate.

LAYOUT OF SUPPORTING VALLEY RAFTERS

The layout of valley rafters is similar to that of hip rafters. The unit of run for both is 17 inches. The total run of the supporting valley is the run of the main roof. Its length is found either by the step-off method or by the rafter tables in the same manner as for hip rafters. A single cheek cut is made at the ridge. The rafter is shortened by one-half the 45° thickness of the ridge board.

BIRD'S MOUTH OF THE VALLEY RAFTER

From the plumb line at the plate, measure toward the tail end of the rafter a distance equal to one-half the thickness of the rafter stock, and lay out another plumb line. Measure down, along the last plumb line from the top edge of the rafter, a distance equal to the amount left on the common rafter. Draw a level line at this point, Figure 15-18. The valley rafter is not dropped like the hip because its top corners do not project above the slope of the roof. However, one edge above the shortened valley needs backing.

OVERHANG AND TAIL CUT OF THE VALLEY RAFTER

The length of the valley rafter overhang is found in the same manner as that used for the hip rafter. Step off the same number of times from the plumb cut in line with the outside of the plate as stepped off for the common rafter. However, a unit run of 17 is used instead of 12.

The tail cut of the valley rafter is a double cheek cut, but is exactly the opposite of the tail cut of the hip rafter. From the last plumb line, measure toward the tail a distance that is one-half the thickness of the rafter stock. Lay out another plumb line. Square both plumb lines across the top of the rafter, and draw diagonals from the center, Figure 15-19.

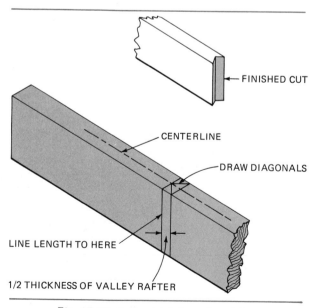

Fig. 15-19 Tail cut of the valley rafter

LAYOUT OF THE SHORTENED VALLEY

The length of the shortened valley is found by using the run of the minor roof. Its bird's mouth, overhang, and tail cut are laid out the same as for the supporting valley.

However, the plumb cut at the ridge is different from that of the supporting valley. Because the two valley rafters meet at right angles, the cut through the thickness of the shortened valley is a square cut. To shorten this rafter, measure back at right angles from the plumb line at the ridge a distance that is one-half the thick-

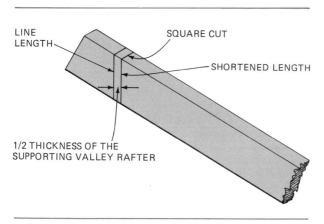

Fig. 15-20 Plumb cut of the shortened valley rafter

ness of the supporting valley rafter stock, and lay out another plumb line, Figure 15-20.

LAYOUT OF THE VALLEY JACK

The length of the valley jack can be found if its total run is known. The total run of any valley jack rafter is equal to the run of the common rafter minus the horizontal distance that it is located in from the corner of the building, Figure 15-21. Remember that all jack rafters are shortened common rafters and that the unit of run is 12.

The ridge cut of the valley jack is the same as a common rafter and is shortened in the same way. The cheek cut against the valley rafter is a single cheek cut. The valley jack is shortened at this end by one-half the 45° angle thickness of the valley rafter stock. The layout of all other valley jack rafters is made by making each set shorter by the common difference found in the rafter tables.

LAYOUT OF THE VALLEY CRIPPLE JACK

As stated before, the length of any rafter can be found, if its total run is known. The run

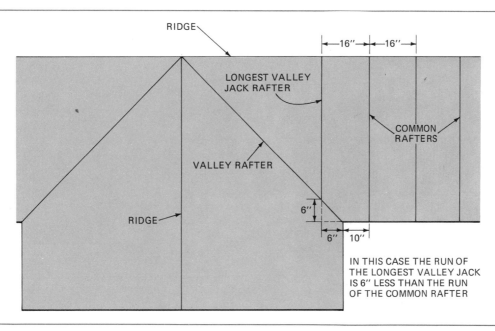

Fig. 15-21 Determining the run of the longest valley jack rafter

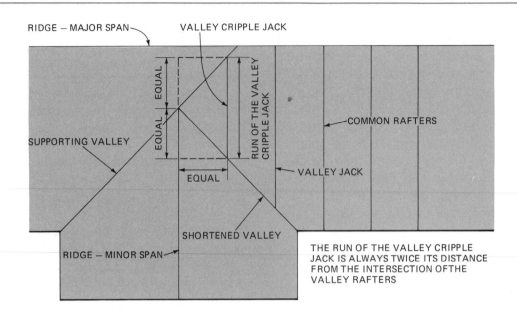

RIDGE — MAJOR SPAN

VALLEY CRIPPLE JACK

EQUAL EQUAL

EQUAL

RUN OF THE VALLEY CRIPPLE JACK

COMMON RAFTERS

SUPPORTING VALLEY

EQUAL

VALLEY JACK

SHORTENED VALLEY

RIDGE — MINOR SPAN

THE RUN OF THE VALLEY CRIPPLE JACK IS ALWAYS TWICE ITS DISTANCE FROM THE INTERSECTION OF THE VALLEY RAFTERS

Fig. 15-22 Determining the run of the valley cripple jack rafter

of the valley cripple jack is always twice its horizontal distance from the intersection of the valley rafters, Figure 15-22. Use the common rafter tables, or step-off in a manner similar to that used for common rafters to find its length. On each end, shorten one-half the 45° thickness of the valley rafter stock, Figure 15-23, and make a single cheek cut.

HIP-VALLEY CRIPPLE JACK RAFTER LAYOUT

Determine the length of the rafter by first finding its total run. The run of a hip-valley cripple jack rafter is equal to the plate line distance between the bird's mouths of the hip and valley rafters, Figure 15-24. Lay out the

DRAW DIAGONAL THROUGH CENTERLINE

LINE LENGTH

ONE HALF THE THICKNESS OF ITSELF

DRAW DIAGONAL THROUGH CENTERLINE

1/2 THE 45 DEGREE THICKNESS OF THE SUPPORTING VALLEY

1/2 THE 45 DEGREE THICKNESS OF THE SHORTENED VALLEY

Fig. 15-23 Layout of the valley cripple jack rafter

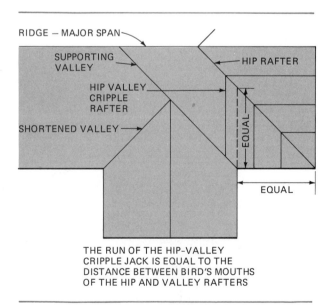

RIDGE — MAJOR SPAN

SUPPORTING VALLEY

HIP RAFTER

HIP VALLEY CRIPPLE RAFTER

EQUAL

SHORTENED VALLEY

EQUAL

THE RUN OF THE HIP-VALLEY CRIPPLE JACK IS EQUAL TO THE DISTANCE BETWEEN BIRD'S MOUTHS OF THE HIP AND VALLEY RAFTERS

Fig. 15-24 Finding the total run of the hip-valley cripple jack rafter

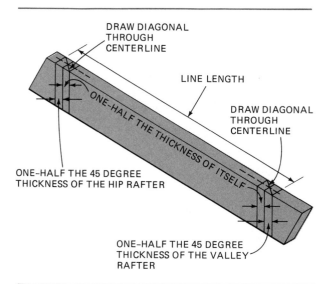

Fig. 15-25 Layout of the hip-valley cripple jack rafter

length using the cut of the common rafter. All hip-valley cripple jacks cut between the same hip and valley rafters are the same length. On each end, shorten by one-half the 45° thickness of the hip and valley rafters, and make single cheek cuts, Figure 15-25.

RIDGE LENGTH OF THE WING

When the ridge heights of both roofs are the same, the line length of the ridge of the wing is equal to the building projection plus one-half the span of the main roof, Figure 15-26. When ridge heights are different, the line length of the ridge is equal to the building projection plus one-half the span of the minor roof, Figure 15-27. The actual length of the ridge in both cases is shorter than the line length. However, it is customary to make the intersecting ridge longer than needed. The end rafters are located by plumbing up from the plate. The surplus end of the ridge is cut off after the roof is framed.

The end of the ridge at the intersection of the valleys is simply a square cut. It is not necessary to make the ridge fit perfectly into the intersection.

ERECTING THE INTERSECTING ROOF

The intersecting roof is raised by framing opposing members of the main span first. Then the valley rafters are installed. To prevent

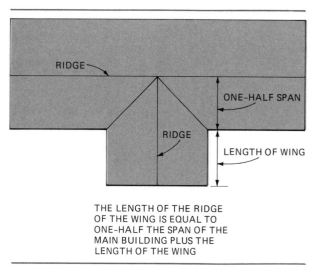

Fig. 15-26 Finding the line length of the ridge of the wing when ridge heights are the same

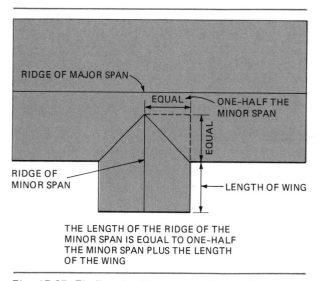

Fig. 15-27 Finding the line length of the ridge of the minor span when ridge heights are different

bowing the ridge of the main span, install rafters to oppose the valley rafters. Install common and jacks in sets opposing each other, and sight members of the roof as framing progresses, keeping all members in a straight line.

ROOF SHEATHING

Roof sheathing is applied after the roof frame is complete. Sheathing gives rigidity to the roof frame and provides a nailing base for

Fig. 15-28 Plank decking is used on plank and beam construction. *(Courtesy of Georgia Pacific)*

the roof covering. Solid lumber or plywood is commonly used to sheath roofs, although many kinds of composition boards are also used.

LUMBER SHEATHING

Lumber sheathing is usually square-edge or matched 1x8 or 1x10 boards. They are laid close together when a solid and continuous base is needed, such as for asphalt shingles. The boards may be spaced when the roof covering is wood shingles or shakes. Spaced sheathing is usually 1x4 strips with the on center spacing the same as the shingle exposure. Joints between boards must be made on a rafter and staggered. Each board must bear on at least two rafters.

Plank decking is used on post and beam construction where the roof supports are spaced farther apart. Plank decking may be of 2-inch nominal thickness or greater depending on the span. Usually both edges and ends are matched. The plank roof often serves as the finish ceiling for the rooms below, Figure 15-28.

PLYWOOD SHEATHING

Plywood roof sheathing is laid with the face grain running across the rafters for greater strength. End joints are made on the rafters and staggered. Nails are spaced 6 inches apart on the ends and 12 inches apart on intermediate supports, Figure 15-29. Adequate blocking,

tongue and grooved edges, or other suitable edge support such as Plyclips must be provided, when spans exceed the indicated value of the plywood roof sheathing. *Plyclips* are small metal pieces shaped like a capital I and are used between adjoining edges of the plywood between rafters, Figure 15-30. Two Plyclips are used for 48-inch or greater spans. One is used for lesser spans.

The ends of sheathing are allowed to run "wild" at the rakes until sheathing is completed. A chalk line is snapped and the sheathing ends are trimmed to the line with a portable electric circular saw.

Fig. 15-29 Sheathing a roof with plywood

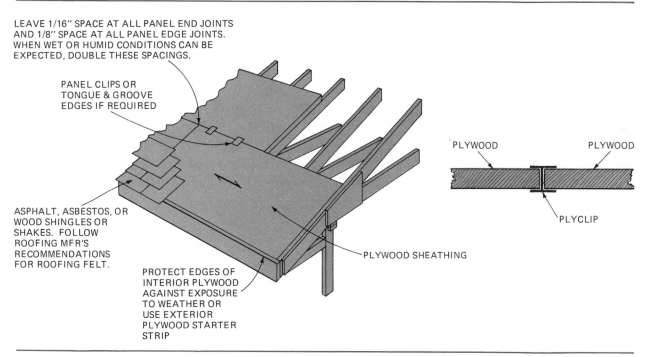

LEAVE 1/16" SPACE AT ALL PANEL END JOINTS
AND 1/8" SPACE AT ALL PANEL EDGE JOINTS.
WHEN WET OR HUMID CONDITIONS CAN BE
EXPECTED, DOUBLE THESE SPACINGS.

PANEL CLIPS OR
TONGUE & GROOVE
EDGES IF REQUIRED

ASPHALT, ASBESTOS, OR
WOOD SHINGLES OR
SHAKES. FOLLOW
ROOFING MFR'S
RECOMMENDATIONS
FOR ROOFING FELT.

PROTECT EDGES OF
INTERIOR PLYWOOD
AGAINST EXPOSURE
TO WEATHER OR
USE EXTERIOR
PLYWOOD STARTER
STRIP

PLYWOOD SHEATHING

PLYWOOD   PLYWOOD

PLYCLIP

Fig. 15-30 Plyclips can be used to support the edges of the plywood sheathing

REVIEW QUESTIONS

Select the most appropriate answer.

1. The unit of run of hip and valley rafters is
   a. 12 inches.   c. 17 inches.
   b. 16 inches.   d. 18 inches.

2. When the hip rafter is framed against the ridge, the amount of shortening is
   a. the thickness of the ridge.
   b. one-half the 90° thickness of the ridge.
   c. one-half the 45° thickness of the ridge.
   d. one-half the 45° thickness of the common rafter.

3. How much is laid out on the blade of the square for 6 inches of run when stepping-off a hip rafter?
   a. 6 inches   c. 8 1/2 inches
   b. 6/17 foot   d. 17 inches minus 6 inches

4. Using the rafter tables, the line length of a hip rafter with a rise of 6 inches per foot of run and a total run of 12 feet is
   a. 16'-6".   c. 18'-0".
   b. 17'-4".   d. 18'-8".

5. The unit of run of any jack rafter is
   a. 12 inches.
   b. 16 inches.
   c. 17 inches.
   d. the same as hip and valley rafters.

6. The length of any rafter in a roof of specified pitch can be found if
    a. its total rise is known.
    b. its total run is known.
    c. the rise per foot of run is known.
    d. its length per foot of run is known.

7. The total run of any hip jack rafter is equal to
    a. its distance from the corner.
    b. one-third of its total rise.
    c. its length minus the total rise.
    d. its common difference in length.

8. The total run of the shortened valley rafter is equal to
    a. the total run of the minor span.
    b. the total run of the major span.
    c. the total run of the common rafter.
    d. the total run of the hip rafter.

9. The total run of the hip-valley cripple jack rafter is equal to
    a. one-half the run of the hip jack rafter.
    b. one-half the run of the longest valley jack rafter.
    c. the distance between bird's mouths of the hip and valley rafters.
    d. the difference in run between the supporting and shortened valley rafters.

10. The line length of the ridge of a hip roof is
    a. the length of the building minus its width.
    b. two times the width of the building.
    c. the width of the building minus its length.
    d. one-half the length of the building.

STAIR FRAMING 16

OBJECTIVES

After completing this unit, the student will be able to

- *describe several stairway designs.*
- *define terms used in stair framing.*
- *determine riser height and tread run of a stairway.*
- *determine the length of and frame a stairwell.*
- *lay out a stair carriage and frame a straight stairway.*
- *lay out and frame a stairway with a landing.*
- *lay out and frame a stairway with winders.*
- *lay out and frame basement stairs.*

Stairs must be constructed so they are comfortable to ascend and safe to descend. A great degree of skill is necessary for their design and construction. The carpenter must know how to lay out and erect well-designed staircases that are comfortable, safe, and of first-class workmanship.

STAIRWAY DESIGN

Stairs, stairway and *staircase* are terms used to designate one or more flights of steps leading from one level of a structure to another. Stairs are further defined as *main,* or *finish,* stairs; *basement* stairs; and *service* stairs. Finish stairs extend from one habitable level of a house to another. Basement and service stairs extend from a habitable to a nonhabitable level. Stairways in residential construction should not be less than 36 inches wide, and preferably wider, to allow the passage of two persons at a time and for the moving of furniture, Figure 16-1.

Fig. 16-1 Recommended stair widths

A *straight* stairway is continuous from one floor to another without any turns or landings. *Platform* stairs have intermediate landings between floors. Platform-type stairs usually change direction at the landing. An L-type stairway changes direction 90 degrees. A stairway that changes direction 180 degrees is called a U-type stairway. Platform stairs are installed to reduce the length of space occupied by the stairs. They also provide a temporary resting place while ascending or descending and are a safety feature in case of falls.

A *winding* staircase gradually changes direction as it ascends from one floor to another. In many cases, only a part of the staircase winds to change direction 90 degrees. Winding stairs solve the problem of a short run, but their use is not recommended. Besides being relatively complex to construct, they pose a danger because of their tapered treads.

Stairways entirely walled-in on both sides are called *closed* stairways. Closed stairways are more economical to build, but add little charm or dignity to the house.

Stairways that have one or both sides open to a room are called *open* stairways. Open stairways require a *balustrade* (guardrail) which lends beauty and grace to the staircase and the house, generally affecting the character of the entire interior.

A staircase may be open and closed in the same flight. The flight may be open for partway and closed the remainder. Also, one side of the staircase may be closed while the other side is open for all or part of the flight. Figure 16-2 illustrates some stairway designs.

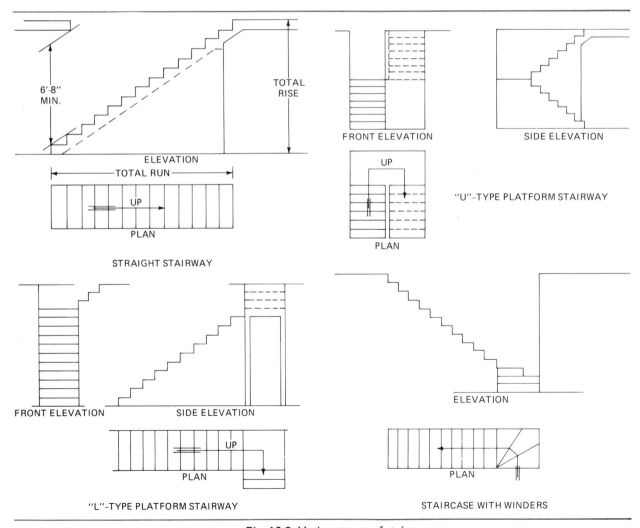

Fig. 16-2 Various types of stairways

272

STAIR FRAMING TERMS

The terms used in stair framing are defined in the following material. Figure 16-3 illustrates the relationship of the various terms to each other and to the total staircase.

- The *total rise* of a stairway is the vertical distance between finish floors.
- The *total run* is the total horizontal distance which the stairway covers.
- A *rise* is the vertical distance from one step to another. A *riser* is the finish material that covers this distance.
- A *tread* is the horizontal member on which the feet are placed when climbing or descending the stairs. The *tread run* is the horizontal distance between the faces of the risers. The *nosing* is that part of the tread that extends beyond the face of the riser.
- A *stair carriage*, sometimes called a *stair horse* or *rough stringer*, is usually a nominal

2x10 framing member which has been cut out to receive the treads and risers.

- A *stairwell* is an opening in the floor for the stairway to pass through and to provide adequate headroom for persons using the stairs.
- The *finish stringers*, also called *skirts*, are the finished sides of a staircase. When placed against a wall, they are called *closed stringers*. On the open side of a staircase, they are called *open stringers* or *mitered stringers*.

Finish stringers support the treads and risers in one method of construction. In this case, stair carriages are not used, and the stringers are grooved and dadoed to receive the treads and risers. Stringers milled in this manner are called *housed stringers*. The layout for housing stringers and for cutting stair carriages is made in a similar manner.

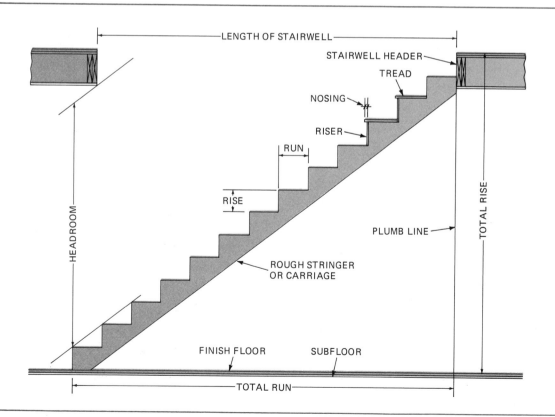

Fig. 16-3 Stair framing terms

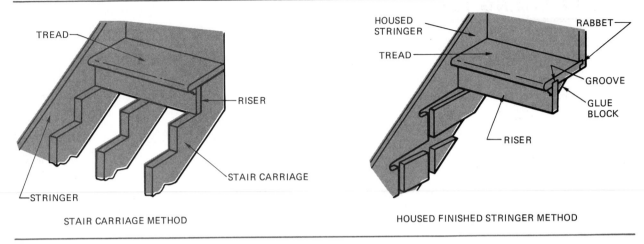

STAIR CARRIAGE METHOD

HOUSED FINISHED STRINGER METHOD

Fig. 16-4 Two principal methods of stair construction

METHODS OF STAIR CONSTRUCTION

Two principal methods of stair construction are used: the stair carriage and housed finish stringer methods, Figure 16-4. Some areas use one method only; others use both.

Carriages or rough stringers are not required when housed finish stringers are used. The two housed finish stringers are sole supports of the stairs. Generally, the housed stringer staircase is prefabricated in the mill. The framing crew frames only the stairwell, and the stairs are installed when the house is ready for finishing.

Sometimes, the finish carpenter builds a housed stringer staircase on the job site by using a router and jig to house the stringers. Then the treads, risers, and other stair parts are cut to size and assembled.

When stair carriages are used, they are installed when the structure is being framed. Rough treads are installed for easy access to upper levels until the stairs are ready for finishing.

DETERMINING THE RISE AND RUN

It is very important that staircases be constructed at a proper angle for maximum ease in climbing and for a safe descent. The relationship of the rise and run determine this angle, Figure 16-5. The preferred angle is between 30° and 35°.

Although building codes differ, many specify that the height of a riser shall not exceed

7 3/4 inches. No codes specify a minimum height riser. However, stairs with a slope of less than 20° become awkward to climb and, therefore, dangerous, besides wasting much valuable space in their run. Stairs with a slope that is excessively steep are difficult to climb and dangerous to descend.

To determine the riser height, the total rise of the stairway must be known. This height is divided by 8 inches. The result is usually a whole number and a fraction. The next largest whole number is the total number of risers required.

For example, a total rise of 9'-0'' (108 inches) is divided by 8 inches. The result is

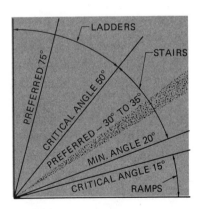

Fig. 16-5 Recommended angles for stairs

13 1/2. The next largest whole number is 14, which represents the number of risers required.

Dividing the total rise by the number of risers gives the height of each riser. For example, dividing the total rise (108 inches) by the number of risers (14) gives a riser height of 7 5/8 inches.

If it is desired to decrease the riser height, increase the number of risers by one and divide the total rise by that number. For example, increasing 14 by one and dividing 108 inches by 15 equals slightly over 7 3/16 inches.

Another method of determining the riser height is by the use of a story rod. Stand the rod, usually a 1x2, vertically on the floor and mark on it the finish floor to finish floor height (total rise of the stairs). Set the dividers at a comfortable rise and divide the rod into equal spaces, adjusting the dividers as necessary.

Once the riser height is known, the tread width can be determined. The tread width is measured from the face of one riser to the next and does not include the nosing. To find the tread width apply the following rule: The sum of one riser and one tread should equal between 17 and 18. For example, if the riser height is 7 5/8 inches, then the minimum tread width may be 17 inches minus 7 5/8 inches, which equals 9 3/8 inches. The maximum tread width may be 18 inches minus 7 5/8 inches, which equals 10 3/8 inches.

Another formula to find the tread width, found in many building codes, states that the sum of two risers and one tread shall not be less than 24 inches nor more than 25 inches. With this formula, a rise of 7 5/8 inches calls for a minimum tread width of 8 3/4 inches and a maximum of 9 3/4 inches. However, many codes specify a minimum tread width of 9 inches.

Remember that decreasing the riser height increases the run of the stairs using up more space. Increasing the riser height decreases the run but makes the stairs steeper and more difficult to climb. The carpenter must use good judgment and adapt the rise and run dimensions to the space in which the stairway is to be constructed in conformance with the building code. A riser height of 7 1/2 inches and a tread run of 10 inches makes a safe, comfortable stairway.

FRAMING A STAIRWELL

A stairwell is framed in the same manner as any large floor opening. Unit 10 Floor Framing describes methods of framing floor openings. Several methods of framing stairwells are illustrated in Figure 16-6.

The width of the stairwell depends on the width of the staircase. Additional width must be provided for an open staircase to allow the finish stringers and balustrade to be returned around the landing above. Stair finish is described later in Unit 28.

The stairwell must be long enough to provide adequate headroom and depends upon the slope of the stairs. Headroom is the vertical distance measured from the outside corner of a tread and riser to the ceiling above. Most building codes require a minimum of 6'-8'' for headroom. However, 7'-0'' headroom is preferred.

To find the length of the stairwell, add the thickness of the floor above (including subfloor, floor joists, furring, and ceiling finish) to the desired headroom and divide by the riser height. The result will probably be a whole number and a fraction. Use the next largest whole number and multiply by the tread width to find the length of the stairwell.

For example, a stairway has a riser height of 7 3/4 inches and a tread width of 10 inches. The total thickness of the floor above is 11 3/4 inches, which when added to the desired headroom of 84 inches equals 95 3/4 inches. Dividing this total by 7 3/4 inches equals 12.4. Changing 12.4 to 13 and multiplying by 10 inches equals 130 inches or 10'-10'' (length of the stairwell).

This length is correct if the header of the stairwell acts as the top riser. If the carriage is framed flush with the top of the opening, add another tread width to the length of the stairwell, Figure 16-7.

LAYING OUT A STAIR CARRIAGE

Place the stair carriage stock on a pair of sawhorses. Sight the stock for a crowned edge. Stand on the side of the crowned edge

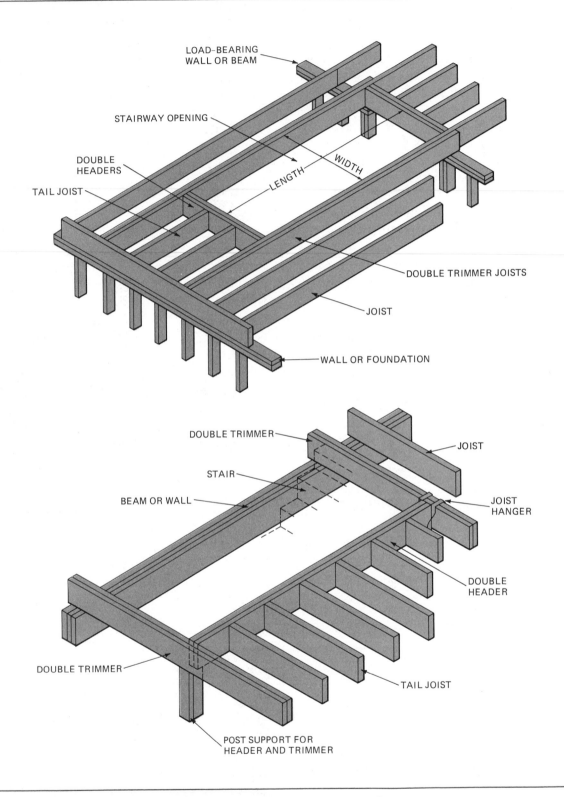

Fig. 16-6 Methods of framing stairwells

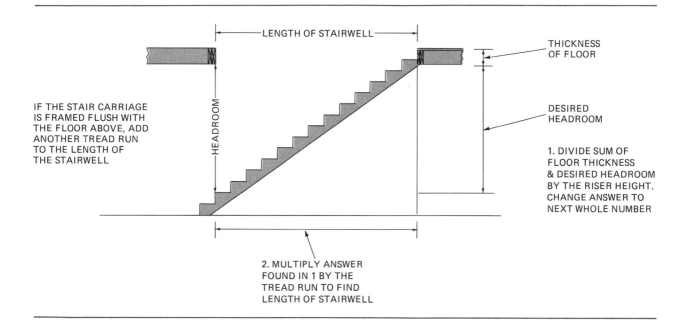

LENGTH OF STAIRWELL

HEADROOM

THICKNESS OF FLOOR

DESIRED HEADROOM

IF THE STAIR CARRIAGE IS FRAMED FLUSH WITH THE FLOOR ABOVE, ADD ANOTHER TREAD RUN TO THE LENGTH OF THE STAIRWELL

1. DIVIDE SUM OF FLOOR THICKNESS & DESIRED HEADROOM BY THE RISER HEIGHT. CHANGE ANSWER TO NEXT WHOLE NUMBER

2. MULTIPLY ANSWER FOUND IN 1 BY THE TREAD RUN TO FIND LENGTH OF STAIRWELL

when laying out. This will be the top edge of the carriage. Set gauges on the framing square with the rise on the tongue and the tread run on the blade or use a pitch block. Step-off the necessary number of times, marking along the outside of the tongue and blade. These lines are the back sides of the risers and treads, Figure 16-8.

EQUALIZING THE BOTTOM RISER

The stair carriage must be laid out so that each step has the same riser height and tread run. If the bottom of the carriage rests on the subfloor and the finish floor and tread stock are the same thickness, the height of the first riser is the same as all the rest.

- If the tread stock is thicker than the finish floor, and the carriage rests on the subfloor, then the first riser height is made less by the difference.

- If the carriage rests on the finish floor, the first riser height is made less by the thickness of the tread stock.

Draw a level line where the carriage rests on the floor. Cutting the bottom end off to equalize the bottom riser is called "dropping" the stair carriage, Figure 16-9.

EQUALIZING THE TOP RISER

If the header of the stairwell acts as the top riser, fasten the carriage a distance down from the top, taking into consideration any difference in the thicknesses of the tread stock and the finish floor above.

For example, if the tread stock is 1 1/16 inches and the finish floor above is 3/4 inch, the difference is 5/16 inch. Fasten the carriage so that the top tread cut is 5/16 inch more than the riser height below the top of the subfloor above, Figure 16-10. It is important that every riser in the staircase be exactly the same height.

Fig. 16-8 Laying out a stair carriage

277

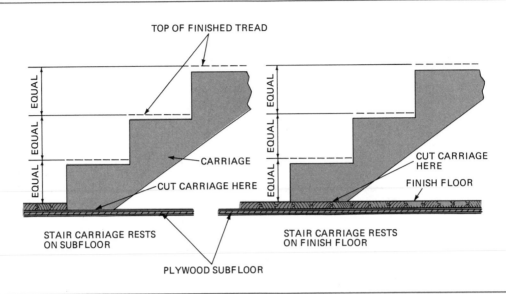

Fig. 16-9 "Dropping" the stair carriage to equalize the first riser

WIDTH OF THE TOP TREAD

The width of the top finished tread must be the same as all others in the staircase. If the carriage is framed against the stairwell header and the same thickness of finish is applied to

the header as the thickness of the riser stock, then the tread cut at the top of the carriage is made the same as all others.

If the carriage is framed to the stairwell in a different manner, allowances must be made so that the top finished tread width will be the

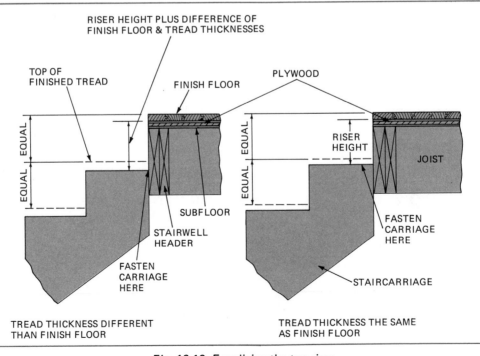

Fig. 16-10 Equalizing the top riser

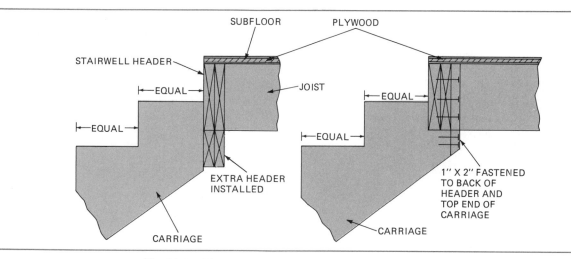

Fig. 16-11 Methods of framing the stair carriage to the stairwell

same as all others in the staircase, Figure 16-11. Draw a plumb line for the correct tread width at the top.

CUTTING THE STAIR CARRIAGES

After the first carriage is laid out, cut it, using a portable electric circular saw and following the layout lines carefully. Finish the cuts at the intersection of the riser and tread with a handsaw.

> **CAUTION: In some cases, when making a cut at a sharp angle to the edge, the guard of the saw may not retract. In this case, retract the guard by hand until the cut is made a few inches into the stock. Then release the guard and continue the cut. Do not wedge the guard in an open position to overcome the difficulty.**

Using the first carriage as a pattern, lay out and cut as many other carriages as needed. The number of carriages depends on the width of the stairway. Carriages are approximately spaced not over 18 inches on center. However, this spacing may be increased depending on such factors as whether risers are used and the thickness of the tread stock.

BUILT-UP STAIR CARRIAGE

In order to conserve materials, the triangular blocks cut out of the stair carriage are fastened to the edge of a 2x4 to form a *built-up stair carriage*, Figure 16-12. The blocks are fastened

by using the first stair carriage as a pattern. Built-up stair carriages are usually used as intermediate carriages in the stairway.

FRAMING A STRAIGHT STAIRWAY

If the stairway is closed or partly closed, the walls must be prepared before the stair carriages are fastened in position.

PREPARING THE WALLS OF THE STAIRCASE

If a wall is to be plastered, a ground must be installed. To find the location of the ground,

Fig. 16-12 A built-up stair carriage

279

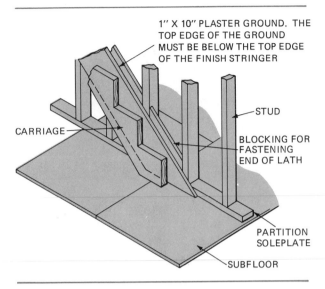

Fig. 16-13 Applying a ground to the wall
of a staircase for plastering

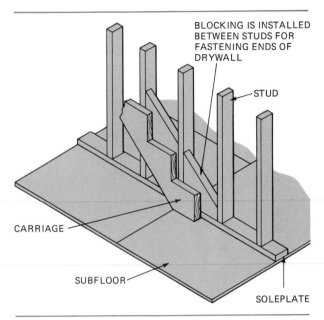

Fig. 16-14 Block installed between studs
for fastening drywall

lay the stair carriage against the wall in its proper position. Check several treads for level. If the carriage is not level, determine the reason and make the necessary adjustments.

Mark the studs in the wall near the top and bottom of the carriage about 1 inch above the outside corner of the tread and riser cut. Remove the carriage and snap a chalk line between the marks. This line must be below the top edge of the finish stringer. Fasten a nominal 1x10 board so its top edge is to the line. Use three 8d common nails into each stud. Nail also into the bottom plate of the wall. Provide blocking at the top, if necessary, Figure 16-13.

If the wall is to be covered with gypsum board (drywall), this is sometimes applied before the carriages are installed. This eliminates the need to cut the drywall around the cutouts of the stair carriage. This method also requires no blocking between the studs.

If the drywall is to be applied after the stairs are framed, blocking is required between studs in back of the stair carriage. Snap a line in the same manner as for plaster grounds. Install 2x6 or 2x8 blocking, on edge, between and flush with the edges of the studs, so their top edge is to the chalk line, Figure 16-14.

Housed staircases may be installed after the walls are finished. However, adequate blocking must be provided for fastening the staircase. Sometimes they are installed before the walls are

finished. In this case, they must be furred out away from the studs to allow the wall finish to extend below the top edge of the finished stringer.

INSTALLING THE STAIR CARRIAGE

When installing the stair carriage, fasten the first carriage in position on one side of the stairway by toenailing it at the top into the stairwell header. Make sure it is fastened so the distance from the subfloor above to the top tread is correct.

Toenail the bottom end of the stair carriage to the subfloor in its proper position. Fasten the carriage to the wall by spiking into the studs. Drive spikes near the bottom edge of the carriage to avoid splitting the triangular sections.

Fasten a second carriage on the other wall in the same manner as the first. If the stairway is to be open on the side, toenail the top and bottom only. The location of the stair carriage on the open end of a stairway is in relation to the position of the handrail. Locate the carriage so that its outside face will be in line with the centerline of the handrail when it is installed.

Fasten intermediate carriages in position by toenailing them at the top into the stairwell

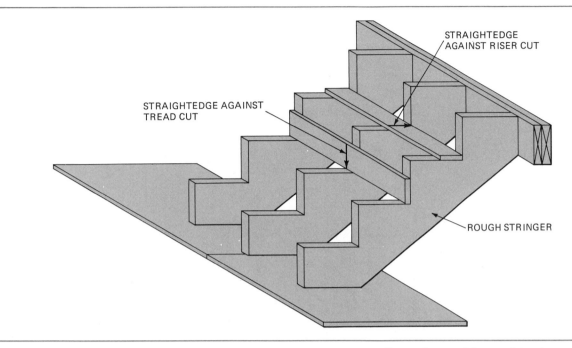

Fig. 16-15 Check the position of tread and riser cuts on intermediate stringers with a straightedge.

header and toenailing them at the bottom into the subfloor. Check their position by testing the tread and riser cuts with a straightedge placed across the outside carriages, Figure 16-15. About halfway up the flight, or as necessary, fasten a riser board to straighten the carriages and tie them together, Figure 16-16.

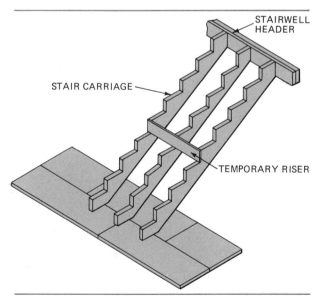

Fig. 16-16 Fasten a temporary riser about halfway up the flight to straighten and tie the stair carriages together.

If a wall is to be framed under the stair carriage at the open side, fasten a bottom plate to the subfloor plumb with the outside face of the carriage. Lay out the studs on the plate. Cut and install studs under the carriage in a manner similar to that used to install gable studs. Be careful to keep the carriage straight and not crown it up in the center, Figure 16-17. Install rough treads on the carriages until the stairway is ready for finishing.

MAKING A HOUSED STRINGER

The method of determining the riser height and tread width for a housed staircase is the same as for any staircase. After these dimensions have been determined, the layout is made on the stringer stock.

A pitch board is often used instead of a framing square for laying out a stair stringer. A pitch board is a piece of stock, usually 3/4 inch thick, cut to the rise and run of the stairs. A strip of wood, usually a 1x2, is fastened to the *rake* edge (long edge) of the pitch board. This is used to hold the pitch board against the edge while laying out the stringer, Figure 16-18.

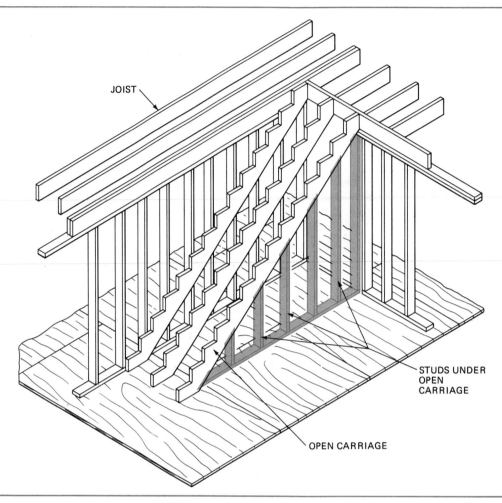

Fig. 16-17 Install studs under the stair carriage on the open side of the stairway.

On the face side of the stringer stock, draw a line parallel to and about 2 inches down from the top edge. This line is the intersection of the tread and riser faces. Using the pitch board, lay out the risers and treads for each step of the staircase. Use a sharp knife to scribe the lines.

These lines show the location of the face side of each riser and tread and are the outside edges of the housing, Figure 16-19.

To mark the inside edges of the housing, wedge-shape templates may be used. The templates are tapered so the wedges will drive

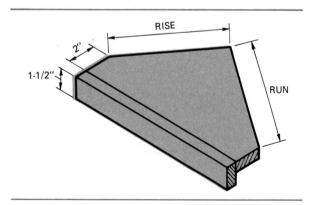

Fig. 16-18 Pitch board for a housed stringer

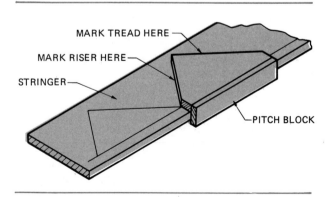

Fig.16-19 Laying out a housed stringer, using a pitch board

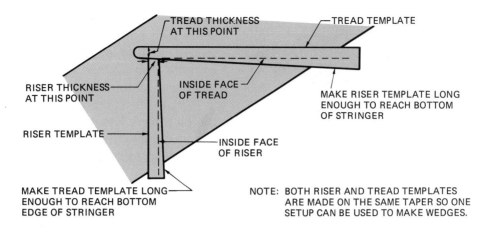

Fig. 16-20 Templates for laying out the inside edges of the housed stringer

the risers and treads tight against the outside edge of the housing. One template is made for the tread housing and one for the riser housing. Both templates are tapered the same to permit the use of stair wedges of the same pitch, Figure 16-20.

To mark the inside edges of the housing, the templates are held to the lines previously scribed. The bottom edge of the template is marked outlining the cut for the bottom edge of the housing. The housing is now ready to be cut.

One method of cutting a housed stringer is to use a routing template and router after the housing is laid out. Stair routing templates are adjustable for different risers and runs. Templates may also be made by cutting out thin plywood or hardboard to the shape of the housing.

Finished stringers may need extra length. The extra length provides sufficient stock for fitting against the baseboard at the top and bottom of the stairway.

LAYING OUT AN OPEN STRINGER

The layout of an open (or *mitered*) stringer is similar to that of a housed stringer. However the open stringer is not housed. The riser layout line is to its outside face and is mitered. The tread layout is to the face side of the tread.

The pitch board is held against the top edge of the stringer, however, instead of against a line in from the edge. The risers and treads are marked lightly with a sharp pencil, Figure 16-21.

To lay out the miter cut for the risers, measure in at right angles from the riser layout line a distance equal to the thickness of the riser stock. Draw another plumb line at this point. Square both lines across the top edge of the stringer stock. Draw a diagonal line on the top edge to mark the miter angle, Figure 16-22.

To mark the tread cut on the stringer, measure down from the tread layout line a distance equal to the thickness of the tread stock. Draw a level line at this point for the tread cut. The tread cut is square through the thickness of the stringer. Allow a little extra length at top and bottom for fitting to existing trim.

LAYING OUT AND FRAMING A STAIRWAY WITH A LANDING

A stair landing is an intermediate platform between two flights of stairs. The landing is usually floored with the same materials as the main floors of the structure. Some codes require that the minimum dimension of a landing be not less than 2'-6''. Other codes require the minimum dimension to be the width of the stairway. L-type stairs may have the landing near the bottom or the top. U-type

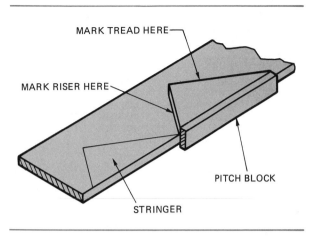

Fig. 16-21 Layout of an open stringer

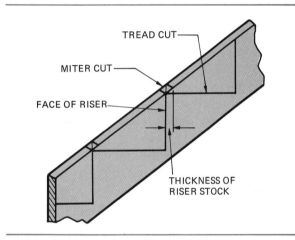

Fig. 16-22 Laying out the miter angle
of an open, mitered stringer

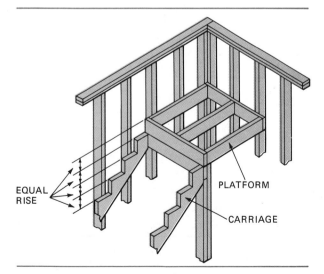

Fig. 16-23 The platform is located so that its top is in line with one of the tread runs.

stairs usually have the landing mid-way on the flight. Many codes state that no flight of stairs shall have a vertical rise of more than 12 feet. Therefore, any staircase with a vertical rise of more than 12 feet must have an intermediate landing or platform.

Platform stairs are built by first erecting the platform. The platform is located so that its top is in line with one of the tread runs. This is to allow an equal riser height for the total flight, Figure 16-23. The stairs are then framed to the platform as if two straight flights were being framed. Either the stair carriage or the housed stringer method of construction may be used, Figure 16-24.

LAYING OUT WINDERS

Although winders are not recommended for safety reasons, some codes state that they may be used in individual dwelling units, if the required tread width is provided at a point not more than 12 inches from the side of the stairway where the treads are narrower. However, in no case shall any width of tread be less than 6 inches at any point, Figure 16-25. Other codes permit a narrower tread.

To lay out a winder, draw a full-size layout on the floor by first drawing a square that has sides equal to the width of the stairway. Using one corner of the square as the center, swing an

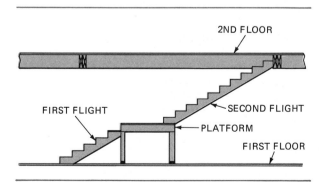

Fig. 16-24 Stair carriages are framed to the platform in a way similar to that used for two straight flights.

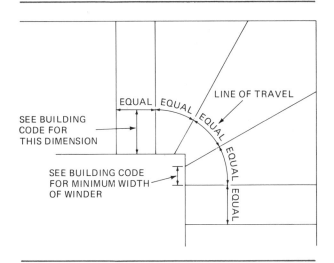

Fig. 16-25 Dimensions of a winder

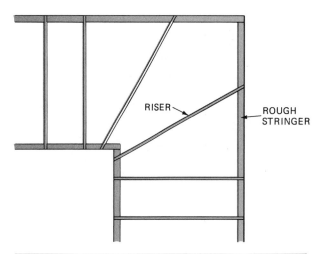

Fig. 16-26 Draw a full-size layout of a set of winders to find the cuts on the rough stringer.

arc showing the line of travel. The radius of the arc may be 12 to 18 inches as the code permits.

Divide the arc into equal parts so that each part equals or exceeds the tread width of the rest of the stairs. Draw lines from the center of the arc through the points dividing the arc. These lines represent the faces of the risers. Draw lines parallel to these to indicate the riser thickness, Figure 16-26.

The cuts on the stair carriage for the winding steps are obtained from the full-size layout. Lay out and cut the carriage and fasten it to the wall. Install rough treads until the stairs are ready for finishing.

If one side of the staircase is to be open, the newel post is installed. Then, the risers are mitered to or mortised into the post, Figure 16-27.

FRAMING BASEMENT STAIRS

Basement stairs are sometimes built without riser boards. Two carriages or stringers are used, one on each side. Nominal 2x10 treads are used.

The carriages are not cut out, but may be dadoed to receive the treads. Or, *cleats* may be fastened to the stringers to support the treads, Figure 16-28.

Lay out the stringers in the usual manner. Cut the bottoms to fit the floor and the tops to fit against the header of the stairwell. "Drop" the carriages as necessary to provide equal risers.

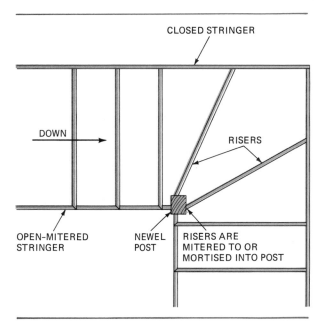

Fig. 16-27 On the open side, winder risers are mitered against or mortised into the newel post.

DADOED STRINGERS

If the treads are to be dadoed into the stringers, lay out the thickness of the tread on the stringer. Mark the depth of the dado on both edges of the stringer. Set the portable electric saw for the depth. Make cuts for the top and bottom of each tread. Then make a series of cuts between each. Chisel from both edges toward the center, removing the excess to

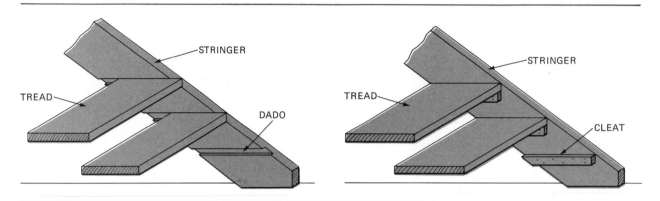

Fig. 16-28 Basement stair stringers may be dadoed or cleated to support the treads.

make the dado. Spike all treads into the dadoes and assemble the staircase. Fasten the assembled staircase in position. The lower end of a basement stairway is sometimes anchored by installing a kicker plate which is fastened to the floor, Figure 16-29.

CLEATED STRINGERS

If the treads are to be supported by cleats, measure down from the top of the tread a distance equal to its thickness and draw another level line for each tread. Fasten 1x3 cleats to the carriages so their top edges are to the bottom line. Fasten the carriages in position. Cut the treads to length and install the treads between the carriages so the treads rest on the cleats.

CALCULATING ROUGH STRINGER LENGTHS

The rough length of a stringer may be determined by scaling across the framing square. Let the tongue of the square represent the total rise of the stairway. Let the blade represent the total run. Using a scale of 1 inch equals 1 foot, scale across the square to find the length of the stringer, Figure 16-30.

For example, the total rise of a stairway is 9'-0'' and the total run is 11'-0''. From the 9-inch mark on the tongue, scale across to the 11-inch mark on the blade. The scaled-off length is found to be 14'-2 1/2''. Sixteen-foot lengths must be ordered.

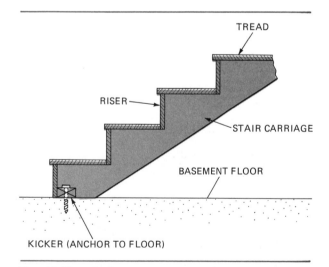

Fig. 16-29 A kicker plate is installed to fasten the bottom end of basement stairs.

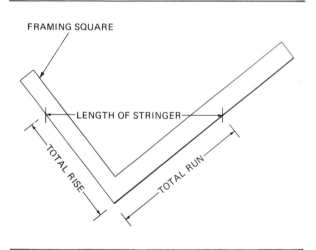

Fig. 16-30 Scale across the framing square to find stringer rough lengths.

REVIEW QUESTIONS

Select the most appropriate answer.

1. Stairways in residential construction should have a minimum width of
   a. 30 inches.   c. 36 inches.
   b. 32 inches.   d. 40 inches.

2. A *nosing* is
   a. a horizontal member of a step.
   b. a vertical member of a step.
   c. the top molded edge of a finish stringer.
   d. the overhang of the tread beyond the riser.

3. The maximum riser height stated in this unit is
   a. 8 1/4".   c. 7 1/2".
   b. 7 3/4".   d. 7 1/4".

4. The riser height for a stairway with a total vertical rise of 8'-9" is
   a. 8 inches.   c. 7 5/8 inches.
   b. 7 3/4 inches.   d. 7 1/2 inches.

5. Applying the rules stated in this unit, what is the tread run if the riser height in a stairway is 7 1/4 inches?
   a. 8 3/4 to 9 3/4 inches   c. 9 3/4 to 10 3/4 inches
   b. 9 to 10 inches   d. 10 to 11 inches

6. Many building codes specify a minimum tread width of
   a. 8 1/2 inches.   c. 9 1/2 inches.
   b. 9 inches.   d. 10 inches.

7. A flight of stairs has a riser height of 7 1/2 inches and a tread run of 10 inches. The total thickness of the upper floor is 10 inches. What is the length of the stairwell if the stairwell header acts as the top riser and the desired minimum headroom is 6'-8"?
   a. 9'-0".   c. 10'-0"
   b. 9'-6".   d. 10'-2"

8. Most building codes specify a minimum headroom clearance of
   a. 6'-6".   c. 7'-0".
   b. 6'-8".   d. 7'-6".

9. The stair carriage of a stairway with a riser height of 7 1/2 inches rests on the finish floor. What is the riser height of the first step if the tread thickness is 3/4 inch?
   a. 6 3/4 inches   c. 7 3/4 inches
   b. 7 1/2 inches   d. 8 1/4 inches

10. A stairway has a riser height of 7 1/2". The tread stock thickness is 1 1/16". The finish floor thickness of the upper floor is 3/4 inch. What distance down from the top of the upper subfloor must the rough carriage be fastened if the stairwell header acts as the top riser?
    a. 7 3/16 inches   c. 7 1/2 inches
    b. 7 3/8 inches   d. 7 13/16 inches

INSULATION AND VENTILATION

17

OBJECTIVES

After completing this unit, the student will be able to

- *describe how insulation works and define insulating terms and requirements.*

- *describe the commonly used insulating materials and state where insulation is placed.*

- *properly install various kinds of insulation.*

- *explain the need for ventilating a structure, describe types of ventilators, and state the minimum recommended sizes.*

- *state the purpose of and install vapor barriers.*

- *describe various methods of construction to resist the transmission of sound.*

Thermal insulation is a material used to resist the passage of heat through the walls, floors, or ceilings of a structure. Most materials used in construction have some insulating value, Figure 17-1. Even the air spaces between studs resist the passage of heat. After insulation is installed, however, the stud space has many times the resistance to the passage of heat than the air space alone had.

HOW INSULATION WORKS

Air is an excellent insulator if confined to a small space in which the air is still. This is commonly called a *dead air space*. Millions of tiny air cells, entrapped in the unique cellular structure of wood, make it a natural barrier to the transmission of heat.

Most insulating materials are manufactured by trapping dead air in tiny spaces to provide resistance to the flow of heat. Among the materials used for insulating are glass fibers, mineral fibers (rock), organic fibers (paper), and plastic foam.

Another type of insulation, called *reflective insulation,* works by reflecting heat. Aluminum foil is high in reflective properties and is used extensively for this type of insulation. Reflective insulation is perhaps more effective in preventing summer heat flow through walls and ceilings than it is in preventing the flow of winter cold. It should be considered more for use in warm climates than cold.

R is the measured resistance of a material to the flow of heat. The higher the R-value, the more efficient the material is in retarding the

THERMAL PROPERTIES OF VARIOUS BUILDING MATERIALS, PER INCH OF THICKNESS	
Material	Thermal Resistance R
Wood	1.25
Air Space	0.97
Cinder Block	0.28
Common Brick	0.20
Face Brick	0.11
Concrete (Sand and Gravel)	0.08
Stone (Lime or Sand)	0.08
Steel	0.0032
Aluminum	0.00070

Fig. 17-1 R-values of various building materials per inch of thickness

Fig. 17-2 Insulation is rated according to its thermal resistance and is stamped with an "R" number.

passage of heat. Insulation is rated according to its efficiency and is stamped with an R number, Figure 17-2.

INSULATION REQUIREMENTS

Extensive studies of heating and air-conditioning requirements have resulted in a number of R-value design standards suitable for new construction. Average winter low-temperature zones of the United States are shown in Figure 17-3. This information is used to determine the amount of insulation for walls, ceilings, and floors.

Perhaps the most widely used recommendations are in the *All-Weather Comfort Standard*,

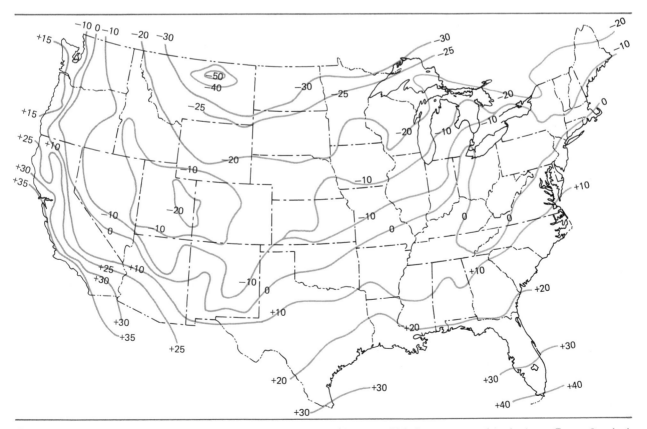

Fig. 17-3 Average low temperature zones of the United States *(Courtesy U.S. Department of Agriculture, Forest Service)*

All-Weather Comfort Standard	
	Insulation "R" Number
Ceilings	R19
Walls	R11
Floors over unheated spaces	R13
Moderate Comfort and Economy Standard	
	Insulation "R" Number
Ceilings	R13
Walls	R8
Floors over unheated spaces	R9
Minimum Comfort Standard	
	Insulation "R" Number
Ceilings	R9
Walls	R7
Floors over unheated spaces	R7

Fig. 17-4 R-values and insulation requirements

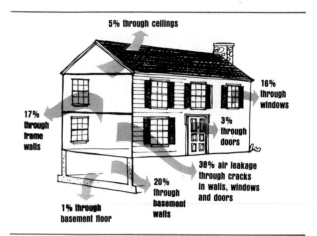

Fig. 17-5 Areas and amounts of heat loss for a typical house with moderate insulation *(Courtesy of Dow Chemical Company)*

Figure 17-4. In areas where the temperatures typically reach –10°F to –15°F., the recommendations of the *All-Weather Comfort and Economy Standard* or the *Minimum Comfort Standard* are often followed.

In warm climates, less insulation is needed to provide comfort in the heating season. However, air-conditioned homes should receive more insulation in walls, ceilings, and floors to assure economy in the operation of cooling equipment.

Comfort and operating economy are dual benefits. Insulating for maximum comfort automatically provides maximum economy of heating and cooling operations. It also reduces the initial costs of heating and cooling equipment because smaller units are required.

WHERE HEAT IS LOST

The amount of heat lost from the average house varies with the type of construction. The principal areas and approximate amount of heat loss for a typical house with moderate insulation are shown in Figure 17-5.

Houses of different architectural style vary in their heat loss characteristics. A single-story house, for example, contains less wall area than a two-story house and a proportionally greater ceiling area. Greater heat loss through floors is

experienced in homes erected over concrete slabs or unheated crawl spaces unless well insulated in these areas.

The transfer of heat through uninsulated ceilings, walls, or floors can be reduced almost any desired amount by installing insulation. Maximum quantities in these areas can cut heat losses by up to 90 percent. The use of 2x6 studs in exterior walls permits installation of 6-inch-thick insulation achieving an R- value of 19.

Windows and doors are generally sources of great heat loss. By reducing the heat transfer through other sections of the house, discomfort caused by these sources can be offset. Weather-stripping around windows and doors reduces a great amount of heat loss. In colder sections, heat losses through glass surfaces can be reduced 50 percent or more by installing double glazed windows or by adding storm sash and storm doors.

TYPES OF INSULATION

Insulation is manufactured in a variety of forms and types. These are commonly grouped as *flexible, loose-fill, rigid, reflective,* and *miscellaneous* types.

FLEXIBLE INSULATION

Flexible insulation is manufactured in *blanket* and *batt* form. Blanket insulation,

Fig. 17-6 Blanket insulation comes in rolls.

Figure 17-6, is furnished in rolls with widths suited to 16- and 24-inch stud and joist spacing. The usual thickness is from 1 inch to 3 1/2 inches. The body of the blanket is made of fluffy material of rock or glass wool, wood fiber, or cotton. Organic insulation materials are treated to make them resistant to fire, decay, insects, and vermin. Most blanket insulation is covered with paper or sheet material with tabs on the sides for fastening to studs or joists. The covering sheet serves as a vapor barrier to resist the movement of water vapor and should always face the warm side of the wall. Aluminum foil and asphalt or plastic laminated paper are commonly used as vapor barrier materials.

Batt insulation, Figure 17-7, is made of the same material as blanket insulation in thicknesses up to 12 inches. Widths are for standard stud

and joist spacing. Lengths are either 24 or 48 inches. Batts are supplied with or without covers.

LOOSE-FILL INSULATION

Loose-fill insulation is usually composed of materials in bulk form. It is supplied in bags or bales and is placed by pouring, blowing, or packing by hand. Materials include rock or glass wool, wood fibers, shredded redwood bark, cork, wood pulp products, and vermiculite.

Loose-fill insulation is suited for use between ceiling joists in unheated attics. It is also used in the sidewalls of existing houses that were not insulated during construction.

RIGID INSULATION

Rigid insulation is usually a fiber or foamed plastic material manufactured in sheet or board forms, Figure 17-8. The material is available in a wide variety of sizes with widths up to 4 feet and lengths to 12 feet. The most common types are made from polystyrene, polyurethane, glass fibers, processed wood, and sugar cane or other fibrous vegetable materials.

Structural insulating boards in densities ranging from 15 to 31 pounds per cubic foot are used as building boards, roof decking, sheathing, and wallboard. Their primary purpose is structural while insulation is secondary, however. In house construction, perhaps the most common

Fig. 17-8 Rigid insulation being installed between floor joists over a crawl space

Fig. 17-7 Batt insulation is made up to 12 inches thick.

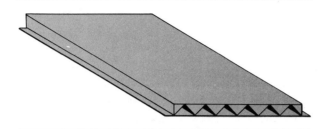

Fig. 17-9 Accordian-type reflective insulation is made up of several spaced layers of aluminum foil.

forms are sheathing and decorative coverings in sheet or tile forms.

REFLECTIVE INSULATION

Reflective insulation usually consists of aluminum foil and paper products with a reflective coating. These reflective insulations are effective only when the reflective surface faces an air space with a depth of 3/4 inch or more. Where a reflective surface contacts another material, the reflective properties are lost and the material has little or no insulating value. The most common type is aluminum foil, which comes in rolls of various widths and lengths. Another type called *accordion* insulation is considered to be the most effective of the reflective insulations, Figure 17-9. It is made up of several spaced layers of foil. As it is applied, the material is stretched out to form dead air spaces.

Reflective insulation of the foil type is sometimes applied to other types of insulation to give it added insulating qualities. The metal foil also acts as an excellent vapor barrier. The foil may be applied to blankets, batts, and rigid insulation. It is sometimes applied to the stud-surface side of gypsum lath.

MISCELLANEOUS INSULATION

Foamed-in-place insulation is sometimes used under certain conditions. A urethane foam is produced by mixing two chemicals together. It is blown into place and expands on contact with the surface.

Sprayed insulation is usually inorganic fibrous material which is blown against a surface that has been primed with an adhesive coating. This type is often left exposed for acoustical as well as insulating properties.

Other types of insulating material such as lightweight *vermiculite* and *perlite aggregates* are sometimes used in plaster as a means of reducing heat flow.

WHERE AND HOW TO INSULATE

To reduce heat loss all walls, ceilings, roofs, and floors that separate heated from unheated spaces should be insulated. Insulation should be placed in all outside walls and in the ceiling. In houses with unheated crawl spaces, it should be placed between the floor joists. A ground cover of roll roofing or plastic film such as polyethylene should be placed on the soil of crawl spaces to decrease the moisture content of the space as well as of the wood members. Provision should also be made for ventilation of unheated areas. (Vapor barriers, condensation, and ventilation are discussed in greater detail later in this unit.)

In houses of flat or low-pitched roofs, insulation should be used in the ceiling area with sufficient space left above the insulation for ventilation. Insulation is used along the perimeter of houses built on slabs where air-conditioning systems are used, and insulation is placed in all exposed ceilings and walls in the same manner as when insulating against cold weather heat loss. Figure 17-10 shows the placement of insulation in various types of structures.

INSTALLING FLEXIBLE INSULATION

Cut the batts or blankets with a knife. Make lengths slightly oversize so stapling flanges can be formed at the top and bottom. Measure out from one wall a distance equal to the desired lengths of insulation and draw a line on the floor. Roll out the material from the wall and cut on the line. Compress the insulation with a straightedge and cut with a sharp knife, Figure 17-11. Cut the necessary number of standard lengths.

Place the batts or blankets between the studs. Staple the tabs of the vapor barrier to the

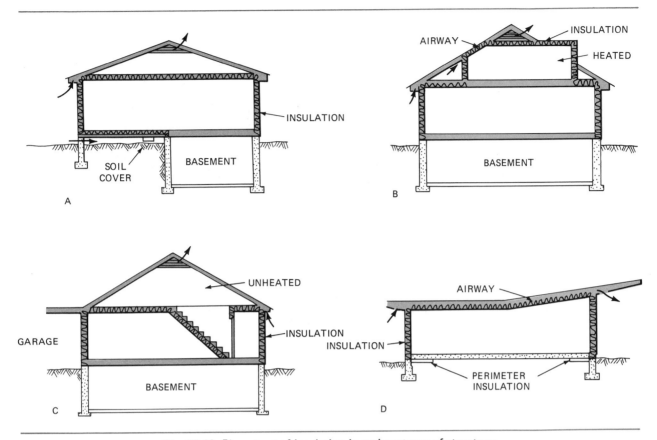

Fig. 17-10 Placement of insulation in various types of structures

Fig. 17-11 Compress flexible insulation with a straightedge and cut it with a sharp knife.

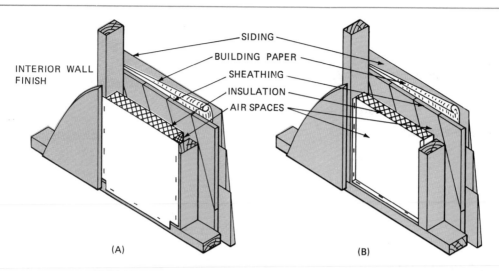

INTERIOR WALL
FINISH

SIDING
BUILDING PAPER
SHEATHING
INSULATION
AIR SPACES

(A)   (B)

IN (A), THE INSULATION FLANGE IS STAPLED TO THE INSIDE EDGE
OF THE STUD. IN (B), THE INSULATION IS RECESSED TO PROVIDE
AN AIR SPACE, OF AT LEAST 3/4-INCH, ON EACH SIDE.

Fig. 17-12 Methods of installing wall insulation

inside edges of the studs as well as the top and bottom plates, Figure 17-12. Use a hand stapler to fasten the insulation in place, Figure 17-13.

Fill any spaces around doors and windows but do not pack the insulation tightly. Squeezing or compressing it reduces its effectiveness.

It is good practice to use narrow strips of vapor barrier material along the top, bottom, and sides of wall openings to protect these areas, Figure 17-14. Ordinarily these are not covered too well by the barrier on the blanket or batt. If the insulation has no covering or stapling tabs, it is friction fitted between the framing members.

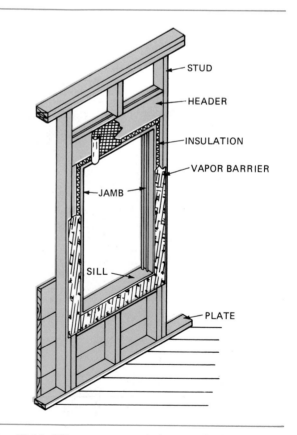

STUD
HEADER
INSULATION
VAPOR BARRIER
JAMB
SILL
PLATE

Fig. 17-13 Use a hand stapler to fasten wall insulation in place.

Fig. 17-14 Fill spaces around doors and windows with insulation and install vapor barrier material over them.

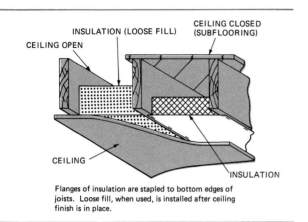

Flanges of insulation are stapled to bottom edges of joists. Loose fill, when used, is installed after ceiling finish is in place.

Fig. 17-15 Method of installing insulation between ceiling joists under unheated attic

CEILING INSULATION

Ceiling insulation is installed by stapling it to the ceiling joists or by friction fitting it between them, Figure 17-15. If furring strips have been applied to the ceiling joists, the insulation is simply laid on top of the furring strips. Extend the insulation across the top plate. However, if the eaves are vented, care must be taken not to block the flow of air into the attic. It may be necessary to compress the insulation against the top of the wall plate to permit a free flow of air, Figure 17-16.

INSULATING FLOORS OVER CRAWL SPACES

Flexible insulation installed between floor joists over crawl spaces may be held in place by wire mesh or pieces of heavy gauge wire wedged between the joists. The vapor barrier on the insulation faces the ground, Figure 17-17, instead of the heated side to prevent moisture from entering from below.

INSULATING MASONRY WALLS

Masonry walls require furring strips. Fasten 1x2 furring strips 16 inches O.C. and insulate as for wood frame walls, Figure 17-18.

INSTALLING LOOSE-FILL INSULATION

Loose-fill insulation is poured into place by hand or blown into place with special equipment. Level the surface to the desired depth, Figure 17-19.

CAUTION: Use a dust mask to prevent inhaling fine dust particles.

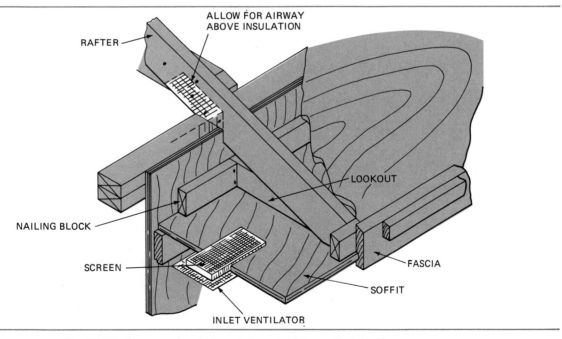

Fig. 17-16 Compress insulation at the top of the wall plate, if necessary, to permit a free flow of air from the eaves.

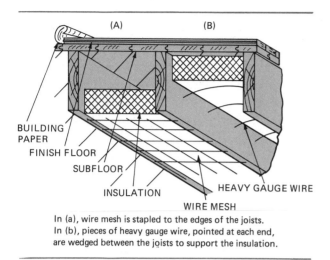

In (a), wire mesh is stapled to the edges of the joists.
In (b), pieces of heavy gauge wire, pointed at each end,
are wedged between the joists to support the insulation.

Fig. 17-17 Methods of installing insulation between floor joists

INSTALLING RIGID INSULATION

Rigid insulation is easily cut with a knife or saw. It is usually applied by friction fitting between the framing members. However, it may be applied to masonry walls with special adhesives, Figure 17-20. Polystyrene foam plastic, commonly called Styrofoam, has excellent insulating qualities and is also waterproof. That is, it does not absorb moisture and, therefore, needs no vapor barrier. It is frequently used to insulate existing structures before new siding is applied, Figure 17-21. It is also used extensively on roofs either below or above the roof sheathing, Figure 17-22.

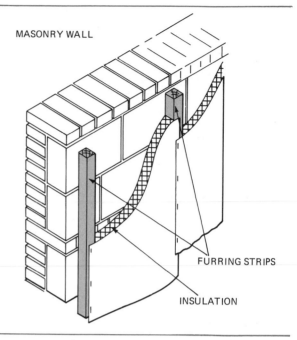

Fig. 17-18 Method of insulating masonry walls

INSTALLING REFLECTIVE INSULATION

The accordion-type reflective insulation is installed between framing members in the same manner as blanket insulation. It is attached to either the face or side of the studs. However, it is important that an air space of at least 3/4 inch is maintained between its surface and the inside edge of the stud.

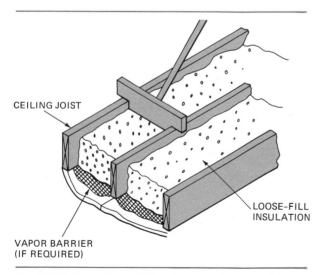

Fig. 17-19 Level loose-fill insulation to the desired depth.

Fig. 17-20 Rigid insulation is applied to basement walls before backfilling. *(Courtesy of Dow Chemical Company)*

Fig. 17-21 Rigid insulation applied over old existing siding *(Courtesy of Dow Chemical Company)*

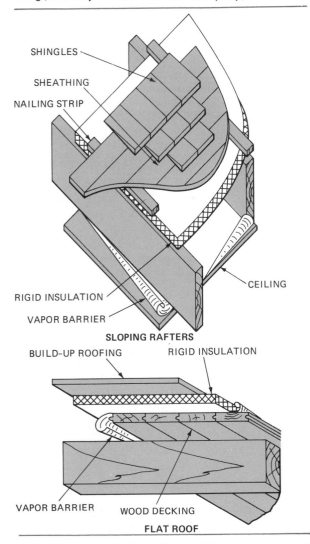

Fig. 17-22 Method of installing rigid insulation on roofs

The roll type reflective insulation is stapled to the inside edges of the studs. As it is rolled out across the studs, it is bowed into the stud spaces by at least 3/4 inch.

PROPERLY INSTALLED INSULATION

Insulation that is not properly installed can cause serious problems, such as rendering the material useless and causing wood to decay and paint to blister, among others. It is extremely important that the carpenter be aware of the consequences of improper application and not take its installation lightly. The carpenter must know how to install insulation and vapor barriers in the recommended fashion and how to provide proper and adequate ventilation. Follow the manufacturer's directions carefully when installing any type of insulation.

CONDENSATION

Warm air can hold more moisture than cold air. As the warm air in a structure moves through the walls, ceilings, and floors, the moisture may condense under certain conditions when it comes in contact with a cold surface, Figure 17-23. In a wall, this contact is made on the inside surface of the exterior wall sheathing. Condensation of moisture in walls, attics, roofs, and floors leads to serious problems.

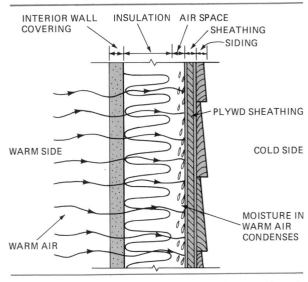

Fig. 17-23 Moisture in warm air condenses when it comes in contact with a cold surface.

- If insulation absorbs moisture, the dead air spaces become filled with water, which is an excellent conductor of heat. This would render the insulation practically useless.

- Condensed moisture is absorbed by the wood frame. This raises the moisture content of the wood and improves the conditions for fungi to grow, which causes the wood to rot.

- Moisture absorbed by exterior trim causes the exterior paint to blister and peel. Ex-

cessive condensation may even damage the interior finish.

- A warm attic that is inadequately ventilated may cause the formation of ice dams at the eaves. During cold weather after a heavy snowfall, heat causes the snow next to the roof to melt. Water running down the roof freezes on the cold roof overhang, forming an ice dam. This causes the water to back up on the roof, under the shingles, and into the walls and ceiling, Figure 17-24.

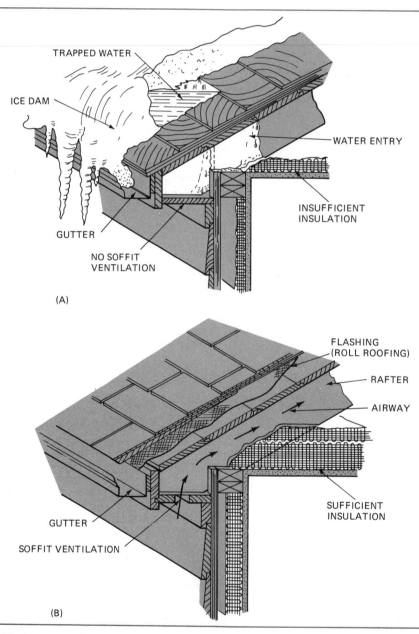

Fig. 17-24 A warm attic may cause the formation of ice dams on the roof during cold weather.

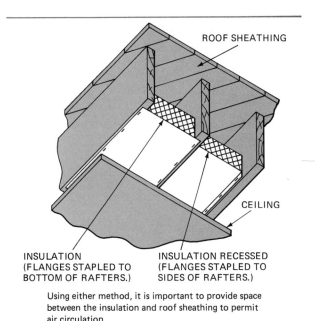

INSULATION
(FLANGES STAPLED TO
BOTTOM OF RAFTERS.)

INSULATION RECESSED
(FLANGES STAPLED TO
SIDES OF RAFTERS.)

Using either method, it is important to provide space
between the insulation and roof sheathing to permit
air circulation.

**Fig. 17-25 Methods of installing flexible
insulation between rafters**

VENTILATION

Ventilation provides part of the answer to combating condensation of moisture. In a well ventilated area, any condensed moisture is removed by evaporation. It is important that areas on the cold side of the insulation be well ventilated. One of the areas where this can be effectively accomplished is in the attic. With a well insulated ceiling and adequate ventilation, attic temperatures are low and melting of snow on the roof over the attic space is greatly reduced.

In crawl spaces under floors, ventilation can be easily provided to evaporate any condensation of moisture that may occur in these areas. In walls, ventilation of the air space on the cold side of the insulation is often difficult especially if the wall is full of insulation. It is for this reason that some contractors object to filling the stud space with insulation and insist on leaving at least 3/4 inch of air space between the insulation and the inside of the exterior wall sheathing. However, with a properly installed vapor barrier on the inside surface of the wall and a type of exterior finish that allows the passage of vapor through it — possibly using siding vents — filling the stud space completely with insulation should pose no problem.

On flat roofs, or on other roofs where the ceiling finish is attached to the roof frame, insulation is usually installed between the roof frame members. It is imperative that an adequate air space of at least 1 1/2 inches is maintained between the insulation and the roof sheathing and that this air space is properly vented, Figure 17-25. Failure to do so is likely to result in decay of the roof frame, causing expensive major repairs.

TYPES OF VENTILATORS

There are many types of ventilators. Louvered openings are generally provided in the end walls of gable roofs and should be as close to the ridge as possible. More positive movement of air can be obtained if additional openings are provided in the soffit (roof overhang) area, Figure 17-26.

The minimum free-air area for attic ventilators is based on the ceiling area of the rooms below. The free-air area for the openings should be 1/300 of the ceiling area. For example, if the ceiling area is 1200 square feet, the minimum total free-air area of the ventilators should be four square feet.

The ratio of ventilator openings are free-air areas. The actual area must be increased to allow for any restrictions such as louvers or wire screen mesh. Use as coarse a screen as conditions permit because lint and dirt tend to clog fine-mesh screens. When painting, be careful not to contact the screen and clog the mesh with paint. Ventilators should never be closed. In cold climates, it is especially important to leave them open in the winter as well as summer.

VENTILATING HIP ROOFS

Hip roofs should have air inlets in the soffit and outlets at or near the peak of the roof. The most efficient type of inlet opening is the continuous slot, Figure 17-27. The air outlet opening near the peak can be a globe-type metal ventilator, Figure 17-28, or several smaller roof ventilators located near the ridge. They can be located below the peak on the rear slope of the roof so they are not visible from the front of the house. Gabled extensions of a hip roof

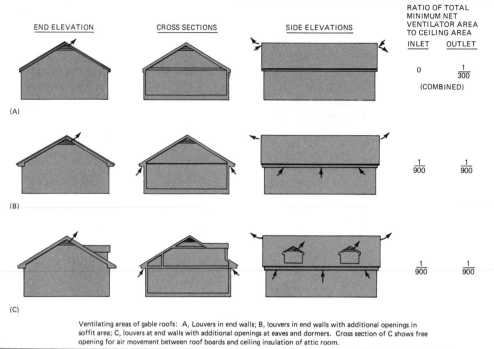

	END ELEVATION	CROSS SECTIONS	SIDE ELEVATIONS	RATIO OF TOTAL MINIMUM NET VENTILATOR AREA TO CEILING AREA	
				INLET	OUTLET
(A)				0	$\frac{1}{300}$ (COMBINED)
(B)				$\frac{1}{900}$	$\frac{1}{900}$
(C)				$\frac{1}{900}$	$\frac{1}{900}$

Ventilating areas of gable roofs: A, Louvers in end walls; B, louvers in end walls with additional openings in soffit area; C, louvers at end walls with additional openings at eaves and dormers. Cross section of C shows free opening for air movement between roof boards and ceiling insulation of attic room.

Fig. 17-26 Methods of ventilating gable roofs

house are sometimes used to provide efficient outlet ventilators, Figure 17-29. If the ridge is long enough, a continuous ridge vent may be used.

VENTILATING CRAWL SPACES

For crawl spaces, provide at least four foundation wall vents near the corners of the building, Figure 17-30. The total net area of the ventilators should be 1/1600 of the ground area when a vapor barrier ground cover is used. The use of a ground cover is recommended under all conditions. It not only protects wood framing members from ground moisture, but also allows the use of small, inconspicuous ventilators.

The practice of closing vents in cold weather should be emphatically discouraged. Ventilation is needed as much, if not more, in winter as in summer.

Fig. 17-27 The continuous slot-type vent in the soffit is a most efficient type of air inlet.

Fig. 17-28 Globe-type ventilators are sometimes used to vent hip roofs.

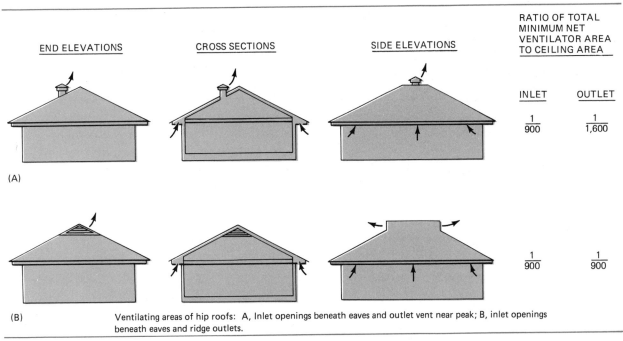

END ELEVATIONS   CROSS SECTIONS   SIDE ELEVATIONS

RATIO OF TOTAL MINIMUM NET VENTILATOR AREA TO CEILING AREA

	INLET	OUTLET
(A)	$\frac{1}{900}$	$\frac{1}{1,600}$
(B)	$\frac{1}{900}$	$\frac{1}{900}$

Ventilating areas of hip roofs: A, Inlet openings beneath eaves and outlet vent near peak; B, inlet openings beneath eaves and ridge outlets.

Fig. 17-29 Methods of ventilating hip roofs

VAPOR BARRIERS

As stated previously, ventilation is part of the answer to combating moisture condensation. The use of *vapor barriers* is the answer to the rest. One way to combat condensation is to evaporate it. The other way is to block any moisture from entering the cold air space. The use of vapor barriers accomplishes this.

Some discussion of vapor barriers has been included in previous sections because vapor barriers are often part of flexible insulation. However, further information is necessary.

Water vapor passes through most building materials. A considerable amount of water vapor is generated in a house from cooking, dishwashing, laundering, bathing, and other sources. To help prevent vapor from entering cold air spaces, a vapor barrier is used.

Most commonly used as a vapor barrier is polyethylene film. Other materials are asphalt laminated papers and aluminum foil. Most blanket and batt insulations are provided with a vapor barrier on one side.

When a more positive seal is desired, wall-height rolls of plastic film vapor barriers are applied over studs, plates, and window and door headers. Some contractors apply the film over the entire opening and then cut the film after

2" X 6" STUDS 24" O.C.

FINISH FLOOR

VAPOR BARRIER

SUBFLOOR

SCREENED VENT

INSULATION

FLOOR JOIST

SOIL COVER (VAPOR BARRIER)

Fig. 17-30 Crawl space ventilator and soil cover

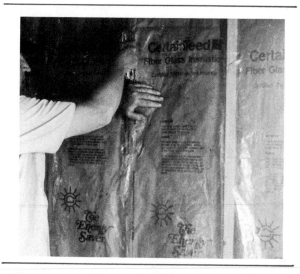

Fig. 17-31 Installing polyethylene film vapor barrier

the finish is applied to assure a more positive seal.

This system is used over insulation having no vapor barrier or to insure excellent protection over any type of insulation. The barrier is fitted tightly around outlet boxes. A ribbon of sealing compound around outlets or switch boxes minimizes vapor loss at these areas.

Vapor barriers are not usually installed on ceilings. This is to allow the vapor to pass through the ceilings where it can be easily evaporated by the great movement of air in the attic. By allowing the vapor to pass through ceilings, its entrance into wall cavities, where the movement of air is not as great, is reduced. However, where temperatures commonly fall below -20°F or where roof slopes are less than 3 in 12, vapor barriers are recommended in ceilings.

Because no type of vapor barrier can be considered 100 percent impervious and some vapor leakage into the wall may be expected, the flow of vapor to the outside should not be impeded by materials of high vapor resistance on the cold side of the wall. For example, sheathing paper should be of a type that is not highly vapor resistant. This also applies to exterior siding and exterior finishes. Bathroom and kitchen vents and fans should be used to expel generated moisture to the outside.

INSTALLING PLASTIC FILM VAPOR BARRIER

Plastic film vapor barriers come in rolls large enough to cover entire wall surfaces. Unroll a length long enough to cover the wall and allow a few inches extra in length. Unfold the section and staple it along the top plate, letting the rest drape down to the floor. Smooth out wrinkles and staple about 6 to 12 inches apart to every stud, Figure 17-31. Cut the material and carefully fit it around all openings. Lap all joints and repair all tears. Cut off the excess at the floor line.

SOUND INSULATION

Sound, or acoustical, insulation resists the passage of noise through a building section. In the past, the reduction of sound transfer between rooms was more important in apartments and motels. More attention is now being paid to insulating for sound in the home. Excessive noise is not only annoying, but it can be harmful. It not only causes fatigue and irritability, but can damage the sensitive hearing nerves of the inner ear. Sound insulation between active areas of the home and the quiet areas is now desirable as is insulation from outdoor sounds. Sound insulation between the bedroom area and the living area and isolation of the bathrooms and lavatories is also desirable. Sound control is an important part of residential and commercial construction. The carpenter should be familiar with the methods of sound insulation.

SOUND TRANSMISSION

Noises such as a loud radio, barking dog, or shouting create sound waves which radiate outward from the source until they strike a wall, floor, or ceiling. These surfaces are set in vibration by the pressure of the sound waves in

25	Normal speech can be understood quite easily
30	Loud speech can be understood fairly well
35	Loud speech audible but not intelligible
42	Loud speech audible as a murmur
45	Must strain to hear loud speech
48	Some loud speech barely audible
50	Loud speech not audible

This chart from the Acoustical and Insulating Materials Association illustrates the degree of noise control achieved with barriers having different STC numbers.

Fig. 17-32 Approximate effectiveness of walls with varying STC ratings

the air. Because the wall vibrates, it conducts sound to the other side in varying degrees, depending on the wall construction.

The resistance of a building section, such as a wall, to the passage of sound is rated by its *Sound Transmission Class* (STC). The higher the number, the better the sound barrier. The approximate effectiveness of walls with varying STC numbers is shown in Figure 17-32.

Sound travels readily through the air and through some materials. When airborne sound strikes a wall, the studs act as conductors unless they are separated in some way from the covering material. Electrical outlet boxes placed back to back in a wall easily pass sound. Faulty construction, such as poorly fitted doors, often allows sound to pass through. Therefore, good construction practices aid in controlling sound.

WALL CONSTRUCTION

A wall that provides sufficient resistance to airborne sound should have an STC rating of 45, or greater. At one time, the resistance was usually provided only by double walls, which

resulted in increased costs. However, a system of using sound-deadening *insulating board* with a gypsum board outer covering has been developed that provides good sound resistance. Figure 17-33 shows various types of wall construction and their STC rating.

FLOOR AND CEILING CONSTRUCTION

Sound insulation between an upper floor and the ceiling below involves not only the resistance of airborne sounds, but also that of impact noises. *Impact noise* is caused by objects, such as dropped objects, footsteps, or moving furniture, striking or sliding along the floor surface. The floor is set into vibration by the impact and sound is radiated from both sides of the floor. Impact noise control must be considered as well as airborne sounds when constructing floors sections.

An *Impact Noise Rating* shows the resistance of various types of floor-ceiling construction to impact noises. The higher the positive value of the INR, the more resistant to impact noise transfer. Figure 17-34 shows various types of floor-ceiling construction with their STC and INR ratings.

SOUND ABSORPTION

The amount of noise in a room can be minimized by the use of *sound absorbing materials.* Perhaps the most commonly used sound absorbing material is *acoustical tile.* These are most often used in the ceiling where they are not subjected to damage. The tiles are soft and are made of wood fiber or similar materials. The tile surface consists of small holes or fissures or a combination of both, Figure 17-35. These holes or fissures act as sound traps in which the sound waves enter, bounce back and forth, and finally die out. A more complete description and methods of installing ceiling tile are found later in Unit 25.

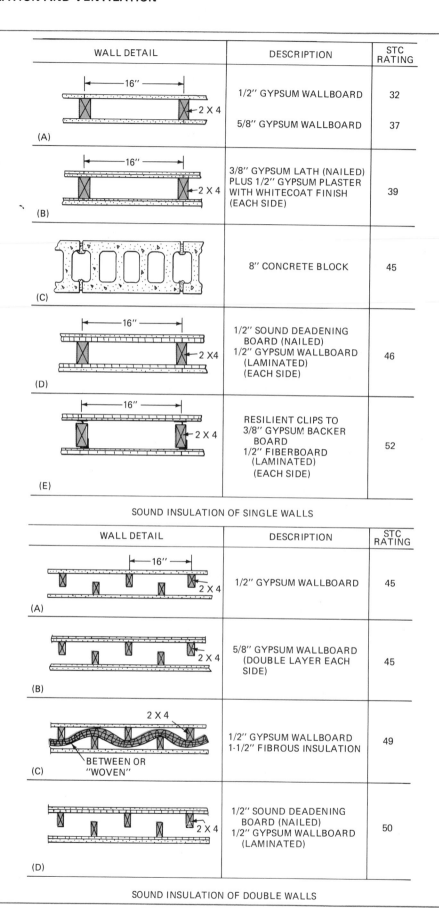

SOUND INSULATION OF SINGLE WALLS

SOUND INSULATION OF DOUBLE WALLS

Fig. 17-33 Sound transmission rating of various types of wall construction

DETAIL	DESCRIPTION	ESTIMATED VALUE	
		STC RATING	APPROX. INR
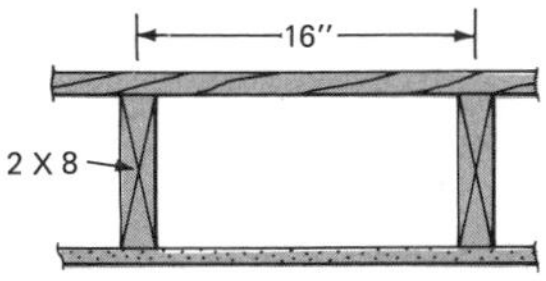 (A) — 16" — 2 X 8	FLOOR 7/8" T. & G. FLOORING CEILING 3/8" GYPSUM BOARD	30	−18
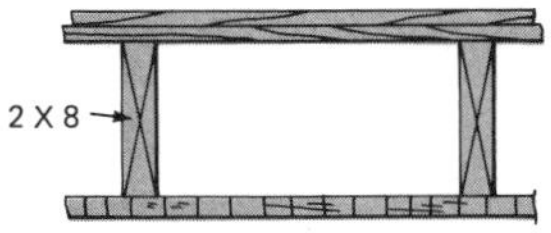 (B) 2 X 8	FLOOR 3/4" SUBFLOOR 3/4" FINISH FLOOR CEILING 3/4" FIBERBOARD	42	−12
 (C) 2 X 8	FLOOR 3/4" SUBFLOOR 3/3/4" FINISH FLOOR CEILING 1/2" FIBERBOARD LATH 1/2" GYPSUM PLASTER 3/4" FIBERBOARD	45	−4

RELATIVE IMPACT AND SOUND TRANSFER IN FLOOR–CEILING COMBINATIONS
(2- BY 8-IN. JOISTS)

DETAIL	DESCRIPTION	ESTIMATED VALUE	
		STC RATING	APPROX. INR
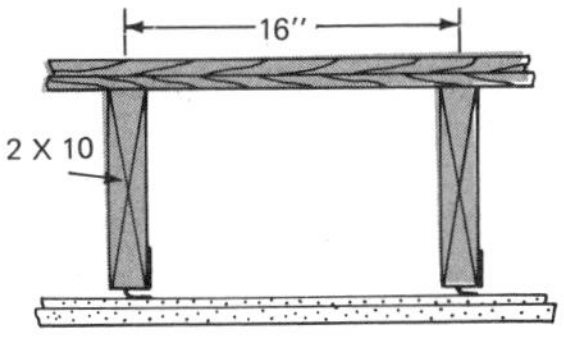 (A) — 16" — 2 X 10	FLOOR 3/4" SUBFLOOR (BUILDING PAPER) 3/4" FINISH FLOOR CEILING GPYSUM LATH AND SPRING CLIPS 1/2" GYPSUM PLASTER	52	−2
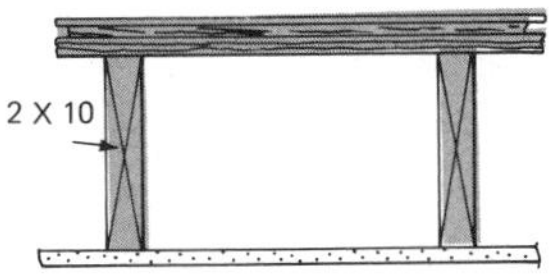 (B) 2 X 10	FLOOR 5/8" PLYWOOD SUBFLOOR 1/2" PLYWOOD UNDERLAYMENT 1/8" VINYL-ASBESTOS TILE CEILING 1/2" GYPSUM WALLBOARD	31	−17
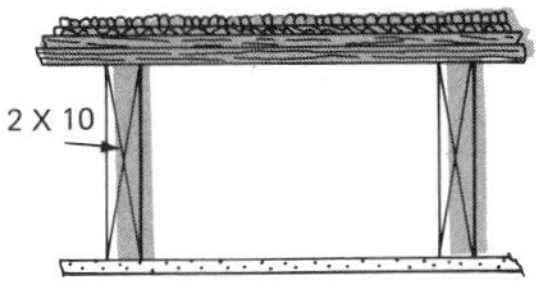 (C) 2 X 10	FLOOR 5/8" PLYWOOD SUBFLOOR 1/2" PLYWOOD UNDERLAYMENT FOAM RUBBER PAD 3/8" NYLON CARPET CEILING 1/2" GYPSUM WALLBOARD	45	+5

RELATIVE IMPACT AND SOUND TRANSFER IN FLOOR-CEILING COMBINATIONS
(2- BY 10-IN. JOISTS)

Fig. 17-34 Floor and ceiling construction rated for STC and INR

Fig. 17-35 Sound absorbing ceiling tile

REVIEW QUESTIONS

Select the most appropriate answer.

1. Of the common building materials listed below the most efficient insulator is
   a. cinder block.　　　　　　　　c. stone.
   b. concrete.　　　　　　　　　　d. wood.

2. The insulating term *R* is defined as
   a. resistance of a material to the flow of heat.
   b. the heat loss through a building section.
   c. the conductivity of a material.
   d. the total heat transfer through a building section.

3. If insulation is covered on both sides, only one side acts as a vapor barrier and should face the
   a. warm side.　　　　　　　　　c. inside.
   b. cold side.　　　　　　　　　　d. outside.

4. Reflective insulation usually consists of
   a. foamed plastic.　　　　　　　c. aluminum foil.
   b. glass wool.　　　　　　　　　d. vermiculite.

5. Reflective insulation is effective only when
   a. it is placed on the warm side.
   b. it comes in contact with the wall covering.
   c. it faces an air space of at least 3/4 inch deep.
   d. it is used under the exterior finish.

306

6. For a more effective seal, the tabs of vapor barrier covering flexible insulation are stapled to
   a. the sides of the framing members.
   b. the inside edges of the framing members.
   c. the inside surface of the wall sheathing.
   d. the top and bottom plates.

7. Squeezing or compressing flexible insulation tightly into spaces
   a. reduces its effectiveness.
   b. increases its efficiency since more insulation can be installed.
   c. is necessary to hold it in place.
   d. helps prevent air leakage by sealing cracks.

8. When insulation is placed between roof framing members, there should be an air space between the insulation and the roof sheathing of at least
   a. 3/4 inch.
   b. 1 inch.
   c. 1 1/2 inches.
   d. 2 inches.

9. Ventilators
   a. are closed in winter to conserve heat.
   b. are left open year round.
   c. are closed in stormy weather.
   d. are closed when the house is to be left vacant for long periods of time.

10. Vapor barriers
    a. should be installed in exterior walls and ceilings.
    b. should be installed on ceilings only.
    c. should not be installed on ceilings unless the roof pitch is less than 3 in 12.
    d. "envelope" the house and are installed on all walls, floors and ceilings.

SECTION

3

Exterior Finish

CORNICE CONSTRUCTION 18

OBJECTIVES

After completing this unit, the student will be able to

- *describe various types of cornices and name their parts.*
- *build box, snub, and open cornices.*
- *apply rake trim and build a cornice return.*
- *install gutters.*

That part of the exterior finish where the walls meet the roof is called the *cornice*. This section is commonly called the *eaves*. In some parts of the country, the exterior finish work at this section is called a *coving*. On hip or mansard roofs, the cornice extends around the entire building. The trim that extends up the slope of a gable roof on the end walls is called the *rake cornice*. A short section extending around the corner at the end wall is called the *return cornice*. The return cornice provides a means of stopping the rake trim and adds to the design of the building at this point, Figure 18-1. The cornice is a major part of the architectural design of a building. Much care should be taken

Fig. 18-1 A cornice return and the rake trim

in its construction. It is important that all cornice parts be installed straight and true with well-fitted joints.

TYPES OF CORNICES

Cornices are generally classified in three types called the box, the snub, and the open cornice.

- The *box cornice*, Figure 18-2, is probably most commonly used. It not only presents a finished appearance but, because of its overhang, also helps to protect the side-walls from the weather.

Box cornices may be narrow or wide. A narrow box cornice is one in which the cuts on the rafters serve as nailing surfaces for cornice members. A wide box cornice requires additional members for fastening the trim of the cornice, Figure 18-3.

- The *snub cornice*, Figure 18-4, does not provide as attractive an appearance nor give as much protection because of its small overhang. In this type, there is no rafter projection beyond the wall. The snub

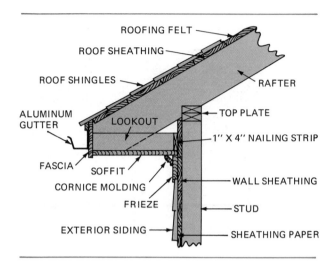

Fig. 18-2 The box cornice and its members

cornice is chosen primarily to cut down the cost of material and labor.

- The *open cornice*, Figure 18-5, is used when the exposed ends of the rafters add appearance to the cornice. It is often used when the rafters are large laminated or solid beams with a wide overhang that exposes the roof decking on the bottom side. Open cornices give contemporary or

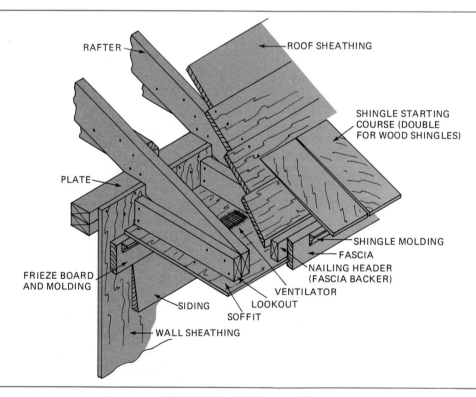

Fig. 18-3 A wide box cornice

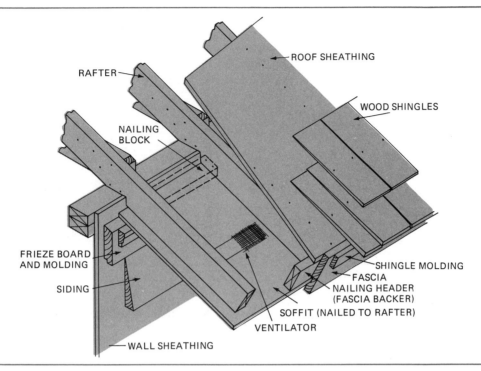

RAFTER

ROOF SHEATHING

WOOD SHINGLES

NAILING BLOCK

FRIEZE BOARD AND MOLDING

SIDING

SHINGLE MOLDING

FASCIA

NAILING HEADER (FASCIA BACKER)

SOFFIT (NAILED TO RAFTER)

VENTILATOR

WALL SHEATHING

Fig. 18-3 (Continued)

rustic design to post-and-beam framing and still provide protection to sidewalls at low cost. This type cornice might also be used for conventionally framed low-cost homes, utility buildings, or cottages.

The bottom side of the roof sheathing on open cornices must present an attractive appearance. Nails or staples must not be driven com-pletely through the exposed sheathing. To present a more attractive appearance, tongue-and-grooved stock, either V-grooved or beaded, may be used with its face down as roof sheathing on that part where the rafters overhang the wall. If the thickness of the sheathing covering the roof is not the same as the sheathing covering the over-hang, the difference is feathered out by using wood shingles as wedges on top of the rafters.

Fig. 18-4 The snub cornice

Fig. 18-5 The open cornice

PARTS OF A CORNICE

The *subfacia*, sometimes called *false* fascia or *rough* fascia, Figure 18-6, is a horizontal framing member attached to the rafter tails to provide an even surface for nailing of other cornice members. It is not always used; when used, it is generally a nominal 1- or 2-inch-thick piece. The width depends on the slope of the roof, the tail cut of the rafters, or the type of cornice construction.

The *soffit*, sometimes called the *planchir*, is the finish member on the underside of the box cornice. Soffit material may be 1-inch nominal thickness solid lumber, matched boards, 1/4- or 3/8-inch plywood, or corrugated aluminum (usually with perforations for ventilation). Special soffit material is also available, usually with precut screened openings for attic ventilation. Soffits may be installed horizontally or fastened to the bottom edge of the rafter tails to the slope of the roof.

Lookouts are horizontal framing members used to support the soffit. They may be of 1- or 2-inch nominal thickness and run from the wall to the end of the rafter. They may be installed on every rafter or spaced up to 48 inches O.C. depending on the construction and materials used.

The *fascia* is a horizontal member of the cornice fastened to the subfacia or rafter tails. Usually of 1-inch nominal thickness it is sometimes grooved to receive the soffit. The fascia usually extends below the soffit 1/4 to 3/8 inch. That portion of the fascia that extends below is called the *drip*. The fascia encloses the outside end of the cornice and provides a nailing surface for the gutter. The fascia may be built up of one or more members to provide more overhang to the roofing or to enhance the beauty of the cornice. In some cases, the fascia may be built up of large crown molding for decorative purposes.

The *frieze* is a horizontal member of the cornice that is fastened to the side wall directly below and against the soffit. Usually of 1-inch-nominal-thickness solid lumber, its bottom edge is sometimes rabbeted to receive the sidewall finish. In other cases, the frieze may be furred away from the sidewall to allow the exterior

Fig. 18-6 The subfacia, or rough fascia, is fastened to the rafter tails.

siding to extend above and behind its bottom edge. However, the frieze is not always used and the sidewall finish is allowed to come up to the soffit. The joint between the siding and the soffit is then covered by a molding.

The *cornice molding* is usually a bed molding used to cover the joint between the frieze and the soffit. Its position remains the same even if the frieze is not used.

Figure 18-7 shows several cross sections of different styles of cornices. Study the location and names of the parts.

BUILDING A NARROW BOX CORNICE

A narrow box cornice requires no lookouts for support of the soffit. The first member of the narrow box cornice to be installed is the soffit. The level cut on the tail of the rafter to which the soffit is to be attached has already been made.

Assuming that the narrow box cornice soffit material is nominal 1-inch solid lumber, prepare it by slightly back-beveling its outside edge with a jack plane. This is done to assure a tight fit between the fascia and the soffit on the

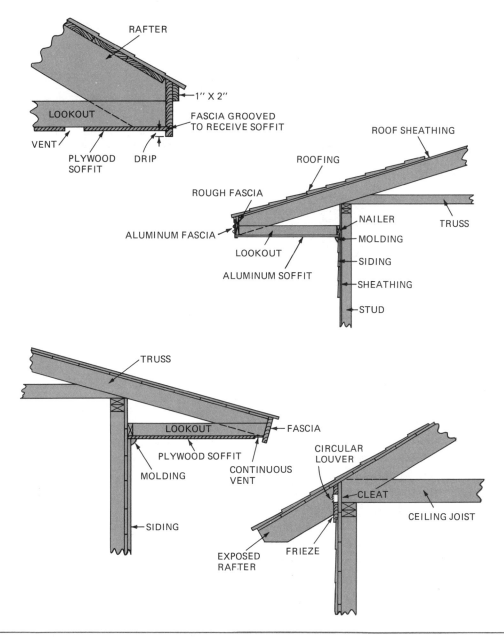

Fig. 18-7 Several styles of cornices

bottom — or where it is seen from the outside, Figure 18-8. When back-beveling, make sure the outside edge is straight and has no crook. If necessary, straighten any crook while planing. Sight the edge to make sure it is straight from one end to the other.

Square and cut one end of the soffit material if the cornice is not returned around the corner. If the cornice is to be returned, miter the end at a 45° angle. Measure and cut the other end square so it falls on the center of

one of the rafter tails. Tack this piece in position against the level cut of the rafter tails, keeping the outside edge even with the plumb cut on the rafter tail. Use a short straightedge on the plumb cut of the rafter tail to determine this, if necessary, Figure 18-9. The position of the first piece of soffit depends on the type of rake cornice. Position the piece so its end will be compatible with the rake finish or the return cornice. In the case of a hip or mansard roof, position it so the soffit material can be joined by

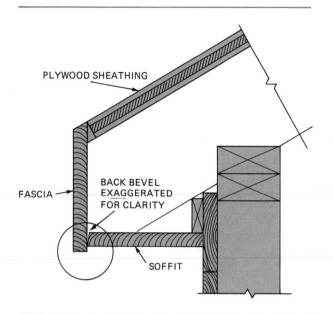

Fig. 18-8 The outside edge of the soffit is back-beveled to assure a tight fit with the fascia.

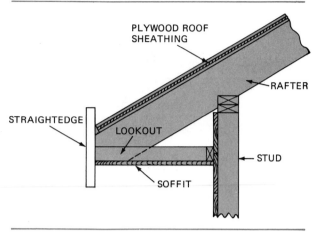

Fig. 18-9 Use a short straightedge to position the soffit

the finish on the other side. Square, cut, and tack successive lengths of soffit material in position until the end is reached. Make the end cut so it can be returned or stopped against the cornice finish on that end.

Make a slight *undercut* on ends of finish that are to be joined when only the face side shows. If one or both edges are exposed, make a square cut for a short distance from the edge and then undercut slightly. Undercutting ends allows a tight joint to be more easily made on the face side. Undercutting also assures that a standing cut is not made on the end. A *standing cut* is the opposite of an undercut where the ends on the back side butt up before the face side comes up tight, Figure 18-10.

After all lengths of soffit material are tacked in place, sight the entire length for straightness in and out (horizontally). Remove nails, as necessary, and move the soffit boards in or out until the outside edges of all boards are in a straight line from one end of the building to the other. Sight by eye from both ends to make sure the outside edges are in a perfectly straight line.

Once the soffit is straight horizontally, sight it for straightness up and down (vertically). Shim any high spots with wood shingles between the soffit and the rafter.

When all boards are straight, both up and down and in and out, nail them into position. Boards that are 8 inches wide or over require three nails. Those that are 6 inches wide or less, require only two nails into each rafter tail. Use galvanized common or casing nails that are at least three times longer than the thickness of the soffit material. Set each nail below the surface.

Any rafter tails that extend beyond the outside edge of the soffit must be trimmed. Install soffits in the same way wherever they are required on the rest of the building.

INSTALLING THE FASCIA

Cut a square or mitered end of the fascia material as may be necessary. Cut the other end square so it rests on the center of a rafter tail. Tack the first piece in position. Tack succeeding pieces in position in a way similar to that used in installing the soffit. When all pieces are tacked

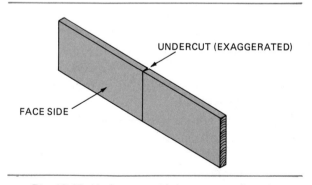

Fig. 18-10 Undercut end joints on exterior trim

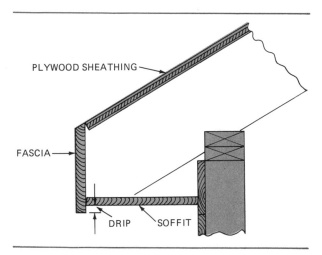

Fig. 18-11 The fascia drips below the soffit by at least 1/4 inch.

into position, sight the fascia for straightness up and down. There is no need to sight for straightness horizontally because nailing the fascia against the soffit will straighten it in this plane. The fascia should extend below the soffit at least 1/4 inch, Figure 18-11. Move the fascia boards as necessary by pulling the nails and retacking to straighten them.

When all of the fascia material is in a straight line, nail it to the rafter tails, Figure 18-12. Use two or three nails into each rafter tail as required. Use galvanized nails and set

them. Nail through the fascia at about 16-inch intervals into the edge of the soffit. Use additional fasteners where required to assure a tight joint between the fascia and soffit.

If the top edge of the fascia extends above the roof sheathing, it is chopped with a hatchet and then planed to the slope of the roof flush with the top side of the roof sheathing.

Often a strip of 1x2 or 1x3 is added at the top edge of the fascia to provide more overhang and to give it a more massive appearance. Snap a line on the fascia and nail the strip to the fascia so its bottom edge is to the line. Use galvanized nails and set them below the surface. Plane the top edge of the strip to the slope of the roof, if necessary.

INSTALLING THE FRIEZE

Before installing the frieze, first apply a narrow strip of sheathing paper on the wall with its top edge against the soffit and so its bottom edge will project below the frieze about 2 to 3 inches, Figure 18-13. This is done to assure a more weatherproof joint between the bottom edge of the frieze and the exterior siding. When the rest of the sheathing paper is applied prior to the exterior siding, the top edge is tucked under this strip.

Fig. 18-12 The fascia is fastened to rafter tails.

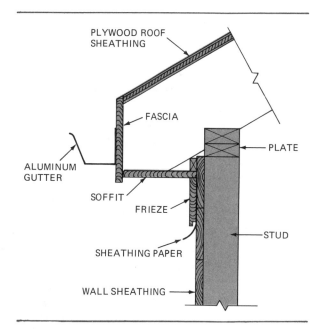

Fig. 18-13 Fasten sheathing paper in back of and extending below the frieze.

Apply furring strips if the frieze is to be furred away from the wall sheathing; or, rabbet the bottom edge of the frieze stock to receive the exterior siding. Fasten the frieze to the sidewall by keeping its top edge against the soffit. Use 2 or 3 galvanized common or casing nails into each stud according to the width of the frieze. Set all nails below the surface. The ends of the frieze are mitered if they are to be returned around the corner of the building. The ends are cut square if they are to be butted against corner boards. If butted against corner boards, the corner boards are installed before the frieze. Corner boards are discussed in a later unit.

INSTALLING CORNICE MOLDING

The cornice molding is applied and covers the joint between the frieze and the soffit. The molding is cut square on the ends if butted against the rake trim or mitered if it is to be returned around the corner. Fasten the molding by driving galvanized finish, casing or common nails through its center at about a 45° angle, approximately 16 inches O.C., or as necessary, to make a tight joint between the frieze and soffit.

If no frieze is used, the cornice molding is installed after the exterior siding has been applied.

BUILDING A WIDE BOX CORNICE

A wide box cornice differs from a narrow box cornice in that lookouts must be installed to support the soffit. To install lookouts, first level from the end of the rafter tail to the wall sheathing. Do this on both ends of the building. Snap a line against the wall sheathing between the two marks. Apply a piece of 1x3 strapping to the wall sheathing so its bottom edge is to the line.

Stretch a line from each end rafter at the intersection of the plumb cut and level cut at its tail end. Install lookouts by fastening the outside end to the rafter tail and the inside end by toenailing into the strapping. At the outside, keep the bottom edge as close as possible to the

stretched line but without touching it. At the inside, keep the bottom edge of the lookout flush with the bottom edge of the strapping.

Installation of other members of the cornice is then done in a manner similar to that used for building a narrow box cornice. If the outside edge of the soffit fits into a groove in the fascia, the soffit edge is extended beyond the tail cut of the rafter a little less than the depth of the groove.

BUILDING A SLOPING BOX CORNICE

In a sloping box cornice, Figure 18-14, the soffit is applied to the bottom edges of the rafter overhang. No lookouts are required. If the fascia is to be installed plumb, the outside edge of the soffit must be beveled to the slope of the roof. If the fascia is installed square to the roof slope, then the outside edge of the soffit is cut square. This soffit is installed and straightened in the same way as horizontal soffits. The fascia and other members of the cornice are then applied in a manner similar to that described previously.

BUILDING A SNUB CORNICE

In a snub cornice, Figure 18-15, the rafter tail cut is made flush with the outside edge of the wall framing. The wall sheathing is applied up to the top edge of the rafter. The roof sheathing is then applied over the top edge of the wall sheathing.

In a snub cornice, no soffit is applied and the fascia acts also as the frieze. The bottom edge may be rabbeted to receive the exterior wall siding. The fascia may also be furred away from the wall to allow the siding to be applied under it.

To install the fascia, first apply a narrow strip of sheathing paper to the top of the wall so its bottom edge extends a few inches below the fascia when installed. Mark the bottom edge of the fascia on both ends of the building. Snap a chalk line between the two marks. Install the fascia so its bottom edge is to the line. For ventilation, bore holes in the fascia centering them on the rafter spaces to receive circular

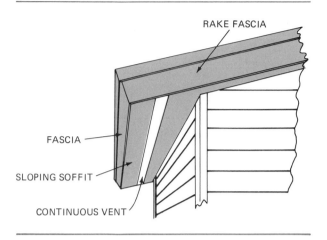

Fig. 18-14 Sloping box cornice

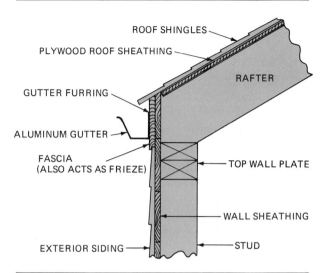

Fig. 18-15 Cross section of a snub cornice

aluminum louvers. These louvers are available in sizes from one to three inches in diameter and press fit into the bored holes.

Use two or three galvanized nails, as necessary, through the fascia into the ends of each rafter. Set all nails below the surface. Make end cuts to be compatible with the rake finish or for returning around the corner of the building. Make square cuts between fascia pieces. Block plane the joints as necessary to make a tight fit. Make a slight undercut on end joints to facilitate a tight joint on the face of the boards. If necessary plane the top edge of the fascia flush with and to the slope of the roof.

BUILDING AN OPEN CORNICE

The open cornice is almost the same as a wide box cornice except that the soffit is eliminated, Figure 18-16. The frieze must also be cut between the rafters. Rafter tails must be cut in an absolutely straight line because any shimming of the fascia to straighten it will show.

Test straightness of rafter tail cuts with a straightedge or by stretching a line over small blocks from one end rafter to the other. Trim any rafter tails that do not lie in a straight line. Ordinarily if an open cornice is to be used on a building, extra care is taken when making rafter tail cuts.

INSTALLING THE FRIEZE

The frieze is usually installed first. Cut and apply some sheathing paper so it will lie behind

the frieze when installed and extend below its bottom edge by about 2 or 3 inches. The frieze is sometimes installed so its bottom edge is flush with the bottom edge of the rafters. In this case, separate pieces are cut between rafters.

In other cases, the bottom of the frieze may extend a few inches below the bottom edge of the rafters. When installed in this manner, the frieze is installed in long lengths with notches in the pieces to fit around each rafter.

In both methods, the frieze must be marked and cut to fit snugly against the sides of the rafters. These cuts are not necessarily square to the edge of the frieze because the rafters may lie tilted slightly one way or the other.

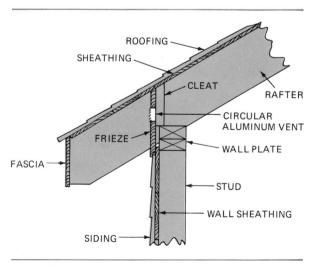

Fig. 18-16 Cross section of an open cornice

To lay out the frieze to assure tight fits, tack a section of frieze stock to the wall with its top edge (previously beveled to the roof slope) against the bottom edge of the rafters. Hold a straightedge that extends across the face of the frieze against the side of the rafter and mark a line across the frieze. Lay out for the next rafter in the same manner. Square lines from a point where the top of the frieze meets the side of the rafters to the bottom edge of the frieze. From a point where the squared line meets the bottom edge of the frieze, draw lines parallel to the first lines laid out. These are the cutting lines. Make cuts slightly under, Figure 18-17.

If the frieze is to be notched around the rafters, the squared line is brought down to the bottom of the notch. The depth of the notch is determined by measuring from a chalk line snapped on the exterior wall for the location of the bottom edge of the frieze to the bottom edge of the rafter.

The frieze is fitted between rafters in this manner along the entire length of the wall and fastened in position. Nailing surfaces for the frieze are provided by nailing cleats to the side of each rafter or by installing blocking between the rafters. The top edge of the frieze, when properly installed, should fit snugly against the bottom side of the roof sheathing. The frieze may be vented in the same manner as the snub

cornice. Small metal circular louvers may be installed between rafters by boring holes of the specified size and simply tapping the louvers in the bored holes.

INSTALLING THE FASCIA

Installing the fascia on an open cornice is done in much the same manner as in a closed cornice. Since no soffit is used, the fascia usually drips below the bottom edge of the rafter tail cut by at least 1/4 inch. However, designs differ and, in some cases, the bottom part of the tail cut of the rafter is left exposed. In any case, the fascia must be installed in a straight, true line with well-fitted joints between the sections.

RAKE CORNICES

The simplest type of rake trim is the *snub rake cornice*. This usually consists of a 1x6 or 1x8 board, furred away from the wall sheathing by the thickness of the exterior siding. Usually a 1x2 or 1x3 is added flush with the top edge to give it added dimension and to provide more overhang, Figure 18-18. This type may be used in combination with any kind of sidewall cornice. If a box cornice is used on the sidewall, a dutchman is added to the bottom edge of the

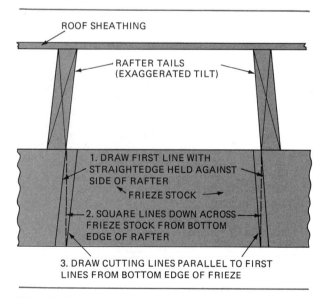

Fig. 18-17 Laying out the frieze of an open cornice to fit between the rafters

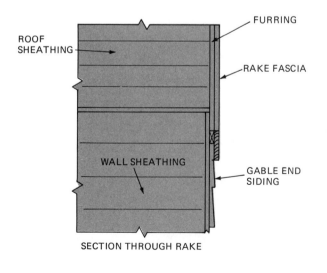

Fig. 18-18 Cross section showing a snub rake cornice

318

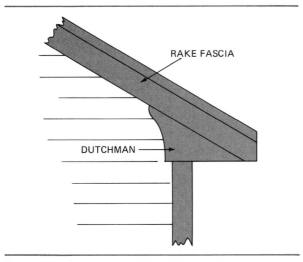

Fig. 18-19 A "dutchman" is added to the rake trim to close in the end of the wall box cornice.

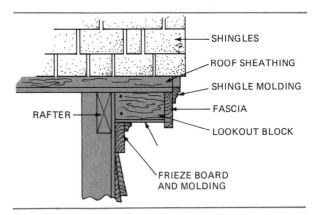

Fig. 18-20 A narrow box cornice on the rakes

rake fascia to close in the box cornice, Figure 18-19. (*Dutchman* is the term given to any odd piece inserted to fill an opening or cover a defect.)

INSTALLING A SNUB CORNICE ON THE RAKES

Determine the thickness of the gable end exterior siding. Apply a furring strip of the same thickness to the gable end with its top edge flush with the top side of the roof sheathing.

Tack the rake fascia to the furring strip, keeping the top edges flush with each other. Let the top end of the fascia extend above the roof on the opposite side. Let the lower end extend by the fascia of the wall cornice by an amount sufficient to stop the gutter against, plus a few inches. Mark the intersection at the peak of the roof by holding a level at the centerline of the ridge and marking a plumb line on the rake fascia. Remove the fascia, cut to the line, and fasten in position so that the plumb cut is in the centerline of the ridge. Fasten the fascia only along the top edge about 16 inches on center. The bottom edge is fastened after the wall siding is applied. Mark, cut, and install the furring strip and rake fascia on the opposite side in the same manner. Build up the fascia with 1x2 or 1x3 strips as may be specified.

The lower end of the rake fascia is usually trimmed after the gutter is installed. Usually

this end is trimmed with a combination of plumb and level cuts. The plumb cut is made about an inch outside of the gutter. The level cut is usually made flush with the bottom edge of the fascia of the sidewall cornice. The dutchman is usually applied after the wall siding.

BOX CORNICES ON THE RAKES

Box cornices on the rakes are built in about the same manner as they are on sidewalls except that they are built to the slope of the roof. When the rake extension is only 6 to 8 inches, the members of the rake cornice can be nailed to a series of short lookout blocks toenailed to the gable end sheathing and also fastened to the projecting roof sheathing, Figure 18-20.

In a moderate overhang of up to 20 inches, both the extending sheathing and a fly rafter aid in supporting the rake cornice. The *fly rafter*, or *barge rafter*, extends from the extended ridge board to the subfascia which connects the ends of the rafters. The roof sheathing boards or plywood should extend from inner rafters to the end of the rake projection to provide rigidity and strength, Figure 18-21. An extra wide overhang at the rake requires a ladder type frame for support, Figure 18-22. (See Unit 14, Figure 14-21, also.)

The wall cornice in a gable roof structure must be joined to the rake cornice in an attractive way. The type of joining depends on the kind of wall and rake cornice.

When the soffit of the wall cornice is installed at the same slope as that of the gable end overhang, the members are returned up the

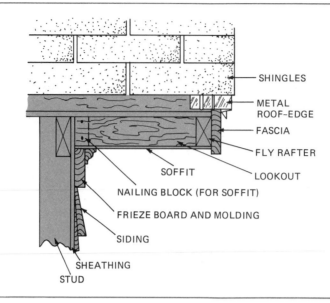

SHINGLES

METAL
ROOF-EDGE

FASCIA

FLY RAFTER

SOFFIT

LOOKOUT

NAILING BLOCK (FOR SOFFIT)

FRIEZE BOARD AND MOLDING

SIDING

SHEATHING

STUD

Fig. 18-21 A moderate overhang at the rakes requires a fly rafter.

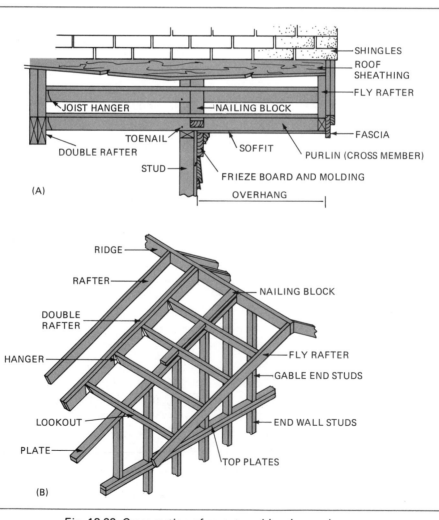

SHINGLES
ROOF
SHEATHING

FLY RAFTER

JOIST HANGER

NAILING BLOCK

FASCIA

TOENAIL

DOUBLE RAFTER

SOFFIT

PURLIN (CROSS MEMBER)

STUD

FRIEZE BOARD AND MOLDING

OVERHANG

(A)

RIDGE

RAFTER

NAILING BLOCK

DOUBLE
RAFTER

HANGER

FLY RAFTER

GABLE END STUDS

LOOKOUT

END WALL STUDS

PLATE

TOP PLATES

(B)

Fig. 18-22 Cross section of an extra wide rake overhang

slope of the roof. This is a simple system and is often used when there are wide overhangs at the sides and ends of the building.

If the rake cornice has no overhang, the wall cornice members are cut square at the end wall and the rake trim is allowed to butt up against and extend by them, Figure 18-23.

In colonial type structures, the wall cornice is returned around the end wall for a short distance. The return projects beyond the building line the same distance on the end of the building as the overhang of the wall cornice. A small sloping roof is constructed on top of the return. Rake cornice members butt against the roof of the return, Figure 18-24.

Prime and paint all exposed cornice members before installation of the gutter. One prime coat and two top coats should be applied.

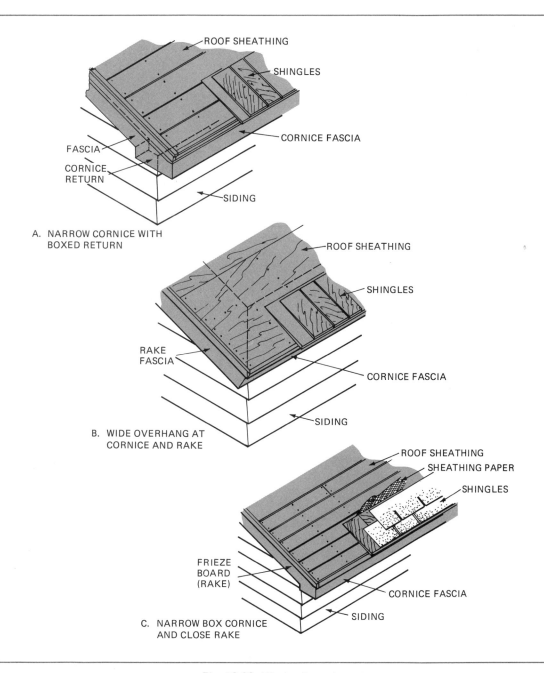

A. NARROW CORNICE WITH BOXED RETURN

B. WIDE OVERHANG AT CORNICE AND RAKE

C. NARROW BOX CORNICE AND CLOSE RAKE

Fig. 18-23 Kinds of cornice returns

Fig. 18-24 A colonial type cornice return

GUTTERS

Gutters may be made of wood, metal, or plastic and are used to collect and drain water from the roof and direct it to *downspouts*, also called *conductors*. The downspouts carry the water downward and away from the building into storm sewers or drywells. In many areas, this system is required to prevent water that is falling from the roof from seeping into basement walls.

WOOD GUTTERS

Wood gutters are usually made of Douglas Fir in sizes 3x4, 4x5, 4x6, and 5x7 and lengths of over 30 feet. Wood gutters are made of cedar, cypress, and redwood, and are rarely used in new construction. The carpenter may be required to replace sections on older homes, however.

METAL GUTTERS

Metal gutters are made in rectangular, beveled, ogee, or semicircular shapes, Figure 18-25. They come in a variety of sizes from 2 1/2 inches to 6 inches in height and from 3 inches to 8 inches in width. Stock lengths run

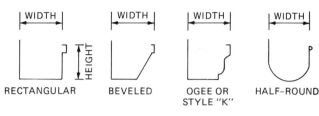

Fig. 18-25 Metal gutter shapes

from 10 to 40 feet. Forming machines are often brought to the job site to form the gutters to practically any desired length.

Besides straight lengths, metal gutters have components comprised of inside and outside corners, joint connectors, outlets, end caps, and others. Metal brackets of various styles are attached to the fascia and support the gutter sections, Figure 18-26.

SLOPE OF THE GUTTER

Most gutters are run level for appearance, however a slight slope toward the downspout is desirable for drainage. Downspouts should be spaced 20 feet minimum and 50 feet maximum. On a short building, the gutter may be installed to slope in one direction with a downspout on one end. On longer buildings, the gutter is

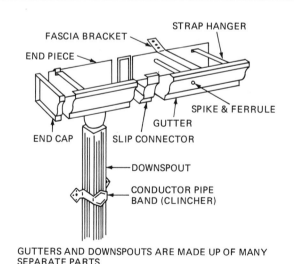

GUTTERS AND DOWNSPOUTS ARE MADE UP OF MANY SEPARATE PARTS

Fig. 18-26 Components of a metal gutter system

PIECE NEEDED	DESCRIPTION
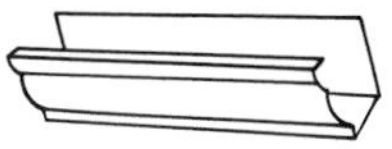	GUTTER COMES IN 10' LENGTH
	SLIP JOINT CONNECTOR USED TO CONNECT JOINTS OF GUTTER
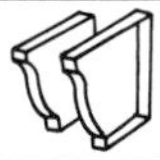	END CAPS — WITH OUTLET USED AT ENDS OF GUTTER RUNS
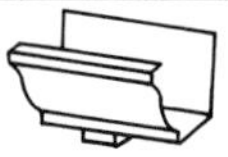	END PIECE — WITH OUTLET USED WHERE DOWNSPOUT CONNECTS
	OUTSIDE MITRE USED FOR OUTSIDE TURN IN GUTTER
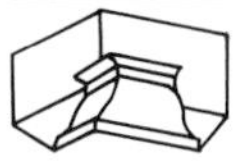	INSIDE MITRE USED FOR INSIDE TURN IN GUTTER
	FASCIA BRACKET USED TO HOLD GUTTER TO FASCIA ON WALL
	STRAP HANGER CONNECTS TO EAVE OF ROOF TO HOLD GUTTER
	STRAINER CAP SLIPS OVER OUTLET IN END PIECE AS A STRAINER
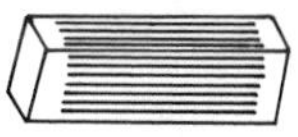	DOWNSPOUT COMES IN 10' LENGTHS
	ELBOW — STYLE B FOR DIVERTING DOWNSPOUT IN OR OUT FROM WALL
	ELBOW — STYLE A FOR DIVERTING DOWNSPOUT TO LEFT OR RIGHT
	CONNECTOR PIPE OR CLINCHER USED TO HOLD DOWNSPOUT SECURELY TO WALL
	SHOE USED TO THROW WATER TO SPLASHER BLOCK
	MASTIC USED TO SEAL ALUMINUM GUTTERS AT JOINTS
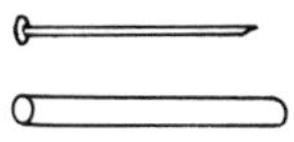	SPIKE & FERRULE USED TO HOLD GUTTER TO EAVE OF ROOF

Fig. 18-26 (Continued)

crowned in the center to allow water to drain to both ends.

LAYING OUT THE GUTTER POSITION

On both ends of the fascia, mark the location of the bottom side of the gutter. The top outside edge of the gutter should be in line with a projection of the top surface of the roof, Figure 18-27. Hold a chalk line to the two marks. Move the center of the chalk line up about 1/4 inch and snap the line on each side. This provides for a 1/4-inch crown in the center. For a slope in one direction only, snap a straight

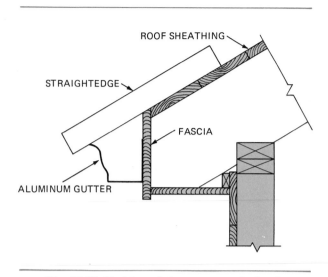

Fig. 18-27 Position of the gutter on the fascia

line about 1/4 inch lower on one end than the other.

INSTALLING METAL GUTTERS

Fasten the brackets to the line on the fascia with screws. Install the gutter sections in the brackets. Use slip-joint connectors to join the sections. Fill joints with the recommended gutter *mastic* (adhesive) before joining.

Install outlet sections at each end or where required. Join with a connector and add the end cap. Use either inside or outside corners where gutters make a turn.

PRACTICAL TIPS FOR INSTALLING CORNICES

The cornice is usually the first part of the exterior trim to be applied. Install the members in a straight and true line where required with well-fitting, tight joints. Use hand tools for cutting, fitting, and fastening exterior trim. Because of its cutting action, a power saw splinters out the face side. A sharp handsaw does the job faster and makes a neater job than a power saw does. Joints may be fitted by block planing or by running a few handsaw cuts through the joint when possible. All fasteners should be well set or countersunk and then puttied to prevent the heads from rusting or corroding. Paint or otherwise protect all exterior trim as soon as possible after installation. Properly installed and protected wood exterior trim will last indefinitely.

REVIEW QUESTIONS

Select the most appropriate answer.

1. That part of the exterior finish where the walls of a building meet the roof is called the
   a. cornice.
   b. fascia.
   c. ridge.
   d. soffit.

2. The exterior trim that extends up the slope of the roof on the gable ends is called the
   a. box finish.
   b. rake finish.
   c. return finish.
   d. snub finish.

3. A member of the cornice generally fastened to the rafter tails is called the
   a. drip.
   b. fascia.
   c. planchir.
   d. soffit.

4. A member of the cornice usually applied under the soffit and against the sidewall is the
   a. fascia.
   b. frieze.
   c. planchir.
   d. rake.

5. The member of a box cornice usually installed first is the
   a. fascia.   c. soffit.
   b. frieze.   d. gutter.

6. The amount the fascia extends below the soffit is called the
   a. drip.   c. projection.
   b. nosing.   d. return.

7. The top edge of the fascia is flush with
   a. the top edge of the rafter.
   b. the top of the roof sheathing.
   c. the bottom of the roof sheathing.
   d. the top edge of the lookout.

8. A cornice that has no soffit is called a
   a. box cornice.   c. rake cornice.
   b. closed cornice.   d. snub cornice.

9. A decorative piece added to the bottom edge of the rake finish to close in the end
   of a wall box cornice is sometimes called a
   a. dutchman.   c. scab.
   b. furring.   d. splice.

10. A device that carries water downward from the roof and away from the building
    is called a
    a. drip.   c. downspout.
    b. gooseneck.   d. gutter.

ROOFING 19

OBJECTIVES

After completing this unit, the student will be able to

- *define roofing terms.*
- *describe kinds of roofing felt, asphalt shingles, and roll roofing and apply them.*
- *describe the various grades and sizes of wood shingles and shakes and apply them.*
- *flash valleys, sidewalls, chimneys, and other roof obstructions.*
- *estimate roofing materials.*

The roof covering is part of the exterior finish. It adds beauty to the exterior and protects the interior. Before roofing is applied, the roof deck must be securely fastened with no loose or protruding nails. All cornice trim should be in place and primed. Properly applied roofing gives years of satisfactory service.

ROOFING TERMS

Some of the terms used in connection with roofing are listed:

- *Square* refers to the amount of roofing required to cover 100 square feet of roof surface.
- *Coverage* refers to the number of overlapping layers of roofing that may occur at any one place on the roof. Roofing may be called single, double, or triple coverage for example.
- *Shingle butt* means the bottom exposed end or edge of the shingle. Wood shingles are tapered from one end to the other. The butt end is the thick end. The thin top end is called the *shingle tip.*
- *Exposure* is the distance between rows (usually called *courses)* of roofing measured from butt to butt. This is the amount that the roofing is exposed to the weather, Figure 19-1.
- *Top lap* refers to the height of the shingle or other roofing minus the exposure. In roll roofing this is also known as the *selvage*

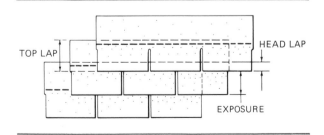

Fig. 19-1 Shingle exposure and lap

- *Head lap* means the distance from the butt of an overlapping shingle to the tip of a shingle two courses below measured up the slope.
- *End lap* is the distance by which adjacent roofing overlaps horizontally.
- *Flashing* refers to a piece of sheet metal (usually lead, zinc, copper, or aluminum) or a piece of asphalt roofing used any place where there is danger of leakage from rain, water, or snow. On a roof, flashing is used at the eaves, around dormers, chimneys, or any other projection.
- *Asphalt cements* and *coatings* are manufactured to various consistencies depending on the purpose for which they are to be used. *Cements* are classified as plastic, lap, and quick setting and will not flow at summer temperatures. They are used as adhesives to bond asphalt roofing products and flashings and are usually troweled on the surface. *Coatings* are usually thin enough to be applied with a brush and are used to resurface old roofing or metal that has become weathered.
- *Electrolysis* is a reaction that occurs when dissimilar metals come in contact with water, causing one of the metals to corrode. When there is a possibility of metals coming in contact with water, a simple way to prevent the corrosion caused by electrolysis is to secure metals with fasteners of the same metal.

ASPHALT ROOFING

Asphalt is commonly used in roofing materials. These materials are generally available in several forms: saturated felts, shingles, and roll roofing. Each form has specific uses and requires specialized methods of application.

SATURATED FELTS

Saturated felts consist of dry felt (heavy paper) saturated with asphalt or coal tar. They are usually made in weights of 15 pounds and 30 pounds per square. One roll of 15 lb felt contains 432 square feet and covers 4 square. A roll of 30 lb felt contains 216 square feet and covers 2 square, Figure 19-2. The rolls are 36 inches wide.

Saturated felts, usually called *asphalt felts*, are used as an underlayment for the finish roofing. This underlayment protects the sheathing from moisture until the roofing is applied and gives additional protection to the roof.

APPLYING THE DRIP EDGE

Before the felt is applied, a metal drip edge or a single course of wood shingles is installed along the roof edges. The metal drip edge, usually made of aluminum or galvanized iron, is applied along the eaves and sometimes along

	Weight per roll	Weight per square	Squares per roll	Roll length	Roll width	Side or end laps	Top lap	Exposure
	60 #	15 #	4	144'	36"	4"	2"	34"
	60 #	20 #	3	108'	36"	to		
Saturated Felt	60 #	30 #	2	72'	36"	6"		

Fig. 19-2 Sizes and weights of saturated felts

Fig. 19-3 Applying a metal drip edge

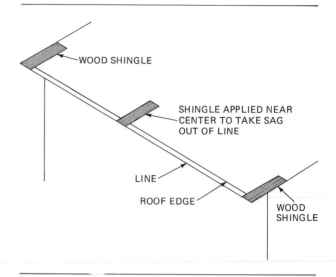

Fig. 19-4 Installing a wood shingle starter course

the rakes, Figure 19-3. It is designed to support the roofing overhang to allow water run-off to drip in the center of the gutter, or free of underlying construction.

Install the metal drip edge by using roofing nails of the same metal spaced 8 to 10 inches along its inner edge. Lap end joints by about 2 inches. If a drip edge is used at the rakes, let it overlap the drip edge at the eaves.

The wood shingle starter course serves the same purpose as the metal drip edge. However, wood shingles are used only along the eaves and not on the rakes.

When using wood shingles for the drip edge, stretch a line from one end of the building to the other to the desired overhang. Sight the line for straightness and install temporary shingles at intermediate points. Fasten the line to butts of these shingles to assure a straight line from one end of the building to the other, Figure 19-4.

Apply the shingles along the edge of the building; have their butts come as close to the line as possible without touching it. Sight each shingle for a bow and fasten it with the crown of the bow up. Use two 3d galvanized nails halfway up the shingle about one-half inch in from each edge. Also fasten two nails through each shingle into the top edge of the fascia. Continue applying shingles until the course is completed.

Although the metal drip edge can be applied faster and is used more, some contractors prefer not using metal as part of the starter course for asphalt shingles. A metal drip edge is not needed on a wood shingle roof.

APPLYING SATURATED FELT

When the drip edge is complete apply a layer of 15 lb asphalt felt over the deck as an underlayment. Lay each course of felt over the lower course at least 2 inches. End lap at least 4 inches. Lap the felt 6 inches from both sides over all hips and ridges, Figure 19-5.

Nail or staple through each lap and through the center of each layer about 16 inches O.C. Roofing nails in combination with metal discs or specially designed felt fasteners hold the underlayment securely in strong winds until the finish roofing is applied.

Fig. 19-5 An asphalt felt underlayment is applied to the roof deck before shingling.

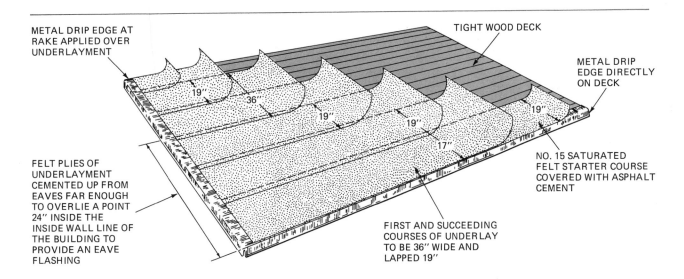

LABELS:
METAL DRIP EDGE AT RAKE APPLIED OVER UNDERLAYMENT

TIGHT WOOD DECK

METAL DRIP EDGE DIRECTLY ON DECK

FELT PLIES OF UNDERLAYMENT CEMENTED UP FROM EAVES FAR ENOUGH TO OVERLIE A POINT 24" INSIDE THE INSIDE WALL LINE OF THE BUILDING TO PROVIDE AN EAVE FLASHING

NO. 15 SATURATED FELT STARTER COURSE COVERED WITH ASPHALT CEMENT

FIRST AND SUCCEEDING COURSES OF UNDERLAY TO BE 36" WIDE AND LAPPED 19"

LOW SLOPE APPLICATION

If asphalt shingles are to be used on slopes lower than 4 inches per foot but not less than 2 inches per foot, a double layer of felt is required. Use a 19-inch course laid along the eaves and completely covering the drip edge or starter course. Then lay 36-inch-wide sheets, each overlapping the preceding course by 19 inches and exposing 17 inches of the underlying sheet, Figure 19-6. Use only enough fasteners to hold the felt securely to the deck until the shingles are applied.

The underlay plies are also cemented together to form an eaves flashing up the roof far enough to reach a point at least 24 inches beyond the inside wall line of the building. The layers are cemented together using a continuous layer of plastic asphalt cement applied at the rate of two gallons per 100 square feet. It is important to spread the cement uniformly with a comb trowel.

ASPHALT SHINGLES

Asphalt shingles are made of saturated felt which is coated with asphalt and surfaced with selected mineral granules. The materials most frequently used for mineral surfacing are naturally colored slate or rock granules in natural form or colored. They come in a wide variety of colors, shapes, and weights, Figure 19-7. The quality of a shingle is determined by its weight. Some shingles are provided with factory-applied adhesive to increase their resistance to wind.

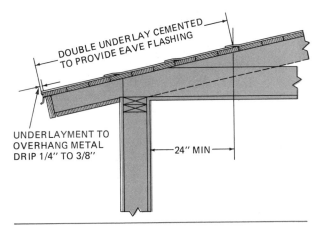

Fig. 19-6 A double coverage of felt is required prior to shingling low slopes. (Courtesy of Asphalt Roofing Manufacturers Association)

Before applying strip shingles, make sure that the roof deck is properly prepared with the underlayment and drip edge applied. Because asphalt roofing products become soft in hot weather, extreme care should be taken not to damage them by digging in with heavy shoes during application or by unnecessary walking on the surface after application.

ROOF SLOPE

The pitch of a roof should not be less than 4 inches per foot for any type of asphalt shingle when conventional methods of application are used. On lower roof slopes, down to 2 inches of rise per foot of run, self-sealing strip shingles are used. If "free tab" strip shingles are used,

PRODUCT	Configuration	Per Square			Size		Exposure	Underwriters' Listing
		Approx. Shipping Weight	Shingles	Bundles	Width	Length		
Wood Appearance Strip Shingle More Than One Thickness Per Strip 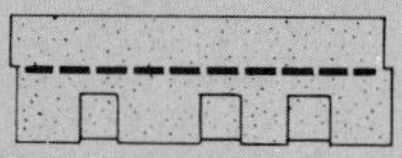 Laminated or Job Applied	Various Edge, Surface Texture & Application Treatments	285# to 390#	67 to 90	4 or 5	11-1/2" to 15"	36" or 40"	4" to 6"	A or C — Many Wind Resistant
Wood Appearance Strip Shingle Single Thickness Per Strip 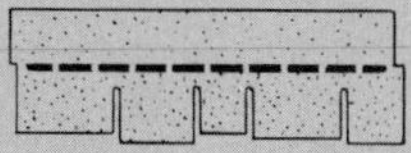	Various Edge, Surface Texture & Application Treatments	Various 250# to 350#	78 to 90	3 or 4	12" or 12-1/4"	36" or 40"	4" to 5-1/8"	A or C — Many Wind Resistant
Self-Sealing Strip Shingle 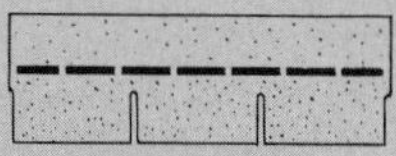	Conventional 3 Tab	205#– 240#	78 or 80	3	12" or 12-1/4"	36"	5" or 5-1/8"	A or C — All Wind Resistant
	2 or 4 Tab	Various 215# to 325#	78 or 80	3 or 4	12" or 12-1/4"	36"	5" or 5-1/8"	
Self-Sealing Strip Shingle 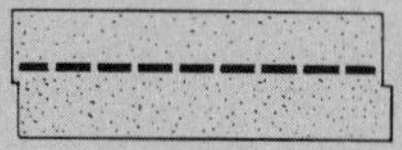 No Cut Out	Various Edge and Texture Treatments	Various 215# to 290#	78 to 81	3 or 4	12" or 12-1/4"	36" or 36-1/4"	5"	A or C — All Wind Resistant
Individual Lock Down 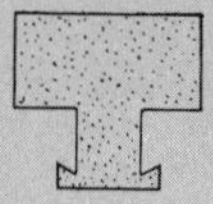 Basic Design	Several Design Variations	180# to 250#	72 to 120	3 or 4	18" to 22-1/4"	20" to 22-1/2"	—	C — Many Wind Resistant

Other types available from some manufacturers in certain areas of the Country. Consult your Regional Asphalt Roofing Manufacturers Association manufacturer.

Fig. 19-7 Asphalt shingles are available in a wide variety of sizes, shapes, and weights.
(Courtesy of Asphalt Roofing Manufacturers Association)

cement each tab down with a spot of quick--setting asphalt cement about 1 1/2 inches in diameter, Figure 19-8. A double coverage felt underlayment must also be provided as previously described.

EAVES FLASHING

Whenever there is a possibility of ice dams forming along the eaves and causing a backup of water, an eaves flashing is needed. For a slope of at least 4 inches rise per foot of run, install a course of 36-inch wide, 90-pound mineral surface roll roofing or a course of smooth roll roofing of not less than 50 pounds. Apply the flashing to overhang the drip edge by 1/4 to 3/8 inch.

The flashing should extend up the roof far enough to cover a point at least 12 inches inside the wall line of the building. If the overhang of

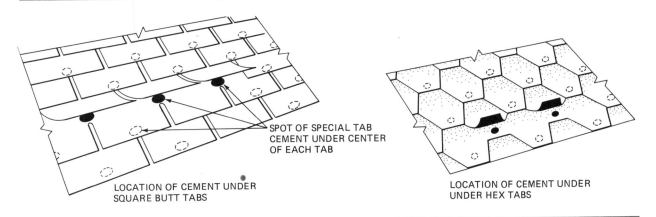

SPOT OF SPECIAL TAB
CEMENT UNDER CENTER
OF EACH TAB

LOCATION OF CEMENT UNDER
SQUARE BUTT TABS

LOCATION OF CEMENT UNDER
UNDER HEX TABS

Fig. 19-8 On low slopes, use self-sealing shingles or spot free tabs with asphalt cement.
(Courtesy of Asphalt Roofing Manfuacturers Association)

the eaves requires that the flashing be wider than 36 inches, the necessary horizontal lap joint is cemented and located on the portion of the roof that extends outside of the wall line, Figure 19-9.

ASPHALT SHINGLE LAYOUT

On small roofs, strip shingles are applied by starting from either rake. On long buildings, a more accurate vertical alignment is insured by starting at the center and working both ways, Figure 19-10.

Mark the center of the roof at the eaves and at the ridge. Snap a chalk line between the marks. Snap a series of chalk lines 6 inches apart on each side of the centerline if the shingle tab cutouts are to break on the halves. Snap lines 4 inches apart if the cutouts are to break on the thirds. When applying the shingles, start the course with the end of the shingle to the chalk line. Start succeeding courses in the same manner, breaking the joints as necessary. Pyramid the shingles up in the center and work both ways toward the rakes.

If it is decided to start at the rakes and cutouts are to break on the halves, start the first full course with a full strip. The second course is started with a strip from which 6 inches has been cut; the third course, with a strip from which the entire first tab is removed; the fourth, with one and one-half tabs removed; and so on,

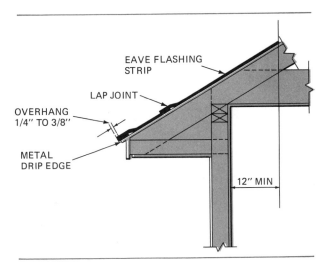

EAVE FLASHING
STRIP

LAP JOINT

OVERHANG
1/4" TO 3/8"

METAL
DRIP EDGE

12" MIN

Fig. 19-9 Eave flashing strip for
a normal slope roof

Fig. 19-10 On long roofs, start shingling in the center
and work toward the rakes.

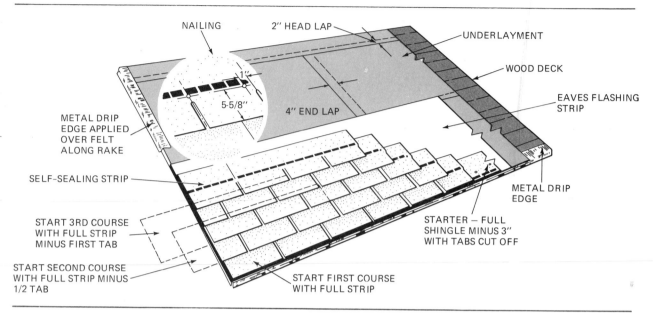

NAILING 2" HEAD LAP UNDERLAYMENT WOOD DECK EAVES FLASHING STRIP

1" 5-5/8" 4" END LAP METAL DRIP EDGE

METAL DRIP EDGE APPLIED OVER FELT ALONG RAKE

SELF-SEALING STRIP

START 3RD COURSE WITH FULL STRIP MINUS FIRST TAB

START SECOND COURSE WITH FULL STRIP MINUS 1/2 TAB

STARTER — FULL SHINGLE MINUS 3" WITH TABS CUT OFF

START FIRST COURSE WITH FULL STRIP

Fig. 19-11 Layout to break cutouts on the halves starting from the rake

Figure 19-11. These starting strips are precut for faster application. Waste from these strips is used on the opposite rake.

If the cutouts are to break on the thirds, cut the starting strip for the second course by removing 4 inches. Remove 8 inches from the strip for the third course, and so on, Figure 19-12.

Cut the shingles by scoring them on the back side with a utility knife. Use a square as a guide for the knife. Bend the shingle; it will break on the scored line.

The layouts just described may have to be adjusted so that tabs on opposite rakes will be of approximate equal widths. It is also recommended that no rake tab be less than 3 inches in width.

STARTER COURSE OF ASPHALT SHINGLES

For the starter course, cut the tab portion off the necessary number of strip shingles and apply with the adhesive strip flush with the roof

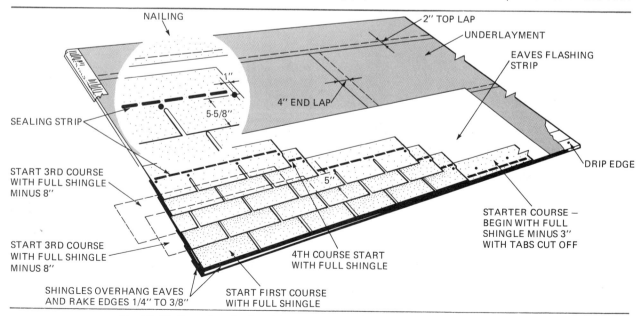

NAILING 2" TOP LAP UNDERLAYMENT EAVES FLASHING STRIP

1" 5-5/8" 4" END LAP

SEALING STRIP

START 3RD COURSE WITH FULL SHINGLE MINUS 8"

START 3RD COURSE WITH FULL SHINGLE MINUS 8"

5"

DRIP EDGE

STARTER COURSE — BEGIN WITH FULL SHINGLE MINUS 3" WITH TABS CUT OFF

4TH COURSE START WITH FULL SHINGLE

SHINGLES OVERHANG EAVES AND RAKE EDGES 1/4" TO 3/8"

START FIRST COURSE WITH FULL SHINGLE

Fig. 19-12 Procedure for breaking cutouts on the thirds starting from the rake

Fig. 19-13 A shingle tab is cut to cover the projecting edge of the rake trim.

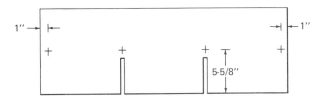

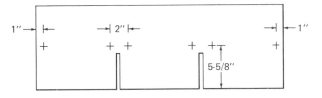

Fig. 19-14 Four to six nails are used to fasten three-tab asphalt strip shingles.

edge. Start the course so that no joint falls in line with a joint or cutout of the course above. Use four galvanized or aluminum roofing nails in each strip spaced equally and slightly above the adhesive.

If the rake trim projects beyond the fascia, a shingle tab about 4 or 5 inches wide is installed to cover the top edge of the rake trim, Figure 19-13. A wood shingle should be first placed under the tab to give it support.

Lay the first full course of shingles on top of the starter course, keeping their bottom edges flush. The number of nails used and their location is specified in the direction sheet provided by the manufacturer. Generally, four or six nails are used, Figure 19-14. The nail length should be sufficient to penetrate nearly the full thickness of the roof sheathing, but should not go entirely through. Drive nails straight so that the edge of the head does not cut into the shingle. Pneumatic staplers are often used to fasten asphalt shingles. Special staples with an extra wide crown are used, Figure 19-15.

Start the course in the layout manner described previously. Cut shingles to overhang the rake by 1/4 to 3/8 inch. A thin strip of wood tacked to the rake trim serves as a guide for marking the end shingles.

EXPOSURE

The maximum exposure of asphalt shingles to the weather depends on the type of shingle. Recommended maximum exposures range from 4 to 6 inches. Most commonly used asphalt shingles have a maximum exposure of 5 inches. Less than the maximum recommended exposure may be used, if desired.

LAYING OUT SHINGLE COURSES

When laying out shingle courses, space up each rake from the top edge of the first course of shingles to gain the desired exposure. Snap

Fig. 19-15 Pneumatic staplers are often used to fasten asphalt shingles.

Fig. 19-16 Lay shingle courses so the top edge is to the chalk line.

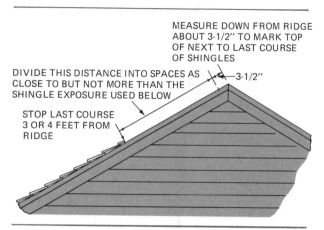

MEASURE DOWN FROM RIDGE ABOUT 3-1/2" TO MARK TOP OF NEXT TO LAST COURSE OF SHINGLES

DIVIDE THIS DISTANCE INTO SPACES AS CLOSE TO BUT NOT MORE THAN THE SHINGLE EXPOSURE USED BELOW

3-1/2"

STOP LAST COURSE 3 OR 4 FEET FROM RIDGE

Fig. 19-17 Space shingle courses evenly to the ridge

5 or 6 lines across the roof. When snapping a long line, it may be necessary to thumb the line down against the roof in the center and snap the lines on both sides of the thumb.

Lay the courses so their top edges are to the chalk line, Figure 19-16. Start each course so the cutouts are broken (staggered) in the desired manner. Continue snapping lines and applying courses until a point 3 or 4 feet below the ridge is reached.

SPACING SHINGLE COURSES TO THE RIDGE

The last course of shingles should be exposed below the ridge cap by about the same amount as all the other shingle courses. Cut a full tab from an asphalt strip. Center it on the ridge and bend it over the ridge. Do this at both ends of the building and mark the bottom edges on the roof. Measure up two inches and snap a line between the marks. This line will be the top edge of the next to last course of shingles.

Space between this line and the top of the last course of shingles applied and mark as close as possible to the exposure of previous courses, Figure 19-17. Do not exceed the maximum exposure. Shingle up to the top.

The line for the last course of shingles is snapped on the face of the course below. Measure up 5 inches from the bottom edge and snap a line. Fasten the last course of shingles by placing their butts to the line. Bend the top edges over the roof and fasten them to the opposite slope.

APPLYING THE RIDGE CAP

Cut hip and ridge shingles from 12x36-inch square butt shingle strips to make approximately 12x12-inch squares. Cut shingles from the cutout to the top edge on a slight taper so the top edge is narrower than the bottom edge, Figure 19-18. Cutting the shingles in this manner keeps the top half of the shingle from protruding when it is bent over the ridge.

The ridge cap is applied after both sides of the roof have been shingled. At each end of the roof, center a shingle on the ridge and bend it over the ridge. Mark its bottom edge on the slope toward the front of the building. Snap a line between the marks.

Beginning at the bottom of a hip or at one end of the ridge, apply the shingles over the hip or ridge exposing each 5 inches. On the ridge,.

Fig. 19-18 Hip and ridge shingles are cut from strip shingles.

shingles are started from the end away from prevailing winds, so the wind blows over the shingle butts, not against them. Keep one edge, from the butt to the start of the taper, to the chalk line. Secure each shingle with one nail on

each side, 5 1/2 inches from the exposed end and 1 inch up from the edge, Figure 19-19. Space the shingles for an equal exposure when nearing the end. The last ridge shingle is cut to size and is face-nailed with one nail on each side of the ridge. The nail heads are covered with asphalt cement to prevent leakage.

NOTE: If shingles are cold, warm them before bending them over the ridge, to prevent cracking them.

ROLL ROOFING

Roll roofing is made of the same materials as asphalt shingles, in a number of different types and weights, Figure 19-20. Some types are applied with exposed nails and have a top lap of 2 to 4 inches. A concealed nail type called *double-coverage* has a top lap of 19 inches.

Roll roofing is recommended for use on roofs with slopes less than 4 inches rise per foot. However, the exposed nail type should

Fig. 19-19 Applying ridge shingles

TYPICAL ASPHALT ROLLS

1	2		3	4		5	6	7
PRODUCT	Approximate Shipping Weight		Sqs. Per Package	Length	Width	Side or End Lap	Top Lap	Exposure	Underwriters' Listing
	Per Roll	Per Sq.							
MINERAL SURFACE ROLL	75# to 90#	75# to 90#	One	36' 38'	36" 36"	6"	2" 4"	34" 32"	C
			Available in some areas in 9/10 or 3/4 Square rolls.						
MINERAL SURFACE ROLL DOUBLE COVERAGE	55# to 70#	55# to 70#	One Half	36'	36"	6"	19"	17"	C

Fig. 19-20 Types of roll roofing

not be used on pitches less than 2 inches rise per foot. Roll roofing applied with concealed nails and having a top lap of at least 3 inches may be used on pitches as low as 1 inch per foot. For maximum life, the concealed nail type roofing is recommended.

GENERAL APPLICATION METHODS

Apply all roll roofing when the temperature is above 45°F to avoid cracking the coating. Cut the roll in 12- to 18-foot lengths and spread in a pile on a smooth surface to allow them to flatten out.

Use only the lap or quick-setting cement recommended by the manufacturer. Store cement in a warm place until ready for use. If necessary to warm it rapidly, place the unopened container in hot water.

> **CAUTION: The materials are flammable and should never be warmed over an open fire or in direct contact with a hot surface.**

Use large head (3/8-inch diameter) galvanized nails long enough to securely hold the roofing, but not so long as to completely penetrate the roof sheathing.

Apply roll roofing only on a solid, smooth, well-seasoned deck where the area below has sufficient ventilation to prevent condensation being absorbed by the deck, causing it to warp.

APPLYING ROLL ROOFING WITH EXPOSED NAILS

Apply the first course of roll roofing with exposed nails so that the lower edge and ends extend over eaves and rakes about 3/8 inch. Nail along the top edge about every 18 inches to temporarily secure it until the second course is laid. Nail along the eaves and rakes about 1 inch in, with nails placed about 2 inches O.C., and slightly staggered to help prevent splitting the roof boards.

Apply the second course by nailing it along its top edge about 18 inches O.C., so it laps the first course by 2 inches. Lift the lower edge and apply lap cement evenly on the upper two inches of the first course. Bed the top course in the cement and nail through the lap about 2 inches O.C., staggering the nails slightly.

Nail along the rakes in the same manner as the first course. Lap end joints by at least 6 inches. Apply lap cement to the joint and nail in a staggered fashion 4 inches O.C., Figure 19-21.

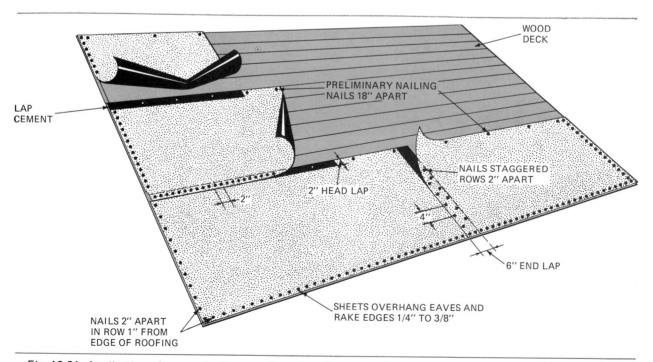

Fig. 19-21 Application of exposed nail-type roll roofing *(Courtesy of Asphalt Roofing Manufacturers Association)*

Continue in like manner, trimming the last sheet even with the ridge.

FINISHING HIPS AND RIDGES

Cut strips of roll roofing 12 inches wide. Snap chalk lines 5 1/2 inches down on each side of the hip or ridge. Apply a 2-inch-wide band of asphalt cement on each side of the hip or ridge just above the chalk line.

Bend the roofing strips in the center and lay them over the hip or ridge and bed them in the cement. Fasten the strips along both edges with nails about 3/4 inch in from the edge and spaced on 2-inch centers, Figure 19-22.

NOTE: It is important that all exposed nails penetrate the cemented portion of the lap in order to seal the nail hole.

APPLYING ROLL ROOFING WITH CONCEALED NAILS

Apply 9-inch wide strips of the roofing along the eaves and rakes overhanging about 3/8 inch. Fasten the strips with two rows of nails 1 inch from each edge and spaced about 4 inches O.C.

Apply the first course with its edges and ends flush with the strips. Secure the upper edge with nails staggered about 4 inches apart. Do not fasten within 18 inches of the rake edge. Apply cement to only that part of the

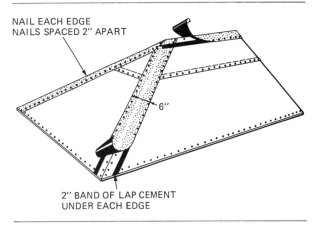

Fig. 19-22 Finishing hips and ridges with exposed nail-type roll roofing *(Courtesy of Asphalt Roofing Manufacturers Association)*

edge strips covered by the first course and press the lower edge and rake edges of the first course firmly in place over the edge strips. Finish nailing the upper edge out to the rakes. Apply succeeding courses in like manner. Make all end laps 6 inches wide and apply cement the full width of the lap.

After all courses are in place, lift the lower edge of each course and apply the cement in a continuous layer over the full width and length of the lap. Press the lower edges of the upper courses firmly into the cement so a small bead appears along the entire edge of the sheet, Figure 19-23.

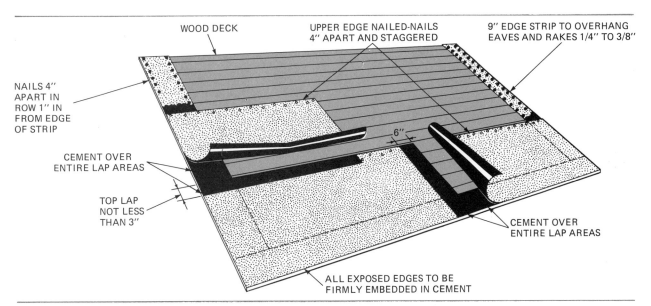

Fig. 19-23 Recommended procedure for applying concealed nail-type roll roofing

COVERING HIPS AND RIDGES

Cut strips of roofing 12x36 inches. Bend the pieces lengthwise through their centers. Snap a chalk line on both sides down 5 1/2 inches from the hip or ridge. Apply cement from one line over the top to the other line.

Fit the first strip over the hip or ridge and press it firmly into place. Start at the lower end of a hip and at either end of a ridge. Lap each strip 6 inches over the preceding one. Nail each strip only on the end which is to be covered by the overlapping piece. Spread cement on the end of each strip that is lapped before the next one is applied. Continue in this manner until the end is reached, Figure 19-24.

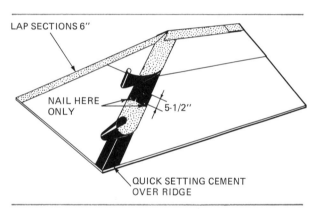

Fig. 19-24 Finishing hips and ridges using concealed nail-type roll roofing

APPLYING DOUBLE COVERAGE ROLL ROOFING

Cut the 19-inch selvage portion from enough roofing to cover the entire length of the roof. Save the surfaced portion for the last course. Apply the selvage portion parallel to the eaves so it overhangs the drip edge by 3/8 inch. Secure it to the roof deck with three rows of nails. Place the top row 4 1/2 inches below the upper edge and bottom row 1 inch above the bottom edge. Place the other row halfway between. Place the nails in the upper and middle rows slightly staggered about 12 inches apart. Place the nails in the lower row about 6 inches apart and slightly staggered. Nail along rakes in the same manner.

Apply the first course and secure it with two rows of nails in the selvage portion. Place one row about 4 3/4 inches below the upper edge and the second about 8 1/2 inches below the first. Space the nails about 12 inches apart in each row and stagger them, Figure 19-25.

Apply succeeding courses in the same manner lapping the full width of the 19-inch selvage each time. Make all end laps 6 inches wide. End laps are made in the manner shown in Figure 19-26. Stagger end laps in succeeding courses.

Lift and roll back the surface portion of each course. Starting at the bottom, apply cement to the entire selvage portion of each

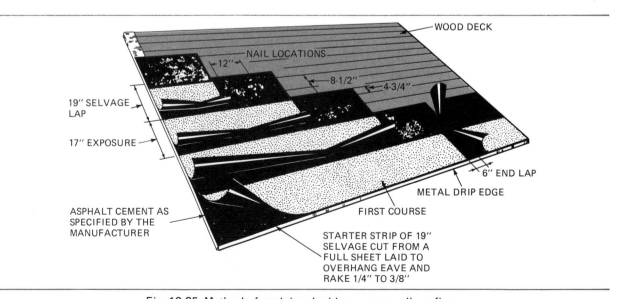

Fig. 19-25 Method of applying double coverage roll roofing

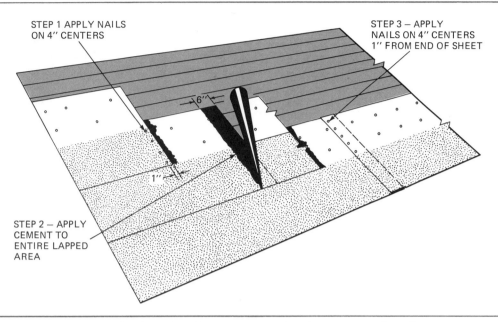

Fig. 19-26 Method of making end laps on double coverage roll roofing

course. Apply it to within 1/4 inch of the surfaced portion. Press the overlying sheet firmly into the cement. Apply pressure over the entire area using a light roller to insure adhesion between the sheets at all points.

Apply the remaining surfaced portion left from the first course as the last course. This type roofing may also be applied in like manner parallel to the rakes, Figure 19-27. Hips and ridges are covered in the same manner shown in Figure 19-28.

FLASHING

Flashing is used to keep water from entering a building. Besides the eaves, other parts of

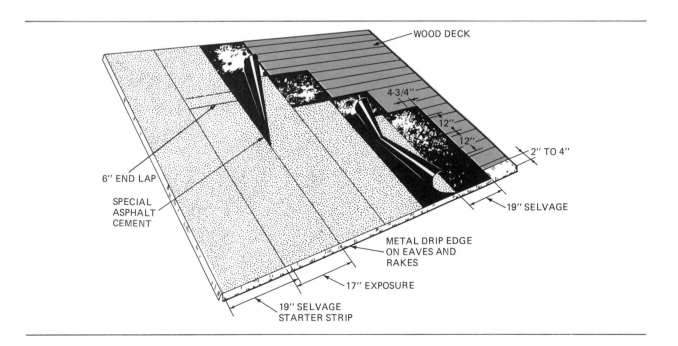

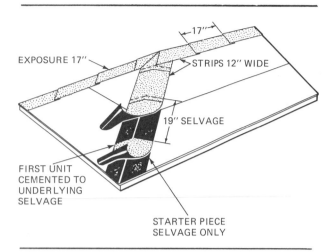

Fig. 19-28 Method of covering hips and ridges with double coverage roofing

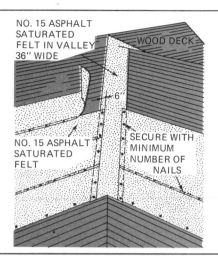

Fig. 19-29 Apply a 36-inch-wide felt underlay in the valley before flashing.

the roof such as valleys, dormers, sidewalls, chimneys, vents, and other projections must be flashed to prevent leakage at the intersections.

VALLEYS

Valleys are formed where two sloping roofs meet. Because drainage concentrates in valleys, they are especially vulnerable to leakage. Valleys must be carefully flashed according to recommended procedures. Valleys are generally flashed in two ways called open and closed valley flashing.

OPEN VALLEY FLASHING

Flashing material for an open valley is of sheet aluminum, copper, zinc, or mineral-surfaced asphalt roll roofing. Advantages of roll roofing are that it is less expensive and that colors that match or contrast with the roof covering can be used. If it is properly applied, roll roofing used as a valley flashing will outlast the main roof covering. Sheet metal, especially copper, may last longer. However, it is good practice to replace all flashing when reroofing.

Prepare the valley by applying a 36-inch-wide strip of 15 pound asphalt felt centered in the valley. Fasten it with only enough nails along its edges to hold it in place. The courses of felt underlay applied to the roof are cut to overlap the valley strip by not less than 6 inches. Seat the felt well into the valley. Be

careful not to cause any break. The eave flashing, if required, is then applied, Figure 19-29.

USING ROLL ROOFING

Lay an 18-inch wide layer of mineral-surfaced roll roofing centered in the valley with its mineral surfaced side down. Use only enough nails spaced 1 inch in from each edge to hold the strip smoothly in place. Press the roofing firmly in the valley when nailing the opposite edge.

On top of the first strip, lay a 36-inch wide strip with its surfaced side up. Center it in the valley and fasten it in the same manner as the first strip.

Snap a chalk line on each side of the valley as a guide for trimming the ends of the courses of the roof covering. These lines are spaced 6 inches apart at the ridge and spread 1/8 inch per foot as they approach the roof edge. Thus, a valley 16 feet long will be 8 inches wide at the eaves. The upper corner of each end shingle is clipped to help keep water from entering between the courses. The roof shingles are cemented to the valley flashing with plastic asphalt cement, Figure 19-30.

USING METAL FLASHINGS

Prepare the valley with underlayment only in the same manner as described previously. Lay a strip of metal centered in the valley. The metal should extend at least 10 inches on each side of the valley centerline for slopes with a 6-inch rise or less and 7 inches on each side for a steeper slope. Carefully press and form it to the valley. Fasten the metal with nails of similar material spaced close to its outside edges with only enough fasteners to hold it smoothly in place. Snap lines on each side of the valley as

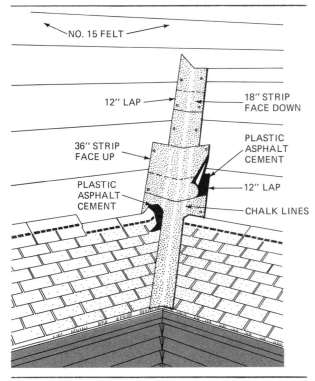

Fig. 19-30 Recommended use of roll roofing for an open valley flashing

described previously. End shingles of each course are applied in the same manner as for a roll roofing flashing.

CLOSED VALLEY FLASHING

Closed valleys are those where the roof shingles meet in the center of the valley and completely cover the valley flashing. Using a closed valley protects the valley flashing, thereby adding to the weather resistance at very vulnerable points. Several methods are used to flash closed valleys.

The first step for any method is to apply the asphalt felt underlayment as previously described for open valleys. Then center a 36-inch width of at least 50-pound roll roofing in the valley over the underlayment. Form it smoothly in the valley and secure it with only as many nails as necessary.

WOVEN VALLEY METHOD

Valleys are commonly flashed by applying shingles on both sides of the valley and weaving each course in turn over the valley. This is called the *woven valley*, Figure 19-31.

Lay the first course of shingles along the edge of the roof up to and over the valley for a distance of at least 12 inches. Lay the first course along the edge of the adjacent roof and extend the shingles over the valley on top of the previously applied shingles.

Fig. 19-31 Valleys are often flashed by weaving shingles. *(Courtesy of Bird and Son)*

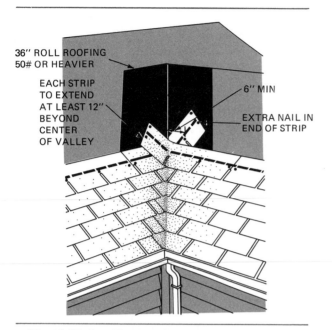

Fig. 19-32 Specifications for a woven valley flashing

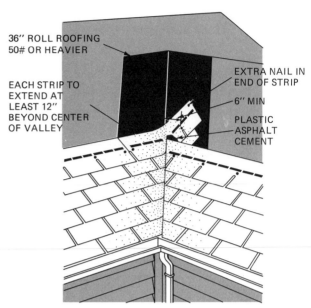

Fig. 19-33 A closed cut valley flashing

Succeeding courses are then applied, weaving the valley shingles alternately, first on one roof and then on the other. When weaving the shingles, make sure they are pressed tightly into the valley and that no nail is located closer than 6 inches to the valley centerline. Fasten the end of the woven shingle with two nails, Figure 19-32. Some contractors prefer to cover each roof area first to a point approximately three feet from the valley and weave the valley shingles in place later.

No end joints should occur in or near the center of the valley. Therefore, it may be necessary to occasionally cut a strip short that would otherwise end near the center. Continue from this cut end with a full length strip over the valley.

APPLYING A CLOSED CUT VALLEY

Apply the shingles to one roof area, letting the end shingle of every course overlap the valley by at least 12 inches. Make sure no end joints occur at or near the center of the valley. Nail at least 6 inches from the center, forming each shingle snugly in the valley. Place two nails at the end of the valley shingle.

Snap a chalk line along the center of the valley. Apply shingles to the adjacent roof area and cut the end shingle to the chalk line. Clip the upper corner of each shingle as described

previously for open valleys. Bed the end that lies in the valley in about a 3-inch wide strip of asphalt cement. Make sure that no nails are located closer than 6 inches to the valley centerline, Figure 19-33.

APPLYING INDIVIDUAL METAL FLASHINGS

When applying individual metal flashings, first estimate the number of shingle courses required to reach the ridge. Using tin snips, cut a piece of metal flashing for each course of shingles. Each piece should be at least 18 inches wide for slopes with at least a 6-inch rise and 24 inches wide for slopes with less than a 6-inch rise. The height of each piece should be at least 3 inches more than the shingle exposure, Figure 19-34.

Prepare the valley with underlayment as described previously. Snap a chalk line in the center of the valley. Apply the starter course and trim the end shingle of each course to the chalk line. Fit and form the first piece of flashing to the valley. Trim its bottom edge flush with the drip edge. Fasten it in the valley with nails of like material to prevent electrolysis. Fasten along the outer edge of the flashing, using only as many nails as are required.

Apply the first course of shingles to both roofs on each side of the valley. Trim the end

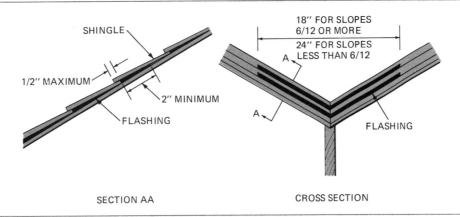

SHINGLE

1/2" MAXIMUM

2" MINIMUM

FLASHING

SECTION AA

18" FOR SLOPES
6/12 OR MORE

24" FOR SLOPES
LESS THAN 6/12

A

A

FLASHING

CROSS SECTION

Fig. 19-34 Specifications for applying individual metal flashings in a valley

shingles to the chalk line. Bed the ends in plastic cement. Do not drive nails through the metal flashing.

Apply the next piece of flashing in the valley over the first course of shingles. Keep its bottom edge about 1/2 inch above the butts of the next course of shingles. Apply the second course of shingles in the same manner as the first. Secure a flashing over the second course. Apply succeeding courses and flashings in this manner. Remember, a flashing is placed over each course of shingles. Do not leave any flashings out. When the valley is completely flashed, no metal surface is exposed.

FLASHING AGAINST A WALL

When roof shingles butt up against a vertical wall, the joint must be made watertight. The most satisfactory method of making the joint tight is with the use of individual metal flashings.

The flashings are cut about 6 inches in width and bent at right angles so they lay about 2 inches on the roof and about 4 inches on the sidewall. The height of the flashings is such that they are about 3 inches more than the exposure of the shingles. When used with strip shingles laid 5 inches to the weather, they would be 8 inches in height. Cut and bend the necessary number of metal flashings.

The roofing is applied and flashed before the siding is applied to the vertical wall. First, apply an underlayment of 15-pound asphalt felt to the roof deck and turn the ends up on the vertical wall by about 3 to 4 inches.

Lay the starter course, working toward the vertical wall. Fasten a metal flashing with its bottom edge flush with the drip edge, and with one nail in each top corner. Lay the first course with its end shingle over the flashing. Apply a flashing over the first course and against the wall. Keep its bottom edge at a point that will be about 1/2 inch above the butt of the next course of shingles. Continue applying shingles and flashings in this manner until the ridge is reached. Do not drive any shingle nails through the flashings. Bed the end of the shingles down with asphalt cement if necessary. Some applicators prefer to cut the shingles back, if necessary, so that no metal flashing is exposed at the shingle cutouts, Figure 19-35.

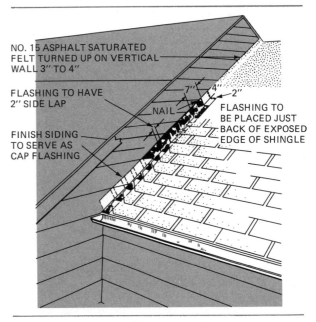

NO. 15 ASPHALT SATURATED FELT TURNED UP ON VERTICAL WALL 3" TO 4"

FLASHING TO HAVE 2" SIDE LAP

FINISH SIDING TO SERVE AS CAP FLASHING

NAIL

7"   4"   2"

FLASHING TO BE PLACED JUST BACK OF EXPOSED EDGE OF SHINGLE

Fig. 19-35 Using individual metal flashings against a wall

FLASHING A CHIMNEY

In many cases, especially on steep pitch roofs, a *cricket* or *saddle* is built on between the upper side of the chimney and roof deck. The purpose of the saddle is to prevent accumulation of water at this point, Figure 19-36.

On chimneys, the flashings are usually of lead or aluminum and are installed by the masons. The upper ends of the flashing are bent around and mortared in between the courses of brick as the chimney is built. The roofer usually finds these flashings in place before starting work.

The underlayment is applied and tucked under the existing flashings. The shingle courses are brought up to the chimney and applied under the flashing on the lower side of the chimney. This is called the *apron flashing*. Shingles are tucked under the apron, cutting the top edge as necessary until the shingle exposure shows below the apron. The apron is then pressed into place on top of the shingles in a bed of plastic cement.

Along the sides of the chimney, the flashings are tucked in the same manner as in flashing against a wall. No nails are used in the flashings. Both shingles and flashings are bedded in asphalt cement. In addition, the standing portions of the chimney flashings are secured to the chimney sides with asphalt cement. The flashing on the upper side of the chimney is cemented to the roof and shingles are applied over it and bedded to it with asphalt cement, Figure 19-37. When necessary to bend the flashing around the corners of the chimney, the metal is gently and carefully tapped with a hammer handle and formed around the corners.

Other rectangular roof obstructions, such as skylights, are flashed in a similar manner. The carpenter usually applies the flashings to these obstructions.

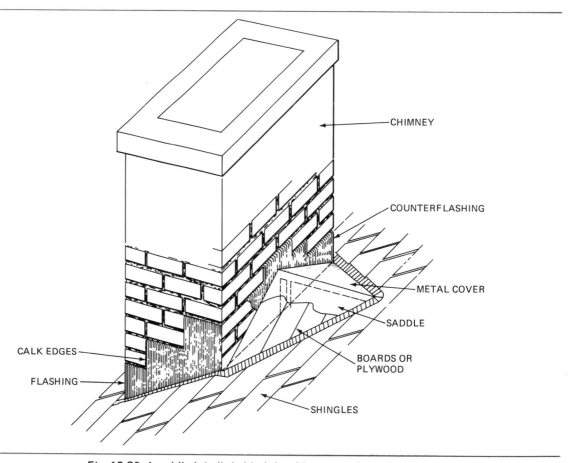

CHIMNEY

COUNTERFLASHING

METAL COVER

SADDLE

BOARDS OR PLYWOOD

CALK EDGES

FLASHING

SHINGLES

Fig. 19-36 A saddle is built behind the chimney to shed water.

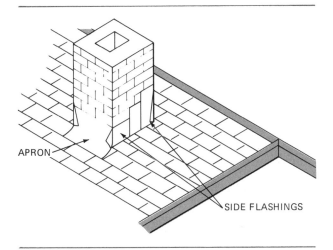

Fig. 19-37 Chimney flashing

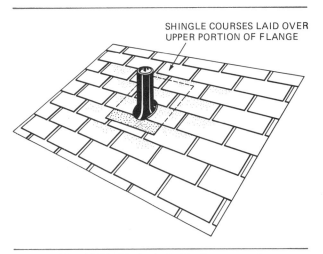

Fig. 19-38 Method of shingling around soil stack flashing

FLASHING SOIL SACKS

Flashings for round pipes such as soil stacks usually come made-up for various roof pitches. They fit around the pipe and have a wide flange on the bottom that rests on the roof deck. All joints are soldered to make them watertight. The top end sometimes is bent and fitted inside the pipe or it may be clamped to the pipe with a neoprene seal. These flashings are usually installed by the plumber before the roofing is applied.

Shingle up to the lower end of the soil stack flashing. Lift the lower part of the flange and apply shingles under it up to the pipe. Replace the flange and bed it in asphalt cement.

Apply shingles around the pipe and over the flange. Do not drive nails through the flashing. Bed shingles to the flashing with asphalt cement, Figure 19-38.

WOOD SHINGLES AND SHAKES

Wood shingles and shakes are used extensively for roof covering as well as for wall covering, Figure 19-39. Shingles are sawed. Shakes are split. Shingles have a relatively smooth surface while shakes have at least one highly textured natural grain split surface.

Fig. 19-39 Wood shingles and shakes are used extensively for roof covering.

CERTIGRADE RED CEDAR SHINGLES

Grade	Length	Thickness (at Butt)	No. of Courses Per Bundle	Bdls/Cartons Per Square		Description
No. 1 BLUE LABEL	16" (Fivex) 18" (Perfections) 24" (Royals)	.40" .45" .50"	20/20 18/18 13/14	4 bdls. 4 bdls. 4 bdls.		 The premium grade of shingles for roofs and sidewalls. These top-grade shingles are 100% heartwood. 100% clear and 100% edge-grain.
No. 2 RED LABEL	16" (Fivex) 18" (Perfections) 24" (Royals)	.40" .45" .50"	20/20 18/18 13/14	4 bdls. 4 bdls. 4 bdls.		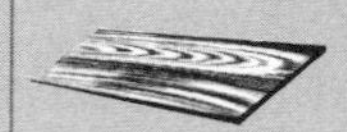 A good grade for many applications. Not less than 10" clear on 16" shingles, 11" clear on 18" shingles and 16" clear on 24" shingles. Flat grain and limited sapwood are permitted in this grade.
No. 3 BLACK LABEL	16" (Fivex) 18" (Perfections) 24" (Royals)	.40" .45" .50"	20/20 18/18 13/14	4 bdls. 4 bdls. 4 bdls.		 A utility grade for economy applications and secondary buildings. Not less than 6" clear on 16" and 18" shingles, 10" clear on 24" shingles.
No. 4 UNDER-COURSING	16" (Fivex) 18" (Perfections)	.40" .45"	14/14 or 20/20 14/14 or 18/18	2 bdls. 2 bdls. 2 bdls.		 A utility grade for undercoursing on double-coursed sidewall applications or for interior accent walls.
No. 1 or No. 2 REBUTTED-REJOINTED	16" (Fivex) 18" (Perfections) 24" (Royals)	.40" .45" .50"	33/33 28/28 13/14	1 carton 1 carton 4 bdls.		 Same specifications as above for No. 1 and No. 2 grades but machine trimmed for parallel edges with butts sawn at right angles. For sidewall application where tightly fitting joints are desired. Also available with smooth sanded face.

Fig. 19-40 Grades, sizes, and description of wood shingles *(Courtesy of Red Cedar Shingle and Handsplit Shake Bureau)*

CERTI-SPLIT RED CEDAR HANDSPLIT SHAKES

Grade	Length and Thickness	18" Pack**			Description
		# Courses Per Bdl.	# Bdls. Per Sq.		
No. 1 HANDSPLIT & RESAWN	15" Starter-Finish 18"x1/2" Mediums 18"x3/4" Heavies 24"x3/8" 24"x1/2" Mediums 24"x3/4" Heavies	9/9 9/9 9/9 9/9 9/9	5 5 5 5 5		 These shakes have split faces and sawn backs. Cedar logs are first cut into desired lengths. Blanks or boards of proper thickness are split and then run diagonally through a bandsaw to produce two tapered shakes from each blank.
No. 1 TAPERSAWN	24"x5/8" 18"x5/8"	9/9 9/9	5 5		 These shakes are sawn both sides.
No. 1 TAPERSPLIT	24"x1/2"	9/9	5		 Produced largely by hand, using a sharp-bladed steel froe and a wooden mallet. The natural shingle-like taper is achieved by reversing the block, end-for-end, with each split.
		20" Pack			
No. 1 STRAIGHT-SPLIT	18"x3/8" True-Edge* 18"x3/8" 24"x3/8"	14 Straight 19 Straight 16 Straight	4 5 5		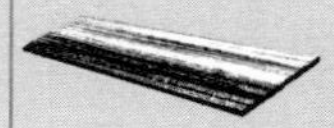 Produced in the same manner as tapersplit shakes except that by splitting from the same end of the block, the shakes acquire the same thickness, throughout.

NOTE: *Exclusively sidewall product, with parallel edges.
**Pack used for majority of shakes.

Fig. 19-41 Types of wood shakes *(Courtesy of Red Cedar Shingle and Handsplit Shake Bureau)*

SHAKE TYPE, LENGTH AND THICKNESS	Approximate coverage (in sq. ft.) of one square, when shakes are applied with 1/2" spacing, at following weather exposures, in inches (h):					
	5 1/2	7 1/2	8 1/2	10	11 1/2	16
18"x1/2" Handsplit-and-Resawn Mediums (a)	55(b)	75(c)	85(d)	100	....	....
18"x3/4" Handsplit-and-Resawn Heavies (a)	55(b)	75(c)	85(d)	100	....	....
18"x5/8" Tapersawn	55(b)	75(c)	85(d)	....	....	....
24"x3/8" Handsplit	....	75(e)	85	100(f)	115(d)	....
24"x1/2" Handsplit-and-Resawn Mediums	....	75(b)	85	100(c)	115(d)	....
24"x3/4" Handsplit-and-Resawn Heavies	....	75(b)	85	100(c)	115(d)	....
24"x5/8" Tapersawn	....	75(b)	85	100(c)	115(d)	....
24"x1/2" Tapersplit	....	75(b)	85	100(c)	115(d)	....
18"x3/8" True-Edge Straight-Split	....	....	....	....	....	112(g)
18"x3/8" Straight-Split	65(b)	90	100(d)	....	....	....
24"x3/8" Straight-Split	....	75(b)	85	100	115(d)	....
15" Starter-Finish Course	Use supplementary with shakes applied not over 10" weather exposure.					

(a) 5 bundles will cover 100 sq. ft. roof area when used as starter-finish course at 10" weather exposure; 6 bundles will cover 100 sq. ft. wall area at 8 1/2" exposure; 7 bundles will cover 100 sq. ft. roof area at 7 1/2" weather exposure; see footnote (h).

(b) Maximum recommended weather exposure for 3-ply roof construction.

(c) Maximum recommended weather exposure for 2-ply roof construction.

(d) Maximum recommended weather exposure for single-coursed wall construction.

(e) Maximum recommended weather exposure for application on roof pitches between 4-in-12 and 8-in-12.

(f) Maximum recommended weather exposure for application on roof pitches of 8-in-12 and steeper.

(g) Maximum recommended weather exposure for double-coursed wall construction.

(h) All coverage based on 1/2" spacing between shakes.

For detailed information on all cedar products, write to: Red Cedar Shingle & Handsplit Shake Bureau, Suite 275, 515-116th Ave. N.E., Bellevue, WA 98004 (In Canada: Suite 1500, 1055 West Hastings Street, Vancouver, B.C., V6E 2H1)

Maximum exposure recommended for roofs:

PITCH	NO. 1 BLUE LABEL			NO. 2 RED LABEL			NO. 3 BLACK LABEL		
	16"	18"	24"	16"	18"	24"	16"	18"	24"
3 IN 12 TO 4 IN 12	3 3/4"	4 1/4"	5 3/4"	3 1/2"	4"	5 1/2"	3"	3 1/2"	....
4 IN 12 AND STEEPER	5"	5 1/2"	7 1/2"	4"	4 1/2"	6 1/2"	3 1/2"	4"	....

Approximate coverage of one square (4 bundles) of shingles based on following weather exposures

Length and Thickness	3 1/2"	4"	4 1/2"	5"	5 1/2"	6"	6 1/2"	7"	7 1/2"	8"	8 1/2"	9"	9 1/2"	10"	10 1/2"	11"	11 1/2"	12"	12 1/2"	13"	13 1/2"	14"	14 1/2"	15"	15 1/2"	16"
16"x5/2"	70	80	90	100*	110	120	130	140	150‡	160	170	180	190	200	210	220	230	240†	....	....	....	....	....	....	....	....
18"x5/2 1/4"	....	72 1/2	81 1/2	90 1/2	100*	109	118	127	136	145 1/2	154 1/2‡	163 1/2	172 1/2	181 1/2	191	200	209	218	227	236	245 1/2	254 1/2‡	....	....	....	....
24"x4/2"	....	....	....	....	....	80	86 1/2	93	100*	106 1/2	113	120	126 1/2	133	140	146 1/2	153‡	160	166 1/2	173	180	186 1/2	193	200	206 1/2	213†

NOTES: *Maximum exposure recommended for roofs.   ‡Maximum exposure recommended for single-coursing No. 1 and No. 2 grades on sidewalls.
†Maximum exposure recommended for double-coursing No. 1 grades on sidewalls.

Fig. 19-42 Maximum exposure and coverage of wood shingles and shakes (Courtesy of Red Cedar Shingle and Handsplit Shake Bureau)

Shingles and shakes are generally used for roofs and walls in the natural, or unstained state.

Most shingles and shakes are made from western red cedar. However, some shingles are made from other durable woods such as eastern white cedar, cypress, and redwood.

Wood shingles are available in lengths of 16, 18, and 24 inches and in four standard grades. A special grade of shingles called *Rebutted and Rejointed* are machine trimmed with parallel edges and butts sawn at right angles, Figure 19-40.

Shakes are available in lengths of 15, 18, and 24 inches. They, also, are available in four different grades, Figure 19-41.

MAXIMUM EXPOSURE AND COVERAGE

The maximum amount of shingle exposed to the weather and the actual area covered by one square depend on the roof pitch, length, and grade of shingle. Figure 19-42 shows the maximum exposure and coverage of wood shingles and shakes.

Shakes are not generally applied to roofs with slopes of less than 4 inches rise per foot unless special construction methods are used. Shingles may be used on slopes down to 3 inches rise with reduced exposures.

SHEATHING AND UNDERLAYMENT

Shingles and shakes may be applied over open or solid sheathing. For spaced sheathing, use 1x6-inch sheathing for shakes and 1x4 inch for shingles spaced on centers equal to the exposure.

No underlayment is required under wood shingles, but it may be desirable for the protection of the roof sheathing. Apply a breather-type paper if one is used. Shakes require a starter course of underlayment and interlayments of 30-pound asphalt felt between courses over the entire roof.

APPLYING WOOD SHINGLES

Apply a starter course of wood shingles as described previously. If a gutter is used, overhang the shingles so that water from the roof will drip in the center of the gutter. If no gutter is installed, project the starter course 1 1/2 inches beyond the fascia. Space adjacent shingles 1/4 inch apart.

Use a shingling hatchet, Figure 19-43. The hatchet should be lightweight, with both a sharp blade and a heel for trimming shingles. A sliding gauge is used for fast and accurate checking of shingle exposure. The gauge permits laying several shingle courses at a time without snapping a chalk line.

Apply each shingle with only two corrosion-resistant nails. Use 3d nails for 16- and 18-inch shingles. Use 4d nails for 24-inch shingles, Figure 19-44. Place each nail about 3/4 inch from the edge of the shingle and not more than

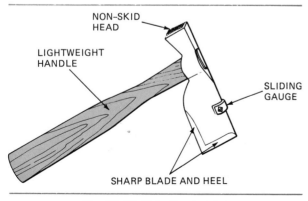

NON-SKID HEAD

LIGHTWEIGHT HANDLE

SLIDING GAUGE

SHARP BLADE AND HEEL

Fig. 19-43 Shingling hatchet

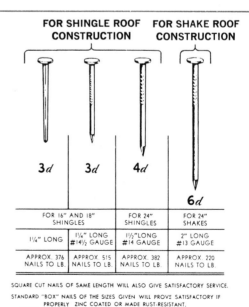

FOR SHINGLE ROOF CONSTRUCTION		FOR SHAKE ROOF CONSTRUCTION	
3d	3d	4d	6d
FOR 16" AND 18" SHINGLES		FOR 24" SHINGLES	FOR 24" SHAKES
1¼" LONG	1¼" LONG #14½ GAUGE	1½" LONG #14 GAUGE	2" LONG #13 GAUGE
APPROX. 376 NAILS TO LB.	APPROX. 515 NAILS TO LB.	APPROX. 382 NAILS TO LB.	APPROX. 220 NAILS TO LB.

SQUARE CUT NAILS OF SAME LENGTH WILL ALSO GIVE SATISFACTORY SERVICE.

STANDARD "BOX" NAILS OF THE SIZES GIVEN WILL PROVE SATISFACTORY IF PROPERLY ZINC COATED OR MADE RUST-RESISTANT.

Fig. 19-44 Recommended nails for wood shingles and shakes

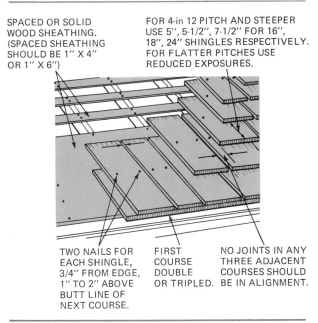

SPACED OR SOLID WOOD SHEATHING. (SPACED SHEATHING SHOULD BE 1″ X 4″ OR 1″ X 6″)

FOR 4-in 12 PITCH AND STEEPER USE 5″, 5-1/2″, 7-1/2″ FOR 16″, 18″, 24″ SHINGLES RESPECTIVELY. FOR FLATTER PITCHES USE REDUCED EXPOSURES.

TWO NAILS FOR EACH SHINGLE, 3/4″ FROM EDGE, 1″ TO 2″ ABOVE BUTT LINE OF NEXT COURSE.

FIRST COURSE DOUBLE OR TRIPLED.

NO JOINTS IN ANY THREE ADJACENT COURSES SHOULD BE IN ALIGNMENT.

Fig. 19-45 Application method for wood shingles

1 inch above the exposure line. Drive the nails flush with the surface, but not so that the head crushes the wood.

Double the starter course of shingles by applying another course on top of it. The first course may be tripled, if desired, for appearance.

Lay succeeding courses across the roof, laying several courses at a time. Stagger the joints in adjacent courses at least 1 1/2 inches, with no joint in any three adjacent courses in alignment, Figure 19-45. Save wide shingles to use along the valleys.

Apply valley shingles first and work out from the valley. Saw shingles extending into the valley to the proper miter. Do not break joints in the valley or lay shingles with the grain parallel to the valley centerline.

After laying several courses, snap a chalk line to straighten the next course. Proceed shingling up the roof. When 3 or 4 feet from the ridge, check the distance on both ends of the roof. Divide the distance as close as possible to the exposure used, so a full course shows below the ridge cap. Shingle tips are cut flush with the ridge.

HIPS AND RIDGES

Hip and ridge caps are usually made the same width as the shingle exposure on the main

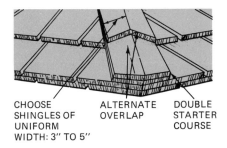

CHOOSE SHINGLES OF UNIFORM WIDTH: 3″ TO 5″

ALTERNATE OVERLAP

DOUBLE STARTER COURSE

Fig. 19-46 When applying hip and ridge shingles, alternate the overlap.

roof. Measure down on both sides of the hip or ridge at each end for a distance equal to the exposure. Snap a line between the marks. Lay a shingle so its bottom edge is to the line and fasten with two nails. Use longer nails so that they penetrate at least 1/2 inch into the sheathing. Trim the top edge flush with the opposite slope.

Lay another shingle on the other side with the butt even and overlapping the first shingle laid. Trim its top edge flush. Double this layer. Apply succeeding layers in the same manner, but alternate the overlap, Figure 19-46. Space the exposure when nearing the end.

SPECIAL APPLICATIONS

Besides being applied in straight single courses, wood shingles are laid in various patterns to achieve certain effects such as emphasizing shadow lines and roof textures. Some of these patterns are called thatch, serrated, weave and ocean wave.

In the *thatch* pattern, shingles are positioned above and below the course line in a random manner. *Serrated* courses are doubled every few courses. In the *weave pattern*, shingles are superimposed at random. The *ocean wave* effect is created by placing a pair of shingles butt to butt under the regular course at intervals, Figure 19-47.

APPLYING WOOD SHAKES

Mark the handle of the shingling hatchet at 7 1/2 and 10 inches from the top of the head.

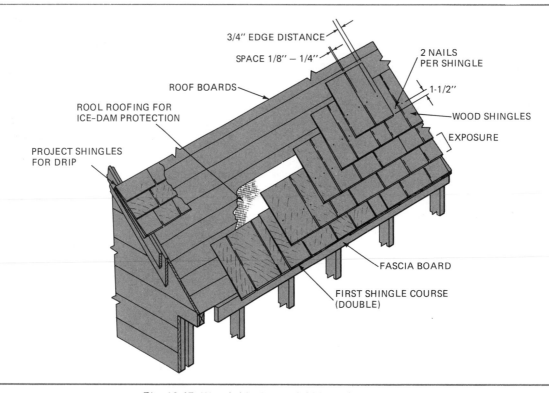

3/4" EDGE DISTANCE

SPACE 1/8" – 1/4"

ROOF BOARDS

ROOL ROOFING FOR
ICE–DAM PROTECTION

PROJECT SHINGLES
FOR DRIP

2 NAILS
PER SHINGLE

1-1/2"

WOOD SHINGLES

EXPOSURE

FASCIA BOARD

FIRST SHINGLE COURSE
(DOUBLE)

Fig. 19-47 Wood shingles are laid in various patterns.

These are the exposures that are used most of the time when applying wood shakes.

Shakes are applied in much the same manner as shingles, Figure 19-48. However, an underlayment and interlayment of felt is required. It is important to lay the felt straight because it serves as a guide when laying shakes.

Fig. 19-48 Applying wood shakes to a roof with a pneumatic stapler

Start by laying out a full 36-inch width of 15-pound asphalt felt along the eaves. Nail it securely with large head galvanized roofing nails.

Next lay an 18-inch wide strip with its bottom edge positioned 20 inches above the butt line of the first course of shakes. Take into consideration the overhang over the fascia. Nail only the top edge of the felt. Fasten successive strips on their top edge only so their bottom edges are 10 inches from the bottom of the previous strip. After the roof is felted, the shakes are tucked under the felt so that the top 4 inches of the shake are covered, with the bottom 20 inches exposed, Figure 19-49.

Check the exposure regularly with the hatchet handle since there is a tendency to angle toward the ground. An easy way to be sure of correct exposure is to look between the cracks of the course below the one being nailed. The bottom edge of the felt will be visible. The butt of the shake being nailed is positioned directly above it.

Adjust the exposure so that tips of shakes in the next-to-last course just come to the ridge. Use an economical 15-inch starter-finish course for the last course. This saves time

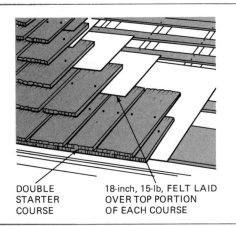

DOUBLE STARTER COURSE

18-inch, 15-lb, FELT LAID OVER TOP PORTION OF EACH COURSE

Fig. 19-49 An interlayment of felt is required when laying shakes.

trimming shakes at the ridge. Use either factory-made hip and ridge caps or cap the ridge with shakes in the same manner as that used with wood shingles.

ESTIMATING ROOFING MATERIALS

Find the area of the roof in square feet. Divide the total by 100 to determine the number of squares needed. Generally add 10 percent for waste and cutting. A simple roof with no dormers, valleys, or other obstructions requires less to be added; a complicated roof requires more.

- For wood shingles, add one square for every 240 linear feet of starter course. Add one square of shakes for 120 linear feet of starter course.

- Add one extra square of shingles for every 100 linear feet of valleys and about two squares for shakes.

- Add an extra bundle of shakes or shingles for every 16 feet of hip and ridge to be covered.

- Figure two pounds of nails per square at standard exposures.

- Remember that a square of roofing will cover 100 square feet of roof surface only when applied at standard exposures. Allow proportionally more material when these exposures are reduced.

REVIEW QUESTIONS

Select the most appropriate answer.

1. A square is the amount of roofing required to cover
   a. 50 square feet.   c. 150 square feet.
   b. 100 square feet.   d. 200 square feet.

2. One roll of 15-pound asphalt felt will cover
   a. 1 square.   c. 3 square.
   b. 2 square.   d. 4 square.

3. When applying asphalt felt on a roof deck as underlayment, lap each course over the lower course by at least
   a. 2 inches.   c. 4 inches.
   b. 3 inches.   d. 6 inches.

4. Under no circumstances should asphalt strip shingles be used on slopes lower than
   a. 1 inch rise per foot of run.
   b. 2 inches rise per foot of run.
   c. 3 inches rise per foot of run.
   d. 4 inches rise per foot of run.

5. When applying asphalt shingles, it is recommended that no rake tab be less than
   a. 2 inches in width.   c. 4 inches in width.
   b. 3 inches in width.   d. 5 inches in width.

6. For slopes with a 6-inch rise per foot of run or less, metal valley flashings should extend on each side of the valley centerline by at least
   a. 6 inches.   c. 10 inches.
   b. 8 inches.   d. 12 inches.

7. When flashing a valley by weaving shingles, do not locate any nails closer to the valley centerline than
   a. 6 inches.   c. 10 inches.
   b. 8 inches.   d. 12 inches.

8. Flashings about 6 inches wide are used when flashing a roof that butts against a vertical wall. They are bent so that
   a. 3 inches lays on the wall and 3 inches lays on the roof.
   b. 4 inches lays on the wall and 2 inches lays on the roof.
   c. 2 inches lays on the wall and 4 inches lays on the roof.
   d. 1 inch lays on the wall and 5 inches lays on the roof.

9. A built-up section between the roof and the upper side of a chimney is called a
   a. saddle.   c. furring.
   b. dutchman.   d. breast.

10. Concealed nail roll roofing may be used on roofs with slopes as low as
    a. 1 inch rise per foot of run.
    b. 2 inches rise per foot of run.
    c. 3 inches rise per foot of run.
    d. 4 inches rise per foot of run.

11. Most wood shingles and shakes are made from
    a. cypress.   c. eastern white cedar.
    b. redwood.   d. western red cedar

12. The longest available length of wood shingles and shakes is
    a. 16 inches.   c. 24 inches.
    b. 18 inches.   d. 28 inches.

13. The maximum exposure for No. 1, 16-inch wood shingles laid on roofs at least as steep as 4 in 12 is
    a. 4 1/2 inches.   c. 5 1/2 inches.
    b. 5 inches.   d. 6 inches.

14. If no gutter is used, wood shingles normally overhang the fascia by
    a. 3/8 inch.   c. 1 1/2 inches.
    b. 1 inch.   d. 2 inches.

15. It is important to lay interlayment straight prior to applying shakes
    a. so there are no wrinkles in the felt.
    b. to obtain the proper lap.
    c. for ease in nailing.
    d. because the felt serves as a guide.

WINDOWS 20

OBJECTIVES

After completing this unit, the student will be able to

- *describe the most popular styles of windows and name their parts.*
- *select desired sizes and styles of windows from manufacturers' catalogs.*
- *install various types of windows in an approved manner.*
- *cut glass and glaze a sash.*

Windows are items of millwork and are made in large factories, fully assembled and ready for installation. They are made of wood, aluminum, and steel. Because wood is an excellent insulator, there is less heat loss through wood windows. However, metal permits the use of smaller frame members allowing more glass area. Metal windows should be installed carefully to prevent condensation and frosting on interior surfaces.

Windows are installed prior to the application of exterior siding. Care must be taken to provide easy-operating, weathertight units of good appearance. Quality workmanship results in a more comfortable interior, saves energy by reducing fuel costs, minimizes maintenance, gives longer life to the units, and makes application of the exterior siding easier.

PARTS OF A WINDOW

When shipped from the factory, the window is complete except for the interior trim. It is important that the person who is installing the windows knows the names, location, and functions of the parts of a window.

SASH

The *sash* is a frame that holds the glass. It may be installed in a fixed position, may move vertically or horizontally, or swing outward or inward. Vertical outside members of the sash are called *stiles*. Horizontal outside members are called *rails*.

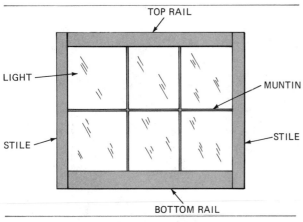

Fig. 20-1 Parts of a sash

MUNTINS

When installed in a sash, the glass is called a light. There may be more than one light of glass in a sash. Small strips of wood that divide the glass into smaller lights are called *muntins*. Muntins divide the glass into rectangular or diamond shapes.

Many windows come with false muntins that do not actually separate or support the glass. They are applied as an overlay to simulate small lights of glass. They are made of wood or plastic and snap in and out of the sash for easy cleaning of the glass, Figure 20-1.

SINGLE GLASS

Float or Plate	Thickness In.	Tolerance	Quality	Maximum Size[1] In.	Approx. Lbs./Sq. Ft.
Parallel-O-Float	1/8	± 1/32″	Glazing	74x120	1.64
	3/16			110x120	2.45
	1/4		Float for Mirrors	75 Sq. Ft.	3.27
				130x252	
Parallel-O-Float or Parallel-O-Plate	5/16		Glazing	122x252	4.08
	3/8			122x264	4.90
	1/2			120x300	6.54
	5/8	+ 1/32″–3/64″		108x300	8.17
	3/4				9.81
	7/8	+ 3/64″–1/16″			11.45
Parallel-O-Grey and Parallel-O-Bronze	3/16	± 1/32″	Glazing	84x120	2.45
	1/4			120x192	3.27
	3/8			119x264	4.90
	1/2			112x300	6.54
Heat Absorbing Float	3/16	± 1/32″	Glazing	110x120	2.45
	1/4			120x192	3.27

Sheet		Range	Quality	Maximum Size[1] In.	Approx. Oz./Sq. ft.
Picture		.043-.053	AA, A, B	50[2]	9-11
		.054-.069 and .070-.080	AA, A	70[2]	12-14 and 15-17
			B	80[2]	
Window	SS	.085-.101	AA, A, B	120[2]	19.5
	DS	.115-.134		140[2]	26.0
Heavy Sheet	3/16	.182-.205	AA, A, B	84x120	40.0

[1]Sizes listed may, in some cases, be too large to meet applicable windload requirements.

[2]United inches (sum of width plus length.)

Fig. 20-2 Kinds, sizes, and grades of construction glass *(Courtesy of Tibby-Owens-Ford Company)*

WINDOW GLASS

Several qualities and thicknesses of sheet glass are used in windows. *Single strength* glass is used for smaller lights. For larger areas *double strength* glass is used. Both of these types may have a slight distortion caused by the manufacturing process and are used when this is not objectionable. However, distortion-free polished *plate glass* is used for large areas, such as picture windows. Many kinds of glass are used in construction, Figure 20-2.

INSULATING GLASS

To help prevent heat loss, and to avoid condensation of moisture on glass surfaces, *insulating glass* is used frequently in the sash in place of a single sheet.

Insulating glass consists of two layers of glass fused together at the edges to make a dead air space of about 3/16 inch between them. The double glass is installed in the sash at the mill and eliminates the need of using storm sash, Figure 20-3. If it is desired to have insulating glass in window units, it must be specified when ordering.

Fig. 20-3 Cutaway of insulating glass glazed in a sash

STORM SASH

Storm sash is commonly used to provide a dead air space. On some windows, it is attached to the outside either on top of or in between the exterior casings of the window. Special storm sash hardware is available for hanging and locking the sash in place.

Wood storm sash is 1 1/8 inches thick and is ordered by specifying the glass size and number of lights that correspond to the window to which it is to be attached. A disadvantage of using storm sash is that they must be removed from windows to provide ventilation in summer.

STORM PANELS

Storm panels are available from manufacturers for certain types of windows. They also provide the second pane of glass and act like a storm sash. However, they fit only those windows for which they are made. They are installed in specially milled sections and become part of the sash and need not be removed to operate the sash when ventilation is desired.

THE WINDOW FRAME

The window sash is held in the *window frame*. The sash may move horizontally or vertically in tracks in the frame. It may also be hinged to and swing from the frame. The frame consists of a number of parts, Figure 20-4.

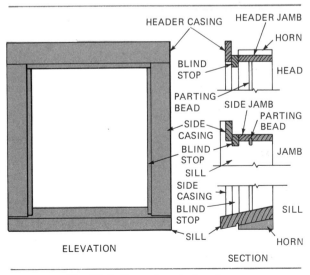

Fig. 20-4 Parts of a window frame

SILL

The bottom horizontal member of the window frame is called a *sill* or, in some locations, a *stool.* It is usually set at an angle to shed water.

JAMBS

The vertical sides of the window frame are called *side jambs* and the top horizontal member is called a *header jamb*.

BLIND STOPS

Blind stops are sometimes applied to certain types of window frames. They are strips of wood attached to the outside edges of the jambs. Their inside edges project about 1/2 inch inside the frame. They provide a weathertight joint between the outside casings and the frame and also act as stops for screens and storm sash. They make the outer edge of the channel for double-hung windows.

CASINGS

Window units usually come with *exterior casings* applied. The side members are called *side casings.* Their lower ends are cut at a bevel and rest on the sill. The top member is called the *header casing.* On flat casings, the joint between them is either rabbeted or tongue-and-grooved to make the joint weathertight. On molded casings, the mitered joint is bedded in compound, Figure 20-5. When windows are installed in multiple units, a *mullion* is formed

where the two side jambs are joined together. The casing covering the joint is called a mullion casing.

DRIP CAP

A *drip cap* comes with some windows and is applied on the top edge of the header casing. It is designed to carry rain water out over the window unit.

WINDOW FLASHING

In some cases, a *window flashing* is also provided. This is a piece of metal as long as the header casing bent to fit over the header casing and against the exterior wall. The flashing prevents the entrance of water at this point. Flashings are usually of aluminum or vinyl.

If window flashings are not provided with the unit, they can be made on the job site. Cut a length of sheet metal from a roll of desired width. The metal should be wide enough to extend up the wall about 2 inches, over the top edge of the header casing, and about 1/4 inch over its face.

Tack the metal along its top edge to a length of nominal 2-inch stock so the location of the bend is even with the edge of the stock. Bend the metal by tapping the overhanging edge with a short 2"x4" block, Figure 20-6.

JAMB EXTENSIONS

In some cases, the jamb width of windows may be specified by standard wall thickness. In other cases, jambs are made narrow and *jamb*

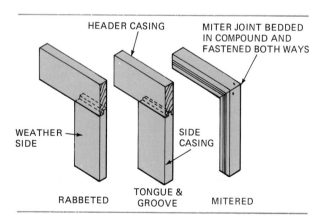

Fig. 20-5 A weathertight joint is made between side and header casings.

Fig. 20-6 Bending a window flashing on the job site

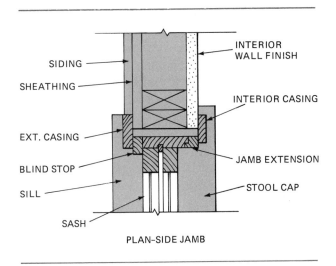

Fig. 20-7 To compensate for varying wall thicknesses, jamb extensions are provided with the window unit.

extensions are provided with the window unit, Figure 20-7. These are installed just before applying the interior trim, so they should be stored for safekeeping until needed. Their installation is discussed later in Unit 27.

WINDOW COATING

Most window units are *primed* (first coat of paint applied) at the factory. Some units have wood parts enclosed in a thick rigid vinyl coat. This vinyl covering is designed not to rust, pit, corrode, or blister and needs no painting.

In case units need priming, this should be done before installation. Store the units under cover and protected from the weather until installed.

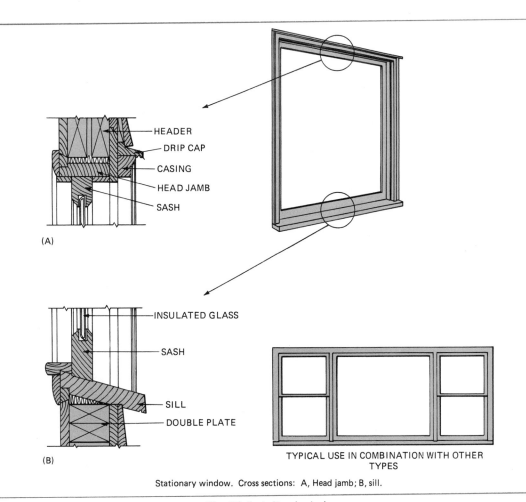

Stationary window. Cross sections: A, Head jamb; B, sill.

Fig. 20-8 A fixed window

TYPES OF WINDOWS

There are several popular types of windows. Each one offers a different combination of functions and styles. They are called fixed, double-hung, casement, sliding, awning, hopper, and jalousie windows.

FIXED WINDOWS

Fixed windows consist of a frame in which a sash is fitted in a fixed position. A good example of a fixed window is what is commonly known as a picture window. This type is often used in combination with other types, Figure 20-8.

DOUBLE-HUNG WINDOW

The *double-hung window* consists of an upper and a lower sash that slide vertically in separate channels in the side jambs, Figure 20-9. A wood strip separating the sash is called a *parting strip.*

In most units, the sash slide in metal channels that are installed in the frames. Each sash is provided with springs, balances, or compression weatherstripping to hold it in place in any location. Compression weatherstripping prevents air infiltration, provides tension, and acts as a counter balance. Some types provide for easy removal of the sash for painting and repair.

When the sash are closed, specially shaped *meeting rails* come together to make a weather-tight joint. *Sash locks* located at this point not only lock the window, but draw the rails tightly together. Other hardware consists of *sash lifts* that are fastened to the bottom rail, although they are sometimes eliminated by providing a finger groove in the rail.

Double-hung windows can be arranged in a number of ways. They can be installed in multiple units side by side or in combination with other types. Figure 20-10 shows them used in a *bay window* unit. Detachable vinyl grille simulates muntins.

CASEMENT WINDOWS

The *casement window* consists of a sash hinged at the side and swinging outward by

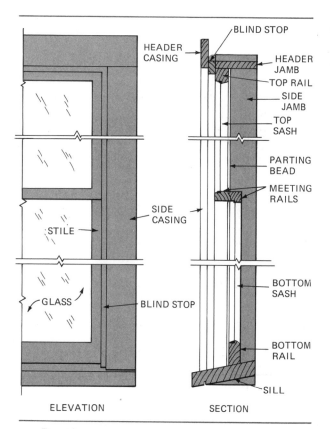

Fig. 20-9 The double-hung window and its parts

Fig. 20-10 Double-hung windows used in a bay window unit

Fig. 20-11 Casement sash used in a bow window unit

means of a crank or lever. Casements are especially practical in kitchens over the sink or countertop where it may be difficult to open other types.

Most casements swing outward because the inswinging type is very difficult to make weather-tight. An advantage of the casement type is that the entire sash can be opened for maximum ventilation. Figure 20-11 shows the use of casement windows in a *bow window* unit.

SLIDING WINDOWS

Sliding windows have sash that slide horizontally in separate tracks located on the header jamb and sill, Figure 20-12. When a window-wall effect is desired, as many units as necessary can be placed side to side. Most units come with all necessary hardware applied.

AWNING AND HOPPER WINDOWS

An *awning window* unit consists of a frame in which a sash hinged at the top swings outward by means of a crank or lever. A similar type, called the *hopper* window, is hinged at the bottom and swings inward.

Each sash is provided with an individual frame so that any combination of width and height can be used. These windows are often used in combination with other types, Figure 20-13.

JALOUSIE WINDOWS

A *jalousie window* consists of a metal frame that holds a series of horizontal glass

Fig. 20-12 Sliding windows have sash that slide horizontally.

Fig. 20-13 Awning windows used in combination with fixed windows

Fig. 20-14 A jalousie window has many horizontal glass slats that operate as a venetian blind does.

slats. The frames are attached to each other so that the glass slats open and close together when operated by a single crank, Figure 20-14. Jalousie windows are very drafty; their use is limited to porches and breezeways or in warm climates.

SCREENS

Most manufacturers provide screens as optional accessories for all kinds of windows. On outswinging and sliding windows the screens are made to be attached to the inside of the frame. On double-hung windows they are mounted on the outside of the frame.

The screen mesh is usually plastic or aluminum. In more expensive screens, bronze is used. The mesh is installed in a lightweight

frame and no fitting is necessary. They are mounted in place with hardware already installed at the factory.

Wood screens are also available in various sizes and are installed full size on windows in the same manner as wood storm sash. Half screens are also used that slide in tracks installed on the inside edges of the side casings.

SELECTING AND ORDERING WINDOWS

The carpenter must study the plans to know the type and location of windows. The floor plan shows the location of each unit. The unit is identified on the plan usually by a number or a letter. These numbers or letters identify the window in detail in a *window schedule* found in a set of plans, Figure 20-15. This schedule includes window style, size, manufacturers' name, and unit number. Rough opening sizes may or may not be shown.

At times a window schedule is not included, and units are only identified by the manufacturer's name and number on the floor plan. In may also be necessary to study details of the window construction. Therefore, in order to gain more information, the builder must refer to the manufacturer's catalog.

The manufacturer's catalog includes a complete description of the units with optional accessories. Also shown are details of the construction of the various units along with the sizes. These sizes show the unit dimension (also masonry opening), the glass size, the sash size, and the rough opening, Figure 20-16. Check all rough openings before ordering windows.

		WINDOW	SCHEDULE	
IDENT.	QUAN.	MANUFACTURER	SIZE	REMARKS
A	6	MORGAN	2'-8" x 3'-10"	D.H. SINGLE
B	3	MORGAN	2'-8" x 3'-10"	D.H. TRIPLE
C	2	MORGAN	3'-4" x 3'-10"	D.H. SINGLE
D	1	ANDERSEN	WIN30	CASEMENT SINGLE
E	1	ANDERSEN	W3N30	CASEMENT TRIPLE
F	1	ANDERSEN	W2N5	CASEMENT DOUBLE

Fig. 20-15 Typical window schedule found in a set of plans

360

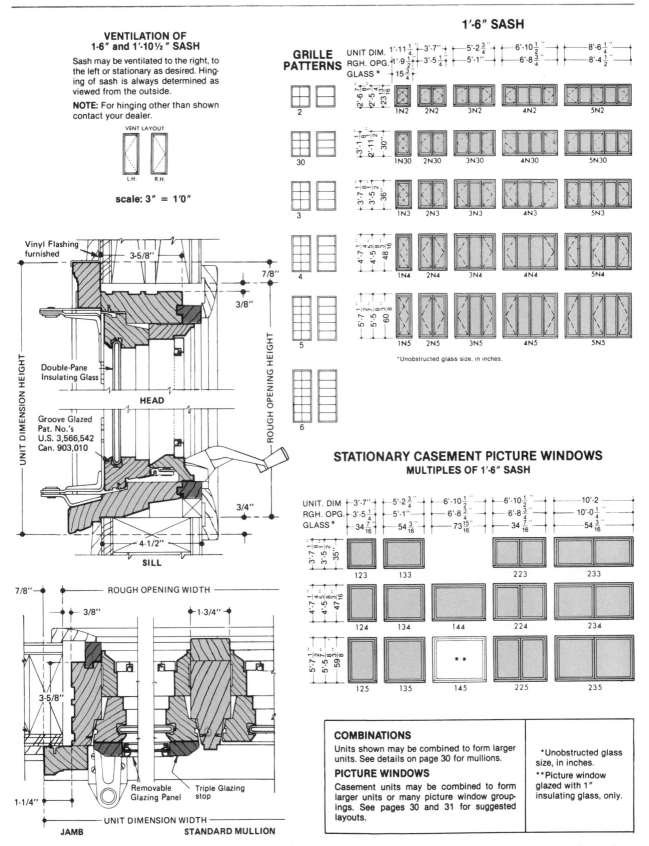

NOTE: Light colored areas are basic parts furnished by Andersen. Dark colored areas are recommended parts to complete unit assembly.

Fig. 20-16 Typical page from a window manufacturer's catalog *(Courtesy of Andersen Corporation)*

INSTALLING WOOD WINDOWS

Prepare the openings by applying strips of 15-pound asphalt felt about 8 inches wide against the wall sheathing on each side of the window openings. Let each end of the strip project above and below the opening about 4 inches. Fasten the strips along each edge with roofing nails or staples about 12 inches apart. Do not fasten the lower end below the window opening, Figure 20-17.

ESTABLISHING WINDOW HEIGHT

Most of the windows in each story of a building are set so their header casings lie in a straight line at the same level even if some of the windows are of different size. One way of establishing this level is to snap a chalk line all around the building to the height of the header casing. Windows are then set to the chalk line.

Another way of establishing this height is by the use of a story pole. Lay out the height of the header jamb from the subfloor on the story pole. When setting windows, hold the story pole with one end against the subfloor and the other end under the header jamb, Figure 20-18.

SETTING THE WINDOW

When setting a window, first cut any *horns* (extensions) from the bottom end of the side jambs so that the end is square and flush with the bottom of the sill. Do not remove any diagonal braces applied at the factory.

If windows are stored inside, they can easily be moved through the openings and set in place. Have sufficient help when setting large units, and handle them carefully. Center the unit in the opening at the bottom.

Place a level on the sill and determine the high end. Shim the opposite end to the correct height with wood shingles between the sill and the bottom header of the rough opening. Tack a nail through the lower end of the casing on that side. Shim the other end of the sill so it is

Fig. 20-17 Flash each side of the opening with felt before setting the window.

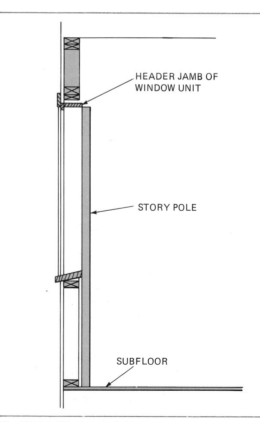

Fig. 20-18 Window heights are sometimes established by using a story pole.

HEADER JAMB OF WINDOW UNIT

STORY POLE

SUBFLOOR

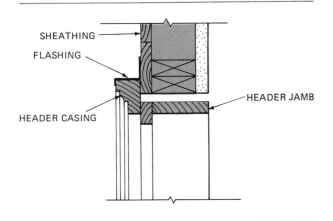

Fig. 20-19 A window flashing is used over the top of the header casing.

SHEATHING

FLASHING

HEADER CASING

HEADER JAMB

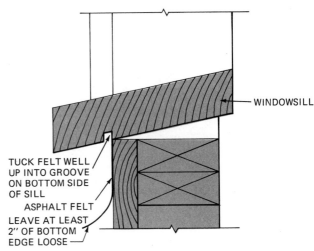

WINDOWSILL

TUCK FELT WELL UP INTO GROOVE ON BOTTOM SIDE OF SILL

ASPHALT FELT

LEAVE AT LEAST 2'' OF BOTTOM EDGE LOOSE

Fig. 20-20 Flashing the sill of a wood window

level and tack the lower end of the casing on that side.

On wide windows with long sills, shim at intermediate points so that the sill is perfectly straight and level with no sag. Use either a long level or a shorter level in combination with a straightedge. Also, sight the sill by eye from end to end.

Plumb the ends of the side jambs with the level and tack the top ends of the side casings. Straighten the side jambs between sill and header jamb and tack through the side casings at intermediate points. Straighten and tack the header casing.

Make sure the sash operates properly and nail the window permanently in place. Use galvanized casing or common nails spaced about 16 inches apart. Keep nails about 2 inches back from the ends of the casings to avoid splitting. Nails should be long enough to penetrate the sheathing and into the framing members. Set the nails so they can be puttied over later.

Some windows have a nailing flange. Large head roofing nails are driven through the flange instead of nailing through the casing.

FLASHING THE HEADER CASING

To flash the header casing, cut the flashing with tin snips so its length is equal to the overall width of the window. Do not let the ends project beyond the side casings because this makes the application of siding more difficult.

If the flashing is applied in more than one piece, lap the joint about 3 inches. Place the flashing firmly on top of the header casing and nail with 3d galvanized lath nails along its top edge and into the wall sheathing, Figure 20-19.

FLASHING THE SILL

Cut a strip of felt about 8 inches wide and long enough to extend to the outside edges of the side flashings. Tuck this strip under the side strips so its ends are even with the outside edges of the side strips. Place the strip so its top edge is well into the groove in the bottom side of the windowsill. Fasten it along its top edge. Leave the bottom edge loose so that sheathing paper can later be applied under it, Figure 20-20.

METAL WINDOWS

Metal windows are available in the same styles as wood windows. The shape and sizes of the parts vary with the manufacturer and the intended use.

In wood construction, the metal window is set in a wood frame, Figure 20-21. In masonry construction, wood *bucks* are fastened to the sides of the opening and the metal window is screwed to the bucks, Figure 20-22.

Metal windows are set in a similar manner to that used for wood windows. Because of the

Fig. 20-21 Installing a metal window

varied methods of manufacture, carefully follow the installation directions provided with the units.

GLAZING

The art of cutting and installing glass in sash is called *glazing.* The carpenter is often required to replace a light of glass if one becomes broken during construction.

Sash are made so that the glass is held in place with glazing points and glazing compound, Figure 20-23. *Glazing points* are small triangular or diamond-shaped pieces of metal driven into the sash parts to hold the glass in place. The glass is installed with the convex or crowned side up in a thin bed of compound. Glass set in this manner is not as apt to break when installed.

CAUTION: Use heavy gloves and handle the broken parts carefully. Broken glass edges are sharp and can cut easily.

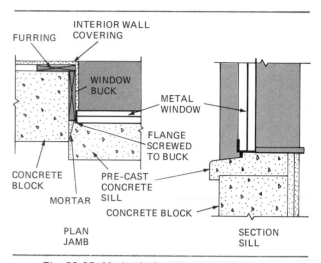

Fig. 20-22 Method of setting a metal window in a masonry opening

To replace a light of glass, first remove the broken glass. Clean all compound and glazier points from the rabbeted section of the sash. Apply a thin bed of glazing compound to the opening. Lay the glass in the sash with its crowned side up. Carefully seat the glass in the bed moving it back and forth slightly.

Fasten the glass in place with glazier points. Slide the driver along the glass. Do not strike the glass with the driver because the glass may break. Special glazier point driving tools prevent glass breakage. If a special driving tool is not available, drive the points with the side of a chisel or a putty knife.

Lay a bead of compound on the glass along the rabbet. Trim the compound by drawing the putty knife along it so that one edge is flush with the inside edge of the sash. The other edge should be flush with the face of the sash. Prime the compound as soon as possible after glazing.

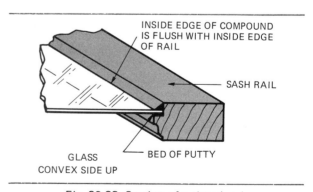

Fig. 20-23 Section of a glazed sash

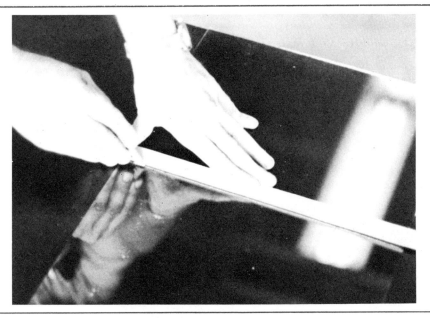

Fig. 20-24 Scoring the glass

CUTTING GLASS

Sometimes it may be necessary to cut a light of glass to size. Lay the glass on a clean, smooth surface. Brush some mineral spirits along the line of cut. Hold a straightedge on the line to be cut. Draw a glass cutter along the line and mark a sharp, uniform score, Figure 20-24. Do not go over the scored line because it will dull the glass cutter. The line must be scored along the whole length the first time with no skips. Otherwise the glass may not break where desired.

Move the glass so the scored line is even with the edge of the workbench. Apply downward pressure on the overhanging glass. If the glass is properly scored, it will break along the scored line, Figure 20-25.

Fig. 20-25 Breaking the glass along the scored line

REVIEW QUESTIONS

Select the most appropriate answer.

1. A frame in a window that holds the glass is called a
   a. light.
   b. mullion.
   c. sash.
   d. stile.

2. Small wood strips that divide the glass into smaller lights are called
   a. mantels.
   b. margins.
   c. mullions.
   d. muntins.

3. When windows are installed in multiple units, the joining of the side jambs forms a
   a. mantel.
   b. margin.
   c. mullion.
   d. muntin.

4. A window that consists of an upper and a lower sash that slide vertically is called a
   a. casement window.
   b. double-hung window.
   c. hopper window.
   d. sliding window.

5. A window that has a sash hinged on one side and swings outward is called
   a. an awning window.
   b. a casement window.
   c. a double-hung window.
   d. a hopper window.

6. The difference between a hopper and an awning window is that the hopper window
   a. swings inward instead of outward.
   b. swings outward instead of inward.
   c. is hung at the top rather than at the bottom.
   d. is hinged on the side rather than on the bottom.

7. Before setting a window in an opening
   a. flash all sides of the opening.
   b. flash the bottom.
   c. flash two sides.
   d. flash the top.

8. Horns on windows are
   a. placed on the header casing.
   b. extensions of the side jambs.
   c. moldings applied to a flat casing.
   d. extensions of the header jamb.

9. The art of cutting and installing glass is called
   a. gauging.
   b. glazing.
   c. gouging.
   d. grouting.

10. Glass is installed
    a. with its crowned side up.
    b. with its concave side up.
    c. with either side up.
    d. on the inside face of the sash.

EXTERIOR DOORS 21

OBJECTIVES

After completing this unit, the student will be able to

- *describe the standard designs and sizes of exterior doors and name their parts.*
- *name the parts of, build, and set an exterior door frame.*
- *hang an exterior door and install a lockset.*

Exterior doors are items of millwork. They are available in many styles and in standard sizes, Figure 21-1. They are classified as entrance, thermal, sash (commonly called French), side-light, combination storm and screen, and screen doors.

DOOR SIZES

Entrance doors are usually 1 3/4 inches thick. Their height is usually 6'-8'' although 7'-0'' is also available. The main entrance door of a residence should be 2'-8'' or 3'-0'' in width. A minimum 2'-6'' width may be used for other residential entrance doors.

Types of Doors	Sizes			Grades
	Thicknesses	Widths	Heights	
	in.	ft and in.	ft. and in.	
Front entrance doors	1-3/4	2'6'' 2'8'' 3'0''	6'8''	Selected
Thermal (insulated-glass) doors	1-3/4	2'6'' 2'8'' 3'0''	6'8''	Selected and Standard
Sash doors, including French doors	1-3/8 1-3/4	2'6'' 2'8'' 3'0''	6'8''	Standard
Side lights	1-3/8	1'0'' 1'2'' 1'4'' 1'6''	6'8-1/2''	Standard
Combination storm and screen doors	1-1/8	2'6'' 2'8'' 3'0''	6'9''	Standard
Screen doors	1-1/8	2'6'' 2'8'' 3'0''	6'9''	Standard

Fig. 21-1 Types, sizes, and grades of exterior doors *(Courtesy of Fir and Hemlock Door Association)*

1-3/4 FRONT ENTRANCE DOORS (EXTERIOR)

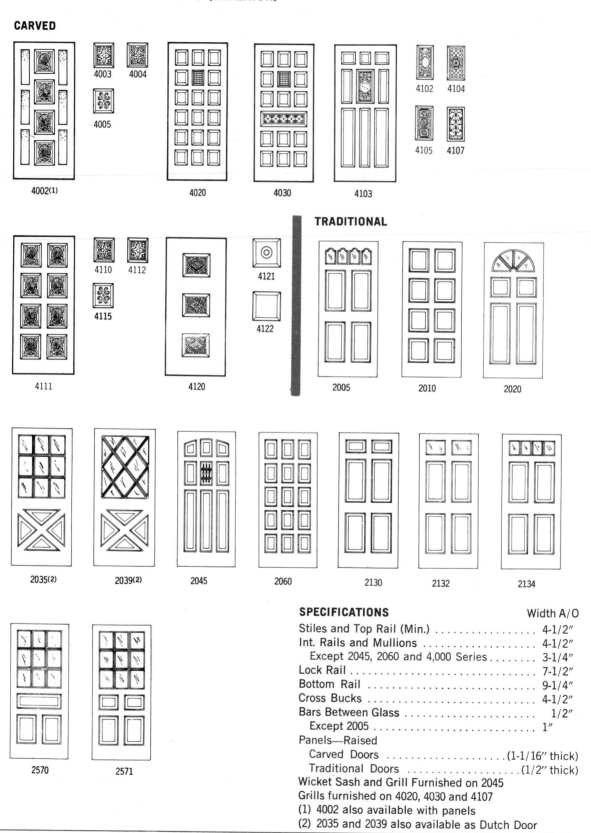

SPECIFICATIONS   Width A/O
Stiles and Top Rail (Min.) .... 4-1/2"
Int. Rails and Mullions .... 4-1/2"
   Except 2045, 2060 and 4,000 Series .... 3-1/4"
Lock Rail .... 7-1/2"
Bottom Rail .... 9-1/4"
Cross Bucks .... 4-1/2"
Bars Between Glass .... 1/2"
   Except 2005 .... 1"
Panels—Raised
   Carved Doors ....(1-1/16" thick)
   Traditional Doors ....(1/2" thick)
Wicket Sash and Grill Furnished on 2045
Grills furnished on 4020, 4030 and 4107
(1) 4002 also available with panels
(2) 2035 and 2039 also available as Dutch Door

Fig. 21-1 (Continued)

1-3/8 and 1-3/4 SASH DOORS (EXTERIOR)

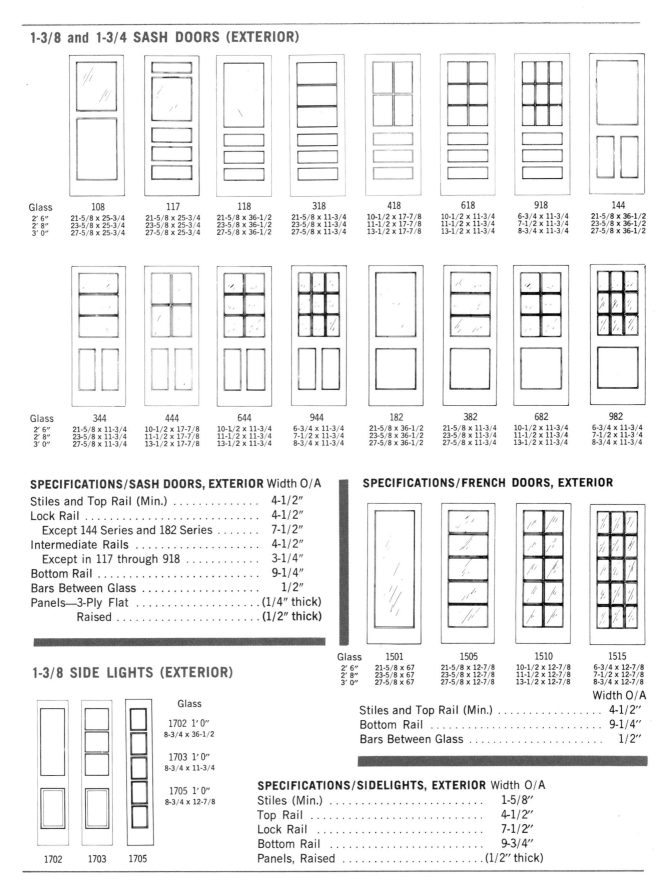

Glass	108	117	118	318	418	618	918	144
2' 6"	21-5/8 x 25-3/4	21-5/8 x 25-3/4	21-5/8 x 36-1/2	21-5/8 x 11-3/4	10-1/2 x 17-7/8	10-1/2 x 11-3/4	6-3/4 x 11-3/4	21-5/8 x 36-1/2
2' 8"	23-5/8 x 25-3/4	23-5/8 x 25-3/4	23-5/8 x 36-1/2	23-5/8 x 11-3/4	11-1/2 x 17-7/8	11-1/2 x 11-3/4	7-1/2 x 11-3/4	23-5/8 x 36-1/2
3' 0"	27-5/8 x 25-3/4	27-5/8 x 25-3/4	27-5/8 x 36-1/2	27-5/8 x 11-3/4	13-1/2 x 17-7/8	13-1/2 x 11-3/4	8-3/4 x 11-3/4	27-5/8 x 36-1/2

Glass	344	444	644	944	182	382	682	982
2' 6"	21-5/8 x 11-3/4	10-1/2 x 17-7/8	10-1/2 x 11-3/4	6-3/4 x 11-3/4	21-5/8 x 36-1/2	21-5/8 x 11-3/4	10-1/2 x 11-3/4	6-3/4 x 11-3/4
2' 8"	23-5/8 x 11-3/4	11-1/2 x 17-7/8	11-1/2 x 11-3/4	7-1/2 x 11-3/4	23-5/8 x 36-1/2	23-5/8 x 11-3/4	11-1/2 x 11-3/4	7-1/2 x 11-3/4
3' 0"	27-5/8 x 11-3/4	13-1/2 x 17-7/8	13-1/2 x 11-3/4	8-3/4 x 11-3/4	27-5/8 x 36-1/2	27-5/8 x 11-3/4	13-1/2 x 11-3/4	8-3/4 x 11-3/4

SPECIFICATIONS/SASH DOORS, EXTERIOR Width O/A

Stiles and Top Rail (Min.) . . . . 4-1/2"
Lock Rail . . . . 4-1/2"
   Except 144 Series and 182 Series . . . . 7-1/2"
Intermediate Rails . . . . 4-1/2"
   Except in 117 through 918 . . . . 3-1/4"
Bottom Rail . . . . 9-1/4"
Bars Between Glass . . . . 1/2"
Panels—3-Ply Flat . . . .(1/4" thick)
   Raised . . . .(1/2" thick)

SPECIFICATIONS/FRENCH DOORS, EXTERIOR

Glass	1501	1505	1510	1515
2' 6"	21-5/8 x 67	21-5/8 x 12-7/8	10-1/2 x 12-7/8	6-3/4 x 12-7/8
2' 8"	23-5/8 x 67	23-5/8 x 12-7/8	11-1/2 x 12-7/8	7-1/2 x 12-7/8
3' 0"	27-5/8 x 67	27-5/8 x 12-7/8	13-1/2 x 12-7/8	8-3/4 x 12-7/8

Width O/A

Stiles and Top Rail (Min.) . . . . 4-1/2"
Bottom Rail . . . . 9-1/4"
Bars Between Glass . . . . 1/2"

1-3/8 SIDE LIGHTS (EXTERIOR)

Glass

1702 1' 0"
8-3/4 x 36-1/2

1703 1' 0"
8-3/4 x 11-3/4

1705 1' 0"
8-3/4 x 12-7/8

1702   1703   1705

SPECIFICATIONS/SIDELIGHTS, EXTERIOR Width O/A

Stiles (Min.) . . . . 1-5/8"
Top Rail . . . . 4-1/2"
Lock Rail . . . . 7-1/2"
Bottom Rail . . . . 9-3/4"
Panels, Raised . . . .(1/2" thick)

Fig. 21-1 (Continued)

1-1/8 SCREEN DOORS (EXTERIOR)

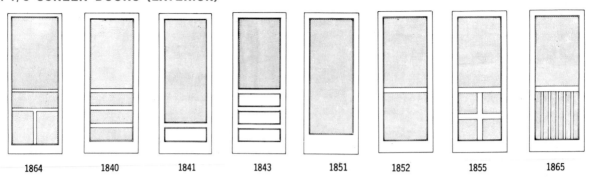

1864   1840   1841   1843   1851   1852   1855   1865

1-3/4 THERMAL (INSULATED-GLASS) DOORS (EXTERIOR)

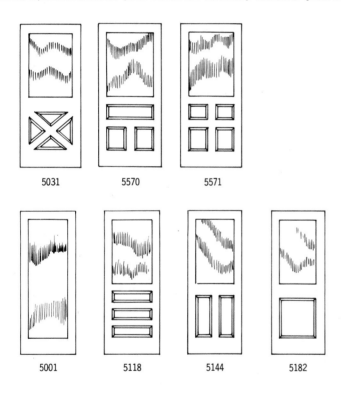

5031   5570   5571

5001   5118   5144   5182

SPECIFICATIONS

	Min. Width
Stiles and Top Rail	4-1/2″
Lock Rail	4-1/2″ or 7-1/2″
Intermediate Rails	3-1/4″ or 4-1/2″
Mullions	4-1/2″
Bottom	9-1/4″
Raised Panels (Thickness)	3/4″

Insulated Glass unit is 1/2″ thick;
  (2 pieces, 1/8″ Tempered Separated by a 1/4″ Air Space)

Glass sizes may vary—check with your supplier

MINIMUM SPECIFICATIONS/
SCREEN DOORS (EXT.)   Width O/A
(Tolerance: plus or minus 1/32″)

	Width O/A
Stiles and Top Rails	3-1/8″
Intermediate Rails	3-1/8″
except 1840	1-5/8″
Mullions	1-1/4″
except 1855	3-1/8″
Bottom Rail	5″
except 1851	9″

1-1/8 COMBINATION STORM
AND SCREEN DOORS (EXTERIOR)

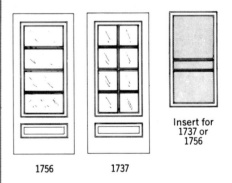

Insert for
1737 or
1756

1756   1737

SPECIFICATIONS/COMB. STORM

	Width O/A
Stiles ....4-1/4″ or 4-1/2″	
Top and Lock Rails	4-1/2″
Bottom Rail	9-1/4″
Panels—Raised .... (1/2″ thick)	

Fig. 21-1 (Continued)

Fig. 21-2 Sidelights are installed on one or both sides of the main entrance door.

Sidelights are installed in a fixed position on one or both sides of the main entrance door, Figure 21-2. Sidelights are made 1/2 inch greater in height than the entrance door to allow for fitting the bottom against the sloping sill of the door frame. Storm and screen doors are made 1 inch greater in height for the same purpose. Sidelights are manufactured 1 3/8 inches in thickness. Storm and screen doors come 1 1/8 inches thick.

PARTS OF A DOOR

The outside vertical members of a paneled door are called *stiles.* The top horizontal member is called the *top rail* and is generally made the same width as the stiles. The bottom horizontal member is called the *bottom rail.* A horizontal member situated at lockset height (38 inches from bottom to the center) is called the *lock rail.* Other rails are called *intermediate rails. Mullions* are vertical members between the stiles. A *bar* is a narrow horizontal, vertical, or diagonal rabbeted member which extends the total length or width of the glass opening and divides the glass into smaller lights. *Panels* fit

between the stiles, rails, and mullions and are usually of wood. They may be raised on one or both sides, Figure 21-3. *Flush doors* have a solid lumber or particle board core covered on both sides with thin face veneers of various kinds of wood.

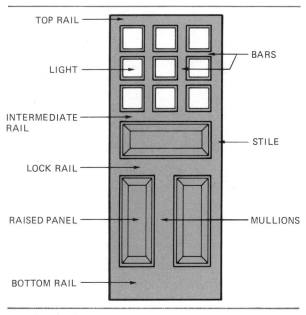

Fig. 21-3 The parts of an exterior paneled door

Fig. 21-4 The door remains in the frame while a prehung unit is being set.

PREHUNG DOORS

Some doors come already fitted, hung, and locked in the door frame. This type is set in the opening in a way similar to that used in setting window units. Center the unit in the opening with the casings removed and the door in a closed position. Level the sill and shim between the side jambs at the bottom. Tack through the side jambs and shim into the stud at the bottom using finishing nails. Plumb the side jambs; shim at the top and tack. Shim at intermediate points along the side jamb, and fasten through the shims, Figure 21-4. Make sure the spacers between the door and the jamb are in place and that the proper joint between the door and jamb is maintained. Open the door to make sure it operates properly. Make any necessary adjustments. Drive all nails home and set them. Do not make any hammer marks on the finish. It is good procedure to drive the nail until it is almost flush and then use a nailset to drive it the rest of the way. When prehung doors are not used, the

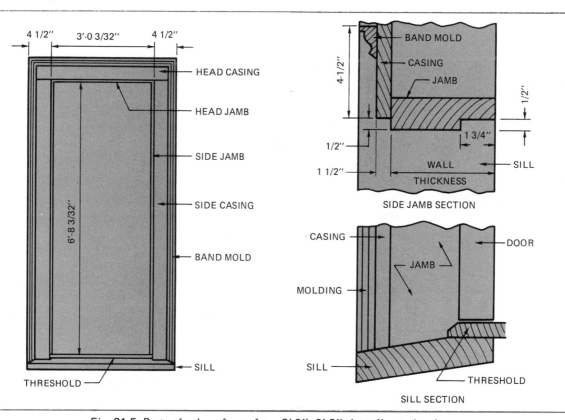

Fig. 21-5 Parts of a door frame for a 3'-0"x6'-8"x1 3/4" exterior door

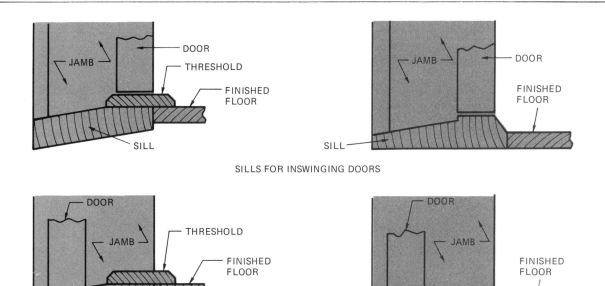

SILLS FOR INSWINGING DOORS

SILLS FOR OUTSWINGING DOORS

Fig. 21-6 Wood sill shapes vary according to the swing of the door
and customs of the geographical location.

carpenter has to build and set the door frame, fit and hang the door, and install the lockset.

PARTS OF A DOOR FRAME

Terms given to members of a door frame are the same as for a window frame. The bottom member is called a *sill* or *stool*. Side members are called *side jambs* and the top a *header jamb*. The exterior casing is also part of the door frame and is usually applied before the door frame is set, Figure 21-5.

SILLS

Sills are made of wood or of metal with vinyl inserts. Both kinds come in various shapes for inswinging or outswinging doors, Figure 21-6. In residential construction, frames are usually made so entrance doors swing inward. It is required that doors swing outward in public buildings.

Wood sills are usually made of oak to resist wearing from the traffic. They are designed so their bottoms rest on the subfloor.

If metal sills are used, the bottom member of the door frame is wood. The metal sill is applied over this piece, Figure 21-7.

JAMBS

Side and header jambs are the same shape. Jambs may be square edge pieces of stock to which a *door stop* is later applied or rabbeted jamb stock may be used. This type may be

Fig. 21-7 Cutaway of an adjustable metal door sill

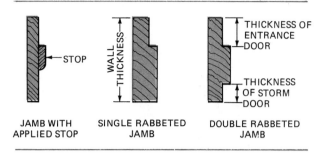

Fig. 21-8 Door jamb shapes

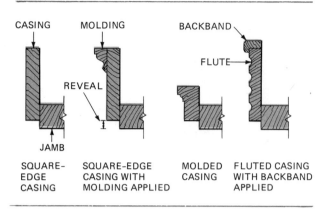

Fig. 21-9 Shapes of exterior door casings

single or double rabbeted. One rabbet is used for a stop for the main door and the other as a stop for combination storm and screen doors, Figure 21-8. Various rabbet widths are available for different door thicknesses. Several jamb widths are available for different wall thicknesses. Clear fir is commonly used for jamb stock because it is straight-grained, less likely to warp, and strong.

EXTERIOR CASINGS

Exterior casings may be plain square edge stock. Applied moldings to the outside edges give a better appearance. Often brick molding is used as casing. Because the main entrance is a distinctive feature of the house, the exterior casing may consist of *fluted*, or otherwise shaped, pieces with appropriate caps and moldings applied, Figure 21-9. Most of these elaborate front entrance frames are purchased in knocked down form and assembled on the job site.

MAKING A DOOR FRAME

One of the most common types of door frames consists of single rabbeted jambs with an oak sill for an inswinging door. To make a door frame, determine the overall wall thickness from the outside surface of the wall sheathing to the inside surface of the interior wall covering. Because the interior wall covering probably has not been applied at this point, its planned thickness must be known.

Rip the jamb stock so its width corresponds to the wall thickness. Rip the edge opposite the rabbet. Smooth and slightly back-bevel the ripped edge with a jack plane. Back-bevel the rabbeted edge if not already done at the mill. Slightly round over (break) any sharp outside

corners. Jamb stock should have no crook or excessive bow or twist.

DADOING THE SIDE JAMBS

Lay the side jambs on a pair of sawhorses so their outside edges are against each other and their ends flush. Square a line across both pieces about 1 inch down from the top end. Hold a piece of jamb stock on end to the squared line and mark its overall thickness toward the bottom of the side jambs. Mark the depth of the rabbet on the edge opposite the rabbet.

Dado out between the lines to the depth of the rabbet. Cut to each line with a fine-toothed hand crosscut saw. Be careful when cutting to the bottom line not to score the face of the rabbet. If the rabbet face is scored at this point, the saw mark will be exposed when the frame is assembled. Chisel in from both edges until the dado depth is reached. The dado bottom should lie in a straight line with no crown. A slight hollow is not objectionable, Figure 21-10.

Place a piece of jamb stock in the dado with its rabbeted face toward the bottom end of the side jamb. Measure from the rabbet face along the side jamb (the door height plus 3/32 inch) and mark. Square a line at this point across the face of the rabbet on the bottom end of the side jamb.

Hold a piece of sill stock so the flat face just below the door lines up with the mark. The top of the bevel on the inside edge of the sill should line up with the inside edge of the side jamb. The lip on the top face of the sill should

Fig. 21-10 The side jamb is dadoed to receive the header jamb.

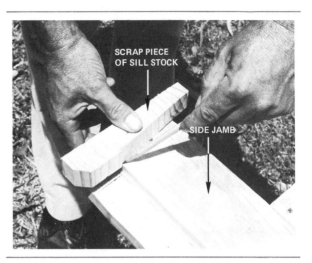

Fig. 21-11 Laying out the dado on the side jamb for the sill

be just inside the rabbet of the side jamb, Figure 21-11. Mark the side jamb on both sides of the sill when held in this position.

Remove the sill stock and place the side jambs back together again. Transfer the marks for the sill from one side jamb to the other. Use a square and a bevel. Dado the side jambs for the sill. Again, be careful not to score the rabbet face of the side jamb when making the upper saw cut.

CUTTING SILL AND HEADER JAMB

Cut the header jamb so it is square on both ends and its length is the width of the door plus 3/32 inch. Lay the header jamb on the sill stock and mark its length. Square these marks across the face of the sill. Measure out from these squared lines along the inside beveled edge a distance of 5/16 inch. The beveled edge is cut to this length. The rest of the sill is cut to the length of the header jamb. This results in notched ends in the sill, Figure 21-12. The projecting ends of the beveled inner edge will butt against the edge of the interior casing when applied.

ASSEMBLING THE DOOR FRAME

Although some have different-shaped jambs and sills, all door frames are assembled in

much the same manner. When assembling the frames, stand the header and sill on end and place the side jamb on them with the ends inserted in the dadoes. Nail through the side jambs into the ends of the sill and header jamb. To assure a tight fit, drive a chisel between the back side of sill and header jamb and the shoulders of the dado before nailing. This will drive the sill and header jamb tight against the dado shoulders on the face side. Use three 8d common nails on each end. Turn the assembly over and install the other side jamb in the same

Fig. 21-12 The ends of the sill are notched on the inside edge.

manner. Keep the edges of side and header jamb flush with each other. The notches in the sill should fit tightly against the inside edges of the side jambs.

APPLYING THE EXTERIOR CASING

The exterior casing is applied so its inside edge is set back 1/2 inch from the inside face of the jambs. This setback is called a *reveal.*

If square-edged casings are used, a rabbeted weathertight joint is made between the side casings and the header casing. If molded casing is used, a bed of caulking compound is applied to the mitered joint, Figure 21-13.

Rabbet the top end of both side casings. Apply both side casings with noncorroding nails spaced about 12 inches apart. Let the bottom ends project below the sill to the bottom of the house frame. Cut and apply the header casing. Make a blind rabbet on the ends of square edge casings to fit the rabbeted ends of the side casings. Miter molded casings and drive nails both ways through the joint. Let the bottom ends of side casings project below the sill to the bottom of the house frame.

SETTING THE DOOR FRAME

When setting a door frame, cut off the horns which project beyond the sill and header jamb from the side jambs. Flash both sides of the opening in the same manner as for window frames. Set the door frame in the opening with the sill on the subfloor. Center the bottom of the frame in the opening and level the sill. Tack the lower end of the casings. Plumb the side jambs and tack the upper end of the casings. Use a six-foot level with metal blocks attached to each end or use a long straightedge to which small blocks of equal thickness are fastened to each end in combination with a level, Figure 21-14.

Straighten the jambs between the top and bottom with a long straightedge and tack the side casings at intermediate points. Straighten the header jamb and tack through the header casing.

Check all parts again for level and plumb. Then drive and set all nails. Apply the metal

Fig. 21-13 A weathertight joint must be made between side and header casings.

Fig. 21-14 Setting an exterior door frame

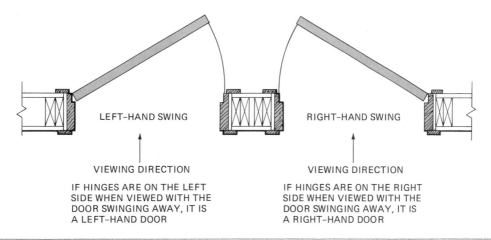

LEFT-HAND SWING

RIGHT-HAND SWING

VIEWING DIRECTION

VIEWING DIRECTION

IF HINGES ARE ON THE LEFT
SIDE WHEN VIEWED WITH THE
DOOR SWINGING AWAY, IT IS
A LEFT-HAND DOOR

IF HINGES ARE ON THE RIGHT
SIDE WHEN VIEWED WITH THE
DOOR SWINGING AWAY, IT IS
A RIGHT-HAND DOOR

Fig. 21-15 Determining the swing of a door

flashing on the header casing and flash the sill in the same manner as for windows.

FITTING THE DOOR

First, determine the outside face of the door. If the design of the door is such that either side may be used toward the outside, sight along the stiles for a bow. The door should be fitted so that the hollow side of the bow is against the door stops. This allows the top and bottom of the door to come up tight against stops when hung. The center comes up tight when the door is latched.

On doors with glass panels, the outside face is that which has the applied molding to hold the glass in the opening. This can easily be determined by looking for the fasteners that hold the molding in place. It is important not to put a door with glass panels on backwards. In such a case, water will seep through the panels during a rainstorm that has heavy winds blowing against the door.

Determine the swing of the door from the plans. A door is designated as having a right-hand or a left-hand swing. When standing on the side of the door so that it swings away, it is a right-hand swing when the hinges are on the right. The conditions are opposite for a left-hand swing, Figure 21-15. Lightly mark the face and hinge edge of the door and the hinge side of the door frame.

Place the door on sawhorses and cut any horns from the stiles. Measure carefully the width and height of the door frame. The frame should be perfectly rectangular; however, this may not be the case and should not be taken for granted. The door must be carefully fitted to the frame so an even joint of 3/64″ (the thickness of a nickel) appears all around between the door and the frame.

Plane the door so that it fits snugly in the door frame. Use a jointer plane, either hand or power, and hold the door in a door jack, Figure 21-16. A *door jack* may be constructed with

Fig. 21-16 When fitting a door, hold it in a door jack.

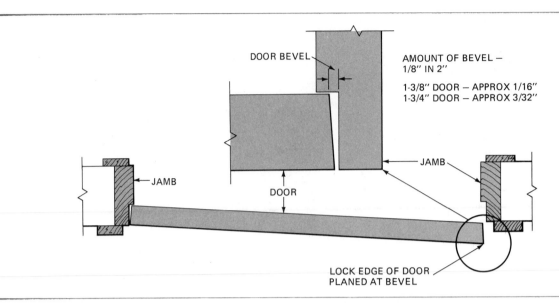

DOOR BEVEL

AMOUNT OF BEVEL –
1/8" IN 2"

1-3/8" DOOR – APPROX 1/16"
1-3/4" DOOR – APPROX 3/32"

JAMB

JAMB

DOOR

LOCK EDGE OF DOOR
PLANED AT BEVEL

Fig. 21-17 The lock edge of a door must be planed at a bevel for clearance.

materials available on the job site or a door jack that is commercially made may be used.

Fit the top end against the header jamb and then fit the bottom against the sill so that a proper joint is acquired. This may require that the door be placed in and then removed from the frame several times.

Next, fit the hinge edge against the side jamb. Finally, plane the lock edge so the desired joint is obtained on both sides. The lock edge must also be planed on a bevel so that when the door is opened, the back corner does not strike the jamb. The amount of bevel is determined by slightly opening the door so that the same amount of joint is between the back corner and the edge of the side jamb. This bevel is often made on the lock edge after the door is hung and before the lockset is installed, Figure 21-17.

Extreme care must be taken when fitting doors not to get them undersize. Check the fit frequently by placing the door in the opening, even if this takes extra time. Speed will come with practice. Handle the door carefully to avoid marring it or other finish. After the door is fitted, break all sharp corners slightly with a block plane and sandpaper.

LOCATION AND SIZE OF DOOR HINGES

On swinging doors, the loose-pin butt hinge is used. The pin is removed and each side applied to the door and frame. The door is hung

by placing the door in the opening and inserting the pins. Extreme care must be taken so that the hinge leaves line up exactly on the door and frame. On 1 3/4" thick doors use three 4"x4" hinges. On 1 3/8" thick doors, use 3 1/2"x3 1/2" hinges. Always use three hinges to a door.

The hinge leaves are cut in flush with the door edge and jamb and only partway across. The distance from the side of the door and the edge of the hinge is called the *backset*. A backset is required so the edge of the hinge will not be exposed when the door is opened. The hinge pin is located beyond the door face, allowing the door to be fully opened and to clear the door casings, Figure 21-18.

On paneled doors, the top hinge is placed with its upper end in line with the bottom edge

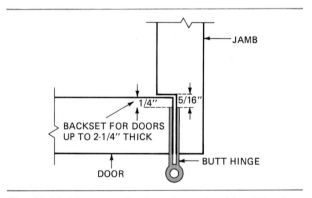

JAMB

1/4"

5/16"

BACKSET FOR DOORS
UP TO 2-1/4" THICK

DOOR

BUTT HINGE

Fig. 21-18 Hinges are backset from the side of the door and the edge of the door stop.

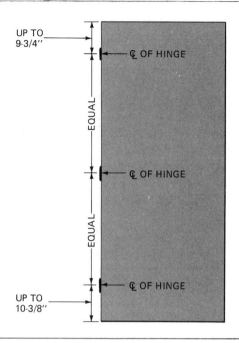

Fig. 21-19 Recommended placement of hinges on flush doors

of the top rail. The bottom hinge is placed with its lower end in line with the top edge of the bottom rail. The middle hinge is centered between the top and bottom hinges. On flush doors, the recommended placement is not more than 9 3/4 inches down from the top and 10 3/8 inches up from the bottom with the middle hinge centered between the two, Figure 21-19.

HANGING A DOOR

To hang a door, place the door in the opening. Shim the top and bottom so the proper joint is obtained. Shim the lock edge so that the hinge edge is tightly against the door jamb.

Use a sharp knife and mark across the door and jamb the location of one end of each hinge. Do not mark either on the door or jamb any more than the hinge thickness. Mark a small X on both the door and the jamb on the side of the knife mark on which the hinge is to be mortised, Figure 21-20. Care must be taken not to cut the hinge in on one side of the mark on the door and the other side of the mark on the frame.

Remove the door from the frame. Place it in a door jack with its hinge-edge up.

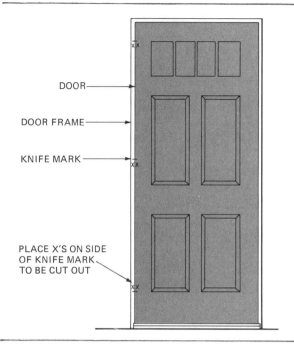

Fig. 21-20 Mark the location of hinges on the door and frame with a sharp knife.

LAYING OUT THE HINGE GAIN

Lay out the ends of the hinge gain by placing the end of a hinge leaf on the knife mark. Make sure the leaf is over the X previously marked. Hold the leaf square with the barrel of the pin against the door face. With a sharp knife mark one end. Tap the hinge until it just covers the knife mark and mark the other end, Figure 21-21.

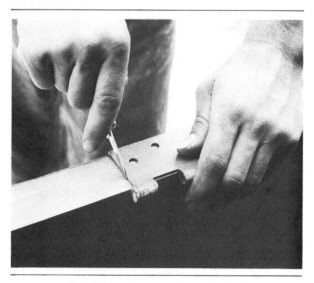

Fig. 21-21 A mortise-type lockset

Fig. 21-22 Laying out the backset of the hinge gain

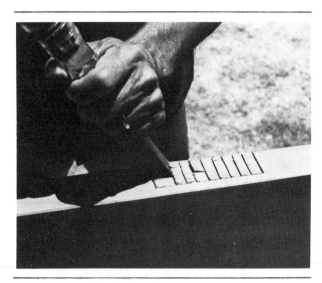

Fig. 21-23 Chiseling out the hinge gain

ADJUSTING THE BUTT GAUGE

The butt gauge has two rods. One rod has two markers. One marker marks the backset on the door, and the other marks the backset on the frame with compensation for clearance between the door and the door stop. Adjust this rod for the desired backset on the door. Between the two end marks, lay out the backset on the door and jamb. Take care not to mark beyond the end marks, Figure 21-22.

The other rod has one marker and is adjusted for the hinge thickness. With this marker, lay out the hinge leaf thickness. Again, be careful not to let the marks extend beyond the ends of the gain.

With a sharp knife deepen the backset layout. Do **not** use a chisel. Using a chisel at this point may split out the edge of the door.

Then, with the bevel of the chisel down, cut a small chip from each end of the gain. The chips will break off at the scored end marks. With the flat of the chisel against the shoulders of the gain deepen the end marks. Make a series of small chisel cuts along the length of the gain. Then with the flat of the chisel down, pare the excess down to the depth of the gain. Be careful not to cut off the backset, Figure 21-23.

The hinge should press fit into the gains and the leaf, flush with the surface. If the hinge leaf is above the surface, deepen the gain until the leaf lies flush. If the leaf is below the surface, shim it with thin pieces of cardboard.

Cardboard from the boxes the hinges come in is usually used.

Press the hinge leaf in the gains and drill pilot holes for the screws. Center the holes carefully on the drilled holes of the hinge leaf. A self-centering punch may be used for this purpose. Drilling off-center will cause the hinge to move from its position when the screw is driven.

Hang the door in the frame by inserting the top pin first. Insert the other pins and try the swing of the door. If the door binds against the jamb, shim the hinges between the hinge pins and the screws. This will move the door toward the lock side, Figure 21-24. Check the joint. Remove and rejoint the door, if necessary. Check the bevel on the lock edge. Plane to the proper bevel if necessary. Break all sharp corners.

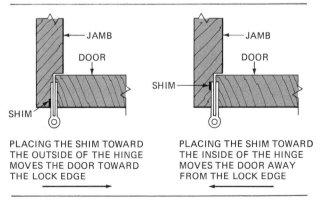

PLACING THE SHIM TOWARD THE OUTSIDE OF THE HINGE MOVES THE DOOR TOWARD THE LOCK EDGE

PLACING THE SHIM TOWARD THE INSIDE OF THE HINGE MOVES THE DOOR AWAY FROM THE LOCK EDGE

Fig. 21-24 Shimming the hinges will move the door toward or away from the lock edge.

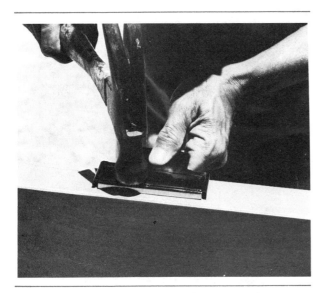

Fig. 21-25 Butt hinge markers are used to
lay out hinge gains.

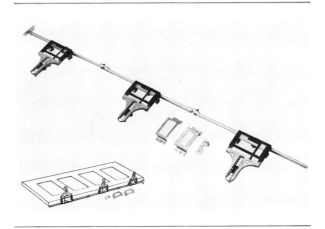

Fig. 21-26 The butt hinge template

OTHER METHODS OF
HANGING DOORS

Instead of laying out hinge gains with a
butt gauge, *butt hinge markers*, Figure 21-25,
are often used. The markers come in different
sizes, and the hinge layout is made by simply
tapping the marker with a hammer. The gain is
then chiseled out in the usual manner, after the
depth is laid out using the butt gauge.

When many doors need to be hung, a *butt
hinge template* is used, Figure 21-26. The
template is laid both on the door and the jamb
and the gains are cut with a portable electric
router. A special hinge mortising bit and template guide is used. The template guide is
attached to the base of the router and guides it
along the template, Figure 21-27.

The template is adjustable for different size
hinges, hinge locations, and door thicknesses. It
is made to give the desired joint at the top of the
door and the proper clearance between the door
and door stop. Because the mortises have
rounded corners, hinges with rounded corners
are used. This saves time by not having to chisel
the corners of the mortise square.

Care must be taken not to clip the template
when removing the router. It is best to let the
router come to a stop before removing it. Both
the template and bit could be damaged beyond
repair.

Fig. 21-27 Using a butt hinge template and
router to cut hinge gains

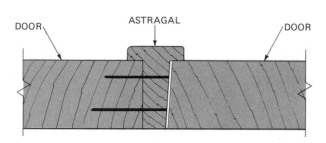

Fig. 21-28 An astragal is required between double doors for weathertightness.

OTHER EXTERIOR DOORS

Other exterior doors, such as *double doors*, are hung in a similar manner. An *astragal* is required between double doors for weathertightness, Figure 21-28.

Sliding glass doors, Figure 21-29, usually come with precut parts. Instructions for assembly are included and should be followed carefully. Installation of the frame is similar to that of setting frames for swinging doors. After the frame is set the doors are installed using special hardware supplied with the unit.

Garage doors come in many styles, kinds, and sizes. Two popular kinds are the swing-up and the roll-up doors. The *swing-up* type is a rigid one-piece door. The *roll-up* type has hinged sections that roll upward and turn to a horizontal position. Special hardware, required for both types, is supplied with the door. Also supplied are the manufacturer's directions for installation. These should be followed carefully.

DOOR LOCKS

The *cylindrical lockset*, often called *key-in-the-knob lockset*, is used extensively on exterior doors in residential construction, Figure 21-30. *Mortise locksets* are used less frequently on residential construction, but are commonly used in commercial construction, Figure 21-31. Cylindrical locksets are quickly and easily installed. Mortise locksets require more time to install. Because of the many different kinds, manufacturer's instructions for installation are included with each set.

Fig. 21-29 Sliding glass door units usually come assembled or with precut parts.

Fig. 21-30 A cylindrical lockset

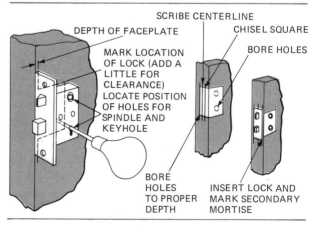

Fig. 21-31 A mortise-type lockset

INSTALLING CYLINDRICAL LOCKSETS

To install a cylindrical lockset, open the package, check the contents, and read the manufacturer's directions carefully, Figure 21-32.

Open the door to a convenient position and wedge the bottom to hold it in place. Measure up, from the bottom of the door to the centerline of the lock, the recommended distance

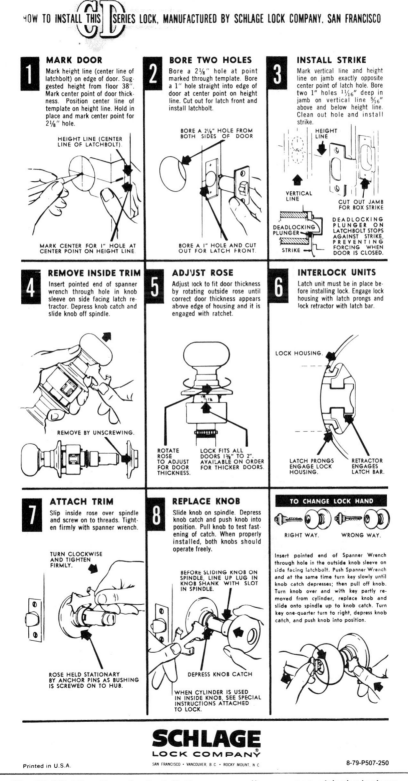

Fig. 21-32 Directions for installation are usually provided with the lockset.

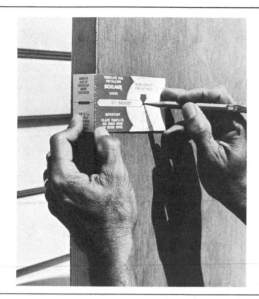

Fig. 21-33 Using a template to lay out the centers of the holes for a lockset

Holes may also be bored by hand using expansive and auger bits and a bit brace.

Place the latch unit in the door edge and mark around the edges of the faceplate with a sharp knife. Remove the latch unit. Deepen the vertical lines with the knife. Do not use a chisel along these lines because this may split out the edge of the door. Then, chisel out the recess so that the latch unit lays flush with the door edge. Faceplate markers, Figure 21-35, may also be used to lay out the mortise. Install the lockset in the door according to directions.

INSTALLING THE STRIKER PLATE

The *striker plate* is installed on the door jamb so when the door is closed it latches tightly with no play. If the plate is installed too far out, the door will not close tightly. If the plate is installed too far in, the door will not latch. To locate the striker plate, place it over the latch in the door. Close the door snugly against the stops. Push the striker plate in against the latch and mark the top end of the jamb. Then hold a pencil against the door edge and draw a line down the face of the striker plate even with the edge of the jamb, Figure 21-36.

(usually 38 inches). At this point, square a light line across the edge and stile of the door.

Position the furnished template on the door and lay out the centers of the holes, Figure 21-33. Usually two holes have to be bored: one through the side and one into the edge. The direction sheet specifies the sizes of the holes.

Boring jigs are frequently used to guide the bits, Figure 21-34. They also help prevent splintering of the door face when the holes are bored. Attach the jig to the door at the layout lines. Power driven bits make boring easier.

Open the door and place the striker plate on the jamb to the layout lines. Mark around the plate with a sharp knife and chisel out the

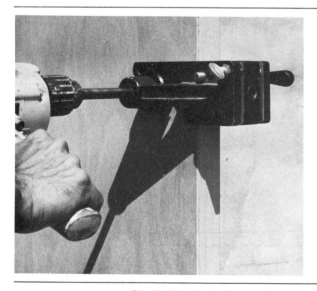

Fig. 21-34 Boring jugs are frequently used when boring holes for locksets.

Fig. 21-35 Using a faceplate marker

Fig. 21-36 Installing the striker plate

mortise so the plate lies flush with the jamb. Screw the plate in place and chisel out the center to receive the latch. **NOTE: If the striker plate has a bent flange, drill a series of small holes vertically in the jamb to receive the flange and proceed with the layout just described.**

MORTISE LOCK

Mortise locks are installed in a similar manner except, instead of boring a hole, a mortise must be made in the edge of the door. This is done by boring a series of holes in the door edge and chiseling the mortise to receive the lock, Figure 21-37. Care must be taken to read the directions carefully because some holes in the side of the door may be bored only partway through.

After the door is fitted and hung, remove all hardware and prime the door and all exposed parts of the door frame. Replace the door and hardware after the prime coat is dry.

Fig. 21-37 A series of holes is bored in the door edge to install a mortise lockset.

REVIEW QUESTIONS

Select the most appropriate answer.

1. The standard thickness of exterior wood doors in residential construction is
   a. 1 3/8 inches.   c. 1 3/4 inches.
   b. 1 1/2 inches.   d. 2 1/4 inches.

2. The minimum width of exterior doors in residential construction is
   a. 3'-0''.   c. 2'-6''.
   b. 2'-8''.   d. 2'-4''.

3. The height of an exterior door in residential construction is generally not less than
   a. 7'-0''.   c. 6'-8''.
   b. 6'-10''.   d. 6'-6''.

4. The thickness of storm and screen doors is usually
   a. 3/4 inch.   c. 1 3/8 inches.
   b. 1 1/8 inches.   d. 1 1/2 inches.

5. A narrow member dividing the glass in a door into smaller lights is called a
   a. bar.   c. rail.
   b. mullion.   d. stile.

6. When casings are installed on door frames, the setback is called a
   a. gain.   c. rabbet.
   b. backset.   d. reveal.

7. Projections of the side jambs beyond the sill and head jamb are called
   a. horns.   c. casings.
   b. ears.   d. rabbets.

8. Before hanging a door, sight along its length for a bow. The hollow side of the bow should
   a. face the outside.   c. face the inside.
   b. face the door stops. d. be straightened with a plane.

9. The joint between the door and door frame should be close to
   a. 3/32''.   c. 1/8''.
   b. 3/64''.   d. 3/16''.

10. On paneled doors, the top end of the top hinge is usually placed
    a. in line with the bottom edge of the top rail of the door.
    b. 7 inches down from the top end of the door.
    c. not more than 10 3/4 inches down from the top end of the door.
    d. in line with the top edge of an intermediate rail.

EXTERIOR WALL SIDING 22

OBJECTIVES

After completing this unit, the student will be able to

- *describe the shapes, sizes, and grades of various exterior wall siding products.*
- *install corner boards.*
- *apply the following siding products: horizontal and vertical lumber siding, wood shingles and shakes, plywood and hardboard siding, and metal and plastic siding.*
- *prepare a base for stucco.*
- *estimate required amounts of siding.*

Exterior wall covering is called *siding* except when a masonry covering such as stucco or brick veneer is used. Siding is available in many different kinds of material, such as solid lumber, plywood, hardboard, metal, and plastic. Siding can be obtained in many different patterns. Many prefinished types are available that eliminate the need to refinish for many years. Siding may be applied according to the type, horizontally, vertically, or in many interesting patterns.

WOOD SIDING

Wood siding is used extensively in both residential and commercial construction, Figure 22-1. For durability and resistance to the weather, cypress, cedar and redwood siding are preferred. Siding is also manufactured from spruce, pine, fir, and other kinds of wood.

Fig. 22-1 Wood siding is used extensively in both residential and commercial construction.
(Courtesy of California Redwood Association)

In some species such as redwood and cedar, siding is available in vertical grain or mixed vertical and flat grain, Figure 22-2. Vertical grain wood siding is the highest quality because it warps less, takes paint and other finishes better, and is easier to work. It also has less knots,

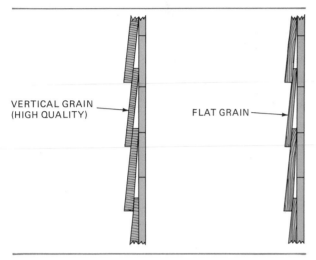

VERTICAL GRAIN (HIGH QUALITY)

FLAT GRAIN

Fig. 22-2 In some species, siding is available in vertical grain or in mixed vertical and flat grain.

pitch pockets, wanes, checks, shakes, or other defects.

It is recommended that wood siding be coated with a water-repellent preservative to serve as a basecoat for other finishes. This treatment resists moisture entry and decay, and results in a longer paint life.

Many boards are cut to length as the siding is put in place. Freshly cut ends should receive a liberal coating of preservative before installation.

Some wood siding patterns are used only horizontally and others only vertically. Some may be used in either manner.

BEVEL SIDING

Bevel siding, commonly called *clapboards*, is thicker on one edge than the other, Figure 22-3. Usually one surface of bevel siding is smooth and the other rough sawn. For a painted surface, the smooth side is exposed. The rough side is exposed when a stain finish is desired because wood stain lasts longer on rough wood surfaces.

	BEVEL	BUNGALOW
	PLAIN 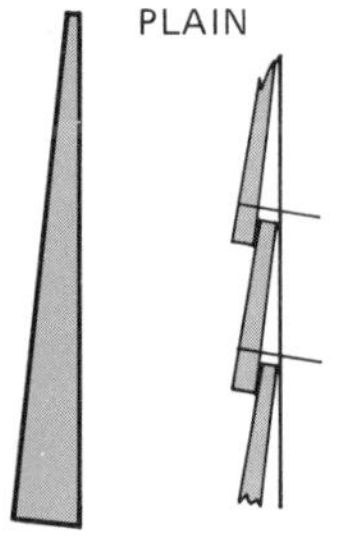	PLAIN 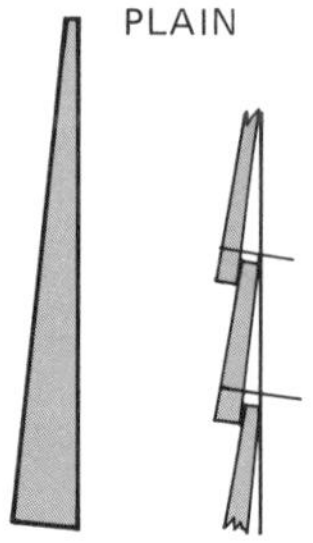
PATTERNS	PLAIN BEVEL MAY BE USED WITH SMOOTH FACE EXPOSED OR SAWN FACE EXPOSED FOR TEXTURED EFFECT.	THICKER AND WIDER THAN BEVEL SIDING. PLAIN BUNGALOW OR "COLONIAL" MAY BE USED WITH SMOOTH FACE EXPOSED OR SAWN FACE EXPOSED FOR TEXTURED TEXTURED EFFECT.
APPLICATION AND NAILING	RECOMMEND 1" MINIMUM OVERLAP ON PLAIN BEVEL SIDING. USE 6d SIDING NAILS AS SHOWN.	SAME AS FOR BEVEL SIDING, BUT USE 8d SIDING NAILS.
AVAILABLE GRADES *Most commonly used	ALL SPECIES EXCEPT WRC SUPERIOR PRIME WRC CLEAR-VG-ALL HEART* A*, B*, C*	SEE BEVEL SIDING GRADES
SEASONING	USUALLY SHIPPED AT 12% OR LESS MOISTURE CONTENT.	USUALLY SHIPPED AT 12% OR LESS MOISTURE CONTENT.

Fig. 22-3 Plain bevel siding

Fig. 22-3 (Continued) *(Courtesy of California Redwood Association)*

Bevel siding comes in butt edge thicknesses of 1/2, 5/8, and 3/4 inch and widths of 4, 6, 8, 10, and 12 inches. Thicknesses depend on the width. The thicker and wider plain bevel siding is also called *bungalow* or *Colonial* siding.

Besides a plain bevel pattern, it is available with the butt edge rabbeted, commonly called

Dolly Varden siding, Figure 22-4. Another type of bevel siding called *Anzac* siding has a back shaped to lie flat against the sheathing at the top, Figure 22-5.

DROP SIDING

Drop siding is available in several patterns. It has matched or *shiplapped* edges and is

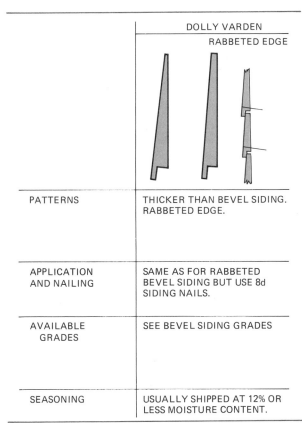

	DOLLY VARDEN
	RABBETED EDGE
PATTERNS	THICKER THAN BEVEL SIDING. RABBETED EDGE.
APPLICATION AND NAILING	SAME AS FOR RABBETED BEVEL SIDING BUT USE 8d SIDING NAILS.
AVAILABLE GRADES	SEE BEVEL SIDING GRADES
SEASONING	USUALLY SHIPPED AT 12% OR LESS MOISTURE CONTENT.

Fig. 22-4 Rabbeted edge bevel siding is called Dolly Varden siding.

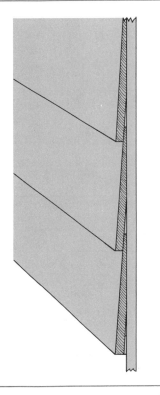

Fig. 22-5 Anzac siding has a back shaped to lie flat against the sheathing.

389

usually available in 1x6 or 1x8 nominal sizes, Figure 22-6. Because drop siding, like Anzac siding, lies flat against the studs, both are used frequently for garages and similar buildings without sheathing. However, these types having rabbeted, matched, or shiplap edges must be applied with a constant exposure distance. Applying a full width course of siding above and below windows with these types is not always possible. The exposure distance of plain beveled siding may be varied slightly, however, so that full courses can be maintained above and below windows and doors by adjusting the exposure.

LOG CABIN SIDING

An unusual type of siding, called *log cabin*, gives a rustic log cabin appearance, Figure 22-7. The surface is convex; it is 1 1/2 inches at its thickest point. The edges are shiplapped for weathertightness. It is available in widths of 6, 8, and 10 inches.

MATCHED SIDING

Matched siding may be used for horizontal, vertical, or diagonal application, Figure 22-8. It may also be applied in a combination of directions to achieve very attractive patterns. This type is manufactured in nominal 1-inch thickness and in widths from 4 to 12 inches. The edges are usually beveled to form a V-groove when placed together. Other designs are available.

BOARD AND BATTEN

A type of vertical siding used frequently for some architectural styles consists of rough sawn boards with small strips called *battens* used to cover the joint between the boards. This type of siding can be arranged in several ways: *board and batten*, *batten and board*, and *board and board*, Figure 22-9.

	DROP
PATTERNS	AVAILABLE IN 13 DIFFERENT PATTERNS. SOME T&G (AS SHOWN), OTHERS SHIPLAPPED
APPLICATION AND NAILING	6d FINISH NAILS FOR T&G, 8d SIDING NAILS FOR SHIPLAP.
AVAILABLE GRADES *Most commonly used	NO. 1 COMMON* NO. 2 COMMON* NO. 3 COMMON OR SUPERIOR*, PRIME*, E
SEASONING	SHIPPED 15% MOISTURE CONTENT OR LESS WHEN SPECIFIED.

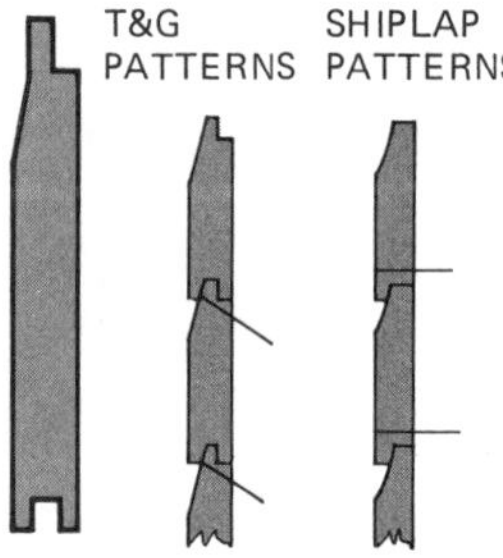

Fig. 22-6 Drop siding is available in several patterns.

	LOG CABIN
PATTERNS	1-1/2″ AT THICKEST POINT.
APPLICATION AND NAILING	NAIL 1-1/2″ UP FROM LOWER EDGE OF PIECE. USE 10d CASING NAILS.
AVAILABLE GRADES *Most commonly used	NO. 1 COMMON* NO. 2 COMMON* NO. 3 COMMON
SEASONING	SHIPPED 15% MOISTURE CONTENT OR LESS WHEN SPECIFIED.

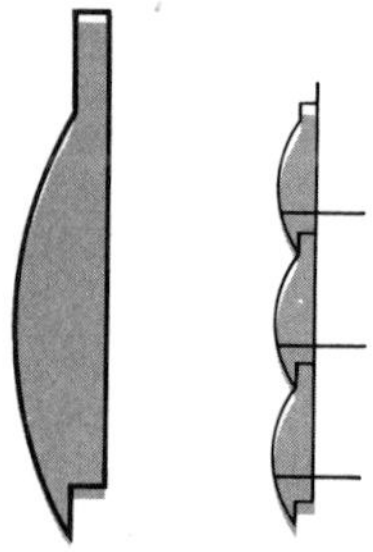

Fig. 22-7 Log cabin siding

	TONGUE & GROOVE PLAIN
PATTERNS	AVAILABLE IN SMOOTH SURFACE OR ROUGH SURFACE.
APPLICATION AND NAILING	USE 6d FINISH NAILS AS SHOWN FOR 6'' WIDTHS OR LESS. WIDER WIDTHS, FACE NAIL TWICE PER BEARING WITH 8d SIDING NAILS.
AVAILABLE GRADES *Most commonly used	NO. 1 COMMON* NO. 2 COMMON* NO. 3 COMMON OR SUPERIOR*, PRIME*, E
SEASONING	SHIPPED 15% MOISTURE CONTENT OR LESS WHEN SPECIIFED.

Fig. 22-8 Matched siding is also called
tongue and groove siding

Unless 5/8'' to 3/4'' sheathing has been applied, nailing blocks between the studs must be installed for fastening the siding. Nailing strips of 1x4 stock, laid horizontally and spaced 16 to 24 inches O.C., can be used over thin or fiberboard sheathing. However, in this case, thicker casings may have to be used around windows and doors.

WOOD SHINGLES AND SHAKES

Wood shingles and shakes are frequently used as siding, Figure 22-10. A complete description of wood shingles and shakes is found in Unit 19 Roofing. Wood shakes are also available in panels. A number of shakes are glued to a plywood or fiberboard base. This form is easier and quicker to apply than individual shakes. Wood shingles and shakes are available prefinished in a number of colors.

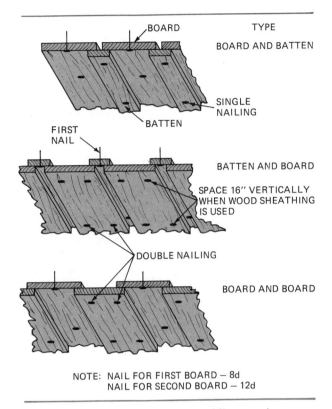

NOTE: NAIL FOR FIRST BOARD – 8d
NAIL FOR SECOND BOARD – 12d

Fig. 22-9 Board and batten siding may be
applied in several ways.

PLYWOOD SIDING

Plywood may be used in a number of ways for siding. It may be used in panel form either vertically or horizontally, Figure 22-11. These sidings are available in a wide variety of surface textures ranging from textured and grooved 303 siding to smooth paintable MDO plywood,

Fig. 22-10 Wood shingles are frequently used as siding
(*Courtesy of Western Wood Products Association*)

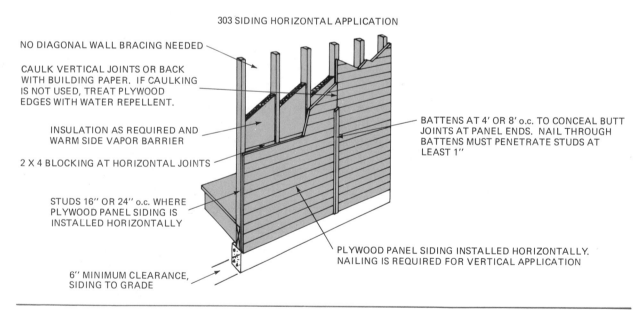

APA STURD-I-WALL CONSTRUCTION

LEAVE 1/16" SPACE AT ALL PANEL END AND EDGE JOINTS.

303 SIDING VERTICAL APPLICATION

INSULATION AS REQUIRED AND
WARM SIDE VAPOR BARRIER

JOINTS — NO CAULKING REQUIRED WHERE
EDGES ARE SHIPLAPPED, BACKED WITH
BUILDING PAPER, OR BATTEN. IF CAULKING
IS NOT USED WITH UNBATTENED SQUARE
BUTT JOINTS, TREAT PLYWOOD EDGES
WITH WATER REPELLENT.

NO DIAGONAL WALL BRACING REQUIRED
WITH PLYWOOD PANEL SIDING

PLYWOOD PANEL SIDING

LEAVE 1/16" SPACE AT
END AND EDGE JOINTS

6" MINIMUM CLEARANCE,
SIDING TO GRADE

303 SIDING HORIZONTAL APPLICATION

NO DIAGONAL WALL BRACING NEEDED

CAULK VERTICAL JOINTS OR BACK
WITH BUILDING PAPER. IF CAULKING
IS NOT USED, TREAT PLYWOOD
EDGES WITH WATER REPELLENT.

INSULATION AS REQUIRED AND
WARM SIDE VAPOR BARRIER

2 X 4 BLOCKING AT HORIZONTAL JOINTS

STUDS 16" OR 24" o.c. WHERE
PLYWOOD PANEL SIDING IS
INSTALLED HORIZONTALLY

BATTENS AT 4' OR 8' o.c. TO CONCEAL BUTT
JOINTS AT PANEL ENDS. NAIL THROUGH
BATTENS MUST PENETRATE STUDS AT
LEAST 1"

PLYWOOD PANEL SIDING INSTALLED HORIZONTALLY.
NAILING IS REQUIRED FOR VERTICAL APPLICATION

6" MINIMUM CLEARANCE,
SIDING TO GRADE

Fig. 22-11 Plywood is used in panel form for siding both horizontally and vertically. *(Courtesy of American Plywood Association)*

Figure 22-12. Selecting plywood siding is mostly a matter of preference because so many surface textures are available. Consideration should also be given to the siding face grade desired. (See Unit 2, Figure 2-8, for a guide in the selection of plywood.)

USING PLYWOOD TO SIMULATE BOARD AND BATTEN SIDING

Plywood may be used to give a board and batten appearance to the exterior. Smooth-faced plywood panels are first applied vertically

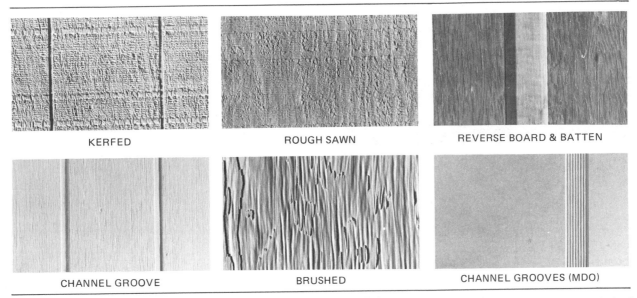

KERFED   ROUGH SAWN   REVERSE BOARD & BATTEN

CHANNEL GROOVE   BRUSHED   CHANNEL GROOVES (MDO)

Fig. 22-12 Plywood siding is available in a wide variety of surface textures. *(Courtesy of American Plywood Association)*

to the wall. Then, narrow wood strips are applied to the plywood directly over each stud at 16- to 24-inch intervals.

PLYWOOD LAP SIDING

Plywood panels may be ripped into strips of practically any width up to 24 inches and used as *lap siding,* Figure 22-13. Either the smooth, paintable medium-density overlay exterior plywood or some of the 303 plywood sidings are suitable for use as lap siding. Grooved panels are not recommended for cutting into lap siding.

HARDBOARD SIDING

Hardboard siding is available in two general types: panel siding and lap siding. *Hardboard panels* are available in widths of four feet and in lengths up to ten feet. The usual thickness is 7/16 inch. *Hardboard lap siding* is available in lengths up to 16 feet and in widths of 6, 8, and 12 inches. Some hardboard siding comes prefinished in a variety of colors; others may be primed or unprimed. Many surface textures are available. Some textures simulate stucco, wood shingles and shakes, pecky cypress, and rough cedar. Because of the number of surface textures and designs, it is advisable to research manufacturers' catalogs.

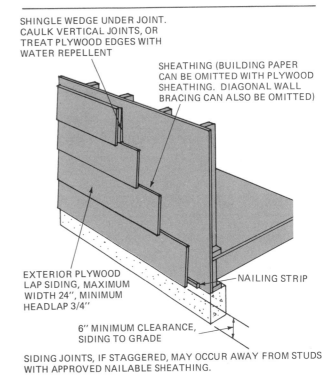

SHINGLE WEDGE UNDER JOINT. CAULK VERTICAL JOINTS, OR TREAT PLYWOOD EDGES WITH WATER REPELLENT

SHEATHING (BUILDING PAPER CAN BE OMITTED WITH PLYWOOD SHEATHING. DIAGONAL WALL BRACING CAN ALSO BE OMITTED)

EXTERIOR PLYWOOD LAP SIDING, MAXIMUM WIDTH 24", MINIMUM HEADLAP 3/4"

NAILING STRIP

6" MINIMUM CLEARANCE, SIDING TO GRADE

SIDING JOINTS, IF STAGGERED, MAY OCCUR AWAY FROM STUDS WITH APPROVED NAILABLE SHEATHING.

Fig. 22-13 Plywood panels may be ripped into strips and used as lap siding. *(Courtesy of American Plywood Association)*

METAL SIDING

Most metal siding is made of aluminum and usually has a backing of rigid insulation. The siding comes prefinished with a baked-on enamel, giving a smooth surface. Other types are manu-

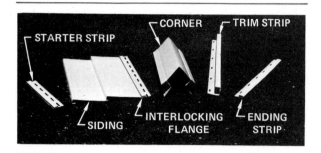

Fig. 22-14 A number of accessories are used with metal and plastic siding.

factured with various surface textures giving the appearance of rough sawn lumber, wood shingles, shakes, and many other kinds. It also comes in styles for both horizontal and vertical application and can be used on new construction or over existing siding.

The edges of metal siding interlock each other. A constant exposure distance is maintained. Elongated nailing holes are prepunched in the interlocking tab that allow for contraction and expansion of the siding.

A number of accessories are used when applying a metal siding such as starter strips, inside and outside corners, and special pieces used around windows and doors, Figure 22-14. Backer strips support the ends at joints.

PLASTIC SIDING

Plastic siding, usually vinyl, is similar to metal siding in shapes, accessories, and method of application. Like aluminum siding, it is available in a number of surface textures and colors. The color is in the material itself, not coated on; therefore, this type of siding never needs finishing. If desired, it can be washed with a garden hose. Also, it resists dents.

STUCCO

Stucco is a coating of specially formulated plaster for exterior walls in which cement is largely used. It is put on wet, but when dry becomes very hard and durable. Stucco is spread by masons; however, it is the responsibility of the carpenter to prepare the wall to receive the stucco.

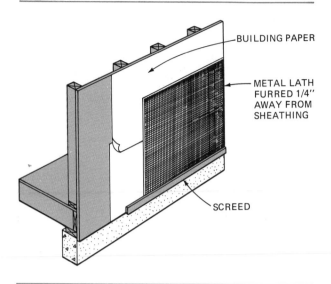

Fig. 22-15 Wall preparation for a stucco finish

WALL PREPARATION FOR STUCCO

The wall is prepared by first applying a layer of sheathing paper. Metal lath is then applied over the paper. The metal lath should be galvanized to prevent rusting and is kept 1/4 inch away from the sheathing so that the stucco is forced through the mesh of the lath. Special furring nails, metal furring strips, or self-furring lath may be used for this purpose. Lath fasteners should penetrate the sheathing at least 3/4 inch and be spaced 18 to 24 inches apart in both directions, Figure 22-15. Screeds to control the thickness of stucco are placed in appropriate locations where they can later be removed or covered by the finish.

INSTALLING SIDING

Before siding is applied to the walls, all window and door frames must be installed. Apply flashings over doors, windows, and other locations as necessary. Install the corner boards, if required, and apply the frieze and cornice molding. All of the exterior trim, except the corner boards, have been previously discussed.

CORNER TREATMENT

One method of finishing the exterior at the corners is with the use of *corner boards*. In other cases, the siding may be mitered around

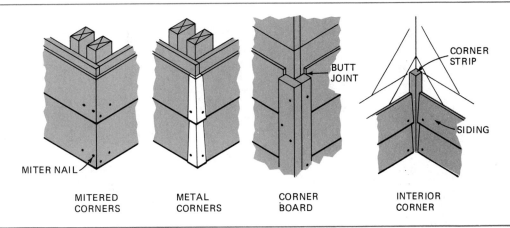

MITER NAIL

BUTT JOINT

CORNER STRIP

SIDING

MITERED CORNERS

METAL CORNERS

CORNER BOARD

INTERIOR CORNER

Fig. 22-16 Methods of returning and stopping siding

the corner, or metal corners are used on each course of siding. Interior corners are finished by butting the siding against a square corner board, Figure 22-16.

When siding returns against a roof surface, such as at a dormer, there should be a clearance of about 2 inches. Siding cut tight against roof shingles retains moisture after rains and usually results in paint peeling and wood decay.

The thickness of corner boards depends on the type of siding used. The corner boards should be thick enough so that the siding does not project beyond the face of the corner board.

The width of the corner boards depends on the effect desired. However, one of the two pieces making up an outside corner should be narrower by the thickness of the stock. When the pieces are joined together, the same width is exposed on both sides of the corner. The joint between the two pieces should be on the side of the house that is least viewed.

When installing corner boards, flash the corner by installing a strip of 15-pound felt vertically on each side so that one edge extends beyond the edge of the corner board at least 2 inches. Tuck the top end under any previously applied felt.

With a sharp plane, slightly back bevel the edge of the narrow piece that faces the corner. This is done to assure a tight fit between the two boards. Cut and fit the narrow piece and fasten to the wall with its beveled edge flush with the corner. Fasten with galvanized or other noncorroding nails spaced about 16 inches apart. Fasten by starting at one end and working toward the

other, keeping the edge flush with the corner while fastening at each location.

Cut and fit the wide piece and fasten it to the corner in a similar manner. The outside row of nails are driven into the edge of the narrow piece. Plane the outside edge of the wide piece wherever necessary to make it come flush. Break any sharp corners by planing a slight bevel. Set all nails so they can be puttied over later. Drive additional nails where necessary so that a tight joint is obtained between the two pieces.

APPLYING BUILDING PAPER

A water-repellent building paper should be applied to board sheathing. If no sheathing is used, the building paper should be placed across the studs. If weather-proof panel sheathing such as plywood is used, the building paper may be omitted. Some contractors prefer the use of building paper no matter what kind of sheathing is used.

The purpose of the building paper is to block water and wind penetration. It does not serve as a vapor barrier. A paper with a permeability of at least 5 perms should be used. Thus, there is no danger of entrapping moisture entering from the inside within the walls, Figure 22-17. Materials such as 15-pound asphalt felt, rosin, and similar papers are considered satisfactory.

Apply the paper horizontally starting at the bottom of the wall. Make sure the paper lies flat with no wrinkles, and staple or nail it in position. If nailing, use large head roofing nails. Fasten with single rows near the bottom, the center, and

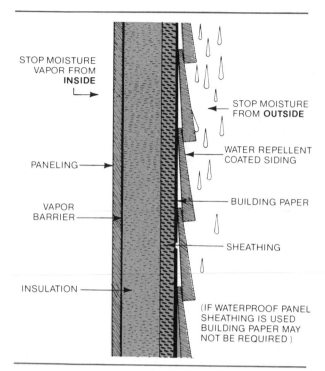

Fig. 22-17 Building paper under siding should not block the passage of vapor from the interior *(Courtesy of California Redwood Association)*

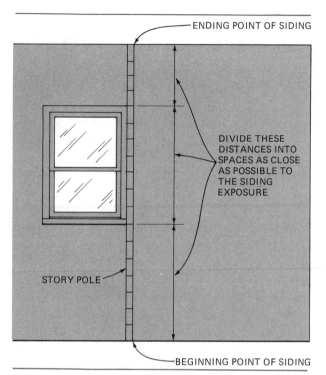

Fig. 22-18 Laying out a story pole

the top about 16 inches apart. Succeeding layers should lap the lower layer by about four inches.

The sheathing paper should lap over any flashing applied at the sides and tops of windows and doors and at corner boards. It should be tucked under any flashings applied under the bottoms of windows or frieze boards. In any case, all laps should be over the paper below.

INSTALLING HORIZONTAL WOOD SIDING

One of the important differences between *plain bevel siding* and other types with matched or shiplap edges is that the exposure of plain beveled siding can be varied. Other types of siding maintain the same exposure to the weather with every course.

The ability to vary the exposure is a decided advantage. When applying horizontal siding, it is desirable from the standpoint of weathertightness and ease of application to have a full course of siding above and below windows and over the tops of doors.

DETERMINING SIDING EXPOSURE

To determine the siding exposure so that it is about equal both above and below the win-

dowsill, divide the overall height of the window frame by the recommended exposure distance. The recommended exposure is 4 inches for 6-inch siding, 6 inches for 8-inch siding, 8 inches for 10-inch siding and 10 inches for 12-inch siding. Dividing this distance results in the number of courses between the top and the bottom of the window.

For example, the overall height of a window from the top of the drip cap to the bottom of the sill is 55 inches. If 8-inch siding is used, the number of courses would be 55 divided by 6 or slightly over nine courses. To obtain the exact exposure distance, divide 55 by 9. The result is 6 1/9 inches or a slack 6 1/8 inches.

The next step is to determine the exposure distance from the bottom of the windowsill to just below the top of the foundation wall. If this is 37 inches then six courses are required at approximately 6 1/8 inches strong. (The term "strong" means slightly more, about 1/32", than the measurement indicated. The term "slack" means slightly less.) Thus, the exposure distance above and below the windowsill are almost the same. The difference is not noticeable to the eye.

The same procedure is used to lay out courses above the window. The layout lines are then transferred to a story pole which is used to lay out the courses all around the building, Figure 22-18. Because of varying window

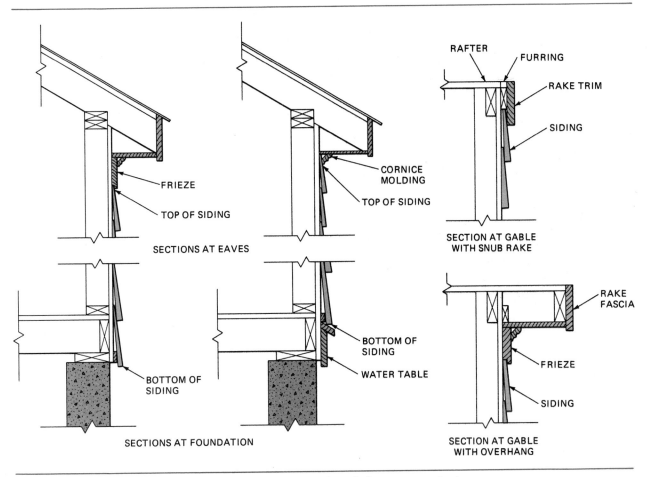

Fig. 22-19 Siding finish at foundation, eaves, and rake

heights, it may not always be possible to lay out full siding courses above and below every window or door.

Another factor in laying out a story pole is determining the point at which the siding begins and ends. Various methods are used at the foundation, at the eaves, and at the rakes according to the overall design of the structure.

Treatment at the Foundation. An effective barrier must be made to keep the weather from entering between the top of the foundation wall and the frame. One simple method is to apply the bottom course of siding so the bottom edge extends below the top of the foundation about 3/4 inch. Another method is by using a *water table*. This consists of a board and a drip cap installed around the structure in such a way that the bottom edge is below the top of the foundation.

At the Eaves. At the eaves the siding may end against the bottom edge of the frieze. The width of the frieze may vary so it is necessary to know its width when laying out the story pole. The siding may also terminate against the soffit. The joint between is covered by a cornice molding. The size of the molding must also be known.

At the Rakes. At the rakes, the siding may be applied under a furred-out rake fascia. In other cases, where the rake overhangs the sidewall, the siding may be fitted against the rake frieze or against the rake soffit and covered with a molding, Figure 22-19.

MATERIAL TRANSITION

Sometimes a different kind of siding is used on the gable ends than on the sidewalls. The joint between the two types must be weathertight. One method is to use a drip cap and flashing.

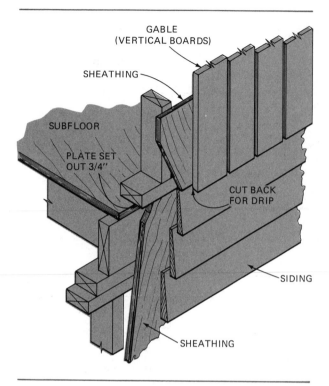

GABLE
(VERTICAL BOARDS)

SHEATHING

SUBFLOOR

PLATE SET
OUT 3/4"

CUT BACK
FOR DRIP

SIDING

SHEATHING

Fig. 22-20 Gable-end finish

In other cases the plate and studs of the gable end are extended out from the wall a short distance allowing the gable end siding to overlap the siding below, Figure 22-20. Furring strips may also be used on the gable end in place of extending the gable plate and studs.

STARTING FROM THE TOP

Another advantage of using plain bevel siding is that application may be made starting at the top and working toward the bottom. With this method, a number of chalk lines may be snapped without being covered by a previous course — thus saving time. Any scaffolding already erected may be used and then dismantled as work progresses toward the bottom.

STARTING FROM THE BOTTOM

Siding may be installed starting from the bottom. Nail a *furring strip* of the same thickness and width of the headlap of siding along the bottom edge of the sheathing, Figure 22-21.

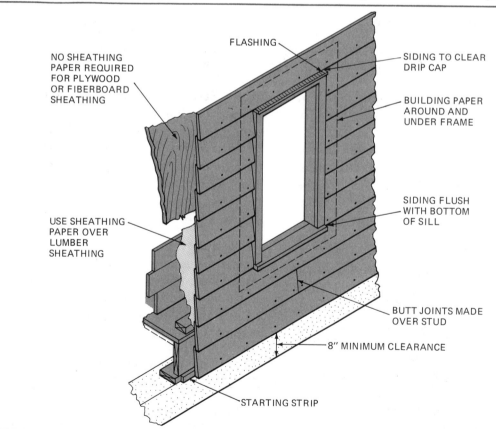

FLASHING

NO SHEATHING
PAPER REQUIRED
FOR PLYWOOD
OR FIBERBOARD
SHEATHING

SIDING TO CLEAR
DRIP CAP

BUILDING PAPER
AROUND AND
UNDER FRAME

USE SHEATHING
PAPER OVER
LUMBER
SHEATHING

SIDING FLUSH
WITH BOTTOM
OF SILL

BUTT JOINTS MADE
OVER STUD

8" MINIMUM CLEARANCE

STARTING STRIP

Fig. 22-21 Installation of bevel siding

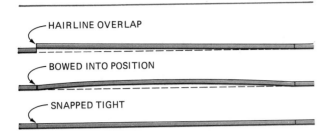

HAIRLINE OVERLAP

BOWED INTO POSITION

SNAPPED TIGHT

Fig. 22-22 Cut bevel siding with a hairline overlap, bow into position, and snap tight.

Snap a line on the wall to lay the top edge of the first course to it.

Snap lines for succeeding courses so that the bottom edge of those courses are laid to it. Each succeeding course overlaps the upper edge of the lower course. The minimum lap for plain bevel siding should not be less than one inch. Siding is nailed to each stud or on 16-inch centers. On beveled siding, nail only through the butt edge just above the top edge of the course below. This allows for slight movement of the siding due to moisture changes without causing splitting. Such an allowance is especially needed for the wider sidings of 8 to 12 inches. Care must be taken to nail as low as possible to avoid splitting the siding in the center. Stagger joints between courses as much as possible.

CAREFUL FITTING

Tight fitting butt joints are obtained by cutting the last piece of each course approximately 1/16 inch too long. Use a fine tooth crosscut saw. Try the fit, letting the other end overlap the last piece installed. Use a block plane to fit the joint. Bow the piece slightly to get the ends in position and snap into place, Figure 22-22.

If corner boards are used, siding ends should butt snugly to the boards. The sealing of all joints helps provide protection against rain, snow, fog, and wind. It is particularly important at the butt joint of short length siding to caulk the joints, Figure 22-23.

When fitting siding under windows, it is necessary that the siding fits snugly in the groove on the underside of the windowsill for weathertightness, Figure 22-24.

Fig. 22-23 Caulk the ends of siding where it butts against trim.

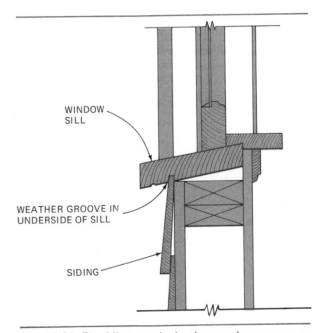

WINDOW SILL

WEATHER GROOVE IN UNDERSIDE OF SILL

SIDING

Fig. 22-24 Fit siding snugly in the weather groove on the underside of the windowsill.

HORIZONTAL SIDING WITH A CONSTANT EXPOSURE

Some sidings such as rabbeted bevel siding and drop siding maintain a constant exposure to the weather. With this type, it is not always possible to provide a full course below and above windows. However, there is no need to snap lines for every course. Take extreme care to start the first course in an absolutely straight line. Succeeding courses maintain the exposure because of their shaped edges. However, it is wise procedure to sight by eye along the butt edge after installing several courses and straighten the course as much as possible without disturbing the joint too much.

Horizontally applied, matched siding in narrow widths is blind nailed at the tongue only. For widths greater than 6 inches, an additional nail should be used, Figure 22-25.

INSTALLING VERTICAL MATCHED SIDING

When installing vertical matched siding, start at one end of the building with the first board plumb. Remove the grooved edge and keep it flush with the corner. Face nail the edge nearest the corner, and blind nail into the tongue edge. Nails should be placed from 16 to 24 inches apart. Unless wood or plywood sheathing is used, wood nailing blocks should be provided between studs.

Apply succeeding pieces by toenailing into the tongue edge. Make sure the edges between boards come up tight. If they do not come up tight by nailing alone, drive a chisel with its beveled edge against the tongue into the sheathing and use it as a pry to force the board up tight; nail close to the chisel. If this is not successful, toenail a short block of the siding with its grooved edge into the tongue of the board, Figure 22-26. Drive the nail home until it forces the board up tight. Drive nails into the siding on both sides of the scrap block. Remove the scrap block. Continue applying pieces in the same manner, keeping the bottom ends in a straight line. Avoid making horizontal joints between lengths. If necessary to apply vertical siding in more than one length, use a mitered or rabbeted joint on the ends for weathertightness.

FITTING AROUND DOORS AND WINDOWS

Because of varying locations and widths of windows and doors, it is necessary to fit vertical matched siding to and around the window and door casings. Cut and fit the piece just before the one to be fitted against the casing. Then remove it, and set it aside for the time being. Cut, fit, and tack the piece to be fitted against the casing in its place. Level from the top of the window casing and the bottom of the sill and mark the piece.

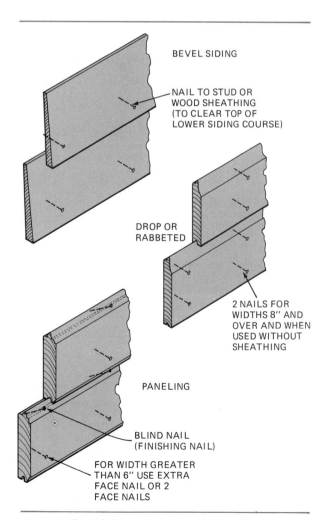

BEVEL SIDING

NAIL TO STUD OR WOOD SHEATHING (TO CLEAR TOP OF LOWER SIDING COURSE)

DROP OR RABBETED

2 NAILS FOR WIDTHS 8" AND OVER AND WHEN USED WITHOUT SHEATHING

PANELING

BLIND NAIL (FINISHING NAIL)

FOR WIDTH GREATER THAN 6" USE EXTRA FACE NAIL OR 2 FACE NAILS

Fig. 22-25 Methods of nailing siding

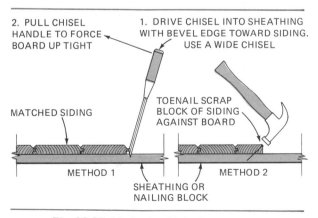

2. PULL CHISEL HANDLE TO FORCE BOARD UP TIGHT

1. DRIVE CHISEL INTO SHEATHING WITH BEVEL EDGE TOWARD SIDING. USE A WIDE CHISEL

MATCHED SIDING

TOENAIL SCRAP BLOCK OF SIDING AGAINST BOARD

METHOD 1

METHOD 2

SHEATHING OR NAILING BLOCK

Fig. 22-26 Methods of bringing matched siding edges up tight

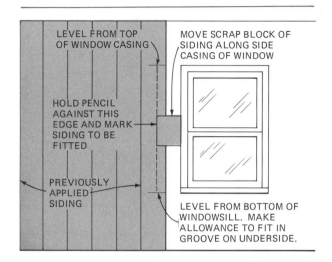

Fig. 22-27 Method of fitting vertical matched siding to a window casing

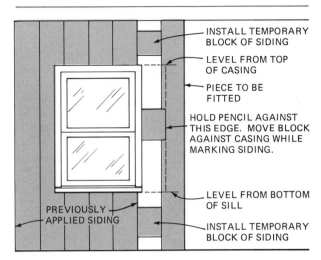

Fig. 22-28 Fitting vertical matched siding to the other side of the window

To mark the piece so it will fit snugly against the side casing, first cut a scrap block about 6 inches long of the siding material and remove the tongue from one edge. Be careful to remove all of the tongue, but no more.

Hold the block so its grooved edge is against the side casing and the other edge is on top of the siding to be fitted. Mark the piece by holding a pencil against the outer edge of the block while moving the block along the length of the side casing, Figure 22-27.

Cut the piece, following the layout lines carefully. When laying out to fit against the bottom of the sill, make allowance to rabbet the siding to fit in the weather groove on the bottom side of the windowsill.

Place the two pieces in position by lifting the joint between the two away from the wall. Then press at the joint, snap the two in position, and fasten. Face nail where it is not possible to blind nail.

Continue to apply the siding with short lengths across the top and bottom of the window. Each length under the window must be rabbeted at the top end to fit in the weather groove at the sill.

On the other side of the window a full length must also be fitted to the casing. To mark the piece, first tack a short length of scrap siding above and below the window and against the last pieces of siding installed. Tack the length of siding to be fitted against these blocks.

Level from the top and bottom of the window and mark the piece for the horizontal cuts.

To lay out the piece for the vertical line to fit against the side casing, use the same block with the tongue removed, as used previously. Hold the grooved edge against the side casing and with a pencil against the other edge, ride the block along the side casing while marking the piece to be fitted, Figure 22-28.

Remove the piece and the scrap blocks from the wall. Carefully cut the piece to the layout lines and then fasten in position. Continue applying the rest of the siding in a like manner until the end of the wall is reached.

INSTALLING BOARD AND BATTEN SIDING

Board and batten siding is installed in a way similar to that used with matched vertical siding. However, this type of siding sometimes is applied in random widths.

In the *standard pattern* the boards are applied first with about 1/2-inch spaces between them. The spaces are covered with battens. In the *reverse pattern*, the battens are applied first and spaced the width of the board plus 1/2 inch O.C. When boards are also used as battens, the spacing between boards is the width of one board less 1 1/2 inches.

The undercourse of each pattern is fastened with one row of nails in the center, spaced about

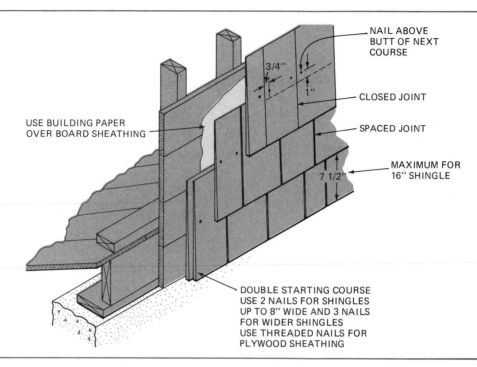

Fig. 22-29 Single-coursed wood shingle siding

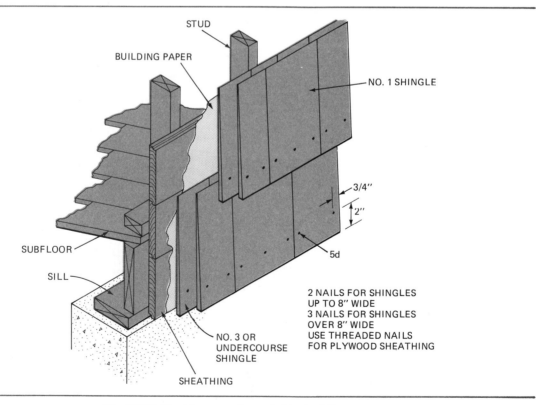

Fig. 22-30 Double-coursed wood shingle siding

402

16 to 24 inches vertically. Nails in the top course do not penetrate the undercourse. See Figure 22-9 for the location of nails. Nails should be of sufficient length to penetrate at least 1 inch into nailing blocks.

APPLYING WOOD SHINGLES AND SHAKES

Wood shingles are applied to sidewalls in either single-layer courses or double-layer courses. In single coursing, shingles are applied in the same manner as on roofs, Figure 22-29. However, greater weather exposures are permitted on sidewalls. Shingle walls are covered by two thicknesses of shingles, while roofs require three.

In double coursing, the top layer is applied with its butts slightly lower than the first layer, Figure 22-30. Double coursing allows for greater weather exposure and gives deep, bold shadow lines. See Figure 22-31 for maximum allowable exposure.

If shingles are to be applied with their butts in a straight line, snap a chalk line for each course of shingles. Lay out a story pole as for bevel siding and plan for full courses above and below windows and doors, where possible.

SINGLE COURSING

To apply single coursing, stretch a chalk line along the bottom from one corner to the other. Sight the line for straightness. Temporarily fasten at intervals along the wall, if necessary, to bring the line straight. Even a tightly stretched line will sag in the center over a long distance.

Apply a single course of shingles so the butts are as close to the chalk line as possible without touching it. Remove the line and apply another course on top of the first course. Joints between shingles may be open about 1/4 inch or closed depending on the desired effect. No joint should be closer, to the right or left, than 1 1/2 inches, of joints in the course below. If the joints are to be closed, use a shingling hatchet and fit each joint by trimming its edge, keeping the butt end to the chalk line. Butt ends are not trimmed.

Rebutted and rejointed shingles are frequently used on sidewalls. These are shingles that have been machine-trimmed to provide parallel edges and square butts. They are used on sidewalls when tight-fitting joints are desired without having to fit each individual shingle.

After the starter course has been applied, apply successive single-layer courses in the same manner. Use two 3d galvanized nails in each shingle, driven about 1 inch above the butt line of the next course. Drive the nails flush but not so the head crushes the wood.

For ease in applying courses, many contractors prefer to tack a straightedge to the chalk line so that the butts of the shingle course rest on the straightedge. Use 1x3 lengths of

SINGLE COURSE APPLICATION			
Wood Shingles		Wood Shakes	
Size	Maximum Exposure	Size	Maximum Exposure
16''	7 1/2''	18''	8 1/2''
18''	8 1/2''	24''	11 1/2''
24''	11 1/2''	32''	15''

DOUBLE COURSE APPLICATION					
Wood Shingles		Wood Shakes			
Size	Maximum Exposure	Hand or Taper Split		Straight Split	
		Size	Exposure	Size	Exposure
16''	12''	18''	14''	18''	16''
18''	14''	24''	20''	24''	22''
24''	16''				

Fig. 22-31 Maximum allowable exposures for wood shingle and shake siding

strapping and tack in position using 6d finish nails. Use only as many nails as necessary to straighten the piece and hold it firmly in position.

When applying an extra wide shingle (10 inches or more), it is the custom in some localities to apply the shingle with one nail in each edge. Then the shingle is scored from tip to butt with the shingling hatchet so the score is offset from a joint in the course below. Then it is nailed with one nail on each side of the score. This is done because wide shingles will eventually split. When it does split, it will split along the scored line rather than elsewhere, with the possibility of the split occurring in line with the joint below.

CORNERS

Shingles may be butted to corner boards like any horizontal wood siding. On outside corners, they may also be applied by alternately overlapping each course in the same manner as in applying a wood shingle ridge. Inside corners may be woven by alternating the corner shingle first on one wall and then the other, Figure 22-32.

DOUBLE COURSING

When double coursing, the starter course is tripled. A lower grade, called undercoursing grade, is doubled and then the face grade course is applied. The outer course is applied 1/2 inch lower than the undercourse.

The undercourse is fastened in the same manner as single-coursed shingles. Each outer course shingle is face-nailed with two 5d galvanized nails driven 1 to 2 inches above the butts and about 3/4 inch in from each edge plus additional nails in the same line about 4 inches apart across the face of the shingle.

For ease in application, use a rabbeted straightedge. Tack the straightedge on the chalk line for the undercourse. Lay the undercourse shingles with their butts to the top edge and the outer course shingles with their butts against the edge of the rabbet, Figure 22-33. Remove the straightedge when the shingling is completed.

Wood shakes are also applied to sidewalls in either single or double layer courses. Maximum exposure with single-course application is 8 1/2 inches for 18-inch shakes and 11 1/2 inches for 24-inch shakes. Double-course application requires an undercourse of shakes or shingles. When double coursing, weather exposures up to 14 inches with 18-inch shakes and 20 inches with 24-inch shakes are allowed. If straight-split shakes are used, exposures of 16 inches for 18-inch shakes and 22 inches for 24-inch shakes are allowed.

Wood shakes are applied in much the same manner as wood shingles. The nailing normally is concealed in single-course applications. Butt

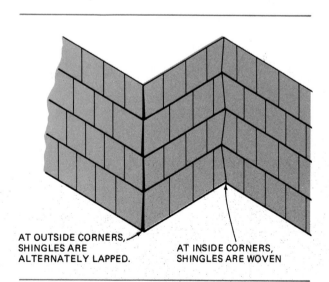

Fig. 22-32 Method of returning shingles at inside and outside corners

AT OUTSIDE CORNERS, SHINGLES ARE ALTERNATELY LAPPED.

AT INSIDE CORNERS, SHINGLES ARE WOVEN

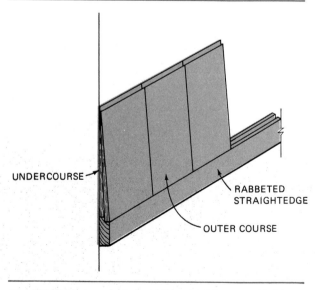

UNDERCOURSE

RABBETED STRAIGHTEDGE

OUTER COURSE

Fig. 22-33 For ease in applying double-coursed shingle siding, use a rabbeted straightedge.

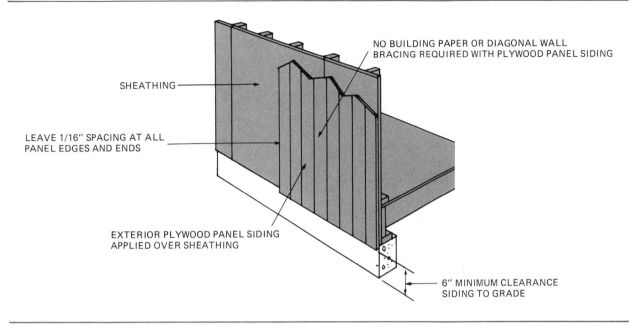

SHEATHING

NO BUILDING PAPER OR DIAGONAL WALL
BRACING REQUIRED WITH PLYWOOD PANEL SIDING

LEAVE 1/16" SPACING AT ALL
PANEL EDGES AND ENDS

EXTERIOR PLYWOOD PANEL SIDING
APPLIED OVER SHEATHING

6" MINIMUM CLEARANCE
SIDING TO GRADE

Fig. 22-34 Plywood panel siding over nailable plywood or lumber sheathing *(Courtesy of American Plywood Association)*

nailing is required with double-course applications. Use 6d galvanized nails or 8d where required.

INSTALLING PLYWOOD AND HARDBOARD SIDING

Plywood, hardboard, and other sheet forms are installed as panels either vertically or horizontally or as lap siding applied horizontally.

INSTALLING VERTICAL PANELS

To install vertical panels start at one end of the wall. Cut the outer edge of the first panel so the inner edge falls on the center of a stud. Nail the sheet in a plumb position. Fasten according to nail specifications for various kinds of plywood panel siding. Apply successive sheets leaving 1/16 inch at panel ends and edges. As with all other types of siding, the bottom edge must be a minimum of 6 inches above the finished grade line, Figure 22-34. Caulk all panel joints unless shiplapped joints or battens are used.

Panel siding is sometimes applied direct to studs in order to cut construction costs. Recommended plywood siding joints at various locations on the building are shown in Figure 22-35.

INSTALLING HORIZONTAL PANELS

When installing horizontal panels, mark on each end of the building what will be the top edge of the panel. Snap a line between the marks. Fasten a full length panel with its top edge to the line and its inner end on the center of a stud. Trim the outer end flush with the corner.

Apply the remaining sheets in the first course in like manner. Trim the end of the last sheet flush with the corner.

Start the next course with approximately one-half a sheet to break the joint with the course below. Carefully fit around doors and windows. Caulk or seal the edges and joints of the cutout section of siding. Fasten plywood panel siding according to the recommendations given in Figure 22-36. Horizontal wall joints may be butted and flashed, offset and lapped, or rabbeted. Battens may be used to cover vertical joints.

APPLYING LAP SIDING

Plywood and hardboard lap siding is applied in much the same manner as plain bevel siding with a few exceptions. To apply lap siding, first install a 1-inch strip of the siding along the bottom of the wall. Apply the first

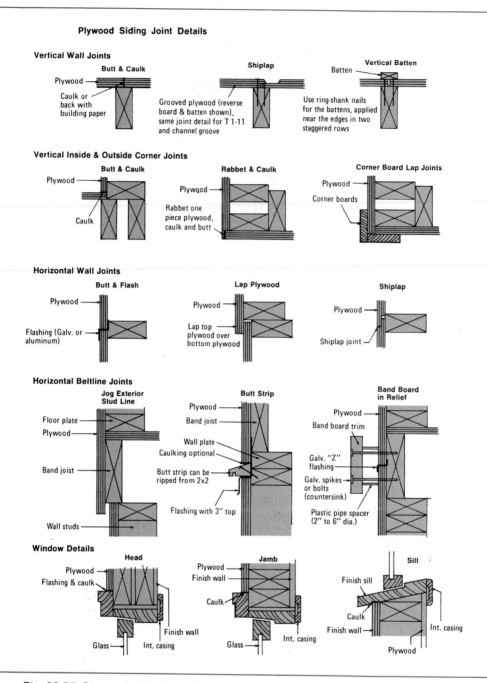

Fig. 22-35 Plywood siding joint details *(Courtesy of American Plywood Association)*

course so the bottom edge overhangs the strip. Nail along the bottom edge at about 16-inch intervals.

End joints may be butted and caulked with a strip of 15-pound felt placed behind the joint. In some types, a 1/8-inch gap is allowed between the ends and a special joint molding is inserted in the gap, Figure 22-37. Where siding butts against trim, the joint is caulked and sealed.

CORNERS

Individual outside corners may be installed as work progresses or can be installed after the entire wall is sided. Slide the corners up into

Exterior Plywood Panel Siding Over Nailable Plywood or Lumber Sheathing						
Description	Nominal Thickness (inch)	Maximum Spacing of Vertical Rows of Nails (inches)		Nail Type & Size	Nail Spacing (inches)	
		Face Grain Vertical	Face Grain Horizontal		Panel Edges	Intermediate Studs
MDO EXT-APA	11/32, 3/8	16	24	6d nonstaining box, siding or casing nails for panels 1/2" thick or less; 8d for thicker panels.	6	12
	1/2 and thicker	24	24			
303-16 o.c. Siding EXT-APA Including T 1-11	5/16 and thicker	16	24			
303-24 o.c. Siding EXT-APA	7/16 and thicker	24	24			

Fig. 22-36 Recommendations for fastening plywood panel siding *(Courtesy of American Plywood Association)*

place with the top slipping under the course above. Seat tabs under the lower edge. When the corner is in position, drive a nail through the exposed hole at the top, Figure 22-38.

Outside corners may also be trimmed with corner boards. The use of lapped or mitered corners is not recommended for this type siding because of the tendency of the corner to open.

Interior corners may be butted against a corner board or alternately woven.

INSTALLING METAL AND PLASTIC SIDING

When installing metal and plastic siding, first apply all necessary accessories around

Fig. 22-37 A special joint molding is sometimes used when applying hardboard lap siding.
(Courtesy of Masonite Corporation)

Fig. 22-38 Installing individual outside metal corners *(Courtesy of Masonite Corporation)*

Fig. 22-39 Installing aluminum siding over existing siding *(Courtesy of Robert Morency, photographer)*

windows, doors, outside and inside corners, and other places where needed. Then, apply the starter strip in a straight line at the bottom of the building. Care must be taken to make sure the starter strip is in as straight a line as possible because this determines the straightness of the siding courses.

Beginning at the corner, fit the first length of siding into the starter strip. Nail through the slots in the upper flange. Do not drive nails tightly against the flange. There should be some play between the nail and flange to allow for expansion and contraction of the siding.

Complete the first course, lapping the end joints. Laps are made so that the exposed end is toward the direction from which it is least viewed. Install successive courses by interlocking them with the course below, Figure 22-39. Allow a gap of 1/8 to 1/4 inch where the siding meets corner, window, and other trim. Cut the siding with a power saw and metal cutting

blade. Tin snips may be needed for other than cutting square ends.

Over window and door casings, add furring strips under the siding to support it at the same slope as adjoining panels. Caulk joints where siding meets trim.

Care must be taken when handling metal siding not to dent it. A dented piece is not usable unless the dented section can be cut out.

ESTIMATING SIDING

To estimate how much siding is needed to cover the walls of a building, first calculate the total surface area. To calculate the area of a gable, multiply the length by the height and divide the result by two. Add the square foot areas together with the areas of other parts that will be sided such as dormers, bays, and porches.

Since windows and doors will not be covered, their total surface areas can be deducted. To do this, multiply the width by the height of each opening. Add the results together. This figure is then subtracted from the total area to be covered by siding, Figure 22-40.

The amount of siding to order depends on the kind of siding used. Determine the coverage of the siding to be used and order on the basis of the coverage and the area to be covered, allowing about 5 percent extra for waste for most kinds of siding.

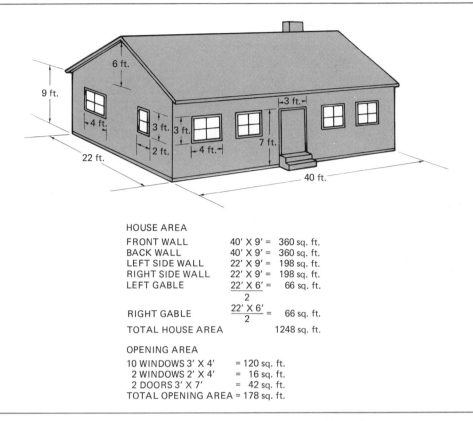

HOUSE AREA

FRONT WALL	40' X 9' =	360 sq. ft.
BACK WALL	40' X 9' =	360 sq. ft.
LEFT SIDE WALL	22' X 9' =	198 sq. ft.
RIGHT SIDE WALL	22' X 9' =	198 sq. ft.
LEFT GABLE	$\frac{22' \times 6'}{2}$	66 sq. ft.
RIGHT GABLE	$\frac{22' \times 6'}{2}$ =	66 sq. ft.
TOTAL HOUSE AREA		1248 sq. ft.

OPENING AREA

10 WINDOWS 3' X 4'	= 120 sq. ft.
2 WINDOWS 2' X 4'	= 16 sq. ft.
2 DOORS 3' X 7'	= 42 sq. ft.
TOTAL OPENING AREA	= 178 sq. ft.

Fig. 22-40 Estimating the area to be covered by siding

REVIEW QUESTIONS

Select the most appropriate answer.

1. Plain beveled siding is applied
   a. horizontally.
   b. vertically.
   c. horizontally or vertically.
   d. horizontally, vertically or diagonally.

2. A type of siding, commonly called Dolly Varden, has
   a. tongue and grooved edges.
   b. rabbeted butt edges.
   c. beveled back sides to lie flat against studs.
   d. a convex shape.

3. A particular advantage of plain bevel siding over other types of horizontal siding is that
   a. it comes in a variety of prefinished colors.
   b. it has a constant weather exposure.
   c. the weather exposure can be varied slightly.
   d. application can be made in any direction.

409

4. Building paper used under siding should
   a. be a vapor barrier.
   b. allow vapor to pass through.
   c. be applied only when no sheathing is used.
   d. be polyethylene film.

5. When applying horizontal siding, it is desirable to
   a. maintain exactly the same exposure with every course.
   b. apply full courses above and below windows.
   c. work from the top down.
   d. use a water table.

6. The height of a window frame is 73 1/2 inches. Dividing this height into spaces as close to 6 inches as possible results in spaces of
   a. 5 7/8".   c. 6 1/16".
   b. 5 15/16".   d. 6 1/8".

7. In order to lay out a story pole for courses of horizontal siding, which of the following must be known?
   a. the width of windows and doors
   b. the kind and size of finish at the eaves and foundation
   c. the location of windows, doors, and other openings
   d. the length of the wall to which siding is to be applied
   e. the size of corner boards

8. Maximum exposure for single coursed 18-inch wood shakes is
   a. 8 inches.   c. 9 inches.
   b. 8 1/2 inches.   d. 9 1/2 inches.

9. When installing metal siding, drive nails
   a. tightly against the flange.
   b. loosely against the flange.
   c. with large heads.
   d. colored the same as the siding.

SECTION

4

Interior Finish

DRYWALL CONSTRUCTION 23

OBJECTIVES

After completing this unit, the student will be able to

- *describe various kinds of gypsum board products and accessories.*

- *handle and stack gypsum board in an approved manner.*

- *describe various drywall attachments and fasteners.*

- *make single-ply and double-ply drywall applications.*

- *conceal fasteners and reinforce and conceal joints.*

Although the application of lumber, plywood, fiberboard, and hardboard paneling to interior walls and ceilings may be considered *drywall construction,* the term is generally accepted as the application of gypsum board products, Figure 23-1. Gypsum board (sometimes also called wallboard, plasterboard, or sheetrock) is used extensively in construction. It makes a strong, high quality wall and ceiling covering; is readily available; is fire-resistant; is easy to apply, decorate, and repair; and is relatively inexpensive.

GYPSUM BOARD

Many types of gypsum board are available for a variety of building needs:

- *Regular gypsum* board is used as a surface layer on walls and ceilings.

- *Type X* has an improved fire resistance made possible through the use of special additives to the core.

- *Water-resistant* has a water-resistant core and paper cover. It serves as a base for application of wall tile in bath, shower, kitchen, and laundry areas.

- *Backing board* is designed to be used as a base in multilayer systems.

- *Coreboard* is available in one-inch thickness and is used in solid gypsum partitions with additional layers applied to the coreboard.

Fig. 23-1 The application of gypsum board to interior walls and ceilings is called drywall construction. *(Courtesy of Gypsum Association)*

- *Predecorated* has coated printed or overlay surfaces that require no further treatment.

- *Veneer plaster base* is used as a base for thin, hard coats of gypsum veneer plaster.

- *Gypsum lath* is used as a base to receive plaster. (See Unit 12 Interior Rough Work.)

- Other gypsum board products are available primarily for exterior work. Standard gypsum boards are 4 feet wide and 8, 10, 12, 14, or 16 feet long.

Gypsum board is made in a number of thicknesses:

1/4 inch — used in multilayer applications or to cover existing walls and ceilings in remodeling

5/16 inch — used primarily in the construction of mobile homes

3/8 inch — used in a double-layer system over wood framing and as a face layer in repair or remodeling

1/2 inch — generally used as single-layer wall and ceiling construction in residential work and in double-layer construction for greater sound and fire resistance

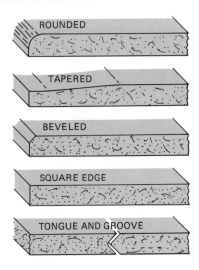

Fig. 23-2 Gypsum board edge shapes

- 5/8 inch — used in quality single-layer and double-layer systems

- 1 inch — used as a core board in gypsum board partitions

The standard edges of gypsum board are *tapered, square edge, beveled, rounded,* and *tongue and grooved,* Figure 23-2. The tapered edge type is most commonly used when the joint is to be reinforced with tape and concealed with compound. Beveled and rounded edges are generally found on predecorated boards. Tongue and grooved edges are given to gypsum sheathing.

METAL TRIM AND CASINGS

Metal trim is often used where gypsum board meets windows, doors, or other openings. It is also used when gypsum board is butted against another material and on exterior corners, Figure 23-3.

METAL FURRING CHANNELS

Open web steel bar joists are not designed to receive wallboard directly. It is necessary to attach *metal furring channels* to the joists to fasten the drywall ceiling.

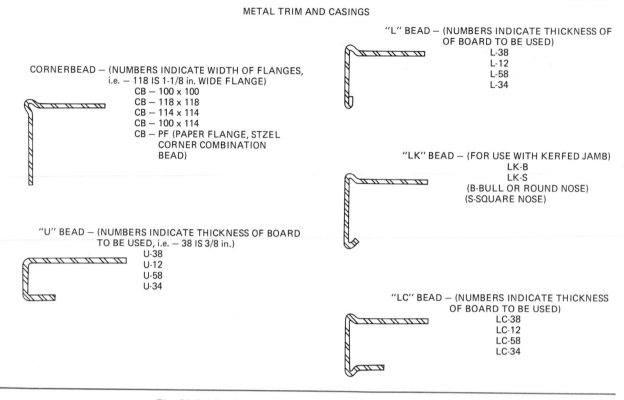

METAL TRIM AND CASINGS

CORNERBEAD — (NUMBERS INDICATE WIDTH OF FLANGES,
i.e. — 118 IS 1-1/8 in. WIDE FLANGE)
CB — 100 x 100
CB — 118 x 118
CB — 114 x 114
CB — 100 x 114
CB — PF (PAPER FLANGE, STZEL
CORNER COMBINATION
BEAD)

"U" BEAD — (NUMBERS INDICATE THICKNESS OF BOARD
TO BE USED, i.e. — 38 IS 3/8 in.)
U-38
U-12
U-58
U-34

"L" BEAD — (NUMBERS INDICATE THICKNESS OF
OF BOARD TO BE USED)
L-38
L-12
L-58
L-34

"LK" BEAD — (FOR USE WITH KERFED JAMB)
LK-B
LK-S
(B-BULL OR ROUND NOSE)
(S-SQUARE NOSE)

"LC" BEAD — (NUMBERS INDICATE THICKNESS
OF BOARD TO BE USED)
LC-38
LC-12
LC-58
LC-34

Fig. 23-3 Metal trim and casings used with gypsum board

Three general types of metal furring channels are available, Figure 23-4. They are attached to joists with appropriate spacing. The minimum width of the furring channel face must be 1 1/4 inches to accommodate abutting edges or ends of the ceiling panels.

DRYWALL FASTENERS

Drywall application requires special fasteners. Ordinary nails or screws are not designed to penetrate the board without causing damage.

NAILS

When nails are used in drywall construction, they should be driven with a hammer that has a crowned head to form a dimple in the board when the nail is driven home, Figure 23-5. Drive the nail at a right angle to the board, taking care not to break the face paper.

Gypsum board nails should have heads that are flat or concave and are thin at the rim. The heads should be at least 1/4 inch and less than 5/16 inch in diameter, Figure 23-6.

Nail penetration into the framing member should be at least 7/8 inch for smooth shank nails and 3/4 inch for ring shank nails. For fire-rated assemblies greater nail penetration is required.

SCREWS

Special drywall screws are used to fasten gypsum board. They have cupped Phillips heads designed to be driven with a power screwdriver (also called a screwgun), Figure 23-7. The screwgun has a clutch controlled by an adjustable nosepiece. When the screw is driven to the proper depth, the clutch is released and the screw can be driven no deeper.

Three basic types of drywall screws are Type *W* for wood, Type *S* for sheet metal, and

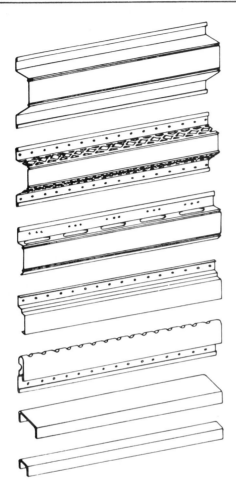

PLAIN DRYWALL CHANNEL (TOP) HAS A HAT SHAPED CROSS SECTION, WHILE RESILIENT DRYWALL CHANNELS (CENTER FOUR) COME IN SEVERAL DESIGNS. ALL ARE OF GALVANIZED STEEL. COLD ROLLED CHANNELS (BOTTOM TWO) ARE "C" SHAPED AND GENERALLY OF 16 GAUGE BLACK ASPHALTUM PAINTED STEEL.

Fig. 23-4 Metal furring channels

Type *G* for solid gypsum, Figure 23-8. Type *W* screws have diamond shaped points and specially designed threads for quick penetration and increased holding power. Minimum penetration in the supporting wood frame is 5/8 inch.

Type *S* screws are self-drilling and have a self-tapping thread. The point is designed to penetrate sheet metal with little pressure.

CAUTION: Wear goggles when driving screws in metal framing. Small particles of drilled metal can fly in all directions.

Type *G* screws have a deeper special thread design for effectively fastening gypsum panels

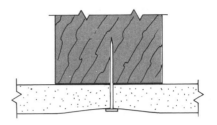

Fig. 23-5 Nails are driven to form a dimple in the board.

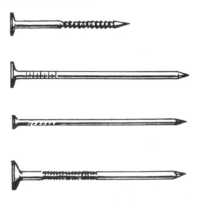

NAILS USED IN DRYWALL CONSTRUCTION HAVE THIN, 1/4 IN. DIAMETER HEADS. SMALLHEADED NAILS SHOWN ARE COLORED FOR USE WITH PREDECORATED WALLBOARD. THESE ARE USED ACCORDING TO INSTRUCTIONS OF THE MANUFACTURERS.

Fig. 23-6 Gypsum board nails

Fig. 23-7 The screwgun is used to drive screws to a desired depth.

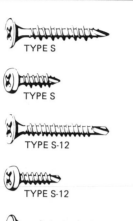

TYPE S SCREWS SHOWN SHOULD PENETRATE AT LEAST 3/8 IN. BEYOND GYPSUM BOARD SURFACE TO HOLD TIGHTLY AGAINST METAL FRAMING. TYPE G SCREWS SHOULD PENETRATE AT LEAST 3/8 IN. INTO SUPPORTING LAYERS OF GYPSUM BOARD TO HOLD TIGHTLY AND TYPE W SCREWS SHOULD PENETRATE AT LEAST 5/8 IN. INTO WOOD FRAMING UNDER GYPSUM BOARD.

Fig. 23-8 Gypsum board screws

Fig. 23-9 Applying adhesive to stud edges *(Courtesy of Gypsum Association)*

together. These screws must penetrate with at least 1/2 inch of the threaded portion into the supporting board.

STAPLES

Staples are used only for attaching the base layer to wood members in a multilayer construction. They should be 16 gauge, flattened, galvanized wire with a minimum wide crown of 7/16 inch.

ADHESIVES

Adhesives are used to bond single layers directly to framing or to laminate gypsum board to base layers. For supplemental support, nails or screws must be used with adhesives. *Stud adhesives* are applied to the framing member with a gun, Figure 23-9.

CAUTION: **Some stud adhesives may contain a flammable solvent. Do not use these types near an open flame or in poorly ventilated areas.**

Drywall joint compound is often used to laminate gypsum boards to each other in addition to being used to conceal joints and fasteners. The compound is applied over the entire board with a suitable spreader. Face boards require fasteners at the perimeter and may require temporary supports until the bond is set.

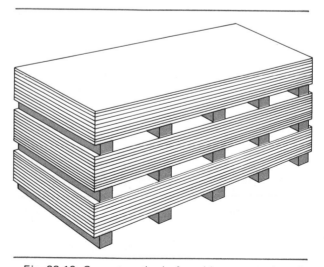

Fig. 23-10 Correct method of stacking gypsum board

HANDLING GYPSUM BOARD

Drywall should be delivered to the job site when application begins. The boards are stored flat and under cover and not exposed to moisture. Supports should be at least 4 inches wide and placed fairly close together, Figure 23-10. Stacking long lengths on too few supports might cause the boards to break. Leaning boards against framing for long periods may cause the boards to warp, resulting in more difficult application. Carry boards, do not drag them, to avoid damaging the edges. Set the boards down gently. Do not drop them.

CUTTING AND FITTING

When walls are less than 8'-1", wallboard is installed at right angles to the studs. Use as long a board as possible to minimize end joints. End joints are difficult to conceal. If possible, use a board of sufficient length to go from corner to corner. If end joints are necessary, stagger them as far apart as possible.

On walls higher than 8'-1", boards are installed vertically with no end joint occurring between floor and ceiling. All edges and ends should fall on a framing member, Figure 23-11.

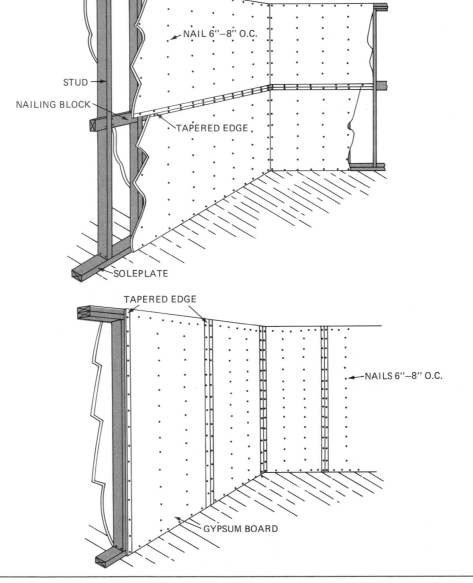

Fig. 23-11 Horizontal and vertical drywall construction

Fig. 23-12 Scoring the sheet and guiding the knife with a drywall square *(Courtesy of Gypsum Association)*

Take measurements accurately and cut the board by first scoring through paper on the face side to the core. Use a utility knife, or the sharp edge of the drywall hammer, and guide it with a drywall T-square, Figure 23-12. The board is then snapped back away from the cut and the back paper broken by snapping the board in the reverse direction or by scoring the back paper with the knife, Figure 23-13.

Smooth ragged edges with a drywall rasp, a coarse sanding block, or the knife. A job-made drywall rasp can be made by fastening a piece of metal lath to a wood block. Cut panels should fit easily into place without being forced. Forcing the panel may cause it to break.

Install ceiling panels first, then the wall panels. Match similar edges and ends; that is, match tapered edges to tapered edges, square ends to square ends.

SINGLE-PLY APPLICATION

Before making a single-ply drywall application, check the framing members for firmness and alignment. A stud with an excessive crook may be straightened by using a handsaw to make a saw cut centered between the top and bottom of the stud. Make the saw cut on the hollow side of the crook and about two-thirds across the width of the stud. While pressing on the crowned edge, drive a wood shingle, tip first, into the saw cut until the stud is straight, Figure 23-14. Trim the shingle flush with the stud. Fasten 1x4 scabs about 16 inches long, on both sides of the stud and centered on the saw cut.

Slightly crooked ceiling joists are sometimes straightened by nailing a strongback across the tops of the joists at about the center of the span, Figure 23-15. Furring strips are straightened by shimming with wood shingles between the furring and the joists. Pry the furring strip until it is straight. Insert shingle shims from both sides and refasten the furring with additional nails.

Fig. 23-13 Snapping and breaking the board along the scored line
(Courtesy of Gypsum Association)

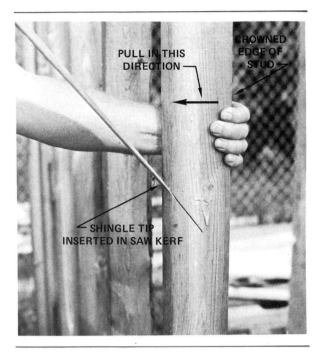

Fig. 23-14 Straightening a stud with an excessive crook

FASTENING

Drywall is fastened to framing members by single or double nailing. Screws and adhesives may also be used.

Single Nailing Method. Single nails are spaced a maximum of 7 inches O.C. on ceilings and 8 inches O.C. on walls into frame members. Nails are first driven in the center of the board and then outward toward the edges.

Double Nailing Method. In double nailing, the first set of nails is spaced 12 inches O.C. with the second set 2 to 2 1/2 inches from the first set. The first nail driven is reseated as necessary after driving the second nail of each set, Figure 23-16.

Attaching with Screws. Screws are spaced 12 inches O.C. on ceilings and 16 inches O.C. on walls when framing members are spaced 16 inches O.C. If framing members are spaced 24 inches O.C., then screws are spaced a maximum of 12 inches O.C. on both walls and ceilings.

Using Adhesives. When using stud adhesives, apply a straight bead about 1/4 inch in diameter to the edge of studs in the center of the panel. Where two panels meet, two parallel beads of adhesive are applied, one near each edge. Zig-

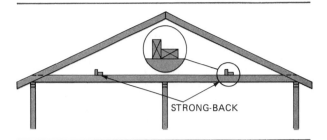

Fig. 23-15 A strongback is sometimes used to straighten crooked ceiling joists.

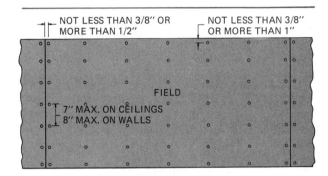

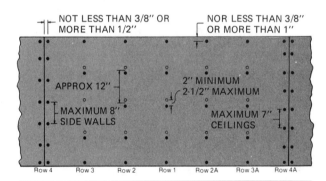

Fig. 23-16 Single and double nailing of gypsum board

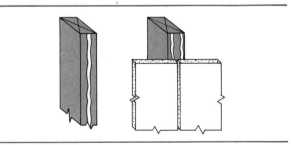

Fig. 23-17 Beads of stud adhesive are applied (left) straight under the field of a board, and (right) parallel under predecorated joints; or joints to be taped.

zag beads should be avoided to prevent the adhesive from squeezing out at the joint, Figure 23-17.

On wall application, supplemental fasteners are used around the perimeter, spaced about 16

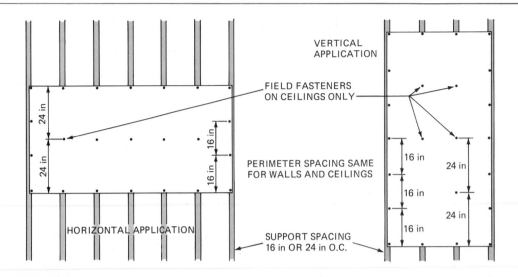

Fig. 23-18 A few temporary fasteners are required when gypsum board
is attached with stud adhesive.

inches apart. On ceilings, in addition to perimeter fastening, the field (center of panel) is fastened at about 24-inch intervals, Figure 23-18.

Gypsum panels may be prebowed to eliminate fasteners at vertical joints. Prebow panels by stacking them face up with ends resting on blocks and centers resting on the floor. Allow them to remain overnight or until the boards have a 2-inch permanent bow. Apply adhesive to the studs and fasten the panel at top and bottom plates. The bow keeps the center of the board in tight contact with the adhesive, Figure 23-19.

CEILING APPLICATION

Gypsum board may be applied parallel or at right angles to joists or furring. If applied parallel, edges and ends must bear completely on

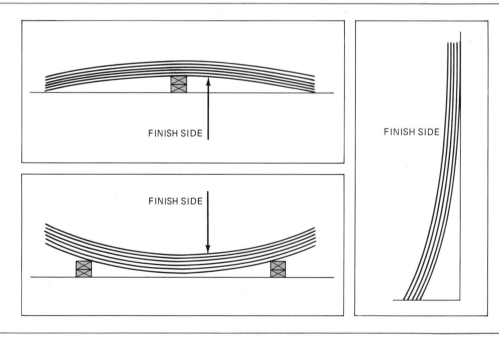

Fig. 23-19 Prebowing keeps the board in tight contact with the adhesive.

framing. If applied at right angles, only the ends need to bear completely on framing. The edges are fastened at each framing member.

Carefully measure and cut the first board. Cut edges should be against the wall. Mark the gypsum board with chalk lines or a straightedge and pencil to indicate the location of the framing for fastening purposes.

Gypsum board panels are heavy. At least two people are needed for ceiling application. Lift the sheet up and over your head. Place it in position and install a deadman under the sheet to hold it in position. *Deadmen* are supports made in the form of a *T*. They are easily made on the job using 1x3 stock with short braces from the vertical member to the horizontal member. The leg of the support is made about 1/4 inch longer than the floor-to-ceiling height. The deadmen are wedged between the floor and the sheet to hold it in position, Figure 23-20.

Fasten the sheet in one of the recommended manners. Hold the board firmly against framing to avoid nail pops or protrusions. Drive nails straight into the member. Nails that miss supports should be removed and the nail hole dimpled so that later it can be covered with joint compound, Figure 23-21.

Once the sheet is fastened, remove the deadmen for use with the next sheet to be

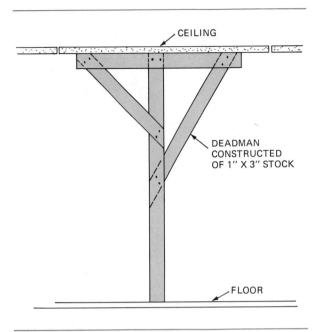

Fig. 23-20 Deadmen are wedged between floor and ceiling to hold the board in place.

applied. Continue applying sheets in this manner, staggering end joints, until the ceiling is covered.

To cut a corner out of a panel in case of a protrusion in the wall, make the shortest cut

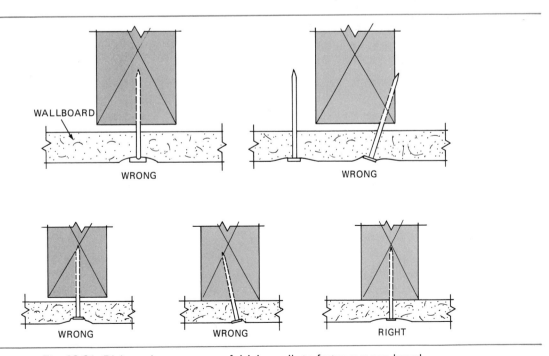

Fig. 23-21 Right and wrong ways of driving nails to fasten gypsum board

Fig. 23-22 Using a knife and drywall saw to cut the corner out of a board

with a drywall saw, Figure 23-22. Then, score and snap the sheet in the other direction. To cut a circular hole, mark the circle with pencil dividers, push a compass saw through the board, and cut it following the circular line.

FLOATING INTERIOR ANGLE CONSTRUCTION

To help prevent nail popping and cracking due to structural stresses where walls and ceilings meet, the *floating angle* method may be used. In this method, the ceiling fasteners are located 7 inches from the wall for single nailing and 11 to 12 inches for double nailing or screw application. However, where joists are parallel to the wall, nailing is started at the intersection.

Gypsum board on sidewalls is applied to provide a firm and level support for the floating edges of the ceiling panels. The top fastener into each stud is located 8 inches down from the ceiling for single nailing and 11 to 12 inches down for double nailing or screw application.

At interior wall corners, the underlying board is not fastened. The overlapping board is fitted firmly against the underlying board to bring it in firm contact with the face of the framing member. The overlapping board is nailed or screwed into the interior corner, Figure 23-23.

WALL APPLICATION

Apply the boards horizontally in rooms where full-length sheets can be used to minimize the number of vertical joints. If joints are necessary, make them at windows or doors. Board ends are cut to fall on studs.

The top panel is installed first. Cut the board to length to fit easily into place without forcing it. Stand the board on edge against the wall and start nails along the top edge at each stud location. Raise the sheet so the top edge is firmly against the ceiling and drive the nails home. Fasten the rest of the sheet in the recommended manner. Apply the top boards completely around the room.

Measure and cut boards for the bottom row. Cut them about 1/4 inch narrower than distance measured. Lay the sheet against the wall and raise it with a foot lifter against the bottom edge of the top sheet, Figure 23-24. (A *foot lifter* is a tool especially designed for this purpose. However, one can be made on the job

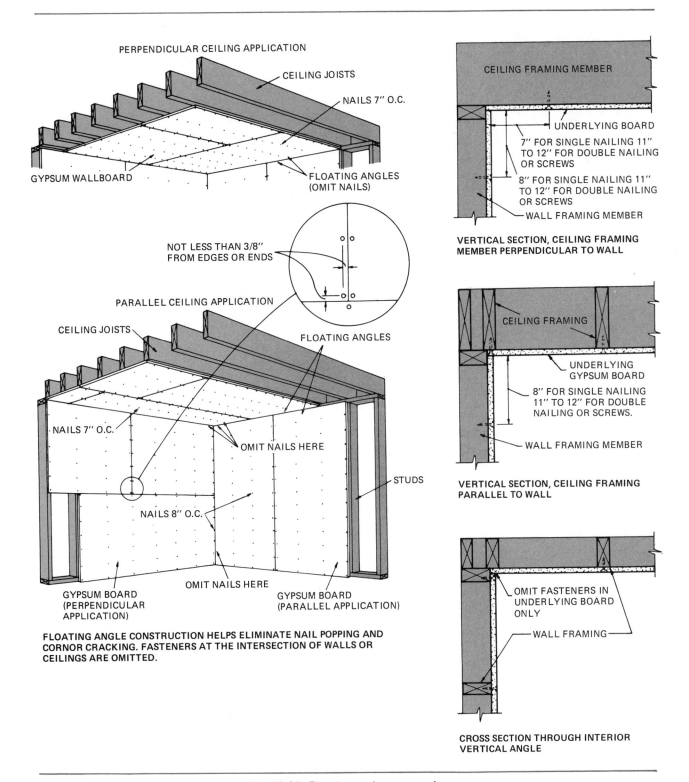

PERPENDICULAR CEILING APPLICATION

CEILING JOISTS

NAILS 7" O.C.

GYPSUM WALLBOARD

FLOATING ANGLES (OMIT NAILS)

NOT LESS THAN 3/8" FROM EDGES OR ENDS

PARALLEL CEILING APPLICATION

CEILING JOISTS

FLOATING ANGLES

NAILS 7" O.C.

OMIT NAILS HERE

STUDS

NAILS 8" O.C.

GYPSUM BOARD (PERPENDICULAR APPLICATION)

OMIT NAILS HERE

GYPSUM BOARD (PARALLEL APPLICATION)

FLOATING ANGLE CONSTRUCTION HELPS ELIMINATE NAIL POPPING AND CORNER CRACKING. FASTENERS AT THE INTERSECTION OF WALLS OR CEILINGS ARE OMITTED.

CEILING FRAMING MEMBER

UNDERLYING BOARD

7" FOR SINGLE NAILING 11" TO 12" FOR DOUBLE NAILING OR SCREWS

8" FOR SINGLE NAILING 11" TO 12" FOR DOUBLE NAILING OR SCREWS

WALL FRAMING MEMBER

VERTICAL SECTION, CEILING FRAMING MEMBER PERPENDICULAR TO WALL

CEILING FRAMING

UNDERLYING GYPSUM BOARD

8" FOR SINGLE NAILING 11" TO 12" FOR DOUBLE NAILING OR SCREWS.

WALL FRAMING MEMBER

VERTICAL SECTION, CEILING FRAMING PARALLEL TO WALL

OMIT FASTENERS IN UNDERLYING BOARD ONLY

WALL FRAMING

CROSS SECTION THROUGH INTERIOR VERTICAL ANGLE

Fig. 23-23 Floating angle construction

Fig. 23-24 Using a foot lifter to raise a gypsum board

Fig. 23-25 Making a cutout for an electrical box *(Courtesy of Gold Bond Building Products)*

by tapering the end of a short piece of 1x3 stock.) Fasten the sheets in the specified manner. Stagger end joints and locate them as far from the center of the wall as possible so they will be less conspicuous.

To make cutouts for electrical boxes and ducts, measure the location carefully and mark it on the sheet. Score the outline of the box, and then score diagonals. Hit the board with a hammer where the diagonals cross. With a sharp knife, trim the opening from the back side, Figure 23-25.

To make cutouts around interior door openings either mark and cut out the sheet before it is applied or make the cutout after it is applied. In this case, use a saw to cut one side, then score the other side on the back, bend it, score it on the face, and break it.

VERTICAL WALL APPLICATION

For vertical wall application, cut the first board (to go in the corner) to length and width. Cut the length about 1/4 inch shorter than measured. Cut the width so the edge away from the corner falls on the center of a stud. The cut edge must be in the corner.

With a foot lifter, raise the sheet so it is snug against the ceiling and the edge is plumb. Fasten in the specified manner. Continue applying sheets around the room, making any necessary cutouts. No horizontal butt joints are

made and vertical joints must have tapered edges for ease in finishing.

MULTI-PLY APPLICATION

Multi-ply application has one or more layers of gypsum board applied over a base layer. The base layer may be gypsum backing board, regular gypsum board, or other base material.

BASE LAYER

The base layer is fastened in the same manner as single-ply construction. On ceilings, it is applied with the long edges at right angles to framing members. On walls, it is applied with the long edges parallel to the framing member.

FACE LAYER

Joints in the face layer are offset at least 10 inches from joints in the base layer. The face layer is applied either parallel to or at right angles to framing. Horizontal application is preferred on walls since it usually results in fewer joints to finish.

The face layer may be attached with nails, screws, or adhesives. If nails or screws are used, the maximum spacing and minimum penetration should be the same as for single-ply application.

Fig. 23-26 In sheet lamination, the entire back of the face layer is covered with adhesive. *(Courtesy of Gypsum Association)*

ADHESIVE ATTACHMENT OF THE FACE LAYER

Adhesive attachment of the face layer employs sheet lamination, strip lamination, or spot lamination. In *sheet lamination* the entire back of the face layer is covered with adhesive, using a notched spreader, Figure 23-26.

In *strip lamination*, the adhesive is applied in ribbons with a special spreader, spaced 16 to 24 inches O.C., Figure 23-27. In *spot lamination*, the adhesive is spotted on the sheet in a regular pattern.

After the adhesive is applied, the face ply is held firmly against the base ply with supplemental fasteners or braces while the adhesive sets. If fasteners are used, they are placed around the perimeter and in the center of ceiling boards. On wall boards, fasteners are used only around the perimeter where they will be concealed by joint treatment or trim.

CONCEALING FASTENERS AND JOINTS

After the gypsum board is installed, it is necessary to conceal the fasteners and to reinforce and conceal the joints. Fasteners are concealed with joint compound and joints are treated with joint tape and joint compound. Ex-

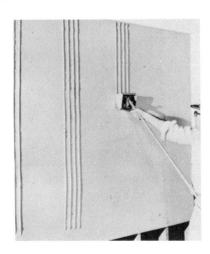

Fig. 23-27 In strip lamination, the adhesive is spaced. *(Courtesy of Gypsum Association)*

terior corners are reinforced with metal corner bead which is concealed with joint compound. Other metal trim is used around doors, windows, and openings and finished with compound.

Install all metal trim before taping and compounding. Use a straightedge to make sure metal trim is applied in a straight line. Use enough fasteners to keep the trim edges tightly against the board.

JOINT TAPE

Joint tape may be made of paper, either perforated or unperforated, glass mesh, or other material and is designed for use with joint compound to reinforce and finish joints between gypsum boards.

JOINT COMPOUNDS

There are three general types of joint compounds. A *bedding compound* is used to adhere the tape to the board. A *finishing compound* is used for subsequent layers. An *all-purpose compound* is used for both bedding the tape and finishing the joint.

JOINT TREATMENT

Fill the recess formed by the tapered edges of the sheets with joint bedding compound,

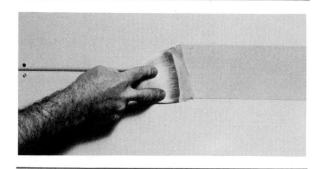

Fig. 23-28 Bedding compound is first applied to the joint. *(Courtesy of Gold Bond Building Products)*

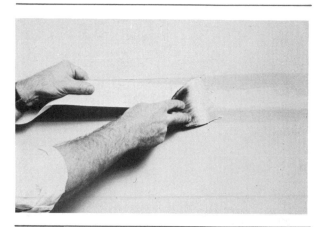

Fig. 23-29 Tape is pressed into the bedding compound. *(Courtesy of Gold Bond Building Products)*

using a joint knife, Figure 23-28. Wipe excess compound applied beyond the groove. Center reinforcing tape and press it down into the joint compound, Figure 23-29. Make sure there are no air bubbles under the tape and that the tape edges are well adhered to the compound. Add additional compound, using a 5- or 6-inch wide knife, and level the compound with the wall surface. Leave the joint completely smooth so that no sanding is necessary prior to the next coat.

A *taping tool* is sometimes used that applies the compound and the tape at the same time, Figure 23-30. Allow the bedding coat to dry thoroughly. This may take 24 hours, or more, depending on temperature and humidity. A quick-set type of compound may be used that reduces drying time to only several hours. It is common to use quick-set compounds to bed the tape and to use slower setting types for finishing coats.

After the bedding coat is thoroughly dry, apply a second coat, feathered out about 2 inches beyond the edges of the first coat, Figure 23-31. On end joints where there are no tapered edges, the tape is higher than the board surface and requires the joint compound to be feathered out about 12 to 16 inches wide. Spot fastener heads and allow to dry. After the second coat is dry, sand lightly.

CAUTION: Wear a dust mask while sanding.

Apply a thin finishing coat to joints and fastener heads. Feather edges of the compound out to about 6 inches from the center of the joint. When the final coat is dry, sand lightly

Fig. 23-30 A taping tool applies tape and compound at the same time. *(Courtesy of Gypsum Association)*

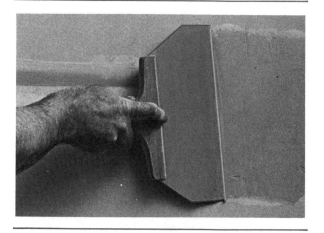

Fig. 23-31 Applying a finishing coat to a taped joint *(Courtesy of Gold Bond Building Products)*

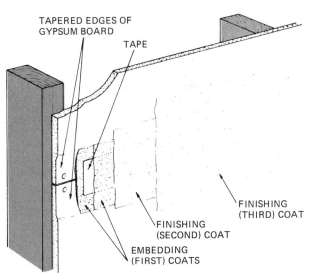

TAPERED EDGES OF
GYPSUM BOARD

TAPE

FINISHING
(THIRD) COAT

FINISHING
(SECOND) COAT

EMBEDDING
(FIRST) COATS

REINFORCING JOINTS WITH TAPE PREVENTS CRACKS FROM
APPEARING AT FILLED GYPSUM BOARD JOINTS. THE JOINT
FILL AND FIRST COAT MAY BE JOINT COMPOUND OR ALL
PURPOSE COMPOUND. THE SECOND AND THIRD COATS
SHOULD BE FINISHING COMPOUND OR ALL–PURPOSE
COMPOUND.

Fig. 23-32 Completed reinforced and concealed drywall joint

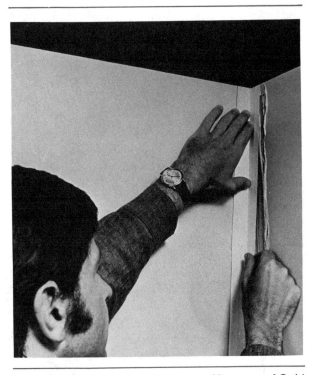

Fig. 23-33 Taping an interior corner *(Courtesy of Gold Bond Building Products)*

but avoid scuffing the face paper of the board, Figure 23-32.

Interior corners are finished in much the same manner except that the tape is folded in the center to fit in the corner, Figure 23-33. The compound is then feathered out from both sides.

APPLYING GYPSUM BOARD IN BATHROOMS

Apply water-resistant gypsum board horizontally with the paper covered edge not less than 1/4 inch above the lip of the shower pan or tub. Do not apply directly over a vapor barrier.

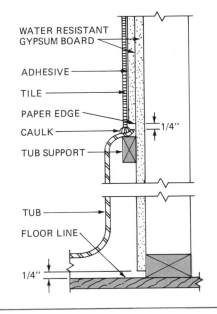

WATER RESISTANT
GYPSUM BOARD

ADHESIVE

TILE

PAPER EDGE

CAULK

TUB SUPPORT

1/4"

TUB

FLOOR LINE

1/4"

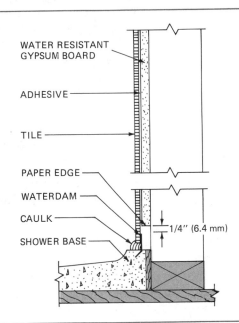

WATER RESISTANT
GYPSUM BOARD

ADHESIVE

TILE

PAPER EDGE

WATERDAM

CAULK

SHOWER BASE

1/4" (6.4 mm)

Fig. 23-34 Installation details around bathtubs and showers

If necessary, the board is furred away from the framing so that the lip of the tub or shower pan is flush with the face of the board, Figure 23-34.

Provide blocking between studs about one inch above the top of the tub or shower pan. Additional blocking is installed between studs behind the horizontal joint of the gypsum board above the tub or shower pan.

Attach water-resistant gypsum board with screws or nails spaced not more than 8 inches apart. When ceramic tile more than 3/8 inch thick is to be applied, the nail or screw spacing should not exceed 4 inches O.C.

After applying the board, tape and compound the joints and nail heads. If the board is to receive tile or wall panels, coat the surface with a compatible sealer before installing tile or wall panels. Openings around pipes are caulked flush with a waterproof, nonhardening caulking compound.

ESTIMATING DRYWALL

To estimate the amount of drywall material needed, determine the area of the walls and ceilings to be covered. To find the ceiling area multiply the length of the room by its width. To find the wall area, multiply the perimeter of the room by the height and subtract all large openings such as windows and doors. Combine all areas to find the total number of square feet of drywall required. Add about 5 percent of the total for waste.

REVIEW QUESTIONS

Select the most appropriate answer.

1. Standard gypsum board width is
   a. 36 inches.
   b. 48 inches.
   c. 54 inches.
   d. 60 inches.

2. Standard gypsum board lengths are
   a. 6, 8, 10, and 12 feet.
   b. 8, 10, 12, and 14 feet.
   c. 8, 10, and 12 feet.
   d. 8, 10, 12, and 16 feet.

3. When fastening drywall, minimum penetration of ring-shanked nails into the framing member is
   a. 1/2 inch.
   b. 3/4 inch.
   c. 7/8 inch.
   d. 1 inch.

4. Gypsum board is installed vertically on walls when the wall height is greater than
   a. 8'-0".
   b. 8'-1".
   c. 8'-4".
   d. 8'-6".

5. Slightly crooked ceiling joists are sometimes straightened by the use of a
   a. deadman.
   b. dutchman.
   c. strongback.
   d. straightedge.

6. In the single nailing method, nails are spaced a maximum of
   a. 8 inches O.C. on walls; 7 inches O.C. on ceilings.
   b. 10 inches O.C. on walls; 8 inches O.C. on ceilings.
   c. 12 inches O.C. on walls; 10 inches O.C. on ceilings.
   d. 12 inches O.C. on walls and ceilings.

7. Screws are spaced
    a. 12 inches O.C. on walls; 10 inches O.C. on ceilings.
    b. 12 inches O.C. on walls and ceilings.
    c. 16 inches O.C. on walls and ceilings.
    d. 16 inches O.C. on walls; 12 inches O.C. on ceilings.

8. Joints in the face layer of a multilayer application are offset from joints in the base layer by at least
    a. 6 inches.   c. 10 inches.
    b. 8 inches.   d. 12 inches.

9. The paper covered edge of water-resistant gypsum board is applied above the lip of the tub or shower pan not less than
    a. 1/4 inch.   c. 1/2 inch.
    b. 3/8 inch.   d. 3/4 inch.

10. When ceramic tile more than 3/8-inch thick is to be applied over water-resistant gypsum board, fasten the board with screws or nails spaced not more than
    a. 4 inches O.C.   c. 8 inches O.C.
    b. 6 inches O.C.   d. 10 inches O.C.

WALL PANELING 24

OBJECTIVES

After completing this unit, the student will be able to

- *describe various types of material used for wall paneling.*
- *apply wall paneling materials in an accepted manner.*
- *estimate quantities of wall paneling materials.*

Except for plaster and plain gypsum board, materials used for finish interior wall covering may be considered as wall paneling. They come in the form of sheets or strips and are classified in several general groups. In each of these groups, many colors, grains, textures, patterns, and shapes are available. These groups include the following:

- Prefinished gypsum board
- Plywood
- Hardboard
- Fiberboard
- Particleboard
- Solid lumber
- Plastic laminates
- Ceramic, metal, and plastic tile

PREFINISHED GYPSUM BOARD

Prefinished gypsum board is plain gypsum board with a decorated vinyl covering. The vinyl comes in solid colors and a variety of patterns. Panels are 3/8 to 5/8 inch thick and 8 and 10 feet long. Some panels come with beveled edges which are left exposed. Joints in square edge panels are covered with matching molding. Molding is also available in matching colors to cover joints, raw edges, and interior and exterior corners, Figure 24-1.

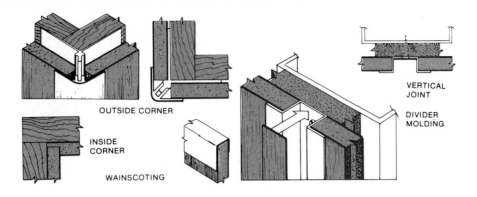

Fig. 24-1 Moldings used to cover joints or raw edges on sheet paneling

Another type of panel, designed for seamless application, has a 2-inch flap of vinyl extending beyond one edge of the board. This flap overlaps the adjacent panel to form a jointless solid surface.

Predecorated gypsum board is usually applied vertically on walls with adhesives. Special nails are used that are colored to match the color of the board. Panels may be applied to open framing or to solid backing.

PLYWOOD

Plywood is widely used for interior wall finish. A tremendous variety is available in both hardwoods and softwoods; plain, textured, and prefinished. The more expensive types have a face veneer of real wood. The less expensive kinds are prefinished with a printed wood grain on a thin vinyl covering. Some sheets are scored to imitate solid wood paneling. Among others, there is always a score 16, 24, and 32 inches from the edge. This is done in case the sheet has to be ripped lengthwise to fall on a stud and join another sheet. The location of the scores facilitates nailing because they indicate the position of hidden studs.

Thicknesses range from 3/16 to 1/2 inch. The most commonly used is 1/4 inch thick. Sheets are 4 feet wide and 7 to 10 feet long with an 8-foot length being the most common.

Edges may be square, beveled, or shiplapped, Figure 24-2. Matching molding is available to treat raw edges and joints. Colored nails are used when exposed fastening is necessary.

HARDBOARD

Hardboard is available in many man-made surface colors, textures, and designs. Some of these designs simulate stone, brick, and stucco; weathered or smooth wood; framed panels; leather; and other materials.

Hardboard is also available for use unfinished. Unfinished hardboard has a smooth, dark brown surface suitable for painting. It may be plain or perforated in a number of designs, Figure 24-3.

Tileboard is tempered hardboard with a baked-on plastic finish. The sheets come in solid colors, marble and other designs, or scored to look like ceramic tile. It is usually used in bathrooms and kitchens, applied over gypsum board.

Hardboard comes in widths of 4 feet and in lengths of from 8 to 12 feet. Thickness ranges

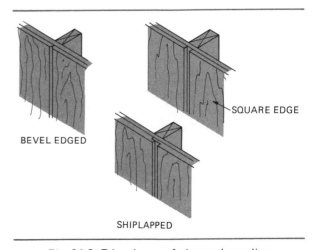

Fig. 24-2 Edge shapes of plywood paneling

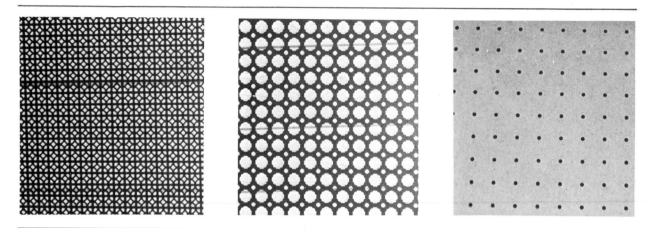

Fig. 24-3 Hardboard is also available in a number of perforated designs.

from 1/8 to 1/2 inch. Solid backing must be provided for boards less than 1/4 inch thick.

FIBERBOARD

Fiberboard is softer than hardboard and plywood. Care must be taken not to mar the finished surface when handling and applying fiberboard. Because of its relative softness, it is usually applied to walls above a wainscoting. A *wainscoting* is a wall finish applied partway up the wall, usually 2 1/2 to 3 feet from the floor, Figure 24-4.

Fiberboard comes in tongue-and-grooved plank form or in sheet form. It is prefinished in a great number of colors and designs and is used primarily for increased sound deadening. The usual thicknesses are 1/2 inch and 3/4 inch.

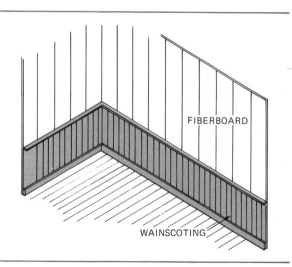

Fig. 24-4 Fiberboard is usually applied to walls above a wainscoting.

Planks are available in a number of widths and in lengths suitable for horizontal application as well as vertical. Sheets usually come in 4-foot widths and in 8-foot lengths. Longer sheets are available. When applied over open framing, fiberboard must be 1/2 inch thick when framing is spaced 16 inches O.C. and 3/4 inch thick for 24-inch spacing.

PARTICLEBOARD

Particleboard comes in sheet form usually with a man-made wood grain finish to imitate random solid wood paneling.

Sheets are usually 1/4 inch thick, 4 feet wide, and 8 feet long. Prefinished particleboard is used when an inexpensive wall covering is desired.

Because the sheets are brittle and break easily, care must be taken in handling and applying. Also, it must be applied only on solid wall backing unless at least 1/2 inch thick.

Unfinished particleboard is not usually used as an interior wall finish. One exception, a particleboard made from aromatic cedar chips, is used to cover walls in closets to repel moths.

SOLID LUMBER PANELING

Solid lumber paneling is used for interior walls in the same manner as siding. It may be applied horizontally, vertically, diagonally, or in various patterns. Wood paneling is desired for its warmth and beauty. It takes paints and stains easily and may be obtained in knotty or clear grades.

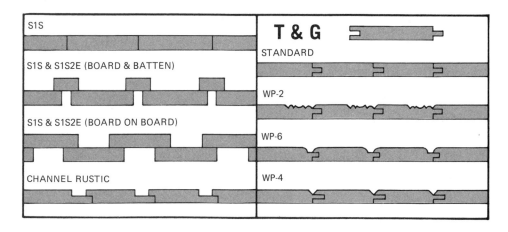

Fig. 24-5 Board paneling is available in a number of designs.

SPECIES

Wood paneling is available in many species each with its own good looks, unique grain, and knot pattern. They may be described as light, medium, and dark-toned woods. Light tones include spruce, white pine, and sugar pine. Some medium tones are hemlock, ponderosa pine, and fir. Darker-toned woods are cypress, cedar, and redwood. For special effects, wormy chestnut, pecky cypress, and white-pocketed Douglas fir may be used.

SURFACE TEXTURES AND PATTERNS

Wood paneling is available either planed for smoother finishing, or rough-sawn for a rustic, informal effect. A number of panel patterns are available. Square edge boards may be used joined edge to edge, spaced on a dark background, or applied in board and batten or board on board patterns. Tongue-and-grooved paneling comes in a number of patterns to show decorative lines when installed, Figure 24-5.

SIZES

Most wood paneling comes in a nominal 1-inch thickness. However, aromatic cedar paneling, used in clothes closets, comes about 3/8 inch thick and is edge and end matched (tongue-and-grooved). Nominal wood paneling widths are 4, 6, 8, 10, and 12 inches.

MOISTURE CONTENT

To avoid shrinkage, wood paneling, like all interior finish, should be dried to a *moisture equilibrium* content. That is, its moisture content should be about the same as the atmosphere in the area which it is to be used. For most parts of the country this is about 8 percent, Figure 24-6.

Interior finish applied with an excessive moisture content will eventually shrink, causing open joints, warping, loose fasteners, and many other problems.

PLASTIC LAMINATES

Plastic laminates are widely used for surfacing kitchen cabinets and countertops. They are also used to cover walls or parts of walls in kitchens, bathrooms, and similar areas where a durable, easy-to-clean surface is desired. Laminates can be scorched by an open flame, but resist heat, alcohol, acids, and stains. They clean easily with a mild detergent.

Laminates are known by such names as Formica, Micarta, Melamite, Pionite, and others. They are manufactured in many colors and designs including many wood grain patterns. Surfaces are available in gloss, satin, and textured finishes, among others. Distributors supply samples or chips to help customers decide which to use.

Plastic laminates come in widths of 24, 36, 48, and 60 inches and in lengths of 5, 6, 8, 10, and 12 feet. Sheets are usually 1 inch wider and

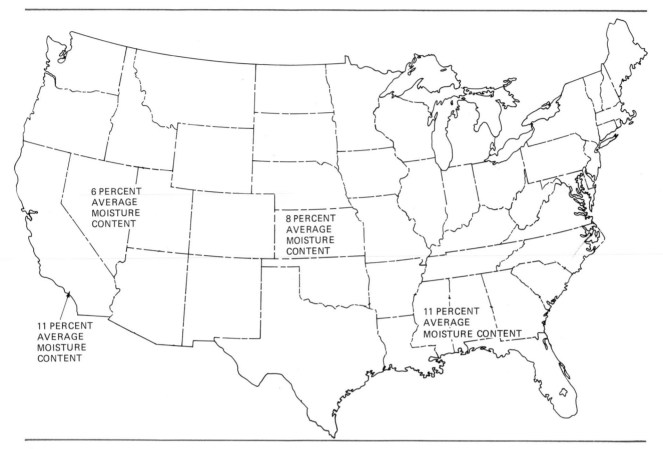

Fig. 24-6 Recommended average moisture content for interior finish woodwork in different parts of the United States

longer than the size indicated. Available sizes depend on the laminate color, pattern, and thickness.

Laminates come in two thicknesses. *Regular laminate* is 0.06 inch thick and is generally used on horizontal surfaces such as countertops and table and desk tops, although it can be used on walls if desired. *Vertical type* laminate is used for vertical surfaces. It is thinner (0.03 inch) because vertical surfaces take less wear than horizontal surfaces.

Some laminates come with a foamlike, soft, spongy backer. This back serves a number of purposes. It makes the sheet more rigid and, therefore, easier to handle and apply. Thin laminate sheets without backers are very brittle and can crack easily if not carefully handled. The backer also provides insulation and absorbs irregularities in the wall to which it is applied, giving a smoother surface to the finished product. Laminates must be applied only to solid wall backing.

OTHER WALL COVERINGS

Other wall coverings such as *ceramic, metal,* and *plastic tile* are usually applied by specialists. However, on small jobs, especially in rural or suburban communities, the general carpenter may be required to install tile.

Tiles are usually rectangular or square and come in a number of different sizes. Special pieces such as base, caps, and inside and outside corners are used to trim the installation, Figure 24-7. Many solid colors, patterns, and designs give a wide choice to achieve the desired wall effect.

INSTALLATION OF SHEET WALL PANELING

Sheet panels such as plywood, predecorated gypsum board, hardboard, fiberboard, and particleboard are applied in a similar manner. Panels may be installed on open studs, to plastered or gypsum board walls, and to masonry walls.

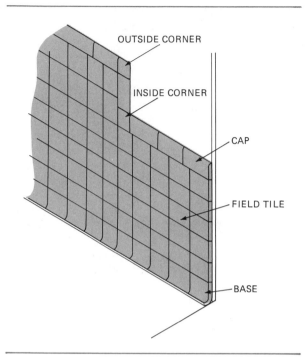

Fig. 24-7 Special pieces of ceramic tile are used to trim the installation.

Fig. 24-8 Applying furring strips to masonry walls *(Courtesy of Masonite Corporation)*

It is desirable to cover open studs with a base layer of 1/2-inch gypsum board. This backing makes a more fire-resistant wall, tends to bring studs in alignment, and provides a rigid finished surface.

Furring strips must be applied to masonry walls before the installation of paneling, Figure 24-8. Instead of furring, a freestanding wood wall close to the masonry wall can be built. Panels may be fastened with *color pins* (nails), or a combination of color pins and adhesive may be used.

Vertical application of sheet wall paneling is done by a standard procedure. On existing walls, find the location of the studs and snap lines on the walls from floor to ceiling to indicate the centerline of each stud. Paneling edges and ends must bear on *solid* backing unless the panels are applied over a backer board with adhesive. However, it is good practice to plan the layout so that panel edges fall on studs in case supplemental nailing is necessary.

Unwrap panels and stand them on their long edge against the wall for at least 48 hours before installation. This allows the panels to adjust to room temperature and humidity, Figure 24-9. Match grain and color by standing

Fig. 24-9 Allow panels to stand in the room for 48 hours to adjust to temperature and humidity. *(Courtesy of Masonite Corporation)*

Fig. 24-10 The first sheet must be set plumb in the corner. *(Courtesy of Masonite Corporation)*

panels around the room to find the most pleasing pattern.

Starting in the corner, cut the first sheet about 1/4 inch less than the wall height. Place the sheet in the corner and plumb the edge, Figure 24-10. Tack it into the plumb position.

Set the dividers for the distance the outside edge of the sheet overlaps the center of the stud. Scribe this amount on the edge of the sheet in the corner, Figure 24-11.

Remove the sheet from the wall and place it on saw horses on which a sheet of plywood or two 2x4s have been placed for support. Cut to the scribed lines with a fine-toothed, hand crosscut saw.

Replace the sheet with the cut edge fitting snugly in the corner. The joint at the ceiling need not be fitted if a molding is to be used. If a tight fit between the panel and ceiling is desired, set

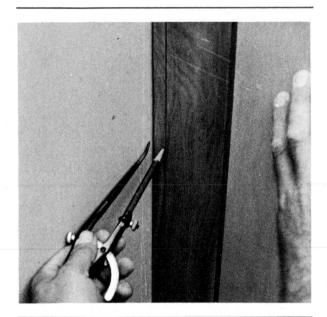

Fig. 24-11 Set dividers and scribe the corner edge of the first sheet.

the dividers to scribe as little as possible at the ceiling line. Remove the sheet again and cut to the scribed line. Replace the sheet and raise it snugly against the ceiling. The joint at the bottom will be covered by baseboard, later.

If sheets are to be nailed, place edge nails about 6 inches apart and 12 inches apart on intermediate studs for 1/4-inch thick paneling. Nails may be spaced farther apart on thicker paneling, Figure 24-12.

If adhesives are used, apply a 1/8-inch continuous bead where panel edges and ends make contact. Apply beads 3 inches long and about 6 inches apart on all intermediate studs, Figure 24-13.

Put the panel in place and tack at the top. Be sure the panel is properly placed in position. Press on the panel to make contact with the adhesive. Use a firm uniform pressure to spread the adhesive beads evenly between the wall and the panel. Then, grasp the bottom of the panel and slowly pull the sheet a few inches away from the wall, Figure 24-14. Press the sheet back into position after about two minutes. After about 20 minutes recheck the panel. Apply pressure to assure firm adhesion and to even panel surface.

Apply successive sheets in the same manner, Figure 24-15. Do not force panels in position. Panels should touch very lightly at

Fig. 24-12 Nailing the edge of sheet paneling. Color pins are used. *(Courtesy of Masonite Corporation)*

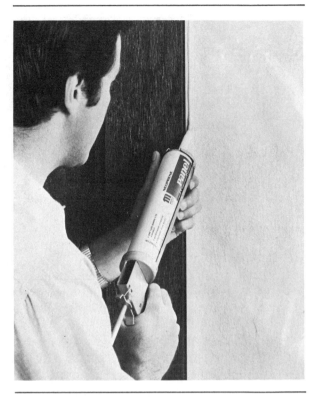

Fig. 24-13 Apply adhesive to existing walls for the installation of sheet paneling. *(Courtesy of Masonite Corporation)*

Fig. 24-14 The sheet is pulled away from the wall to allow the adhesive to dry slightly. *(Courtesy of Masonite Corporation)*

Fig. 24-15 Applying successive sheets in a paneling installation *(Courtesy of Masonite Corporation)*

Fig. 24-16 A saber saw is used to cut an opening for an electrical box. *(Courtesy of Masonite Corporation)*

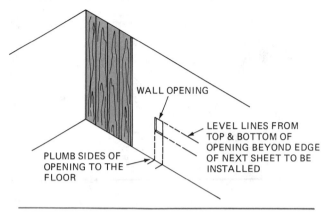

Fig. 24-17 Method of laying out for wall openings

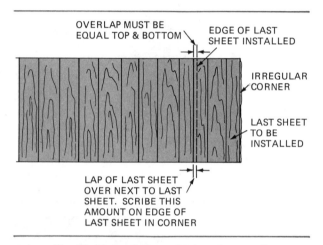

Fig. 24-18 Laying out the last sheet to fit against a finished corner

joints. Measure the location of electrical boxes and other wall openings. A saber saw may be used to cut these openings. When using the saber saw, cut from the back of the panel to avoid splintering the face, Figure 24-16.

A number of methods may be used to accurately mark wall openings on the panel. Plumb both sides of the opening to the floor and mark on the floor. If the opening is close to the ceiling, plumb upward and mark lightly on the ceiling. Level top and bottom of the opening on the wall beyond the edge of the sheet to be installed, or on the adjacent sheet, if closer. Cut, fit, and tack the sheet in position. Level and plumb marks from the wall and floor onto the sheet for the location of the opening, Figure 24-17. Remove the sheet and cut the opening.

Another method is to rub a cake of carpenter's chalk on the edges of the opening. Cut, fit, and tack the sheet in position. Tap on the sheet directly over the opening to transfer the chalked opening to the back of the sheet. Remove the sheet and cut the opening.

Note: Openings for such things as electrical boxes must be cut fairly close to the opening. If the opening is cut oversize, the cover plate may not cover the cutout. This could make it necessary to replace the panel.

The last sheet in the wall need not fit snugly in the corner if the adjacent wall is to be paneled or if interior corner molding is to be used. Take measurements at the top, center,

and bottom and cut the sheet to width. If the last sheet butts against a finished wall and no corner molding is used, the sheet must be cut to fit snugly in the corner.

To mark the panel accurately, first measure the remaining space and rip the panel about 1 inch wider. Place the sheet with the cut edge in the corner and the other edge overlapping the

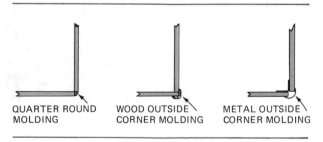

Fig. 24-19 Joint treatment of sheet paneling at exterior corners

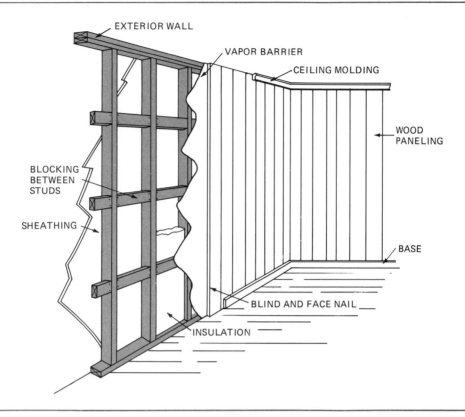

EXTERIOR WALL

VAPOR BARRIER

CEILING MOLDING

WOOD PANELING

BLOCKING BETWEEN STUDS

SHEATHING

BASE

BLIND AND FACE NAIL

INSULATION

Fig. 24-20 Blocking between studs for vertical panels

last sheet installed. Tack the sheet in position so that the amount of overlap is the same at the top and bottom. Set the dividers for this amount of overlap and scribe this amount on the edge in the corner, Figure 24-18.

Cut to the scribed line. If the line is followed carefully, the sheet should fit snugly between the last sheet installed and the corner — regardless of any irregularities in the corner. On exterior corners, the joint may be mitered, covered with a corner molding, or butted against a quarter-round molding, Figure 24-19.

INSTALLING SOLID LUMBER PANELING

If wood paneling is to be installed vertically, Figure 24-20, blocking must be provided between studs not more than 24 inches O.C. If vertical boards are to be applied over existing walls, horizontal furrings strips must be installed.

Allow the material to adjust to room temperature by placing boards around the room. At the same time, match them for grain and color, Figure 24-21.

Fig. 24-21 Place boards around the room to adjust to temperature and humidity and to match color and grain. *(Courtesy of California Redwood Association)*

Fig. 24-22 Tongue-and-grooved paneling is blind nailed.

Select a straight board to start with, and cut it to length about 1/4 inch less than the height of the wall. If tongue-and-grooved stock is used, place it in position with the grooved edge in the corner. Plumb and tack the board in position. If a tight fit is desired, adjust the dividers to scribe an amount a little more than the depth of the groove. Use a hand ripsaw and rip to the scribed line.

Note: A portable, power circular saw cannot be used with satisfactory results for at least two reasons:

1. Because of the irregularity of the scribed line, it is difficult to follow with a power saw.
2. The scribed line is marked on the face side of the board. Cutting from the face side with a power saw splinters the edge on the face side.

Face nail along the cut edge into the corner with finish nails about 16 inches apart. Blind nail the other edge through the tongue, Figure 24-22.

Apply succeeding boards by blind nailing into the tongue only. Make sure the joints between boards come up tightly. (Refer to Figure 22-26 for methods of bringing edge joints of matched boards up tightly if they are slightly crooked.) Severely warped boards should not be used. As installation progresses, check the paneling for plumb. Cut out openings in the same manner as described for sheet paneling.

If the last board in the installation must fit snugly in the corner, proceed as follows:

Plan the layout so that the last board will be as wide as possible. Cut and fit the next to the last board and remove it.

Tack the last board in the place of the next to the last board.

Cut a scrap block about 6 inches long and equal in width to the finished face of the next to the last board.

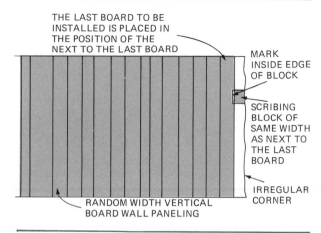

Fig. 24-23 Laying out the last board to fit against a finished corner

- Use this block to scribe the last board by running one edge along the corner and holding a pencil against the other edge, Figure 24-23.

- Cut to the scribed line.

- Fasten the next to the last board in position.

- Fasten the last board in position with the cut edge in the corner. Face nail the edge nearest the corner.

Horizontal application of wood paneling is done in a similar manner except that blocking between studs on open walls or furring strips on existing walls are not necessary. On existing walls, locate and snap lines to indicate the position of stud centerlines. The thickness of wood paneling should be at least 3/8 inch for 16-inch spacing of frame members and 5/8 inch for 24-inch spacing.

Horizontal and vertical application of *fiberboard plank* is done in a manner similar to that used with wood paneling. If wainscoting is applied to a wall, the joint between the different materials is treated in a number of ways, as shown in Figure 24-24.

APPLYING PLASTIC LAMINATES

Plastic laminates are cut in a number of ways. A fine-toothed, hand crosscut saw can be used, but the cut must be made close to the edge of a support because the material is so thin and brittle. After being handsawed, the edge may be

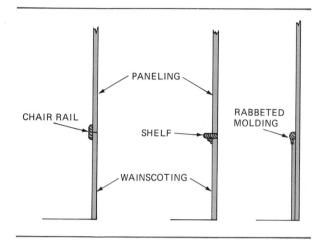

Fig. 24-24 Joint treatment where paneling meets wainscotting

trimmed with a hand plane. However, because of the toughness of the material, ordinary tools dull quickly when used to cut laminates.

An easy and efficient way to cut laminates is with the use of a router with a solid carbide flush trimming bit. Clamp a straightedge to the underside of the sheet along the line of cut with the waste opposite the straightedge. Cut the laminate by guiding the pilot of the router bit along the straightedge, Figure 24-25. Carefully cut the sheet to size. Place the sheet in position to try the fit.

Apply contact cement to the back of the sheet and to the surface on which the sheet is to be placed. Porous surfaces need two coats. Allow one coat to dry before applying a second coat. Contact cement dries rapidly. If no cement sticks to the finger when it is pressed on the surface, the cement is dry even though it may feel tacky. Allow the cement to dry on both surfaces before bonding the sheet.

CAUTION: Some types of contact cement are flammable. Have the area well ventilated, with no open flame. Avoid breathing the fumes.

Bond the sheet by carefully placing it in position. Contact cement bonds on contact. Once contact is made the sheet cannot be repositioned. Small strips of wood or venetian blinds are used to hold the sheet away from the surface while it is positioned. Contact is made and the strips are removed. The entire surface is then rolled out with a 3-inch roller or tapped by using a hammer and a wood block. (The application of plastic laminates is discussed in greater detail in a later unit.)

APPLYING WALL TILE

Ceramic, metal, and plastic wall tiles are frequently used as wall covering in bathrooms.

Fig. 24-25 Cutting plastic laminate using a straightedge and router

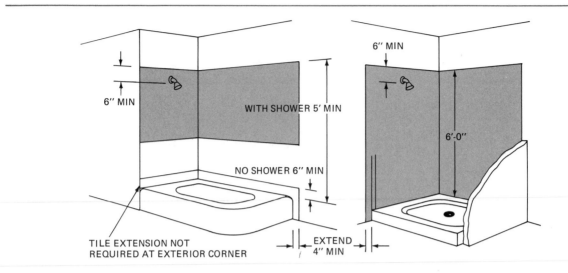

Fig. 24-26 Levels of tile in bath and showers

Tiles are applied down to the top edge of the finished shower floor or tub and overlap the lip. Tiles should be installed to extend to a minimum of 6 inches above the rim of a tub that does not have a showerhead. Over tubs with showerheads, tiles should extend to a minimum of 5 feet above the rim or 6 inches above the showerhead, whichever is higher. In shower stalls, tiles should be a minimum of 6 feet above the shower floor or 6 inches above the showerhead, whichever is higher. A 4-inch minimum extension to the full specified height is required beyond the external face of the tub or shower, Figure 24-26.

BORDER TILES

The first step in applying wall tile is to calculate the border tiles. The border tiles should be the same, and the widest width possible in both corners. Measure the wall from corner to corner. Change the wall measurement to inches and divide it by the width of a tile. Add the width of a tile to the remainder and divide by two to find the width of border tiles.

For example: a room measures 8'-6" or 102 inches. If 4-inch tiles are used, 102" divided by 4" equals 25 with a remainder of 2". Add 4 to the remainder of 2, which equals 6. Divide 6 by 2 to get 3, the width of border tiles in inches.

LAYOUT

Measure out from the corner near the center of the wall and mark a point that will be the edge of a full tile. Mark a plumb line on the wall through this point from the floor to ceiling.

If the floor is level, tiles may be applied by placing the bottom edge of the first row on the floor using plain tile or base tile. If the floor is not level, then base tile cannot be used because the bottom edge of the tiles must be cut to fit the floor while keeping the top edges level.

In this case, place a level on the floor and find the low point. From this point, measure up and mark on the wall the height of a tile. Draw a level line on the wall through the mark. Tack a straightedge on the wall with its top edge to the line. Tiles are then laid to the straightedge. When tiling is completed above, the straightedge is removed, and tiles are cut and fitted to the floor.

APPLICATION

Apply the recommended adhesive to the wall with a trowel. The trowel has grooved edges to allow the recommended amount of adhesive to be applied. Follow the manufacturer's directions for the type of trowel to use and the amount of adhesive to be spread at any one time. Be careful not to spread adhesive beyond the area to be covered. Too heavy a

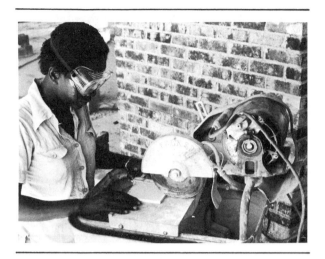

Fig. 24-27 Power tool for cutting ceramic tile

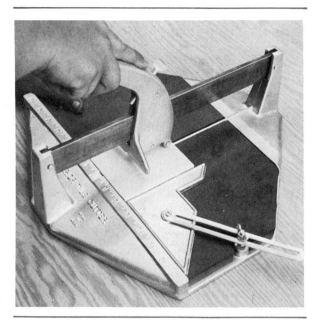

Fig. 24-28 Hand cutter for ceramic tile

coat results in adhesive being squeezed out between the joints of applied tile, causing a mess.

Apply the first tile with its bottom edge to the straightedge or floor, as the case may be, and its vertical edge on the plumb line. Press the tile firmly into the adhesive. Apply other tiles in the same manner, starting from the center plumb line and working toward the corners. As tiles are applied, slight adjustments may need to be made to keep them lined up. Keep fingers clean and adhesive off the face of the tiles. Clean tiles with the solvent recommended by the manufacturer to remove any adhesive from the faces of the tile.

CUTTING BORDER TILES

After all field tiles are applied, it is necessary to cut and apply border tiles. Specialists on large jobs use a power tool for cutting ceramic tile, Figure 24-27. On small jobs, the general carpenter may use a hand-operated ceramic tile cutter, Figure 24-28. This cutter scores and snaps the tile in a way similar to that used in cutting glass. However, this tool cannot cut a small amount from the edge of a tile. For this purpose, a *nibbler,* shown in Figure 24-29, is used. Metal and plastic tiles may be cut with a hacksaw.

To finish the top edges, caps are used. The top edge of the caps are rounded for a finished appearance. Special caps are used at interior and exterior corners.

After all tile has been applied, the joints are filled with tile grout. *Grout* comes in a powder form which is mixed with water to the desired consistency. It is spread over the face of the tile with a rubber trowel to fill the joints. The surface is wiped as clean as possible with the trowel and then the joints are *pointed* (smoothed). A small hardwood stick with a rounded end can be used as a pointing tool. The job is finished by wiping the entire surface with a clean, dry cloth.

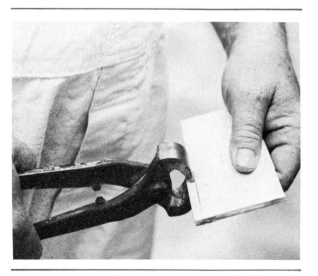

Fig. 24-29 Nibblers are used to cut a small amount from the tile

coverage estimator

The following estimator provides factors for determining the amount of material needed for the five basic types of wood paneling.

Multiply square footage to be covered by factor (length x width x factor).

Nominal Size	WIDTH		Area Factor*
	Dress	Face	
SHIPLAP			
1x6	5 1/2	5 1/8	1.17
1x8	7 1/4	6 7/8	1.16
1x10	9 1/4	8 7/8	1.13
1x12	11 1/4	10 7/8	1.10
TONGUE AND GROOVE			
1x4	3 3/8	3 1/8	1.28
1x6	5 3/8	5 1/8	1.17
1x8	7 1/8	6 7/8	1.16
1x10	9 1/8	8 7/8	1.13
1x12	11 1/8	10 7/8	1.10
S4S			
1x4	3 1/2	3 1/2	1.14
1x6	5 1/2	5 1/2	1.09
1x8	7 1/4	7 1/4	1.10
1x10	9 1/4	9 1/4	1.08
1x12	11 1/4	11 1/4	1.07
PANELING AND SIDING PATTERNS			
1x6	5 7/16	5 1/16	1.19
1x8	7 1/8	6 3/4	1.19
1x10	9 1/8	8 3/4	1.14
1x12	11 1/8	10 3/4	1.12
BEVEL SIDING (1" Lap)			
1x4	3 1/2	3 1/2	1.60
1x6	5 1/2	5 1/2	1.33
1x8	7 1/4	7 1/4	1.28
1x10	9 1/4	9 1/4	1.21
1x12	11 1/4	11 1/4	1.17

*Allowance for trim and waste should be added.

Fig. 24-30 Coverage estimator for board paneling

ESTIMATING SHEET PANELING

To determine the number of sheets of paneling needed, measure the perimeter of the room. Divide this distance by the width of the panels to be used. Deduct from this number any large openings such as doors, windows, or fireplaces. Deduct 2/3 of a panel for a door and 1/2 for a window or fireplace. Round off any remainder to the next highest number.

ESTIMATING BOARD PANELING

Determine the square foot area to be covered. An additional amount, over the area, is needed because of the difference in the nominal size of lumber and its actual size.

Multiply the area to be covered by the area factor shown in Figure 24-30 and add 5 percent for waste in cutting. To cut down waste, order suitable lengths.

ESTIMATING TILE

Determine the square foot area to be covered for the amount of tile to order. Tile usually comes in boxes with a specified number of square feet in each box.

REVIEW QUESTIONS

Select the most appropriate answer.

1. On prefinished plywood paneling that is scored to simulate boards, some scores are always placed in from the edge
   a. 12 and 16 inches.   c. 12 and 24 inches.
   b. 16 and 20 inches.   d. 16 and 32 inches.

2. Tileboard is a type of
   a. tile.   c. fiberboard.
   b. hardboard.   d. plywood.

3. A wainscoting is a wall finish
   a. applied diagonally.
   b. applied partway up the wall.
   c. used as a coating on prefinished wall panels.
   d. used around tubs and showers.

4. For most parts of the country, wood used for interior finish should be dried to a moisture content of about
   a. 8 percent.   c. 15 percent.
   b. 12 percent.   d. 20 percent.

5. The thickest plastic laminate described in this unit is called
   a. vertical type.   c. backer type.
   b. regular type.   d. all-purpose type.

6. The tool recommended to cut thin plywood paneling is the
   a. portable electric circular saw.
   b. fine-tooth hand crosscut saw.
   c. table saw.
   d. saber saw.

7. If wood paneling is to be applied vertically over open studs, wood blocking must be provided between studs for nailing at intervals of
   a. 12 inches.   c. 24 inches.
   b. 16 inches.   d. 32 inches.

8. When applying board paneling vertically to an existing wall
   a. nail paneling to existing studs.
   b. apply horizontal furring strips.
   c. remove the wall covering and install blocking between studs.
   d. use adhesives.

9. An easy and efficient way to cut plastic laminates is by the use of
   a. a table saw.
   b. a portable electric saw.
   c. a router and straightedge.
   d. a handsaw and hand plane.

10. Bathroom tile should extend over the tops of showerheads a minimum of
    a. 4 inches.   c. 8 inches.
    b. 6 inches.   d. 12 inches.

CEILING FINISH

25

OBJECTIVES

After completing this unit, the student will be able to

- *describe the sizes, kinds, and shapes of ceiling tile and lay-in panels.*
- *lay out and install ceiling tile.*
- *lay out and install suspended ceilings.*

Inexpensive and highly attractive ceilings may be created by installing suspended ceilings or ceiling tiles. They may be installed in new construction or in remodeling beneath exposed joists in an unfinished room.

A ceiling that is too high can be replaced simply with a suspended system installed beneath the existing ceiling, Figure 25-1. In addition to improving the appearance, this conserves energy by reducing the amount of space to be heated or cooled and also aids in controlling sound transmission. In basements, where overhead pipes and ducts may make other types of ceiling

Fig. 25-1 An inexpensive and highly attractive suspended ceiling *(Courtesy of Armstrong World Industries)*

application difficult, a suspended type is easily installed. Removable panels make pipes, ducts and wiring accessible.

Ceiling tile may be installed on furring strips fastened to exposed joists. They may also be cemented to existing ceilings provided the ceilings are solid and level. If the existing ceiling is not sound, furring strips may be installed over the existing ceiling and fastened to the joists above.

CEILING TILES AND PANELS

Both tiles and panels are usually made of vegetable or wood fibers. A more fire-resistant type is made of mineral fibers.

Vegetable fiber or wood fiber tiles are pressed into large sheets that are 1/2 to 3/4 inch thick. The sheets are fissured or perforated for acoustical purposes, left plain, or embossed with different designs. Then they are given a factory finish and cut into individual tiles or panels. Mineral fiber tiles are made of rock that is heated to a molten state and then the fibers are sprayed into a light sheet form.

TILES

The most popular sizes of ceiling tile are 12''x12'' and 12''x24'' with thicknesses of 1/2'' and 3/4''. The edges may be tongue-and-grooved for concealed fastening and self-leveling. Other tiles are available with beveled or kerfed edges.

PANELS

Lay-in panels are made of the same type of material as tiles. In addition, they are sometimes made of plastic or glass fibers. They come prefinished in many colors and designs. One special type is coated with a reflective material that gives a mirror effect to the ceiling. The edges may be square, kerfed, or rabbeted, Figure 25-2. The most common sizes are 24''x24'' and 24''x48'' with thicknesses of 1/2'' and 3/4''.

LAYING OUT AND INSTALLING CEILING TILE

Ceiling tiles should be allowed to adjust to normal interior conditions for 24 hours before installation. Some carpenters sprinkle talcum powder on their hands to keep them dry and to avoid fingerprints and smudges on the finished ceiling.

Ceiling tile is sometimes cemented to existing plaster and drywall ceilings. These ceilings must be completely dry, free of dust and dirt, and solid and level. If the existing ceiling is in poor condition or has loose paint, cement application is not recommended.

CALCULATING BORDER TILE SIZES

Before installation begins, it is necessary to calculate the size of the tiles that run along the walls. It is desirable that border tiles be the same and of the widest width possible on opposite walls.

To find the width of the border tiles along the long walls of a room proceed as follows:

- Measure the distance between the opposite long walls.

- In most cases, the measurement will be a number of full feet plus a few inches. Not counting the foot measurement, add 12 more inches to the remaining inch measurement.

- Divide the total number of inches by 2 to find the width of border tiles for the long wall.

Example:

Room length	10' – 6''
Add	12''
Width of border tiles	18'' ÷ 2 = 9''

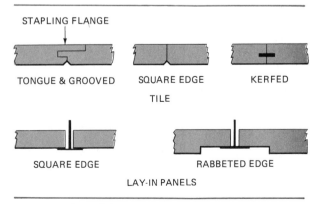

Fig. 25-2 Edge design of ceiling tile and lay-in panels

- Border tiles for the short wall are calculated in the same manner.

Example:

Room length	19' – 8 ''
Add	12''
Width of border tiles	20'' ÷ 2 = 10''

- If 12''x24'' tiles are to be used, the border tile size for the 12'' dimension is done in the same manner as just described, except that the 24'' dimension should be parallel with the long wall.

- To figure the border tile size for the 24'' dimension simply add 24 inches (instead of 12'') to the remaining inch measurement and divide the total number of inches by 2.

Note: **If tiles are to be fastened to furring strips, the 24'' dimension must run parallel to the direction of the furring strips.**

SQUARING THE ROOM

Snap a chalk line on the existing ceiling, parallel to the long wall, as a guide for installing border tiles. Snap the line away from the wall whatever the width of the border tiles was figured to be. If the tiles have a stapling flange add 1/2'' to the measurement. The edge of the stapling flange is laid to the chalk line.

In the example, the long wall border is 9''. If square edge tiles are used, snap the line 9'' away from the wall. If tiles with stapling flanges are used, snap the line 9 1/2 inches away from the wall.

The second chalk line is snapped parallel to the short wall and must be at right angles to the first chalk line or the tiles will not line up properly.

- From the short wall, measure in along the first chalk line the width of the short wall border tiles. Add 1/2 inch if tiles with a stapling flange are used.

- From this point (A) measure along the first chalk line exactly 3 feet and mark point (B).

- Measure parallel to the short wall from point A exactly 4 feet and strike a small arc.

- From point B measure exactly 5 feet toward the arc. Where the 5-foot measurement intersects the arc is point (C).

- Snap a line between and beyond A and C for a guide to install the border tiles along the short wall. With this method the chalk lines are at right angles to each other, as can be seen in Figure 25-3.

INSTALLING TILES

Cut the tiles face up with a sharp utility knife or a fine-toothed handsaw. Cut the first one in the corner so that the outside edge of the flanges line up exactly with both chalk lines. Cut the tile slightly undersize so there is a small gap between it and the wall. The small gap between border tiles and the wall will be covered later by moldings, Figure 25-4.

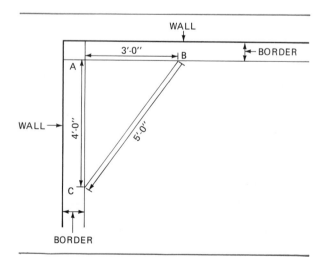

Fig. 25-3 Squaring a ceiling tile layout

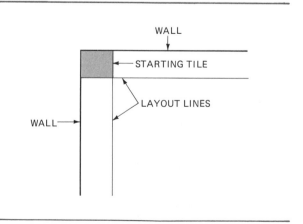

Fig. 25-4 Install the first tile in the corner with its outside edges to the layout lines.

Fig. 25-5 Applying adhesive to the back of ceiling tiles

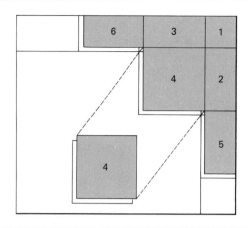

Fig. 25-6 Install a few border tiles, then fill in with full-sized field tiles.

Special tile adhesive is used to cement tiles to existing ceilings. Four daubs of cement are used on 12"x12" tile, Figure 25-5, and six daubs are used on 12"x24" tile.

Before applying the daubs, prime each spot by using the putty knife to force a thin layer into the pores of the back of the tile. Apply the daubs, about the size of a small plum, and press the tile into position. Keep the adhesive away from the edges to allow for spreading when the tile is pressed into place. No staples are required to hold the tile in place.

After the first tile is in place, work across the ceiling, installing two or three border tiles at a time. Then fill in with full-sized field tiles, Figure 25-6. Tiles are applied so they are snug to each other, but not jammed tightly. Continue applying tiles in this manner until the last row is reached.

When the last row is reached, measure and cut each border tile individually. Cut the tiles about 1/4 inch narrower than measured. Do not force the tiles in place. After all tiles are in place, the ceiling is finished by applying molding between the wall and ceiling. The application of wall molding is discussed in a later unit.

APPLYING TILES TO WOOD FURRING

When applying tiles to wood furring, the strips are fastened directly to exposed wood joists or through an existing ceiling into the joists above. They are nailed in the direction that is at right angles to the joists.

If the joists are hidden by an existing ceiling, tap on the ceiling with a hammer until a dull thud is heard. Drive a nail into this spot. Locate other joists by the same method or by measuring. Generally, joists are spaced 16 inches O.C. and usually run parallel to the short dimension of the room. When all joists are located, snap lines on the existing ceiling to indicate the location of the hidden joists for nailing purposes. After locating the joists, calculate the size of the border tiles in the same manner as for cementing tiles.

LAYING OUT FURRING STRIPS

For 12x12 or 12x24 tile, the furring strips must be laid out 12 inches O.C. From the corner, measure out the width of the border tiles plus 1/2 inch for the stapling flange. This measurement is the center of the first furring strip away from the wall. From this mark, measure in either direction, one-half the width of the furring strip; mark and place an *X* on the side of the mark toward the center of the furring strip. From this mark, measure and mark across the room every 12 inches, placing *X*s on the same side of the mark as the first one.

Lay out the other end of the room in the same manner. Snap lines between the marks for the location of the furring strips. The strips are fastened by keeping one edge to the chalk line

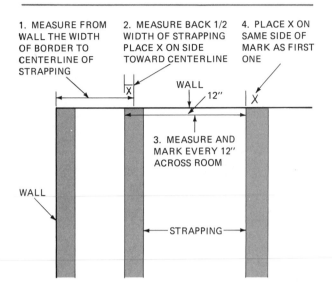

1. MEASURE FROM WALL THE WIDTH OF BORDER TO CENTERLINE OF STRAPPING

2. MEASURE BACK 1/2 WIDTH OF STRAPPING PLACE X ON SIDE TOWARD CENTERLINE

4. PLACE X ON SAME SIDE OF MARK AS FIRST ONE

WALL

12″

3. MEASURE AND MARK EVERY 12″ ACROSS ROOM

WALL

STRAPPING

Fig. 25-7 Furring strip layout for ceiling tile

Fig. 25-8 Installing ceiling tile to wood furring strips below an existing ceiling *(Courtesy of Armstrong Wood Industries)*

with the strip on the side of the line indicated by the *X*, Figure 25-7.

Start by nailing the first furring strip against the wall. Fasten others to the chalk line and finish by nailing the last one against the opposite wall. Use two 8d nails at each exposed joist to keep the strips from warping. Use larger nails through existing ceilings into concealed joists so at least one inch of the nail penetrates into the joist. Butt joints should fall in the center of joists and be staggered so that joints in adjacent rows do not fall on the same joists.

After the furring strips are installed, sight each row by eye for straightness. Use wood shims to straighten the strips when necessary.

SQUARING THE ROOM

Square the room in the same manner as described previously. In this case, the lines for the border tiles are snapped on the furring strips instead of on the existing ceiling.

INSTALLING TILES

Cut and install the first tile in the corner so that the outside edges of the stapling flanges line up exactly with the chalk lines. Work across the ceiling, installing two or three border tiles at a time and filling in between with full-size field tiles, Figure 25-8. Each of the border tiles

should be measured and cut individually because walls are often uneven. Cut each border tile about 1/4 inch less than the actual measurement for easier fitting.

Fasten each tile with four 1/2″ or 9/16″ staples. Place two in each flange using a hand stapler. Use six staples in 12″x24″ tiles.

Measure and cut each tile in the last row individually. Face nail the last tile in place near the wall where the nailhead will be covered by the molding.

LAYING OUT AND INSTALLING SUSPENDED CEILINGS

A *suspended ceiling* consists of panels that are laid into a metal grid system. The *grid system* consists of main runners, cross tees, and wall angles. Members of the grid system come prefinished in a number of colors, wood grains, and patterns.

Main runners are shaped in the form of an upside down T and usually come in 12-foot lengths. They have notches in the side at 6-inch intervals to receive the cross tees. At the ends of the main runners are interlocking devices to join lengths together. Along the top edge, punched holes are spaced at close intervals for hanging the main runner with suspension wire.

Cross tees usually come in 2- or 4-foot lengths and are shaped like main runners. At

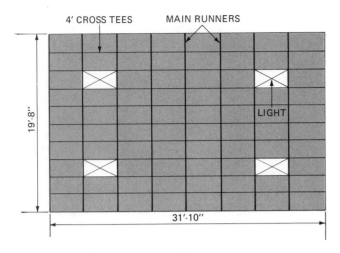

Fig. 25-9 Typical drawing of a suspended ceiling layout

each end of the cross tees are tabs that are inserted and locked into the main runners. Slots are provided in the sides to receive other cross tees.

Wall angles are shaped in the form of an L and come in 12-foot lengths. They are used against the wall to support the ends of main runners, cross tees, and panels.

PLANNING THE ROOM

Measure the dimensions of the room and draw them to scale on paper. Use special care in drawing any odd-shaped walls. Calculate the border panels, and sketch the layout for the planned ceiling. Indicate where any recessed lighting is to be located. Sketching the ceiling layout also helps in estimating materials.

Determine and draw heavier lines to indicate main runners. Main runners should run at right angles to the joists in the room and are spaced 4 feet O.C. Main runners may run parallel to joists, but blocking between the joists from which to suspend the main runners is necessary.

Draw lighter lines to indicate the position of the cross tees. Cross tees may be installed to divide the ceiling into 2′x2′ or 2′x4′ patterns, Figure 25-9.

Space 4-foot cross tees 2 feet apart between main runners. If a 2′x2′ pattern is used, add 2-foot cross tees between the mid points of the

SHORT WALL: 12′ 4″

	2′ x 4′ Panels or	2′ x 2′ Panels
Odd Inches:	4″	4″
Add:	48″	24″
Divide by 2:	$\frac{52″}{2}$ = 26″	$\frac{28″}{2}$ = 14″

26″ is the proper width for **long wall** border panels when using 2′ x 4′ panels and 14″ is the proper width for the long wall border panels when 2′ x 2′ panels are used.

LONG WALL: 14′ x 6″

Odd Inches:	6″
Add:	24″
Divide by 2:	$\frac{30″}{2}$ = 15″

15″ is the proper width for **short wall** border panels.

The 48″ factor is always added to the short wall odd inches; when using 2′ x 4′ panels and a 24″ factor must be used when working with 2′ x 2′ panels. In both cases, the 24″ factor is added to the long wall odd inches.

Fig. 25-10 Calculations for widths of border panels

4-foot cross tees. An alternative is to space main runners 2 feet apart and use 2-foot cross tees throughout. This is a more expensive layout, however, because main runners cost more than cross tees.

CALCULATING BORDER PANELS

As with ceiling tile, it is necessary to determine the proper widths for border panels so that the ceiling will have a well-balanced appearance. A room measuring 12′-4″x14′-6″ requires the calculations shown in Figure 25-10 to arrive at the border panel widths.

INSTALLING WALL ANGLES

To install wall angles, determine the height at which the suspended ceiling is to be placed. Allow at least 3 inches between the new ceiling and existing overhead construction, in order to insert panels in the grid. If recessed lighting is to be used, allow at least 6 inches. Mark the height

Fig. 25-11 Snap chalk lines on the walls
for the location of wall angles.

in each corner of the room. Snap chalk lines between the marks on all walls, Figure 25-11.

Next, nail the wall angle around the room with the bottom side lined up with the chalk line. It may be easier to start the nails in the wall angle by prepunching holes with a center punch or spike. Nail into wall studs wherever possible, placing nails not more than 24 inches apart, Figure 25-12. On masonry walls, use plastic anchors and screws to fasten the wall angles.

At interior corners, the wall angle may be lapped. Miter the wall angle at outside corners. Joints between lengths of wall angle are butted, Figure 25-13. Use tin snips or a hacksaw with a fine-toothed blade. If a hacksaw is used, a temporary wooden miter box can be made to steady the wall angle and to guide the hacksaw. If tin snips are used, lay out the cut using a combination square. Follow the line carefully.

CAUTION: Be careful of the sharp cut ends of the wall angle.

LOCATING AND HANGING SUSPENSION WIRES FOR MAIN RUNNERS

Refer to the sketch of the room previously drawn and border panel calculations for the location of the main runners. Mark their location on opposite walls and snap lines between the marks on the existing joists or ceiling. This line serves as a guide for installing screw eyes to attach suspension wires. Install screw eyes not over 4 feet apart along the main runner chalk lines. Screw eyes should be long enough to penetrate joists a minimum of 1 inch. Stretch a tight line across the room from the top edges of

Fig. 25-12 Installing wall angle
(Courtesy of Armstrong World Industries)

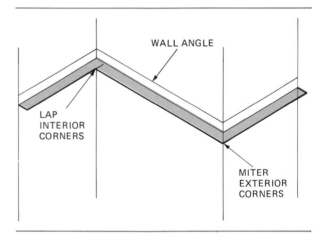

Fig. 25-13 Treatment of wall angle at
interior and exterior corners

the wall angle between the marks indicating the position of the first main runner.

Cut a number of suspension wires to the proper length using wire cutters, Figure 25-14. Suspension wires should be about 12 inches longer than the distance between the overhead construction and the stretched line.

Attach the suspension wires to the screw eyes. Pull on each wire to remove any kinks and

Fig. 25-14 Cutting suspension wire to length with wire cutters

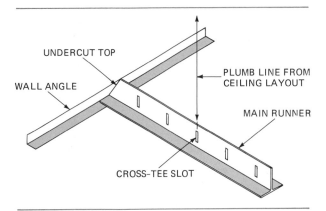

Fig. 25-15 A cross tee slot in the main runner must line up with the layout line.

then make a 90° bend where the suspension wire crosses the stretched line. Stretch the lines, then attach and bend the suspension wires at each main runner location.

INSTALLING MAIN RUNNERS

When ready to install main runners, refer to the sketch and calculations for the width of the short wall border panels (Figure 25-3). Measure out, from the short wall along the stretched line for the main runner, a distance equal to the width of the border panel (point *A*). Mark the line with a string or by pushing a small brad or pin through it. Measure from this point 3 feet and mark the line again (point *B*).

Stretch another line at right angles to the main runner line through point *A*. Measure 4 feet along this line and mark point *C*. Move the ends of this line until the distance between points *B* and *C* equals 5 feet with the two lines intersecting at point *A*.

This method is similar to squaring tile layout except that the measurements are laid out on stretched lines. An alternate method is to lay out the lines on the floor, then plumb the end points up to the ceiling. However, lines can only be laid out on the subfloor, not on the finish floor.

The ends of the main runners must be cut so that one of the slots for the cross tees lines up with the stretched line parallel to the short wall, Figure 25-15. Otherwise, the entire grid system will be out of square.

At each main runner location, measure from the short wall to the stretched line. Trans-

Fig. 25-16 Hanging the main runner *(Courtesy of Armstrong World Industries)*

fer the measurement to the main runner, measuring from a cross tee slot so as to cut as little as possible from the end of the main runner. Cut the main runners about 1/8 inch less to allow for the thickness of the wall angle. Undercut the top slightly for easier installation at the wall.

Hang the main runners by resting the cut end on the wall angle and inserting suspension wires in the appropriate holes in the top of the main runner. Bring the runners up to the bend in the wires. Twist the wires with enough turns to hold the main runners securely, Figure 25-16. More than one section may be needed to

Fig. 25-17 Inserting a cross tee in the main runner
(Courtesy of Armstrong World Industries)

reach the opposite wall. Connect lengths of main runners together by inserting tabs into matching ends. Make sure end joints come up tight. The length of the last section is measured from the end of the last one installed to the opposite wall.

INSTALLING CROSS TEES

Cross tees are installed by inserting the tabs on the ends into the slots in the main runners. These fit into position easily although the method of attaching varies from one manufacturer to another.

Install all full-length cross tees between main runners first. When installed, they effectively straighten the grid. After all full-length cross tees are installed, cut and install cross tees along the border by inserting one end in the main runner and resting the other end on the wall angle, Figure 25-17. After the grid is complete, sight sections by eye and straighten where necessary.

Fig. 25-18 Installing ceiling panels *(Courtesy of Armstrong World Industries)*

INSTALLING CEILING PANELS

Ceiling panels are placed in position by tilting them slightly, lifting them above the grid, and letting them fall into place, Figure 25-18. Be careful when handling panels to avoid marring the finished surface.

Install all full-sized field panels first and then cut and install border panels. Measure each border panel individually and cut them slightly smaller than measured so they can drop into place easily. Cut the panels with a sharp utility knife using a straightedge as a guide. A scrap piece of cross tee material can be used as a straightedge. Always cut with the finished side of the panel up.

CONCEALED GRID SUSPENDED CEILINGS

In cases where it is not desired to expose the grid, a concealed grid system is used. Al-though systems vary with the manufacturer, generally the cross tees are attached below the main runners. Kerfed panels are used, with the kerf inserted into the edges of the cross tees. When complete, all panel or tile edges butt each other completely concealing the grid, Figure 25-19. The only exposed part of the grid is the wall angle.

Locate ceiling joists, install screw eyes and suspension wires, and hang the main runners in the same manner as for exposed grids. Calculate the border sizes and square the room by stretching two lines at right angles to each other.

INSTALLING THE FIRST ROW

To install the first row, cut each border tile carefully to fit so the edges away from the wall are to the stretched lines that square the room. Measure and cut the first tile so that the outside edges line up exactly with the two layout lines and the other edges rest snugly against the wall.

Fig. 25-19 Installing tile in a concealed grid system. Notice that cross tees are attached below main runners. *(Courtesy of Armstrong World Industries)*

After the first four border tiles are in position, snap a cross tee onto the main runner and slide the cross tee into the kerf of the four tiles. Install the remainder of the first row, four or five tiles at a time in the same manner, with the exception of the last tile. Cut the last tile in each row 3/8 inch short of the sidewall. Slide a wall spring between the cut edge and the wall. The spring keeps the row from shifting. Install the last cross tee in each row so that it falls no more than one inch short of the wall.

Continue applying rows of tile in this manner until the last row is reached. To simplify installing the last row, cut the tiles 3/8 inch short from the end wall. To keep the tiles in the end row tight, insert a wall spring between each tile and the end wall.

Because wall springs cannot be used on the last tile, place several drops of white glue on a flange before installing the last tile. Tilt the tile up through the opening and slide it into place.

ESTIMATING MATERIALS

To estimate ceiling tile, measure the width and length of the room to the next whole foot measurement. Multiply these figures together to find the area of the ceiling and the number of square feet of ceiling tile needed.

To estimate furring strips, measure to the next whole foot the length of the room in the direction that the furring strips are to run. Multiply this by the number of rows of furring strips to find the total number of linear feet of furring strip stock needed. To find the number of rows, divide the width of the room by the furring strip spacing and add one.

To estimate suspended ceilings, calculate from the sketch the number of ceiling panels, the length and number of cross tees, the number of linear feet of main runners, and the number of wall angles needed.

REVIEW QUESTIONS

Select the most appropriate answer.

1. The most common sizes in inches of ceiling tile are
   a. 8x12 and 12x12.   c. 12x12 and 12x24.
   b. 12x12 and 16x16.   d. 24x24 and 24x48.

2. The most common sizes in inches of lay-in ceiling panels are
   a. 8x12 and 12x12.   c. 12x12 and 12x24.
   b. 12x12 and 16x16.   d. 24x24 and 24x48.

3. The width in inches of border tiles for a room 14'-9''x22'-3'' is
   a. 9 and 3.   c. 10 1/2 and 7 1/2.
   b. 8 1/2 and 6 1/2.   d. 11 and 8.

4. The width in inches of 24''x24'' border tiles for a room 12'-6''x18'-8'' is
   a. 6 and 8.   c. 14 1/2 and 18 1/2.
   b. 10 and 12.   d. 15 and 16.

5. When squaring a room, use measurements of
   a. 2, 3, and 4 feet.   c. 4, 5, and 6 feet.
   b. 3, 4, and 5 feet.   d. 5, 6, and 7 feet.

6. To locate joists above an existing ceiling
   a. determine the direction of the finished floor above.
   b. tap on the ceiling with a hammer until a dull thud is heard.
   c. look for lines of discoloration on the existing ceiling.
   d. determine the location of studs in the wall below.

7. Nails used to fasten furring to ceiling joists must penetrate into the joists at least
   a. 3/4 inch.   c. 1 1/4 inches.
   b. 1 inch.   d. 1 1/2 inches.

8. Fastening 12″x12″ tiles requires
   a. two staples.   c. four staples.
   b. three staples.   d. six staples.

9. In a suspended ceiling, suspension wire is used to hang
   a. cross tees.   c. wall angle.
   b. main runners.   d. furring strips.

10. The first step in installing a suspended ceiling is to
    a. square the room.
    b. make a sketch of the planned ceiling.
    c. calculate border tiles.
    d. install the wall angle.

INTERIOR DOORS AND DOOR FRAMES 26

OBJECTIVES

After completing this unit, the student will be able to

- *describe the sizes and kinds of interior doors and the methods of operation.*
- *make and set an interior door frame.*
- *hang a swinging interior door.*
- *install locksets on interior doors.*
- *set a prehung door and frame.*
- *install by-passing, folding, and pocket doors.*

Interior doors may be classified as *paneled, sash, flush, louvered,* and *wardrobe* (bifolding) doors. All of these doors except bifold doors are generally manufactured in 1 3/8-inch thickness for interior use. Usually bifold doors are 1 1/8 inches thick.

Most doors are manufactured in 6'-8'' heights, although louvered doors may be obtained in heights of 6'-0'' and 6'-6''. Door widths range from 1'-0'' to 3'-0'' in increments of 2 inches. However, not all sizes are available. Available sizes depend on the kind and style of door, Figure 26-1. Refer to manufacturers' catalogs for available sizes.

PANELED DOORS

Paneled doors consist of a frame with panels set between the members of the frame. Some of the panels may be glass. Names of the door members have been previously discussed in Unit 21 Exterior Doors.

SASH DOORS

Sash doors, commonly called *French doors,* contain many small panels, all of which are glass. They are constructed in about the same manner as doors with wood panels.

FLUSH DOORS

Flush doors are made with solid cores of wood or particleboard, or they may be made with hollow core construction. Hollow core

1-3/8 INTERIOR PANEL DOORS

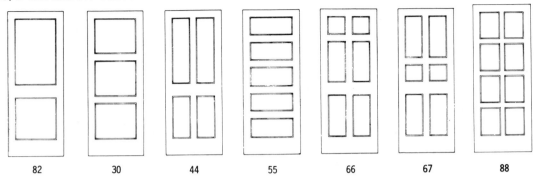

82   30   44   55   66   67   88

SPECIFICATIONS/INT. PAN. DOORS	Width O/A
Stiles and Top Rail (Min.)	4-1/2″
Intermediate Rails	*4-1/2″
Mullions	*4-1/2″
Lock Rail	7-1/2″
Bottom Rail	9-1/4″
Panels—3-Ply Flat	(1/4″ thick)
Raised	(1/2″ thick)

*Except Design 88 shall be 3-1/4″

1-3/8 LOUVER DOORS (INTERIOR)

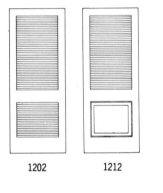

Sticking

1202   1212

Slat and Rout

SPECIFICATIONS/LOUVER DOORS		
FLAT SLAT TYPE		Width O/A
Stiles	1-11/16″—2-9/16″—3-9/16″	
Top Rail	4-1/2″	
Lock Rail and Mullion	4-1/2″	
Bottom Rail	9-1/4″	
Panels—Raised	(1/2″ thick)	
Slats Approx. 1/4″x1-3/8″ (See Detail)		

1-1/8 BI-FOLD WARDROBE DOORS (INTERIOR)

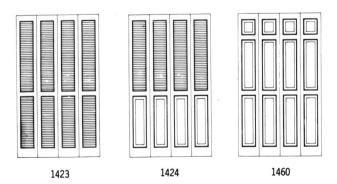

1423   1424   1460

SPECIFICATIONS	Width O/A
Stiles	1-5/16″ to 2-1/16″
Top Rail	3-1/4″
Lock Rail	2-1/4″
Bottom Rail	3-3/4″ to 5-1/8″
Slats Approx.	1/4″ to 1-3/8″
Panels—Raised	(1/2″ thick)

Fig. 26-1 Styles and specifications of commonly used interior doors *(Courtesy of Fir & Hemlock Door Association)*

doors contain a mesh of thin wood or composition material to support the faces of the door. The core is surrounded by a light frame of wood. Solid blocks are placed in the core for the installation of locks. The core is covered with face panels called *skins,* which are generally of 1/8-inch plywood. *Lauan* plywood is used extensively for flush door skins. However, many flush doors are available with faces of birch, gum, oak, mahogany, and other kinds of plywood. Some flush doors are faced with hardboard or plastic laminates, Figure 26-2. Hollow core doors are usually used in the interior. Solid core doors are used on the exterior.

MESH OR CELLULAR CORE
7 PLY CONSTRUCTION ILLUSTRATED

TOP AND BOTTOM RAILS
MINIMUM 2-1/4 INCHES (57.2 mm)

WIDTH OF STILES
MINIMUM 1 INCH (925.4 mm)

WOOD OR WOOD
DERIVATIVE STRIPS

LOCK BLOCKS

FACE VENEER

LADDER CORE
7 PLY CONSTRUCTION ILLUSTRATED

TOP AND BOTTOM RAILS
MINIMUM 2-1/4 INCHES

WIDTH OF STILES
MINIMUM 1 INCH

WOOD OR WOOD
DERIVATIVE STRIPS

LOCK BLOCKS

FACE VENEER

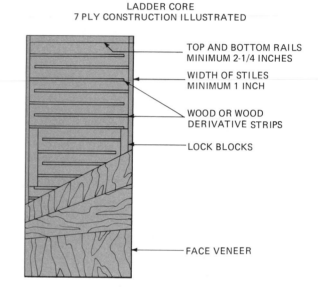

MAT-FORMED COMPOSITION BOARD CORE
7 PLY CONSTRUCTION ILLUSTRATED

TOP AND BOTTOM RAILS
MINIMUM 1 INCH

WIDTH OF STILES
MINIMUM 1 INCH

CORE OF MAT-FORMED
COMPOSITION BOARD

FACE VENEER

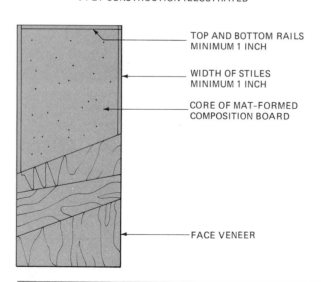

FRAMED BLOCK NON-GLUED CORE
5 PLY CONSTRUCTION ILLUSTRATED

TOP AND BOTTOM RAILS
MINIMUM 1 INCH

WIDTH OF STILES
MINIMUM 1 INCH

WOOD CORE BLOCKS

FACE VENEER

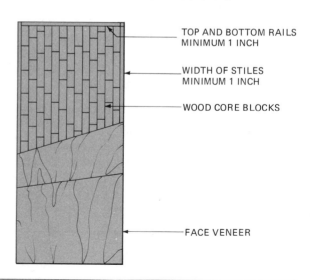

Fig. 26-2 Flush door construction *(Courtesy of National Woodwork Manufacturers Association)*

460

LOUVERED DOORS

Louvered doors are panel doors with spaced horizontal slats called louvers, used in place of panels. The louvers are installed at an angle in such a manner as to obstruct vision but to permit the flow of air through the door. Louvered doors are widely used on clothes closets.

WARDROBE DOORS

Wardrobe doors may be paneled, flush, louvered, or a combination of kinds. They are usually made thinner and narrower than other doors. They are widely used on closets and ordinarily operate in a folding fashion.

METHODS OF OPERATING DOORS

Door operation may be classified as swinging, sliding, or folding. Several types of doors perform these operations.

SWINGING DOORS

Swinging doors are hinged on one edge and swing out of the opening. When closed, they cover the total opening. Swinging doors may swing in one direction (single acting). With special hinges, they may swing in both directions (double acting), Figure 26-3. Swinging doors are the most commonly used type of door; however, the disadvantage is that when they are opened, the doors project into the room.

SLIDING DOORS

There are two kinds of sliding door operations. One type is the bypass door and the other is the pocket door.

The *bypass operation* requires two or more doors that slide on a track mounted on the header jamb of the door frame, Figure 26-4. Rollers are mounted on the doors and ride in the track so that the doors slide by each other.

The disadvantage of bypass doors is, although they do not project out into the room,

Fig. 26-3 Swinging door

Fig. 26-4 Bypassing doors

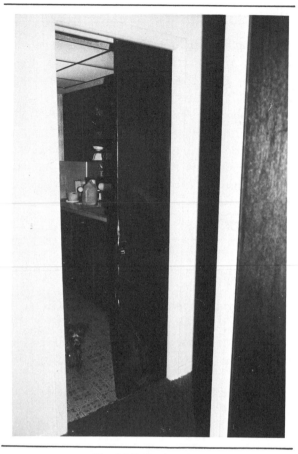

Fig. 26-5 Pocket door

Fig. 26-6 A pocket door frame comes pre-assembled from the factory.

access to the total opening at one time is not possible. However, they are easy to install and are used frequently as wardrobe doors.

The *pocket door* slides in a track at the top into an opening in the partition. When opened, only the lock edge of the door is visible, Figure 26-5. However, pocket door units are more expensive and require more time to install than others. A special pocket door frame unit and track must be installed during the rough framing stage, Figure 26-6. The rough opening in the frame must be large enough to include the finished door opening and the pocket. Pocket doors may be installed as a single unit, sliding in one direction, or as a double unit sliding in opposite directions. The distinct advantages of this type of door operation are that the door does not project into the room and total access to the opening is obtained when the door is opened. Pocket doors are used when these advantages are desired.

BIFOLD DOORS

Bifold doors consist of two or more doors hinged at their edges. The door at the jamb side has pivots installed top and bottom so that it can swing. Other doors fold up against it as it is swung open. The end door has a guide installed at the top that rides in a track to guide the set when opening or closing, Figure 26-7. On very wide openings the guide also acts as a support for the set to keep it from sagging.

Bifolding doors may be installed in double sets, opening and closing from each side. When opened, folding doors do not project too much into the room and almost total access to the opening is obtained. Bifolding doors are frequently installed on closet openings.

MAKING AN INTERIOR DOOR FRAME

The first step in making an interior door frame is to determine if the rough opening size is

Fig. 26-7 Bifold doors

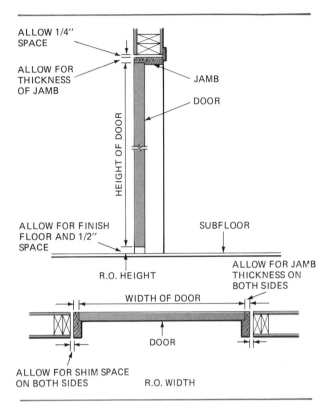

Fig. 26-8 Determining the rough opening size for swinging doors

correct. Measure the door opening to make sure it is the correct width and height. The rough frame width for swinging doors should be the width of the door plus twice the thickness of the side jambs of the door frame plus 1/2 inch for shimming between the door frame and the opening. If the thickness of the door frame stock is 3/4 inch, for example, the rough opening width is 2 inches over the door width.

The rough opening height for swinging doors should be the height of the door, plus the thickness of the header jamb, plus 1/4 inch for clearance at the top, plus the thickness of the finished floor, plus a desired clearance under the door. An allowance of 1/2 inch is usually made for clearance between the finished floor and the door.

For example, if the header jamb and finished floor thickness are both 3/4 inch, the rough opening height should be 2 1/4 inches over the door height, Figure 26-8.

The rough opening size for other than swinging doors, such as sliding and folding doors, should be checked against the manufacturer's installation directions because allowances for hardware may differ with the manufacturer.

CUTTING JAMBS TO WIDTH

Next, measure the finished wall thickness. Door frames are installed after the walls are finished. Cut the door jamb stock to the required width. If rabbeted jamb stock is used, cut the edge opposite the rabbet. Plane and smooth both edges to a slight back bevel. The back bevel permits the door casings, when later applied, to fit tightly against the edges of the door frame in case there are irregularities in the wall, Figure 26-9. Slightly round over all sharp exposed corners.

CUTTING JAMBS TO LENGTH

Cut the side jambs to length a little less than the height of the opening. Cut the header jamb to length. If rabbeted jambs are used, the

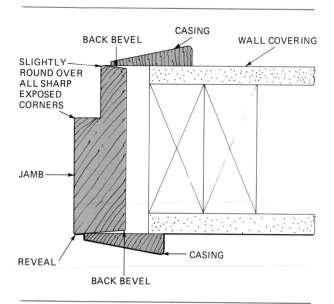

Fig. 26-9 Back bevel jamb edges slightly to permit casings to fit snugly against them.

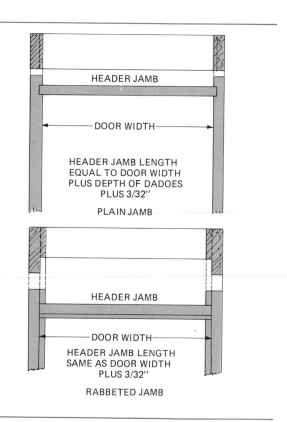

Fig. 26-10 Length of header jambs for plain and rabbeted door frames

length of the header jamb is the width of the door plus 3/32 inch. The extra 3/32 inch is for clearance between the door edges and the side jambs. If plain jambs with applied stops are used, add 1/2 inch for dadoing the header jambs into the side jambs, plus 3/32 inch for clearance, Figure 26-10.

Measure from the bottom ends of the side jambs and mark the location of the face of the header jamb. On plain jambs this would be the height of the door plus 1 1/4 inches. On rabbeted jambs the measurement is 1/2 inch less because of the depth of the rabbet. Square lines across the side jambs at this point. Hold a scrap piece of jamb stock to the squared lines. Mark the depth of the dadoes and cut them to receive the header jamb. On rabbeted jambs, dado depth is to the face of the rabbet. Dado 1/4 inch deep on plain jambs.

ASSEMBLING THE DOOR FRAME

Fasten the side jambs to the header jamb, keeping the edges flush. If there is play in the dado, first wedge the header jamb with a chisel so the face side comes up tight against the dado shoulders before fastening. Cut a narrow strip of wood and tack it to the side jambs a few inches up from the bottom. Tack the strip so that the frame width is the same at the bottom

as it is at the top. Trim off any overhanging ends of the strip. Cut the horns off the top ends of the side jambs, Figure 26-11.

SETTING A DOOR FRAME

A door frame must be set so that the jambs are straight, level, and plumb. They are usually set before the finish floor is laid.

If a rabbeted frame is used, determine the swing of the door so that the rabbet is placed on the correct side. Place the frame in the opening. Wedge the frame in place with wood shingles placed directly opposite the ends of the header jamb. Wedge an equal amount on both sides. The bottom ends of the side jambs should rest on the floor. Keep the edges of the frame flush with the wall.

Level the header jamb by placing shims between the bottom end of the side jambs and the subfloor, if necessary. Tack the frame directly opposite the top wedges, close to the face of the header jamb.

Fig. 26-11 Cutting a horn from the top
end of the side jamb

PLUMBING THE SIDE JAMBS

An accurate and fast method of plumbing side jambs is by use of a *plumb bob.* Measure out from the side jamb along the header jamb a distance of 12 inches. Tack a small finish nail into the edge of the header jamb at this point. Drive the nail to one side so that a line suspended from it falls through the mark. Suspend a plumb bob from the nail just clear of the floor, Figure 16-12. When the plumb bob comes to rest, mark the floor in line with the point of the plumb bob.

Measure 12 inches back from the mark on the floor. Wedge the bottom end of the side jamb to the mark and tack. Wedge the other side so the opening width is the same at the bottom as on the top, and tack in position. Remove the spreader.

Side jambs may also be plumbed by using a 6-foot level or a carpenter's level and straightedge. (This method was described in an earlier unit.)

Sight the door frame and remove any wind. A wind is a twist. The frame must be sighted by eye to make sure that side jambs line up vertically with each other and that the frame is not twisted. This is important when installing rabbeted jambs. Stand to one side of the opening and sight through it. Sight along the edge of the closest side jamb to the opposite edge of the other side jamb to see if they line

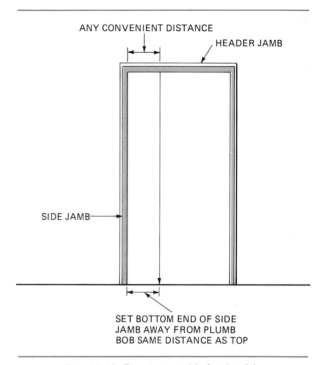

ANY CONVENIENT DISTANCE

HEADER JAMB

SIDE JAMB

SET BOTTOM END OF SIDE
JAMB AWAY FROM PLUMB
BOB SAME DISTANCE AS TOP

Fig. 26-12 Plumbing a side jamb with
the use of a plumb bob

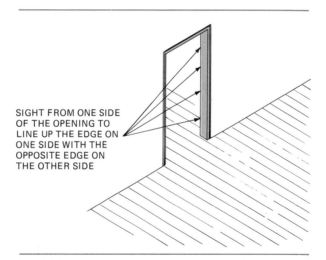

SIGHT FROM ONE SIDE
OF THE OPENING TO
LINE UP THE EDGE ON
ONE SIDE WITH THE
OPPOSITE EDGE ON
THE OTHER SIDE

Fig. 26-13 Sighting a door frame to determine any wind

up, Figure 26-13. If the edges are not in line, move the top or bottom ends of the side jambs until they are. Keep the edges of the jambs as flush with the wall as possible. Fasten top and bottom ends of the side jambs securely. Set all nails.

STRAIGHTENING SIDE JAMBS

Use a 6-foot straightedge against the side jambs and straighten them by shimming at

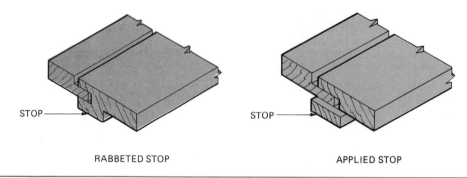

STOP —   STOP —

RABBETED STOP   APPLIED STOP

Fig. 26-14 Some types of prehung door units have a split jamb.

intermediate points. Besides other points, shims should be placed opposite hinge and lock locations. Fasten the jambs opposite and through the shims. Door stops on plain jambs are usually applied after installation of lock sets. A temporary stop is applied at this time on plain jambs.

HANGING AN INTERIOR DOOR

Fitting and hanging interior doors is done in the same way as for exterior doors. See Unit 21 Exterior Doors for the procedure.

SETTING A PREHUNG DOOR

A prehung door unit consists of a door frame with the door and casings already installed. Lock hardware may or may not be installed, but holes are bored for their installation. Shims are stapled to the lock edge and top end of the door to maintain proper clearance between the door and frame. Several frame widths are available to accommodate various wall thicknesses. One type has a split jamb which is adjustable for varying wall thicknesses, Figure 26-14.

A prehung door unit can be set in a matter of minutes. Remove the casings from one side carefully so that you will not split them. Center the unit in the opening. Plumb, level, and fasten the unit by driving nails through the casings, keeping the jambs against the shims between the door for proper clearance.

Some carpenters prefer to install shims between the opening and the frame on the other side. Install wedges at intermediate points

keeping side jambs straight. Nail through the side jambs opposite the wedges. Replace the removed casings.

INSTALLING LOCKSETS

Locksets are installed on interior doors in the same manner as for exterior doors and as described in Unit 21 Exterior Doors. Although their installation is basically the same, some locks are used exclusively on interior doors.

The *privacy* lock is used often on bathroom and bedroom doors. It is locked by pushing or turning a button on the room side, Figure 26-15. A turn of the knob on the room side unlocks the door. On the opposite side, the door can be unlocked only by a pin or key inserted into a hole in the knob. Another type, called a *passage* lockset, has knobs on both sides that are always free to turn and are used where privacy is not a consideration.

Fig. 26-15 A privacy lock on a bathroom door

Fig. 26-16 Bypassing door hangers have an offset. Make sure hangers with the same offset are used on the same door.

Fig. 26-17 Bypassing doors must have flush pulls.

INSTALLING BYPASS DOORS

Bypass doors are installed so they overlap each other by about one inch when closed. Cut the track to length and install it on the header jamb according to the manufacturer's direction.

Install pairs of rollers on each door. The roller hangers may be offset a different amount for each door, according to the door thickness and its position on the track. Make sure that hangers with the same and correct offset are used on the same door, Figure 26-16.

Mark the location and bore holes for door pulls. Bypassing doors have flush pulls, Figure 26-17. The proper size hole is bored partway into the door and the pull tapped into place with a hammer and wood block. The press fit holds it in place. Other kinds of flush pulls are held in place with small screws.

Hang the doors by holding the bottom outward and inserting the rollers in the track. Then gently let the door come to a vertical position. Install the inside door first, then the outside door, Figure 26-18.

Test the door operation and the fit against side jambs. Lower or raise the door as necessary by adjusting one or both roller hangers to make edges fit against side jambs.

Fig. 26-18 Installing bypassing doors

467

Push the doors to one side of the opening and mark the location of the floor guide. The doors ride in the floor guide to prevent them from swaying. The guide is placed in the center of the opening directly below the centerline of the track. Remove the outside door and install part of the guide. Replace the outside door and install the rest of the guide.

INSTALLING BIFOLD DOORS

When installing bifold doors, check the opening for level, plumb, and size specified by the hardware manufacturer. Cut the track to length and fasten it to the header jamb with screws provided in the kit. The track contains adjustable sockets for the door pivot pins. Make sure these are in the track before fastening the track in position. The position of the track on the header jamb is usually optional depending on the way trim is applied, Figure 26-19, or the desired position of the doors. Fasten pivot sockets to the bottom of the side jambs directly in line with the pivot sockets in the track above.

In most cases, bifold doors come hinged together with prebored holes for the pivots and guide. Install pivots in the prebored holes in the top and bottom end of the door closest to the jamb. Note that the *top* pivot is spring loaded and can be depressed. The *bottom* pivot is threaded and can be adjusted for height. The *guide* is installed in the hole provided at the top end of the door farthest away from the jamb. The guide is also spring-loaded and can be depressed with the finger.

Loosen the top pivot socket and slide it along the track toward the center of the opening

and about one foot away from the side jamb. Place the doors in position by inserting the bottom pivot pin in the bottom pivot socket. Tilt the doors to an upright position and at the same time insert the top pivot pin in the top socket while sliding the socket toward the jamb. Adjust both top and bottom pivot sockets so the desired joint is obtained between the door and the jamb. Depress the guide and insert in the track. Support-type guides snap into place in special hardware. Lock top and bottom pivot sockets in position. Try the operation of the door. Adjust the bottom pivot pin to raise or lower the doors, if necessary.

If more than one set of bifold doors are to be installed in an opening, install the other set on the opposite side in the same manner. Install knobs in the manner and location recommended by the manufacturer.

Where sets meet at the middle of an opening, door aligners are installed near the bottom on each of the doors at the centerline. The door aligning hardware keeps the faces of the center doors lined up when closed.

INSTALLING POCKET DOORS

As stated before, the pocket door frame, complete with track, is installed in the rough framing stage. The pocket consists of two ladder-like frames between which the door slides. A steel channel at the bottom is fastened to the floor and keeps the sides of the pocket frame spread the proper distance apart.

The frame, which is usually preassembled at the factory, is made of nominal 1-inch stock. The pocket is covered by the interior wall finish. Care must be taken when covering the pocket frame not to use fasteners that are so long that they penetrate the frame. If fasteners penetrate the pocket deeply, the door will not slide in the pocket. If fasteners penetrate the frame slightly, the door may slide in the pocket, but the fasteners will scratch the side of the door as it is operated. Pocket door frames are set in a manner similar to that used with other door frames.

Attach rollers to the top of the door in the location specified by the manufacturer. Install

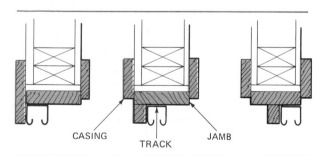

CASING   TRACK   JAMB

Fig. 26-19 The position of the overhead track depends on the way trim is applied.

pulls on the door. On pocket doors an edge pull is necessary in addition to recessed pulls on the sides of the door. A special pocket door pull contains edge and side pulls and is mortised in the edge of the door. In most cases, all the necessary hardware is supplied when the pocket door frame is purchased.

Engage the rollers in the track by holding the bottom of the door outward in a way similar to that used with bypass doors. Adjust the rollers for the correct door height. Test the operation of the door to make sure it slides easily. Install stops around the opening for the door to slide between, Figure 26-20.

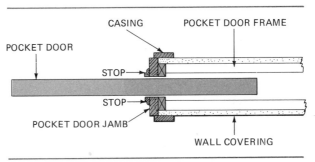

Fig. 26-20 Pocket door details

REVIEW QUESTIONS

Select the most appropriate answer.

1. Most interior doors are manufactured in a thickness of
   a. 1 inch.
   b. 1 3/8 inches.
   c. 1 1/2 inches.
   d. 1 3/4 inches.

2. The height of most interior doors is
   a. 6'-0''.
   b. 6'-6''.
   c. 6'-8''.
   d. 7'-0''.

3. Interior door widths range from
   a. 1'-6'' to 2'-6''.
   b. 2'-2'' to 2'-8''.
   c. 2'-6'' to 2'-8''.
   d. 1'-0'' to 3'-0''.

4. Used extensively for flush door skins is
   a. fir plywood.
   b. lauan plywood.
   c. metal.
   d. plastic laminate.

5. Wardrobe doors are usually made in a thickness of
   a. 1 inch.
   b. 1 1/8 inches.
   c. 1 3/8 inches.
   d. 1 3/4 inches.

6. A disadvantage of bypassing doors is that they
   a. project out into the room.
   b. cost more and require more time to install.
   c. are difficult to operate.
   d. do not provide total access to the opening.

7. If the jamb stock is 3/4 inch thick, the rough opening width for a swinging door should be the door width plus
   a. 3/4".   c. 2".
   b. 1 1/2".   d. 2 1/2".

8. If the jamb stock and the finished floor are both 3/4" thick, the rough opening height for a 6'-8" swinging door should be
   a. 7'-0".   c. 6'-10 1/4".
   b. 6'-11 1/2".   d. 6'-9 1/4".

9. When a rabbeted door frame is made for a swinging door, the header jamb length is the width of the door plus
   a. 3/32".   c. 3/8".
   b. 3/16".   d. 1/2".

10. An accurate and fast method of plumbing side jambs of a door frame is by the use of a
    a. builder's level.
    b. carpenter's 26-inch hand level.
    c. plumb bob.
    d. straightedge.

INTERIOR TRIM 27

OBJECTIVES

After completing this unit, the student will be able to

- *identify standard molding patterns and describe their use.*
- *apply ceiling and wall molding.*
- *apply interior door casings, baseboard, base cap, and base shoe.*
- *install window trim including stools, aprons, jamb extensions, casings, and stop beads.*

install closet shelves and a closet pole.

install a mantel.

The application of specially shaped pieces, called moldings, to cover joints at the intersection of walls, ceilings, floors, around window and door openings, and other places is an important part of the interior trim. These pieces must be cut and fitted accurately with a high level of craftsmanship to present a good appearance.

STANDARD MOLDING PATTERNS

Standard moldings used for interior trim are classified as beds, crowns, coves, full rounds, half rounds, quarter rounds, base, base shoes, base caps, corner molds, back bands, aprons, ply caps, casings, stools, stops, and others. There are a number of sizes and shapes of each kind, Figure 27-1. Molding comes in standard lengths of 8, 10, 12, 14, and 16 feet. Some moldings are available in odd lengths. Door casings, in particular, come in lengths of 7 feet to reduce waste. *Finger-jointed* lengths are made of short pieces joined together with factory-made finger joints. These are used only when a paint finish is to be applied because the joints show through a stained or natural finish.

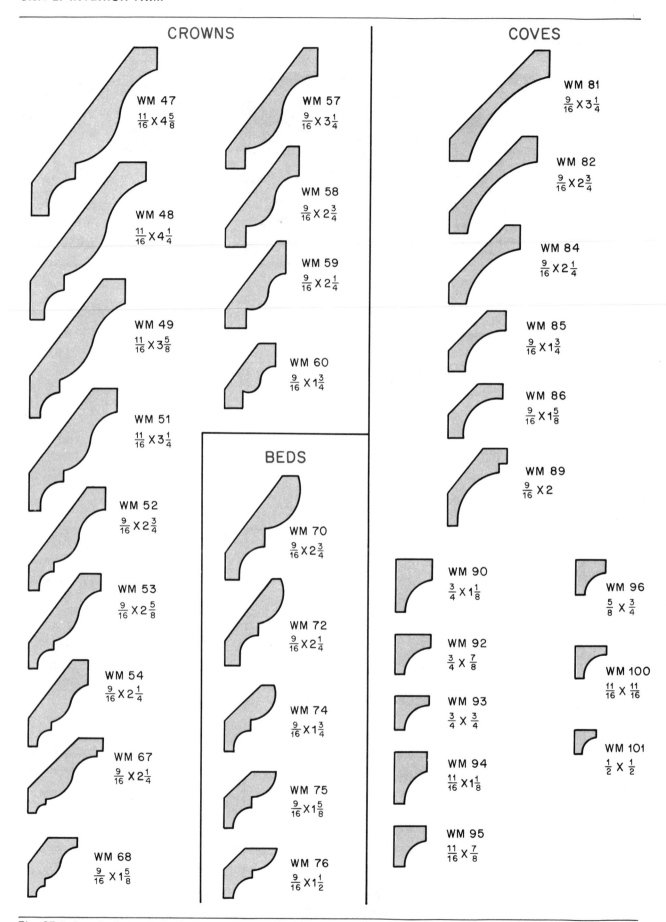

CROWNS

WM 47
$\frac{11}{16} \times 4\frac{5}{8}$

WM 48
$\frac{11}{16} \times 4\frac{1}{4}$

WM 49
$\frac{11}{16} \times 3\frac{5}{8}$

WM 51
$\frac{11}{16} \times 3\frac{1}{4}$

WM 52
$\frac{9}{16} \times 2\frac{3}{4}$

WM 53
$\frac{9}{16} \times 2\frac{5}{8}$

WM 54
$\frac{9}{16} \times 2\frac{1}{4}$

WM 67
$\frac{9}{16} \times 2\frac{1}{4}$

WM 68
$\frac{9}{16} \times 1\frac{5}{8}$

WM 57
$\frac{9}{16} \times 3\frac{1}{4}$

WM 58
$\frac{9}{16} \times 2\frac{3}{4}$

WM 59
$\frac{9}{16} \times 2\frac{1}{4}$

WM 60
$\frac{9}{16} \times 1\frac{3}{4}$

BEDS

WM 70
$\frac{9}{16} \times 2\frac{3}{4}$

WM 72
$\frac{9}{16} \times 2\frac{1}{4}$

WM 74
$\frac{9}{16} \times 1\frac{3}{4}$

WM 75
$\frac{9}{16} \times 1\frac{5}{8}$

WM 76
$\frac{9}{16} \times 1\frac{1}{2}$

COVES

WM 81
$\frac{9}{16} \times 3\frac{1}{4}$

WM 82
$\frac{9}{16} \times 2\frac{3}{4}$

WM 84
$\frac{9}{16} \times 2\frac{1}{4}$

WM 85
$\frac{9}{16} \times 1\frac{3}{4}$

WM 86
$\frac{9}{16} \times 1\frac{5}{8}$

WM 89
$\frac{9}{16} \times 2$

WM 90
$\frac{3}{4} \times 1\frac{1}{8}$

WM 92
$\frac{3}{4} \times \frac{7}{8}$

WM 93
$\frac{3}{4} \times \frac{3}{4}$

WM 94
$\frac{11}{16} \times 1\frac{1}{8}$

WM 95
$\frac{11}{16} \times \frac{7}{8}$

WM 96
$\frac{5}{8} \times \frac{3}{4}$

WM 100
$\frac{11}{16} \times \frac{11}{16}$

WM 101
$\frac{1}{2} \times \frac{1}{2}$

Fig. 27-1 Standard molding patterns *(Courtesy of Wood Moulding and Millwork Producers, Inc., P.O. Box 25278, Portland, Oregon 97225)*

472

QUARTER ROUNDS

WM 103
$1\frac{1}{16} \times 1\frac{1}{16}$

WM 104
$\frac{11}{16} \times 1\frac{3}{8}$

WM 105
$\frac{3}{4} \times \frac{3}{4}$

WM 106
$\frac{11}{16} \times \frac{11}{16}$

WM 107
$\frac{5}{8} \times \frac{5}{8}$

WM 108
$\frac{1}{2} \times \frac{1}{2}$

WM 109
$\frac{3}{8} \times \frac{3}{8}$

WM 110
$\frac{1}{4} \times \frac{1}{4}$

HALF ROUNDS

WM 120
$\frac{1}{2} \times 1$

WM 122
$\frac{3}{8} \times \frac{11}{16}$

WM 123
$\frac{5}{16} \times \frac{5}{8}$

WM 124
$\frac{1}{4} \times \frac{1}{2}$

FLAT ASTRAGALS

WM 133
$\frac{11}{16} \times 1\frac{3}{4}$

WM 134
$\frac{11}{16} \times 1\frac{3}{8}$

WM 135
$\frac{7}{16} \times \frac{3}{4}$

BASE SHOES

WM 126
$\frac{1}{2} \times \frac{3}{4}$

WM 129
$\frac{7}{16} \times \frac{11}{16}$

WM 127
$\frac{7}{16} \times \frac{3}{4}$

WM 131
$\frac{1}{2} \times \frac{3}{4}$

SHELF EDGE/ SCREEN MOULD

WM 137
$\frac{3}{8} \times \frac{3}{4}$

WM 141
$\frac{1}{4} \times \frac{5}{8}$

WM 138
$\frac{5}{16} \times \frac{5}{8}$

WM 142
$\frac{1}{4} \times \frac{3}{4}$

WM 140
$\frac{1}{4} \times \frac{3}{4}$

WM 144
$\frac{1}{4} \times \frac{3}{4}$

GLASS BEADS

WM 147
$\frac{1}{2} \times \frac{9}{16}$

WM 148
$\frac{3}{8} \times \frac{3}{8}$

BASE CAPS

WM 163
$\frac{11}{16} \times 1\frac{3}{8}$

WM 167
$\frac{11}{16} \times 1\frac{1}{8}$

WM 164
$\frac{11}{16} \times 1\frac{1}{8}$

WM 172
$\frac{5}{8} \times \frac{3}{4}$

WM 166
$\frac{11}{16} \times 1\frac{1}{4}$

BRICK MOULD

WM 175
$1\frac{1}{16} \times 2$

WM 176
$1\frac{1}{16} \times 1\frac{3}{4}$

WM 180
$1\frac{1}{4} \times 2$

DRIP CAPS

WM 187
$1\frac{11}{16} \times 2$

WM 196
$\frac{11}{16} \times 1\frac{3}{4}$

WM 188
$1\frac{1}{16} \times 1\frac{5}{8}$

WM 197
$\frac{11}{16} \times 1\frac{5}{8}$

CORNER GUARDS

WM 199
1×1

WM 200
$\frac{3}{4} \times \frac{3}{4}$

WM 201
$1\frac{5}{16} \times 1\frac{5}{16}$

WM 204
$1\frac{5}{16} \times 1\frac{5}{16}$

WM 202
$1\frac{1}{8} \times 1\frac{1}{8}$

WM 205
$1\frac{1}{8} \times 1\frac{1}{8}$

WM 203
$\frac{3}{4} \times \frac{3}{4}$

WM 206
$\frac{3}{4} \times \frac{3}{4}$

BATTENS

WM 224
$\frac{9}{16} \times 2\frac{1}{4}$

WM 229
$\frac{11}{16} \times 1\frac{5}{8}$

ROUNDS

WM 232 $1\frac{5}{8}$

WM 233 $1\frac{5}{16}$

WM 234 $1\frac{1}{16}$

SQUARES

WM 236 $1\frac{5}{8} \times 1\frac{5}{8}$

WM 237 $1\frac{5}{16} \times 1\frac{5}{16}$

WM 238 $1\frac{1}{16} \times 1\frac{1}{16}$

WM 239 $\frac{3}{4} \times \frac{3}{4}$

Fig. 27-1 (Continued)

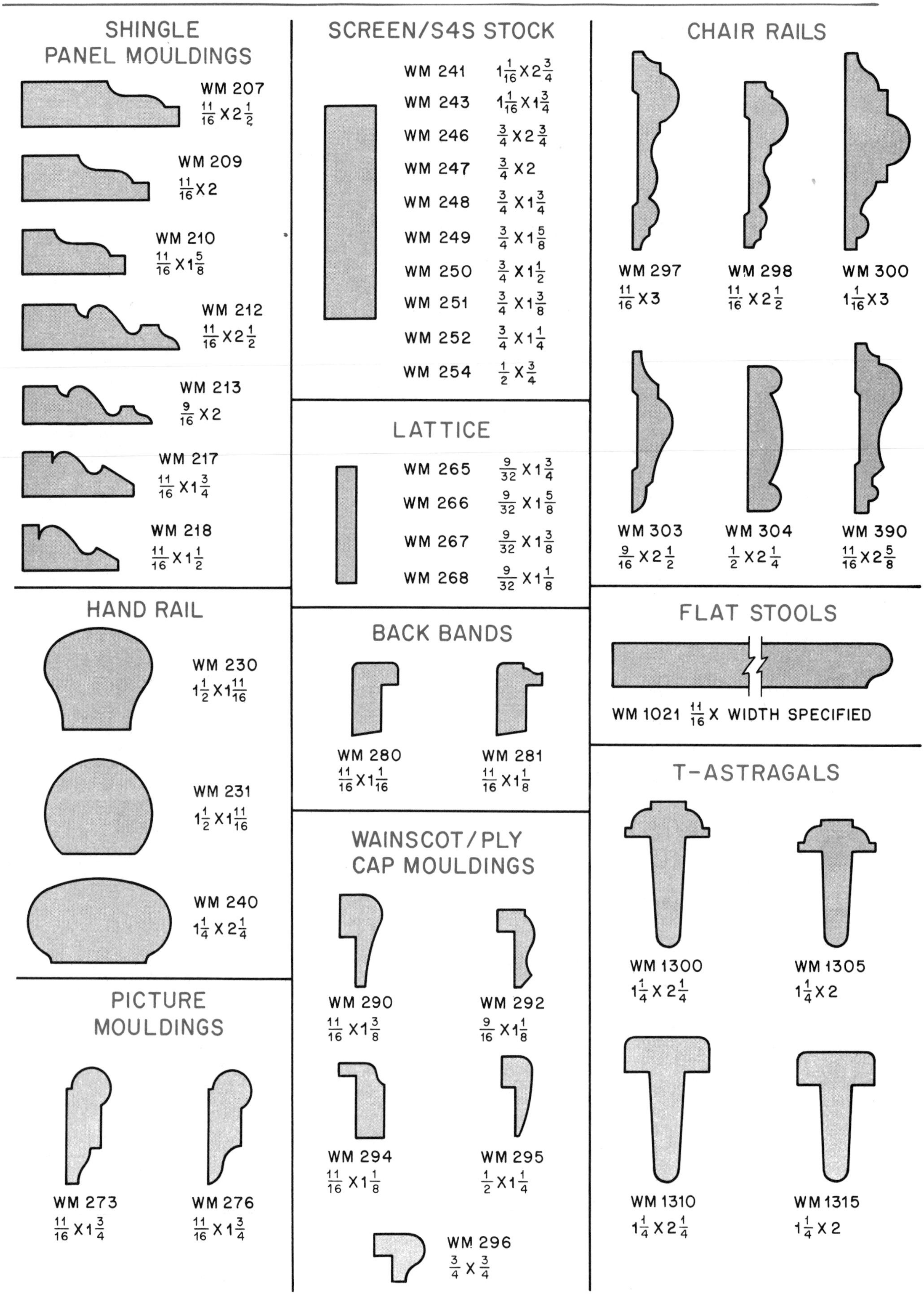

SHINGLE PANEL MOULDINGS

WM 207
$\frac{11}{16} \times 2\frac{1}{2}$

WM 209
$\frac{11}{16} \times 2$

WM 210
$\frac{11}{16} \times 1\frac{5}{8}$

WM 212
$\frac{11}{16} \times 2\frac{1}{2}$

WM 213
$\frac{9}{16} \times 2$

WM 217
$\frac{11}{16} \times 1\frac{3}{4}$

WM 218
$\frac{11}{16} \times 1\frac{1}{2}$

HAND RAIL

WM 230
$1\frac{1}{2} \times 1\frac{11}{16}$

WM 231
$1\frac{1}{2} \times 1\frac{11}{16}$

WM 240
$1\frac{1}{4} \times 2\frac{1}{4}$

PICTURE MOULDINGS

WM 273
$\frac{11}{16} \times 1\frac{3}{4}$

WM 276
$\frac{11}{16} \times 1\frac{3}{4}$

SCREEN/S4S STOCK

WM 241	$1\frac{1}{16} \times 2\frac{3}{4}$
WM 243	$1\frac{1}{16} \times 1\frac{3}{4}$
WM 246	$\frac{3}{4} \times 2\frac{3}{4}$
WM 247	$\frac{3}{4} \times 2$
WM 248	$\frac{3}{4} \times 1\frac{3}{4}$
WM 249	$\frac{3}{4} \times 1\frac{5}{8}$
WM 250	$\frac{3}{4} \times 1\frac{1}{2}$
WM 251	$\frac{3}{4} \times 1\frac{3}{8}$
WM 252	$\frac{3}{4} \times 1\frac{1}{4}$
WM 254	$\frac{1}{2} \times \frac{3}{4}$

LATTICE

WM 265	$\frac{9}{32} \times 1\frac{3}{4}$
WM 266	$\frac{9}{32} \times 1\frac{5}{8}$
WM 267	$\frac{9}{32} \times 1\frac{3}{8}$
WM 268	$\frac{9}{32} \times 1\frac{1}{8}$

BACK BANDS

WM 280
$\frac{11}{16} \times 1\frac{1}{16}$

WM 281
$\frac{11}{16} \times 1\frac{1}{8}$

WAINSCOT/PLY CAP MOULDINGS

WM 290
$\frac{11}{16} \times 1\frac{3}{8}$

WM 292
$\frac{9}{16} \times 1\frac{1}{8}$

WM 294
$\frac{11}{16} \times 1\frac{1}{8}$

WM 295
$\frac{1}{2} \times 1\frac{1}{4}$

WM 296
$\frac{3}{4} \times \frac{3}{4}$

CHAIR RAILS

WM 297
$\frac{11}{16} \times 3$

WM 298
$\frac{11}{16} \times 2\frac{1}{2}$

WM 300
$1\frac{1}{16} \times 3$

WM 303
$\frac{9}{16} \times 2\frac{1}{2}$

WM 304
$\frac{1}{2} \times 2\frac{1}{4}$

WM 390
$\frac{11}{16} \times 2\frac{5}{8}$

FLAT STOOLS

WM 1021 $\frac{11}{16} \times$ WIDTH SPECIFIED

T-ASTRAGALS

WM 1300
$1\frac{1}{4} \times 2\frac{1}{4}$

WM 1305
$1\frac{1}{4} \times 2$

WM 1310
$1\frac{1}{4} \times 2\frac{1}{4}$

WM 1315
$1\frac{1}{4} \times 2$

Fig. 27-1 (Continued)

CASING

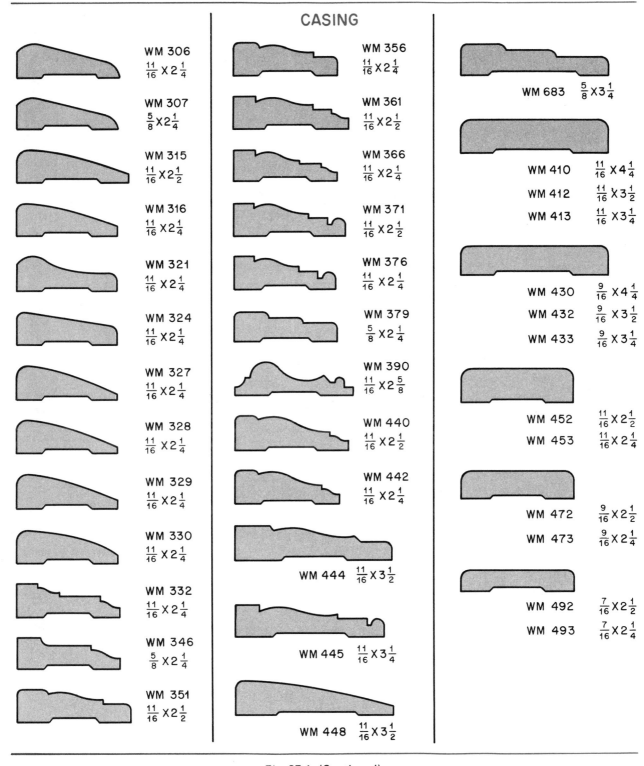

WM 306 $\frac{11}{16} \times 2\frac{1}{4}$

WM 307 $\frac{5}{8} \times 2\frac{1}{4}$

WM 315 $\frac{11}{16} \times 2\frac{1}{2}$

WM 316 $\frac{11}{16} \times 2\frac{1}{4}$

WM 321 $\frac{11}{16} \times 2\frac{1}{4}$

WM 324 $\frac{11}{16} \times 2\frac{1}{4}$

WM 327 $\frac{11}{16} \times 2\frac{1}{4}$

WM 328 $\frac{11}{16} \times 2\frac{1}{4}$

WM 329 $\frac{11}{16} \times 2\frac{1}{4}$

WM 330 $\frac{11}{16} \times 2\frac{1}{4}$

WM 332 $\frac{11}{16} \times 2\frac{1}{4}$

WM 346 $\frac{5}{8} \times 2\frac{1}{4}$

WM 351 $\frac{11}{16} \times 2\frac{1}{2}$

WM 356 $\frac{11}{16} \times 2\frac{1}{4}$

WM 361 $\frac{11}{16} \times 2\frac{1}{2}$

WM 366 $\frac{11}{16} \times 2\frac{1}{4}$

WM 371 $\frac{11}{16} \times 2\frac{1}{2}$

WM 376 $\frac{11}{16} \times 2\frac{1}{4}$

WM 379 $\frac{5}{8} \times 2\frac{1}{4}$

WM 390 $\frac{11}{16} \times 2\frac{5}{8}$

WM 440 $\frac{11}{16} \times 2\frac{1}{2}$

WM 442 $\frac{11}{16} \times 2\frac{1}{4}$

WM 444 $\frac{11}{16} \times 3\frac{1}{2}$

WM 445 $\frac{11}{16} \times 3\frac{1}{4}$

WM 448 $\frac{11}{16} \times 3\frac{1}{2}$

WM 683 $\frac{5}{8} \times 3\frac{1}{4}$

WM 410 $\frac{11}{16} \times 4\frac{1}{4}$

WM 412 $\frac{11}{16} \times 3\frac{1}{2}$

WM 413 $\frac{11}{16} \times 3\frac{1}{4}$

WM 430 $\frac{9}{16} \times 4\frac{1}{4}$

WM 432 $\frac{9}{16} \times 3\frac{1}{2}$

WM 433 $\frac{9}{16} \times 3\frac{1}{4}$

WM 452 $\frac{11}{16} \times 2\frac{1}{2}$

WM 453 $\frac{11}{16} \times 2\frac{1}{4}$

WM 472 $\frac{9}{16} \times 2\frac{1}{2}$

WM 473 $\frac{9}{16} \times 2\frac{1}{4}$

WM 492 $\frac{7}{16} \times 2\frac{1}{2}$

WM 493 $\frac{7}{16} \times 2\frac{1}{4}$

Fig. 27-1 (Continued)

475

BASE MOULDINGS

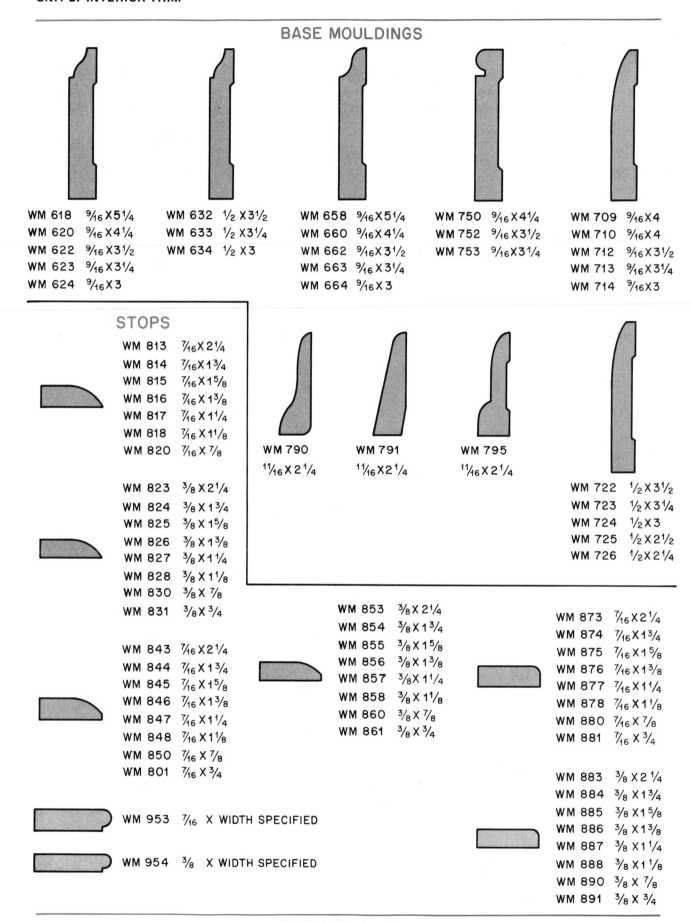

WM 618 $^9/_{16}$ X $5^1/_4$
WM 620 $^9/_{16}$ X $4^1/_4$
WM 622 $^9/_{16}$ X $3^1/_2$
WM 623 $^9/_{16}$ X $3^1/_4$
WM 624 $^9/_{16}$ X 3

WM 632 $^1/_2$ X $3^1/_2$
WM 633 $^1/_2$ X $3^1/_4$
WM 634 $^1/_2$ X 3

WM 658 $^9/_{16}$ X $5^1/_4$
WM 660 $^9/_{16}$ X $4^1/_4$
WM 662 $^9/_{16}$ X $3^1/_2$
WM 663 $^9/_{16}$ X $3^1/_4$
WM 664 $^9/_{16}$ X 3

WM 750 $^9/_{16}$ X $4^1/_4$
WM 752 $^9/_{16}$ X $3^1/_2$
WM 753 $^9/_{16}$ X $3^1/_4$

WM 709 $^9/_{16}$ X 4
WM 710 $^9/_{16}$ X 4
WM 712 $^9/_{16}$ X $3^1/_2$
WM 713 $^9/_{16}$ X $3^1/_4$
WM 714 $^9/_{16}$ X 3

STOPS

WM 813 $^7/_{16}$ X $2^1/_4$
WM 814 $^7/_{16}$ X $1^3/_4$
WM 815 $^7/_{16}$ X $1^5/_8$
WM 816 $^7/_{16}$ X $1^3/_8$
WM 817 $^7/_{16}$ X $1^1/_4$
WM 818 $^7/_{16}$ X $1^1/_8$
WM 820 $^7/_{16}$ X $^7/_8$

WM 823 $^3/_8$ X $2^1/_4$
WM 824 $^3/_8$ X $1^3/_4$
WM 825 $^3/_8$ X $1^5/_8$
WM 826 $^3/_8$ X $1^3/_8$
WM 827 $^3/_8$ X $1^1/_4$
WM 828 $^3/_8$ X $1^1/_8$
WM 830 $^3/_8$ X $^7/_8$
WM 831 $^3/_8$ X $^3/_4$

WM 843 $^7/_{16}$ X $2^1/_4$
WM 844 $^7/_{16}$ X $1^3/_4$
WM 845 $^7/_{16}$ X $1^5/_8$
WM 846 $^7/_{16}$ X $1^3/_8$
WM 847 $^7/_{16}$ X $1^1/_4$
WM 848 $^7/_{16}$ X $1^1/_8$
WM 850 $^7/_{16}$ X $^7/_8$
WM 801 $^7/_{16}$ X $^3/_4$

WM 953 $^7/_{16}$ X WIDTH SPECIFIED

WM 954 $^3/_8$ X WIDTH SPECIFIED

WM 790 $^{11}/_{16}$ X $2^1/_4$

WM 791 $^{11}/_{16}$ X $2^1/_4$

WM 795 $^{11}/_{16}$ X $2^1/_4$

WM 722 $^1/_2$ X $3^1/_2$
WM 723 $^1/_2$ X $3^1/_4$
WM 724 $^1/_2$ X 3
WM 725 $^1/_2$ X $2^1/_2$
WM 726 $^1/_2$ X $2^1/_4$

WM 853 $^3/_8$ X $2^1/_4$
WM 854 $^3/_8$ X $1^3/_4$
WM 855 $^3/_8$ X $1^5/_8$
WM 856 $^3/_8$ X $1^3/_8$
WM 857 $^3/_8$ X $1^1/_4$
WM 858 $^3/_8$ X $1^1/_8$
WM 860 $^3/_8$ X $^7/_8$
WM 861 $^3/_8$ X $^3/_4$

WM 873 $^7/_{16}$ X $2^1/_4$
WM 874 $^7/_{16}$ X $1^3/_4$
WM 875 $^7/_{16}$ X $1^5/_8$
WM 876 $^7/_{16}$ X $1^3/_8$
WM 877 $^7/_{16}$ X $1^1/_4$
WM 878 $^7/_{16}$ X $1^1/_8$
WM 880 $^7/_{16}$ X $^7/_8$
WM 881 $^7/_{16}$ X $^3/_4$

WM 883 $^3/_8$ X $2^1/_4$
WM 884 $^3/_8$ X $1^3/_4$
WM 885 $^3/_8$ X $1^5/_8$
WM 886 $^3/_8$ X $1^3/_8$
WM 887 $^3/_8$ X $1^1/_4$
WM 888 $^3/_8$ X $1^1/_8$
WM 890 $^3/_8$ X $^7/_8$
WM 891 $^3/_8$ X $^3/_4$

Fig. 27-1 (Continued)

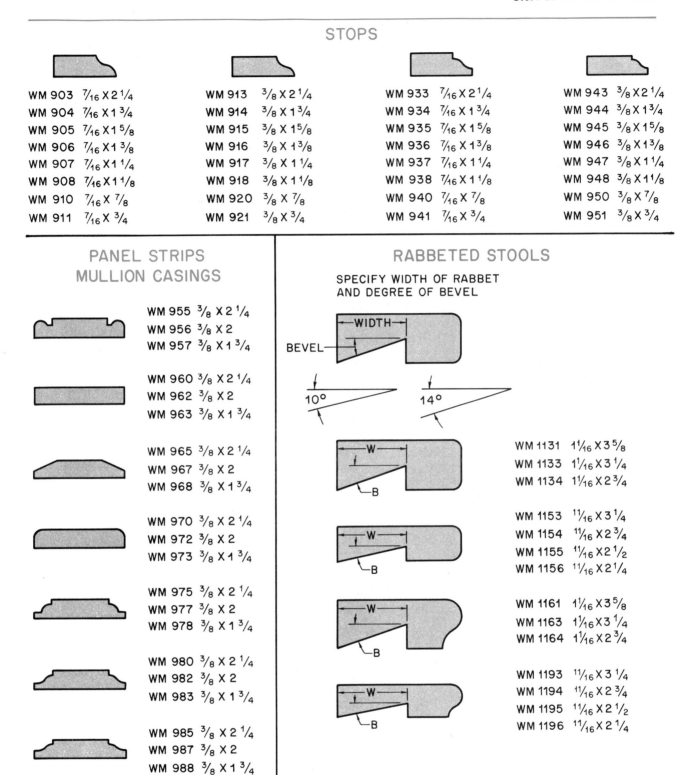

STOPS

WM 903 $7/16 \times 2 1/4$	WM 913 $3/8 \times 2 1/4$	WM 933 $7/16 \times 2 1/4$	WM 943 $3/8 \times 2 1/4$
WM 904 $7/16 \times 1 3/4$	WM 914 $3/8 \times 1 3/4$	WM 934 $7/16 \times 1 3/4$	WM 944 $3/8 \times 1 3/4$
WM 905 $7/16 \times 1 5/8$	WM 915 $3/8 \times 1 5/8$	WM 935 $7/16 \times 1 5/8$	WM 945 $3/8 \times 1 5/8$
WM 906 $7/16 \times 1 3/8$	WM 916 $3/8 \times 1 3/8$	WM 936 $7/16 \times 1 3/8$	WM 946 $3/8 \times 1 3/8$
WM 907 $7/16 \times 1 1/4$	WM 917 $3/8 \times 1 1/4$	WM 937 $7/16 \times 1 1/4$	WM 947 $3/8 \times 1 1/4$
WM 908 $7/16 \times 1 1/8$	WM 918 $3/8 \times 1 1/8$	WM 938 $7/16 \times 1 1/8$	WM 948 $3/8 \times 1 1/8$
WM 910 $7/16 \times 7/8$	WM 920 $3/8 \times 7/8$	WM 940 $7/16 \times 7/8$	WM 950 $3/8 \times 7/8$
WM 911 $7/16 \times 3/4$	WM 921 $3/8 \times 3/4$	WM 941 $7/16 \times 3/4$	WM 951 $3/8 \times 3/4$

PANEL STRIPS
MULLION CASINGS

WM 955 $3/8 \times 2 1/4$
WM 956 $3/8 \times 2$
WM 957 $3/8 \times 1 3/4$

WM 960 $3/8 \times 2 1/4$
WM 962 $3/8 \times 2$
WM 963 $3/8 \times 1 3/4$

WM 965 $3/8 \times 2 1/4$
WM 967 $3/8 \times 2$
WM 968 $3/8 \times 1 3/4$

WM 970 $3/8 \times 2 1/4$
WM 972 $3/8 \times 2$
WM 973 $3/8 \times 1 3/4$

WM 975 $3/8 \times 2 1/4$
WM 977 $3/8 \times 2$
WM 978 $3/8 \times 1 3/4$

WM 980 $3/8 \times 2 1/4$
WM 982 $3/8 \times 2$
WM 983 $3/8 \times 1 3/4$

WM 985 $3/8 \times 2 1/4$
WM 987 $3/8 \times 2$
WM 988 $3/8 \times 1 3/4$

RABBETED STOOLS

SPECIFY WIDTH OF RABBET
AND DEGREE OF BEVEL

WIDTH
BEVEL

$10°$ $14°$

WM 1131 $1 1/16 \times 3 5/8$
WM 1133 $1 1/16 \times 3 1/4$
WM 1134 $1 1/16 \times 2 3/4$

WM 1153 $11/16 \times 3 1/4$
WM 1154 $11/16 \times 2 3/4$
WM 1155 $11/16 \times 2 1/2$
WM 1156 $11/16 \times 2 1/4$

WM 1161 $1 1/16 \times 3 5/8$
WM 1163 $1 1/16 \times 3 1/4$
WM 1164 $1 1/16 \times 2 3/4$

WM 1193 $11/16 \times 3 1/4$
WM 1194 $11/16 \times 2 3/4$
WM 1195 $11/16 \times 2 1/2$
WM 1196 $11/16 \times 2 1/4$

Fig. 27-1 (Continued)

477

MOLDING USE

Some moldings are used only in certain locations. Others may be used in a number of places.

- *Beds, crowns, coves,* and *quarter rounds* are used at the intersection of walls and ceilings and similar locations. Quarter rounds may also be used to trim outside corners on wall paneling or to hold panels in a frame.

- *Base, base shoes,* and *base caps* are usually only used between floor and walls.

- *Casings* are used to trim around windows, doors, and other openings to cover the space between the frame and the wall. *Back bands* are applied to the outside edges of casings for a more decorative appearance.

- *Aprons, stools,* and *stops* are part of the window trim. Stops are also applied to door frames. Casings may also be used as aprons, Figure 27-2.

- *Full rounds* are usually used for closet poles and stair handrails.

- *Half rounds* are often used to band plywood edges and for similar purposes. They may also be used to cover joints between wall and ceiling panels and on screens.

- *Corner molds* are used on outside corners of wall finish.

- *Ply caps* trim the top edge of plywood wainscoting.

APPLYING CEILING MOLDING

Joints between lengths of ceiling molding may be made square or at a 45° angle. Many carpenters prefer to make square joints between moldings because less joint line is shown. Also, the square end acts as a stop when bowing and snapping molding into place at a corner. Often the last piece of molding on a wall is cut slightly

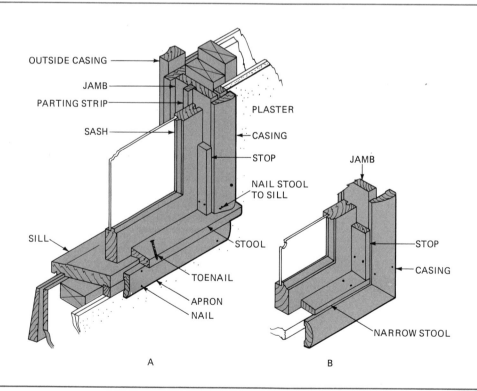

Fig. 27-2 Window trim parts

Fig. 27-3 A coped joint

longer, bowed outward in the center, then pressed into place when the ends are in position. This makes the joints come up tight. In any case, the joint between lengths of molding should be sanded flush except on prefinished moldings.

Joints on exterior corners are mitered. Joints on interior corners are coped. A *coped* joint is made by cutting the piece on one wall with a square end in the corner. The end of the molding on the other wall is cut to fit against the shape or face of the molding on the first wall, Figure 27-3.

Moldings of all types may be mitered with the use of a miter box. A metal miter box can be used, Figure 27-4, or a wood miter box can

be made on the job as needed. The problem with a wood miter box is that the saw cuts used to guide the saw eventually widen with use, resulting in inaccurate cuts. However, it takes only a little time to make another wood miter box.

METHODS OF MAKING MITERS

Miters may also be made with a *miter trimmer*, Figure 27-5. All cuts on the trimmer must be made at or near the end of molding. All softwood molding and trim are usually cut off square, rough mitered with a chop cut, and trimmed with one or two shaving slices. Hardwoods must be sawed at approximately 45° and then trimmed. The harder the wood, the thinner the trim cut. In off-square joinery, paper-thin corrective cuts are made quickly and easily.

Fig. 27-5 The miter trimmer is widely used to cut miters. *(Courtesy of Pootatuck Corporation)*

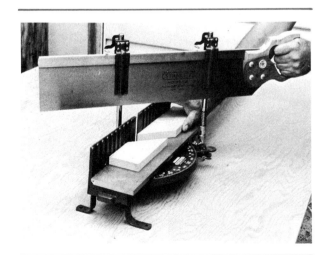

Fig. 27-4 Using a metal miter box

Fig 27-6 Using the table saw and miter guage
(Guard is removed for clarity)

Fig. 27-7 A mitering jig allows left- and right-hand miters to be made quickly and easily. (Guard is removed for clarity)

Miters may also be made by using the table saw with the miter gauge as a guide, Figure 27-6, or by holding the stock in a mitering jig, Figure 27-7. The use of a mitering jig allows both right- and left-hand miters to be cut quickly and easily. The radial arm saw, or the power miter box, may also be used to miter moldings and trim, Figure 27-8.

MAKING A COPED JOINT

To cope the end of molding, first make a back miter on the end. A *back miter* is a miter that starts from the end and is cut back on the face, Figure 27-9. For square corners make the back miter at a 45° angle.

Some beds, crowns, and coves that do not have a square back must be held in the miter box upside down with the face side out. The edge of the cut along the face forms the profile of the cope. Rub the side of a pencil point lightly along the profile to outline it.

Use a coping saw and cut to the profile with a slight undercut, Figure 27-10. Hold the molding so it is over the end of a sawhorse with the side of the molding that goes against the wall lying flat on the top of the sawhorse. **It is important that the molding be held in this position.** Holding it in any other way makes it

Fig. 27-8 The radial arm saw may be used to miter molding.

Fig. 27-9 Making a back miter. The molding is held upside down.

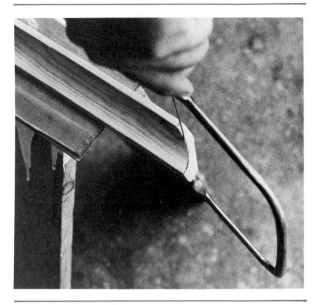

Fig. 27-10 Coping the end of molding

difficult to cut the cope in the proper manner. Cut with the handle of the coping saw above the work and the teeth of the blade pointing away from the handle. Coped joints are used on interior wall corners because they will not open up when the molding is nailed in place, especially if the backing is not solid. Miter joints may open up in interior corners when the ends are fastened.

APPLYING MOLDING

Apply the molding to the first wall with square ends in both corners. If more than one piece is required to go from corner to corner, install the first piece. Fasten the molding by starting a finish nail in the center at about a 45° angle. Use a nail of sufficient length to penetrate into solid wood at least one inch, Figure 27-11. Press the molding in the corner with one hand while driving the nail almost home. Then set the nail below the surface. Nail at about 16-inch intervals and in other locations as necessary to bring the molding tight against wall and ceiling. End nails should be placed 2 to 3 inches from the end to keep from splitting the molding. If it is likely that the molding may split, blunt the pointed end of the nail or drill a hole slightly smaller than the nail diameter.

Install the last piece on the first wall by first squaring one end. Place the square end in the corner and let the other end overlap the first piece. Mark and cut it at the overlap. This method is more accurate than measuring and

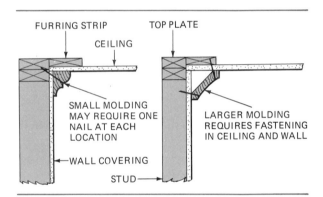

FURRING STRIP   TOP PLATE

CEILING

SMALL MOLDING
MAY REQUIRE ONE
NAIL AT EACH
LOCATION   LARGER MOLDING
                            REQUIRES FASTENING
                            IN CEILING AND WALL

WALL COVERING

STUD

Fig. 27-11 Methods of nailing ceiling molding

then transferring the measurement to the piece. It is advisable to mark all pieces of interior trim for length in this manner whenever possible. Fasten this last piece in the same manner as others.

Cope the first piece on each succeeding wall against the face of the last piece installed. Work around the room in one direction. If the molding is applied in one piece on the last wall, it must be coped on both ends.

APPLYING DOOR CASINGS

Door casings are applied before the baseboard because the base butts against door casing edges. Door casings extend to the floor. The

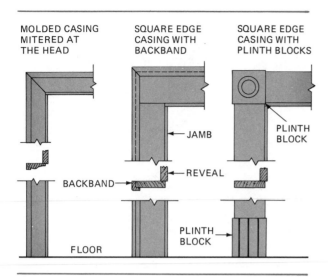

MOLDED CASING MITERED AT THE HEAD

SQUARE EDGE CASING WITH BACKBAND

SQUARE EDGE CASING WITH PLINTH BLOCKS

PLINTH BLOCK

JAMB

REVEAL

BACKBAND

PLINTH BLOCK

FLOOR

Fig. 27-12 Styles of door casing. The setback of the casing on the jamb is often called a reveal.

finish floor may or may not be previously laid, according to the custom of the geographical location.

DESIGN OF DOOR CASINGS

Door casings may have molded faces or plain faces with square edges. On molded casings, the joint at the head is usually mitered. On square edge casings, the joint may be mitered or butted, with the head casing overlapping the side casing. Square edge casings may be further decorated by the use of backbands, Figure 27-12.

APPLYING THE DOOR CASINGS

Door casings are set back from the face of the door frame a distance of about 5/16 inch. This setback is called a *reveal*. Set the blade of the combination square so that it extends 5/16 inch beyond the body of the square. Gauge lines at intervals along the side and head jamb edges. Let the lines intersect where side and head jambs meet.

The following procedure applies to molded door casings mitered at the head. If a number of door openings are to be cased, cut the necessary number of door casings to rough lengths with a miter cut on one end of each piece. For each interior door opening, four side casings and two

header casings are required. Cut side casings in pairs with right- and left-hand miters. Rough lengths are a few inches longer than actually needed.

APPLYING THE HEADER CASING

Mark and cut the header casing. Hold it against the header jamb of the door frame so that the inside edge at a miter cut on one end is on the intersection of the gauged lines. Mark its length on the inside edge at the gauged line on the opposite side. Miter the casing to the mark.

Fasten the header casing in position so its inside edge is to the gauged lines on the header jamb and the ends are in line with the gauged lines on the side jambs. Use finish nails along the inside edge into the header jamb. If the casing edge is thin, use 3d or 4d finish nails spaced about 12 inches apart. Keep the edge of the casing to the gauged lines on the jamb, straightening the casing as necessary as nailing progresses. Drive nails at the proper angle to keep them from coming through the face of the jamb.

Fasten the top edge of the casing into the framing. If the outside edge is thicker, use longer nails spaced farther apart, about 16 inches O.C. Do not drive end nails at this time.

APPLYING THE SIDE CASINGS

Mark the side casing by turning it upside down with the point of the miter touching the floor. Mark the edge of the side casing in line with the top edge of the header casing, Figure 27-13. Make a square cut on the casing to the mark.

Place the side casing in position and try the fit at the mitered joint. If the joint needs fitting, trim the mitered end of the side casing with a block plane until a tight fit is obtained and the inside edges of both side and header casings meet at the corner. Apply a little glue to the joint and nail the side casing in the same manner as you did the header casing.

Avoid sanding the joint to bring the casing faces flush. It is difficult to keep from sanding across the grain on one or the other of the pieces. Cross-grain scratches will be very

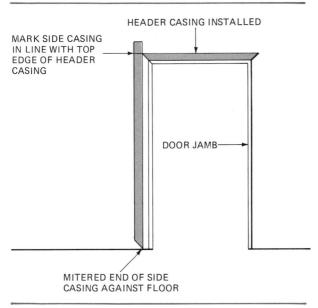

Fig. 27-13 Method of marking the length of side casings

noticeable, especially if the trim is to have a stained finish. Bring the faces flush by shimming, where necessary, between the back of the casing and the wall. Usually the shims needed would be very thin. Any small space between the casing and the wall can be filled with joint filling compound. Most carpenters prefer to do this rather than try to sand the joint.

Drive a 4d finish nail into the edge of the casing and through the mitered joint. Drive end nails and then set all fasteners. Keep nails 2 or 3 inches from the end to avoid splitting the casing. If there is danger of splitting, blunt the pointed end of the nail or drill a hole slightly smaller in diameter than the nail.

APPLYING BASE, BASE CAP, AND BASE SHOE

According to custom and geographic location, the base, Figure 27-14, may be applied

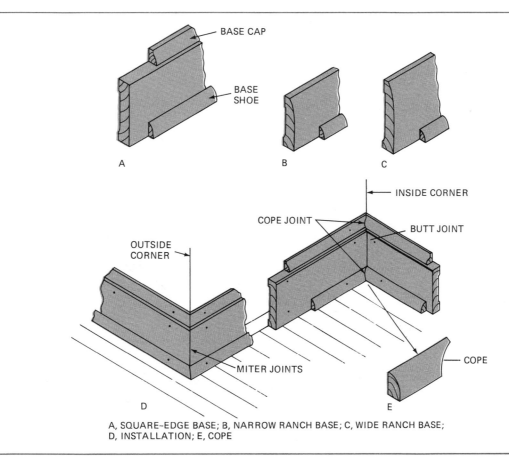

A, SQUARE-EDGE BASE; B, NARROW RANCH BASE; C, WIDE RANCH BASE;
D, INSTALLATION; E, COPE

Fig. 27-14 Styles of base

before or after the finish floor is laid. If the base is applied on top of the finish floor, the base may shrink and cause an open joint between it and the floor, therefore, a base shoe is necessary. If the base is applied before the finish floor, no shoe is required.

The base cap is used to give a decorative effect to the entire base assembly. The base must have a square top edge upon which the cap rests. The base cap conforms easily to irregularities in the wall and, therefore, results in a better fit.

PLANNING THE APPLICATION

Consideration must be given to the order in which the base is applied. The last few pieces of base should end by butting against door casings, if possible, to make application easier.

If the base is applied before the finish floor is laid, another consideration is the direction in which the floor is to be laid. Starting and ending strips of flooring that butt tightly against the base may cause joints in interior corners to open if the joints are not made properly. To avoid this, lay the base that is at right angles to the flooring first, with square ends in the corners. Then, cope both ends of the base that is parallel to the flooring, Figure 27-15.

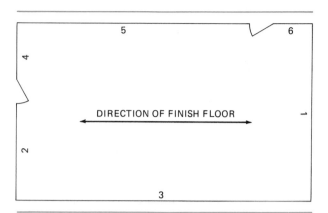

Fig. 27-15 Consideration must be given to the order in which base is applied.

APPLICATION OF BASE

The base is applied in a manner similar to wall and ceiling molding with the exception of laying out the coped ends. Because the base is relatively wide, compared to other moldings, and because of irregularities in the wall, the face of the base may be out of square with the floor. Therefore, it is necessary to lay out the cope by scribing rather than back-mitering it.

Apply the base to the first wall with square ends in each corner. Use two finishing nails at each stud location and set them. Nailing blocks previously installed during framing provide solid

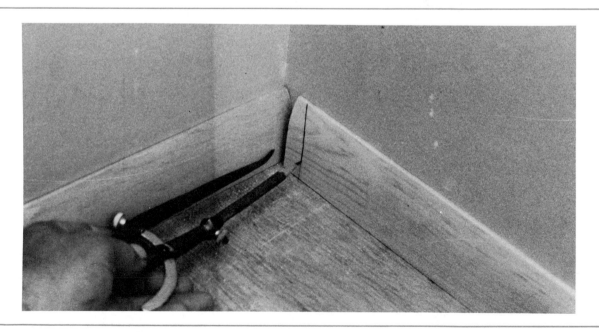

Fig. 27-16 Laying out a coped joint on base in interior corners

484

wood for fastening the ends of the base in interior corners.

Cut the base to go on the next wall about an inch longer than required. Lay the base against the wall by bending it so the end to be scribed lies flat against the wall and against the first base. Set the dividers to scribe about 1/2 inch and lay out the cope by riding the dividers along the face of the base on the first wall. Hold the dividers while scribing so that a line between the two points is parallel to the floor. Twisting the dividers while making the scribe results in an inaccurate layout. Cut the end to the scribed line with a slight undercut, Figure 27-16. Place the base back in position and try the fit. Make adjustments if necessary.

CUTTING THE BASE TO LENGTH

With the base in this position, bend it in against the wall near the center. Place marks on the top edge of the base and the wall where they line up with each other.

Remove the base and put the other end in the opposite corner. Press the base against the wall at the mark. The difference between the two marks is the amount to scribe off the other end. Set the dividers to this distance, mark the scribe, and cut to the scribed line, Figure 27-17.

Place one end in the corner, bow out the center, place the other end in the opposite corner, and press the center against the wall. Fasten in place.

If a little tighter fit is desired, set the dividers slightly less than the distance between the marks. This method of fitting long lengths between walls may be applied to other kinds of trim. However, this works especially well with the base. Continue in this manner around the room in the order previously planned. Make regular miter joints on outside corners.

If both ends of a single piece are to have regular miters for outside corners, it is imperative that it be fastened in the same position as it was marked. To assure this, tack the piece in position with one finish nail in the center. Mark both ends, remove, and make the miters. Replace the piece by first fastening into the original nail hole.

APPLYING THE BASE CAP

The base cap is applied in the same manner as most wall or ceiling molding. Cope interior corners and miter exterior corners.

APPLYING THE BASE SHOE

The base shoe is also applied in a similar manner as other molding. However, the base shoe is nailed into the floor and not into the baseboard itself, Figure 27-18. If the baseboard should shrink slightly, no opening will occur under the shoe.

When the end of the base shoe stops at an opening with no material against which to butt its end, it must be *returned upon itself*. This means that the end is given the same profile as

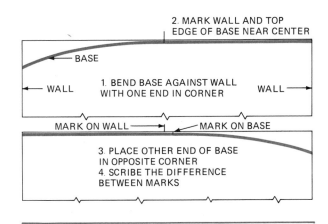

Fig. 27-17 Method of laying out base length to fit between walls

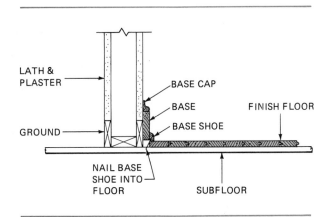

Fig. 27-18 The base shoe is nailed into the floor only.

Fig. 27-19 Returning a molding upon itself

its face. To return an end, use a scrap piece of molding, hold it on end, and draw the profile on the piece to be used. Cut to the line with a coping saw and sand the end smooth, Figure 27-19. Most interior trim whose ends are exposed are returned in this manner.

INSTALLING WINDOW TRIM

Interior window trim, in order of installation, consists of the stool, sometimes called the stool cap, the apron, jamb extensions, casings, and the stops, sometimes called the stop bead. Although the style of the trim may differ with various kinds of windows, the application is basically the same. The procedure described in this unit applies to double hung windows.

INSTALLING THE STOOL

Hold a scrap piece of casing stock on the wall with its inside edge flush with the inside face of the side jamb of the window frame. Draw a light line along its outside edge on the wall at the bottom of the window frame. Do this on both sides.

Cut a piece of stool stock to length equal to the distance between the marks, plus twice the outside edge thickness of the casing. The stool, when installed, projects on both ends beyond the casing by the casing thickness.

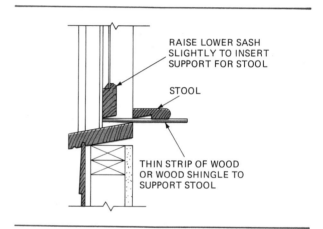

RAISE LOWER SASH SLIGHTLY TO INSERT SUPPORT FOR STOOL

STOOL

THIN STRIP OF WOOD OR WOOD SHINGLE TO SUPPORT STOOL

Fig. 27-20 Method of supporting the stool for layout.

Raise the lower sash slightly and place short, thin strips of wood under it, on both sides, to support the stool while it is laid out, Figure 27-20. Place the stool on the strips so the top of the stool is level, outside edges against the wall, and the ends in line with the marks on the wall previously made.

The stool, in its final position, must be notched between the side jambs with its outside edge against the bottom rail of the sash and the wall. Its inside edge should project inward beyond the casing by the casing thickness.

Set the dividers so that, at the wall, an amount is left on the stool equal to twice the casing thickness. Scribe the stool by riding the dividers along the wall and the bottom rail of the window sash, Figure 27-21.

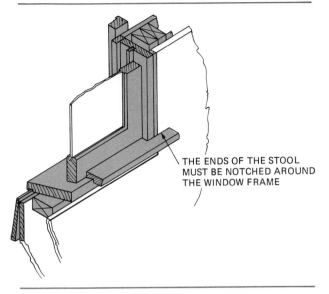

Fig. 27-21 Laying out the stool

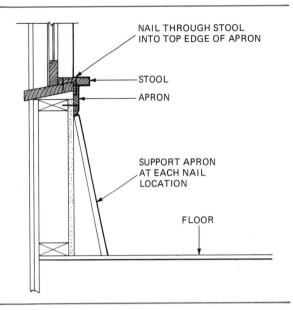

Fig. 27-22 Support the apron when
fastening the stool to it.

Cut to the lines, using a handsaw and return the ends of the stool. Smooth the edge of the stool that goes against the sash. Apply a small amount of caulking compound to the bottom of the stool along its outside edge. Fasten the stool in position by driving finish nails along its outside edge into the sill. Set the nails.

APPLYING THE APRON

The apron covers the joint between the sill and the wall. It is applied with its ends in line with the outside edges of the window casing.

Cut a length of apron stock equal to the distance between the marks laid out on the walls. Return both ends of the apron. Place the apron in position with its upper edge against the bottom of the stool. Be careful not to force the stool upward. Keep the top side of the stool level by holding a square between it and the edge of the side jamb.

Fasten the apron along its bottom edge into the wall. Then drive nails through the stool into the top edge of the apron. When nailing, wedge a short length of 1x6 stock between the apron and the floor at each nail location. This supports the apron while nails are being driven. Failure to support the apron in this manner results in an open joint between it and the stool. Take care not to damage the bottom edge of the apron with the supporting piece, Figure 27-22.

INSTALLING JAMB EXTENSIONS

Many times, windows are installed with narrow jambs. Strips are fastened to the inside edge of the header and side jambs to extend them flush with the inside surface of the finished wall. These strips are called jamb extensions.

Some manufacturers provide jamb extensions with the window unit. However, they are not applied when the window is installed, but when the window is trimmed. Therefore when windows are set, these pieces should be carefully stored and then retrieved when it is time to apply the trim.

Measure the distance from the inside edge of the jamb to the finished wall. Rip the jamb extensions to this width and back bevel the inside edge slightly. Cut the pieces to length and apply them to the header and side jambs. Drive finish nails through the edges into the edge of the jambs.

APPLYING THE CASINGS

Casings are applied with their inside edges flush with the inside face of the window frame. The bottom ends of the side casings rest on the stool. The casing pattern is the same as that used around the interior door frames.

Window casings are applied in the same manner as door casings. Install the header casing first and, then, the side casings. Keep inside edges flush with the face of the jamb.

INSTALLING THE STOPS

Stops are applied to the head and side jambs. Usually the stop applied to the header jamb is fastened with finish nails. Stops applied to side jambs are usually fastened with screws for easy removal in case the sash needs to be taken out of the frame. Joints at the top between stops may be mitered or coped. The bottom ends of the side stops rest on the stool with a square cut.

Cut the stops to width, if necessary. Install the stop at the head first and then the side stops. Fasten the stops so that the sash slides smoothly. The inner edge covers the joint between the jamb and casing and also between the jamb and the extension jamb.

INSTALLING CLOSET TRIM

The simple clothes closet is normally furnished with a shelf and a rod for hanging clothes. Usually a piece of 1x4 stock is installed around the walls of the closet to support the shelf and the rod. This piece is called a *cleat*. The shelf is installed on top of it and the closet pole is installed in the center of it. Shelves are not fastened to the cleat and rods are installed for easy removal in case the closet walls need refinishing.

Shelves are usually 1x12-inch boards. Rods may be 3/4-inch steel pipe, 1 5/16-inch full round wood poles, or chrome plated rods manufactured for this purpose. On long spans, the rod may be supported in its center by special metal closet pole supports. On each end, the closet pole is supported by plastic or metal closet pole *sockets*. In place of sockets, holes and notches are made in the cleat to support the ends of the closet pole.

For ordinary clothes closets, the height from the floor to the top edge of the cleat is 66 inches, Figure 27-23. Measure up from the floor this distance and draw a level line on the back wall and two end walls of the closet. Install

the cleat so its top edges are to the line. The cleat is installed in the same manner as baseboard. Butt the interior corners. Fasten with two finish nails at each stud.

Install the closet pole sockets on the end cleats. The center of the socket should be located at least 12 inches from the back wall and centered on the width of the cleat. Usually each socket is fastened with one screw in its center.

INSTALLING CLOSET SHELVES

For a professional job, fit the ends and back edge of the shelf to the wall. Cut the shelf about 1/2 inch longer than the distance between end walls.

Place the shelf in position by laying one end on the cleat and tilting the other end up and resting against the wall. Scribe about 1/4 inch off the end resting on the cleat. Remove the shelf and cut to the scribed line. Measure the distance between corners along the back wall and transfer to the shelf, measuring from the scribed cut along the back edge of the shelf.

Place the shelf in position tilted in the opposite direction. Set the dividers to scribe the distance from the wall to the mark on the

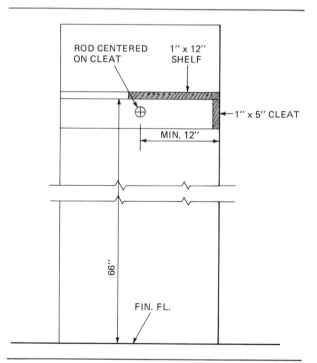

Fig. 27-23 Specifications for an ordinary clothes closet

shelf. Scribe and cut the other end of the shelf. Place the shelf into position resting on the cleats. Scribe the back edge to the wall to take off as little as possible. Cut to the scribed line. Round off the corners on the front edge of the shelf with a hand plane. Sand and place the shelf in position.

INSTALLING THE CLOSET POLE

Measure the distance between pole sockets and cut the pole to this length. Install the pole on the sockets. One socket is closed and the opposite socket has an open end. Place one end of the pole in the closed socket. Then rest the other end on the opposite socket.

LINEN CLOSETS

Linen closets usually consist of a series of shelves spaced 12 to 16 inches apart. Cleats used to support shelves are 3/4"x1" stock, chamfered on the bottom outside corner. A *chamfer* is a bevel that extends only partway through the thickness of the stock.

Lay out level lines for the top edges of each set of cleats. Install the cleats and shelves in the same manner as described for clothes closets.

MANTELS

Mantels are used to decorate fireplaces and to cover the joint between the fireplace and the wall. Most mantels come preassembled from the factory and are available in a number of sizes and styles, Figure 27-24.

Study the manufacturer's installation directions carefully. Place the mantel against the wall and center it on the fireplace. Scribe it to the floor or wall as necessary. Carefully nail the mantel in place and set all nails.

CONCLUDING THE INTERIOR TRIM INSTALLATION

All pieces of interior trim should be sanded smooth after they have been cut and fitted, and before they are fastened. The sanding of interior finish provides a smooth base for the application of stains, paints, and clear coatings. Always sand with the grain, never across the grain.

All sharp, exposed corners of trim should be rounded over slightly. Use a block plane to make a slight chamfer and then round over with sandpaper.

If the trim is to be stained, make sure every trace of excess glue is removed. Excess glue allowed to dry seals the surface and does not allow the stain to penetrate, resulting in a blotchy finish.

Be careful not to make hammer marks in the finish. Occasionally rubbing the face of the hammer head with sandpaper to clean it helps prevent it from glancing off the nailhead while you are driving the nail.

Make sure any pencil lines left on along the edge of a cut are removed before fastening the pieces. Pencil marks in interior corners are difficult to remove after the pieces are fastened in position. Pencil marks show through a stained or clear finish. When marking interior trim make light, fine pencil marks. **Note: Lay-out lines in the illustrations are purposely made dark and heavy only for the sake of clarity.**

Make sure all joints are tight fitting. Measure, mark, and cut carefully. Do not leave a poor fit. Do it over, if necessary!

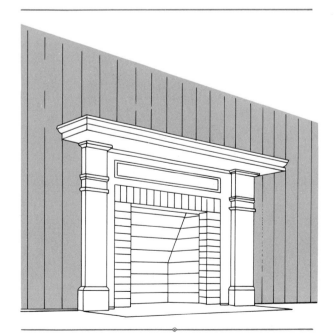

Fig. 27-24 Mantels are available preassembled in a number of styles and sizes.

REVIEW QUESTIONS

Select the most appropriate answer.

1. Bed, crown, and cove moldings are used frequently as
    a. window trim.   c. part of the base.
    b. ceiling molding.   d. door casings.

2. Back bands are applied to
    a. wainscoting.   c. casings.
    b. exterior corners.   d. interior corners.

3. A stool is part of the
    a. soffit.   c. base.
    b. door trim.   d. window trim.

4. The joint between moldings in interior corners is usually
    a. coped.   c. butted.
    b. mitered.   d. bisected.

5. The setback of door casings from the face of the jamb is often referred to as a
    a. gain.   c. reveal.
    b. backset.   d. quirk.

6. Find the length of side casings by
    a. measuring the distance from floor to the header casing.
    b. marking the length on a scrap strip and transferring it to the side casing.
    c. turning the side casing upside down with the point of the miter against the floor.
    d. holding the side casing with the right end up and marking the miter.

7. If the joint between side and head casings needs fitting
    a. sand it.   c. plane it.
    b. fill it.   d. nail it.

8. The cope on baseboard is laid out by
    a. back mitering.   c. a combination square.
    b. returning it.   d. scribing.

9. The base shoe is fastened
    a. to the baseboard only.
    b. to both the base and the floor.
    c. to the floor only.
    d. directly to the wall.

10. When the end of a molding has no material to butt against, its end is
    a. back mitered.   c. returned upon itself.
    b. mitered.   d. coped.

28

OBJECTIVES

After completing this unit, the student will be able to

- *name the parts of stair finish and describe their location and function.*
- *state how a housed stringer staircase differs from a job-built staircase.*
- *apply the following finish to a job-built staircase: risers, newel posts, open and closed finish stringers, treads, return nosings, tread molding, handrail, and balusters.*

The staircase is usually an outstanding feature of a residence. Therefore, all work must be done in a first-class manner. The stair finish members must be tight fitting. Joints on some members must be tight fitting. Joints on some parts, such as handrails and posts, must be made strong enough to provide support and prevent accidents.

STAIR FINISH PARTS

Treads and risers were described and discussed in detail in an earlier unit.

A *nosing* is that part of a stair tread which extends beyond the riser. A *return nosing* is a separate piece mitered to the open end of a tread, Figure 28-1. Return nosings are available in the same thickness as treads, in widths of 1 1/8'' and in lengths of 1'-2''. The tread molding

Fig. 28-1 A return nosing

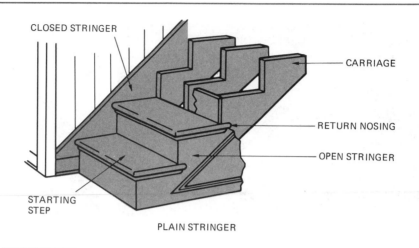

Fig. 28-2 Open and closed finish stringers

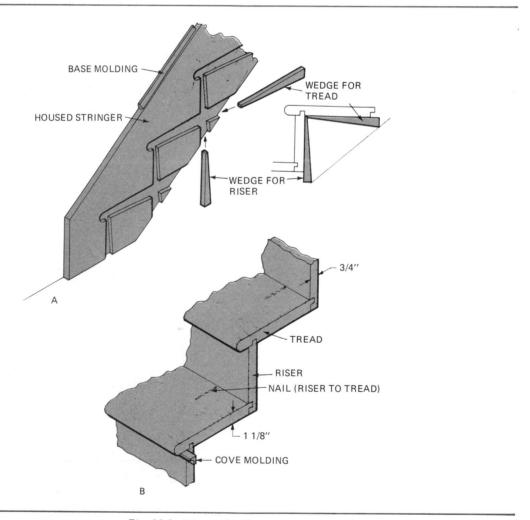

Fig. 28-3 A housed stringer

is usually a small cove used to cover the joint between treads and risers and between treads and open stringers.

Stringers, sometimes called *skirts*, are finish boards used on each end of the risers and treads. Stringers are called *closed* when they are placed against a wall into or against which the treads and risers are fitted. Stringers are referred to as *open* when they face the open end of a stairway. An open stringer is cut so the treads lay over it and the risers are mitered to it, Figure 28-2. Stringers are usually cut from 1x10 or 1x12 softwood lumber. *Housed stringers* are dadoed and grooved to receive the ends of treads and risers, Figure 28-3.

A *starting step* is the first tread-and-riser unit in a stairway. In many cases the starting step is longer than the rest of the treads. Starting steps are usually used when the stair rail curves outward at the bottom. They are available in a number of styles with quarter or semi-circular ends, preassembled, and ready for installation, Figure 28-4.

The *handrail* is the top finishing piece grasped by the hand of the person ascending or descending the stairs. Handrails come in various shapes. They are furnished by mills in straight lengths and in sections of special shapes.

At the start of a stairway, both horizontal and vertical volutes and turnouts may be used.

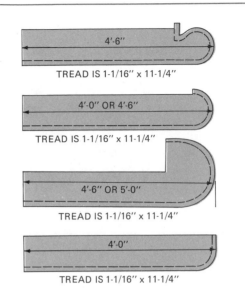

Fig. 28-4 Starting steps are available in a number of styles and sizes.

A gooseneck is generally used at a landing. Many other shapes are available such as quarter and half turns and concave and convex easements, Figure 28-5. These specially shaped sections are joined to straight lengths to obtain the desired shape of the total handrail. Special handrail bolts are used to join sections of handrail together. The bolt has a lag screw thread on one end and a machine screw thread and nut on the

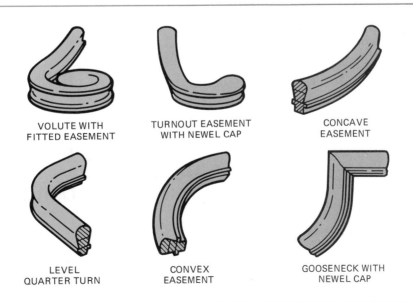

VOLUTE WITH
FITTED EASEMENT

TURNOUT EASEMENT
WITH NEWEL CAP

CONCAVE
EASEMENT

LEVEL
QUARTER TURN

CONVEX
EASEMENT

GOOSENECK WITH
NEWEL CAP

Fig. 28-5 Special handrail parts

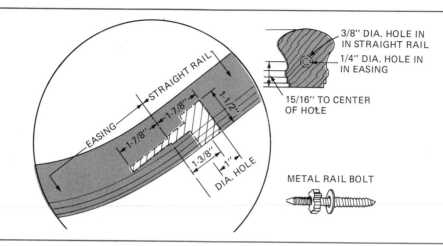

Fig. 28-6 Handrail bolts are used to join sections of handrail.

other end, Figure 28-6. To install a rail bolt to join the easing and the straight rail, drill a 1/4'' hole into the easing 1 7/8'' deep and turn the lag screw bolt in 1/2 the length of the bolt. Then drill a 3/8'' hole in the straight rail 1 7/8'' deep. Drill a 1'' diameter hole in the underside of the straight rail, 1 3/8'' from the end of the rail to the center of the hole. Place a washer and the nut on the threaded bolt end and tighten with a nail set. When perfect alignment of the easement and the straight rail is assured, plug the 1'' hole with a wood plug.

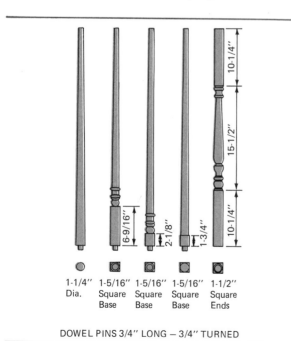

DOWEL PINS 3/4'' LONG — 3/4'' TURNED

Fig. 28-7 Balusters are available in many styles and sizes.

Balusters are usually turned, decorative pieces placed at intervals between the handrail and treads. Usually two or three balusters are used on each tread in a straight run. The bottom end is turned to a smaller diameter to fit in a hole bored in the treads. The top end is cut on a bevel to fit against the underside of the handrail. Balusters are furnished in lengths of 30, 33, 36, 39, and 42 inches so that they can be used in any part of the stairway. Because of the pitch of the handrail, more than one length must be used in a single staircase, Figure 28-7.

Newel posts are posts anchored securely to the staircase that support the ends of the handrail. *Starting newels* are used at the bottom of a staircase. *Landing newels* are used at landings or at the top of the staircase. Newel posts are available in a number of styles and sizes, Figure 28-8.

The entire rail assembly on the open side of a stairway or landing is called a *balustrade*. It includes all the rail details, including the handrail, balusters, and newel posts, Figure 28-9.

DIFFERENCE BETWEEN HOUSED AND JOB-BUILT STAIRCASES

Staircases are either built on the job or made in a mill. A housed staircase is usually made in a mill, preassembled, and delivered to the job for installation. A housed staircase differs from a job-built staircase mainly in the way the treads and risers are supported.

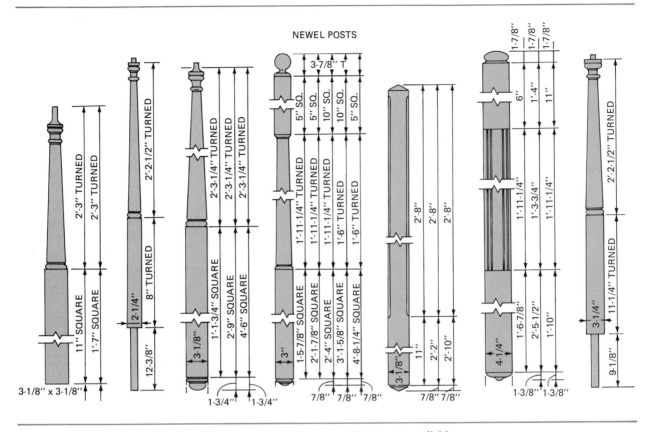

Fig. 28-8 Many kinds of newel posts are available.

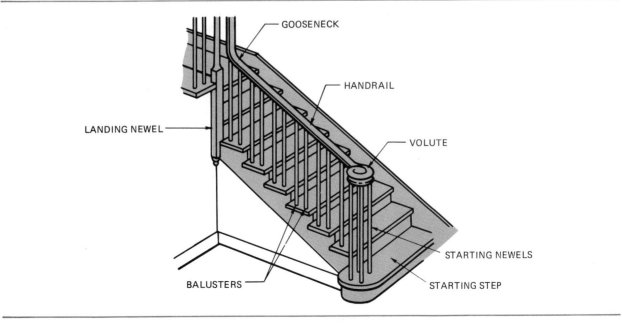

Fig. 28-9 The entire rail assembly is called a balustrade.

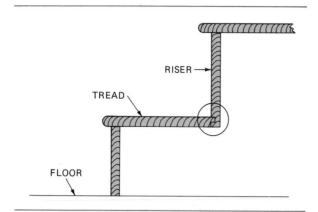

Fig. 28-10 A groove is made in the riser for the rabbeted inner edge of the tread.

In a job-built staircase, the risers and treads are supported by cutout framing members, called *stair horses*, *stair carriages*, or *rough stringers*. See Unit 16 Stair Framing.

In a *housed-stringer staircase*, the risers and treads are supported by being fitted into grooves routed into the stringer. The grooving of the stringer to "let-in" the risers and treads is called *housing the stringer*, which gives this type of staircase its name. The housed staircase is a rigid, self-contained unit which is not affected by shrinkage or movement of the house frame.

FINISHING A JOB-BUILT CLOSED STAIRCASE

A closed staircase is one that runs between two walls. Finishing this type of staircase means working with risers, stringers, treads, tread molding, and the handrail.

RISERS

Usually the first trim applied to the stair framing are the risers. Rip the riser stock to the proper width. The width of the riser stock is the height of the rise on the rough stringer. Cut enough risers for the staircase to a rough length. A *rough length* is a few inches longer than actually needed. Determine the face side of each piece. Then make a groove 3/8"x3/8" near the bottom edge on the face side of all but one piece. The groove is made so its top side is a distance from the bottom edge equal to the

tread thickness. The groove is made so that the the rabbeted inner edge of the tread fits into it, Figure 28-10. The starting riser is not grooved. Cut the riser stock to a length equal to the distance between walls with both ends square. Sand the face side of all risers smooth before fastening them in position. Start at the top and work down. Remove temporary treads as work progresses downward.

CAUTION: Put up positive barriers at the top and bottom of the stairs so that the stairs cannot be used while the finish is applied. A serious accident can happen if a person who does not realize that the temporary treads have been removed, uses the stairs.

Fasten each riser in position with its bottom edge on and its top edge flush with the tread cutout of the stair carriage. The bottom edge of the starting riser rests on the subfloor. If the stairs are built on a finish floor, the starting riser must be cut narrower. Use three 8d finishing nails into each stair carriage. Set all nails.

CLOSED FINISH STRINGER

The closed finish stringer can be compared to a baseboard that runs along the staircase. In fact, the finish stringer must be joined at both ends — to the base on the lower floor and the upper floor. The finish stringer is applied before the base that it meets. It is notched around the previously installed risers and the tread cut of the stair carriage so that its top edge is at least 2 inches above the intersection. Usually 1x10 lumber is sufficient.

Cut a length of lumber so that its bottom edge is resting on the stairs, its bottom end is cut at an angle to fit roughly against the subfloor, and the top end extends about 6 inches beyond the stairway.

With the back side against the wall, tack it in place. Plumb up and across the face of the finish stringer from the face of the risers. If the riser itself is out of plumb, then make the layout from that part of the riser that projects farthest outward.

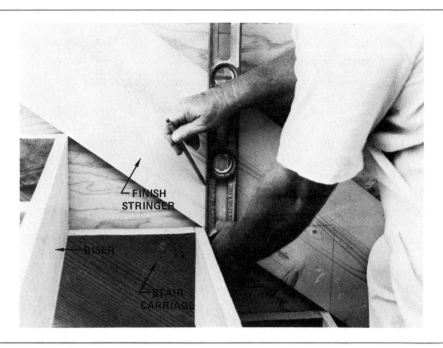

Fig. 28-11 Layout of a closed finish stringer

Level from and about 1/4'' above each tread cut of the stair carriage. Use light, fine lines when making the layout, Figure 28-11.

Remove the stringer and cut to the layout lines. Use a fine-toothed crosscut saw. Follow the plumb lines carefully and make a slight undercut. Plumb cuts butt against the risers, and a fine joint needs to be made here. Not as much care needs to be taken with level cuts. Level cuts are hidden because treads butt against them and cover them.

After the cutouts are made in the finish stringer, tack it back in position and lay out both ends. Lay out plumb lines on both ends so their height is the same as the base to be installed on the walls. Remove the stringer, make the end cuts, sand the board, and place it back in position. Fasten the stringer securely to the wall with finishing nails. Do not nail too low. Care must be taken to avoid splitting the lower end of the stringer.

At each step, and along the wall, drive shims between the back side of the riser and the stair carriage. This drives the riser tightly against the plumb cut of the finish stringer. Usually wood shingle tips are used for shims. Install the finish stringer on the other wall in the same manner.

TREADS

Treads are rabbeted on the back edges to fit into the grooves of the risers. The nosed edge projects beyond the face of the riser by 1 1/8 inches, Figure 28-12. Treads are cut on both ends to fit snugly against the stringers. Rip the tread stock to width and rabbet the back edges.

Along the top edge of the riser, measure carefully the distance between finish stringers. Transfer the measurement and square lines across the tread. Cut in from the nosed edge, square through the thickness for a short distance, and then undercut slightly. Smooth the cut

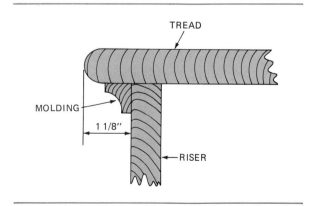

Fig. 28-12 Details of a tread nosing

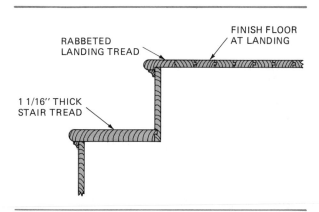

Fig. 28-13 A rabbeted landing tread is used at the top of the stairway when 1 1/16" treads are used.

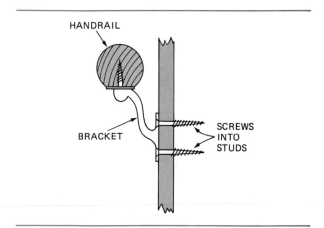

Fig. 28-14 Handrails are attached to walls by means of handrail brackets.

ends with a block plane. Rub one end with wax. (Use a short piece of candle.) Place the other end in position and press on the waxed end until the tread lays flat on the stair carriages.

Place a short block on the nosed edge and tap it until the rabbeted edge is firmly seated in the groove of the riser. Fasten the tread in place with three 8d finish nails into each stair carriage. Set the nails. It may be necessary to drill holes in hardwood treads for the nails to prevent splitting the tread or bending the nail. A little wax applied to the nail makes driving easier and prevents bending the nail.

Start from the bottom and work up, installing the rest of the treads in a similar manner. At the landing, install a 3 1/2" wide landing tread. If 1 1/16" thick treads are used on the staircase, use a landing tread that is rabbeted for 13/16" flooring, Figure 28-13.

TREAD MOLDING

The *tread molding* is usually a small cove made of the same material as the tread. It is installed under the overhang of the tread and against the riser. Cut the molding to length, using a miter box. Predrill holes and fasten the molding in place with 4d finish nails spaced about 12 inches apart. Nails are driven at about a 45° angle through the center of the molding.

HANDRAIL

The *handrail* on a straight closed stairway is usually a straight section attached to the wall by

means of handrail brackets, Figure 28-14. The ends of the handrail are usually returned upon themselves.

Most building codes state that the handrail height shall be not less than 30 inches or more than 36 inches. This height is taken from the top of the tread along a plumb line in line with the face of the riser, Figure 28-15. Handrails are required only on one side in stairways of less than 44 inches in width.

Deduct the distance from the top of the handrail to the bottom screw hole in the handrail bracket from the desired handrail height. Mark the remainder on the wall at each end of the staircase. The measurement is taken along a plumb line from the top of the tread and in line with the riser face. Snap line between the marks.

Space brackets not over 4 feet O.C. and in such a way that bracket screws go into solid wood. Attach brackets to the wall so that the

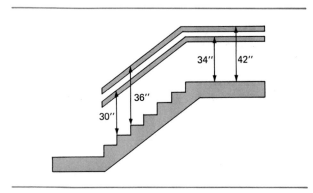

Fig. 28-15 Handrail heights

bottom screw is to the chalk line. Install additional screws into each bracket. Make sure all screws penetrate into solid wood. Apply a little wax to the screw threads for easier driving. (See Unit 3 Fasteners.) Cut the handrail to length, return its ends, and fasten it to the top of the handrail brackets.

FINISHING A JOB-BUILT OPEN STAIRCASE

The order in which trim is applied to an open staircase varies with the design. In most cases the order of application is as follows:

- Install starting riser or starting step and landing riser.
- Set starting and landing newel posts.
- Lay out and install open finish stringer.
- Install remainder of risers.
- Install closed finish stringer if staircase is closed on one side.
- Install treads.
- Apply return nosings.
- Apply tread molding.
- Install handrail including special-shaped sections.
- Lay out and install balusters.

SETTING NEWEL POSTS

Install the bottom and landing risers in the manner previously described. Newel posts must be set plumb, to the proper height, and strong enough to resist lateral pressure applied to persons using the staircase. The height of the newel post depends on the design. Sometimes handrails butt the post and in other cases, the handrail is applied on top of the post. Follow the manufacturer's directions for the height of the newel post above the tread.

STARTING NEWEL

Notch the bottom of the post so that it fits around the outside corner of the bottom step.

Fig. 28-16 The starting newel is notched to fit around the outside corner of the bottom step.

The depth of the notch on the side of the stair carriage should be such that the center of the post is in line with the outside face of the stair carriage. The depth of the notch that lies against the riser is such that the same amount of post is exposed on both sides when finished, Figure 28-16.

The top of the notch (actually it is a blind rabbet) rests on the top edge of the riser and tread cut of the stair carriage. If the post is long enough, cut an opening in the floor to allow the bottom of the post to extend below the floor. By extending the post below the floor, blocking can be installed to which the post is fastened to give it added strength.

Also fasten the post above the floor. Install a short piece of 1x6 with 45° cuts on both ends, on the inside and across the corner made by the riser and the stair carriage. Bore holes of proper size through the block and into the

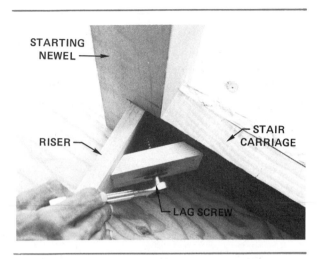

Fig. 28-17 Method of fastening newel posts

post and fasten the post with a lag screw, Figure 28-17.

If the post is slightly out of plumb after it is fastened, loosen the lag screw slightly and install thin shims to plumb the post. On one side, install the shims (near the bottom or top of the notch as necessary) between the post and the stair carriage. On the other side, shim between the riser and the stair carriage. When plumb, retighten the lag screw.

USING A STARTING STEP

If a special starting step which extends beyond the side of the staircase is used, the starting newel is set on top of it. In this case, the bottom end of the newel is turned smaller to go through a hole bored through the tread and the floor.

First, fit the starting step in position and fasten it securely without the tread.

STARTING NEWEL ON A STARTING STEP

Follow the manufacturer's directions for the location and size of the hole to be bored in the tread. Bore a hole in the tread and in the floor directly in a plumb line.

A newel post of this type is not set until the treads have been installed. However, when the treads are installed, the newel post is installed through the bored holes. A through mortise is made in the part that extends below the floor and the post is fastened by driving wedges through the mortise and against the bottom side of the floor, Figure 28-18.

INSTALLING THE LANDING NEWEL

The landing newel is installed in a somewhat similar manner as the starting newel. However, the landing newel is much longer because it accommodates a handrail at the landing, covers the top riser, provides a stop for and extends

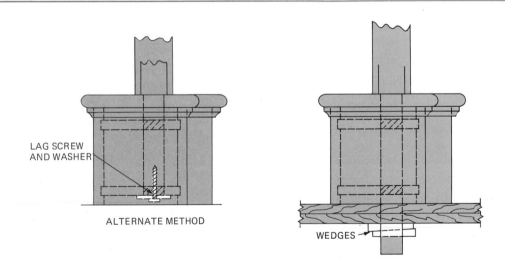

Fig. 28-18 Wedging a newel post below the floor

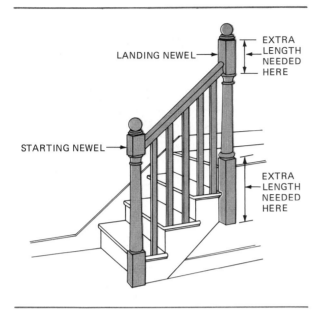

Fig. 28-19 The landing is much longer
than the starting newel.

Fig. 28-20 A "preacher" is used to lay out an
open finish stringer.

below the open finish stringer. The bottom end
is exposed and, therefore, has a decorative
design, Figure 28-19.

It is notched around the corner at the
intersection of the staircase and the landing.
However, in this case, the notch is made the
same as in the starting newel; that is, in such a
manner that when the post is installed, its
centerline lines up with the outside face of the
stair carriage.

The landing newel is fastened with lag
screws from the inside in a manner similar to
that used in fastening the starting newel.

LAYING OUT AND INSTALLING
THE OPEN FINISH STRINGER

Cut a length of 1x10 or 1x12 stock so that
it fits between newel posts and its top edge is
flush with top corners of the stair carriage. Tack
it in this position. If a starting step is used, fit it
between the starting step and the landing
newel. Lay out level lines on the face of the
stringer from the tread cut on the stair carriage.

Lay out plumb lines from the face of the
riser. Since the risers have not yet been installed,
use a preacher. A *preacher* can be made from a

piece of stock about 12" long and with a 4"
nominal width. The thickness must be the
same as the riser stock. It is notched in the
center to fit over the finish stringer and rest on
the tread cut of the stair carriage, Figure 28-20.
Place the preacher over the stringer and against
the riser cut of the stair carriage. Plumb it with
a hand level. Lay out the plumb cut on the
stringer by marking along the side of the preacher
that faces the bottom of the stairway.

Mark the top edge of the stringer along the
side of the preacher that faces the top of the
staircase. Connect both lines diagonally across
the top edge of the stringer for the miter cut,
Figure 28-21. Lay out all plumb lines on the
stringer in this manner.

Remove the stringer and cut to the layout
lines. Make miter cuts along the plumb lines and
cut square through the thickness along the level

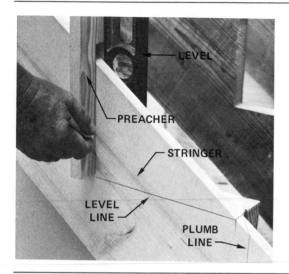 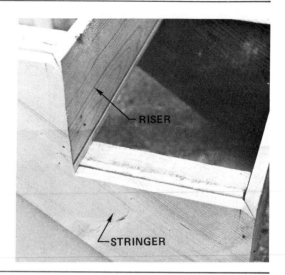

Fig. 28-21 Layout of an open stringer

lines. Sand the piece and fasten it in position. To assure getting the piece in the same position as it was when laid out, first fasten it in the holes in which the piece was originally tacked.

INSTALLING RISERS

Install the remainder of the risers. Cut to length by making a square cut on the end that goes against the wall. Make a mitered cut on the other end to fit to the open finish stringer. Sand all pieces before installation, apply a small amount of glue to the miters, and fasten them in position. Drive finish nails both ways through the miter to hold the joint tight. Wipe off any excess glue and set all nails.

INSTALLING THE CLOSED STRINGER

Lay out and install the closed stringer in the same manner as described for a closed staircase.

INSTALLING THE TREADS

Rip the treads to width and rabbet the back edges. Make allowance for the rabbet when ripping treads to width. Cut one end to fit against the closed finish stringer. Make a cut on the other end to receive the return nosing. This is a combination square and miter cut. The square cut is made flush with the outside face of

the open finish stringer. The miter starts from the nosed edge 1 1/8 inches beyond the square cut, as shown in Figure 28-1. The starting tread is notched around the newel post eliminating the miter. Cut and install all treads in the staircase in this manner.

APPLYING THE RETURN NOSINGS

The return nosings are applied to the open ends of the treads. Miter one end of the return nosing to fit against the miter cut on the tread. Cut the other end square. This end extends beyond the face of the riser 1 1/8 inches, the same amount as the overhang of the nosing beyond the finish stringer. Return the square end on itself. Predrill holes in the nosing for nails. Locate the holes so they are not in line with any balusters later installed. Holes must be bored in the treads to receive the balusters. Any nails in line with the holes will ruin the boring tool.

Apply glue to the joint, and fasten the return nosing to the end of the tread with three 8d finishing nails. Set all nails and sand the joint flush. Apply all other return nosings in the same manner.

APPLYING THE TREAD MOLDING

The tread molding is applied in the same manner as for closed staircases except it is

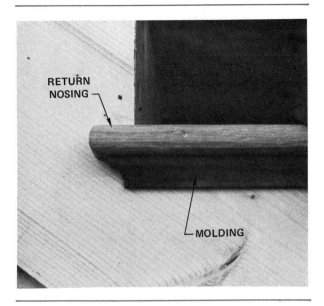

Fig. 28-22 The back end of the return nosing and molding are returned upon themselves.

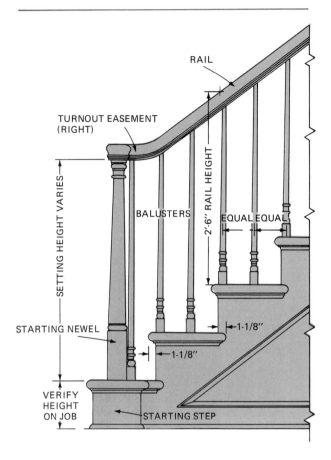

Fig. 28-23 Layout of balusters

mitered and returned around onto the open stringer. The back end of the return molding is cut so that the return nosing projects beyond it the same distance on the end as on the side. The back end of the molding is then returned upon itself, Figure 28-22.

INSTALLING THE HANDRAIL

In most cases, the handrail is cut to fit between newel posts. Mark the height, cut the handrail to fit, and fasten to the posts with a handrail bolt.

In case sections of handrail need to be joined, also use a handrail bolt. Follow the manufacturer's direction for the bevel cut on the ends of the sections. Mark the center of each section carefully. Bore a hole in the end of one section to receive the lag end of the handrail bolt. Bore a hole in the end of the other section to receive the machine-threaded portion. At the proper distance from the end of this section, bore a hole in the bottom to receive and tighten the nut. Install the lag end of the bolt in one section by using double nuts turned against each other on the machine-treaded end.

Put the two sections together, insert and tighten the nut, and plug the bottom hole. Sand the joint flush.

INSTALLING BALUSTERS

Mark the centerline of the balusters on the treads. Each tread usually carries two balusters and sometimes three. The front side of the leading baluster in each tread is aligned with the face of the riser below. Other balusters are spaced so there are equal distances between them, Figure 28-23. They are aligned vertically with the centerline of posts and handrail.

At the centerline of each baluster bore holes in the treads of the required diameter and depth. Plumb from the centerline of the holes up to and mark the handrail for each baluster. Measure, mark, and cut the top end of the baluster on an angle to fit against the bottom of the handrail. Test the baluster for fit and plumb and fasten it in position by drilling and toenailing with finish nails to the under side of the handrail. It is good practice to install the first baluster near the center of the handrail while

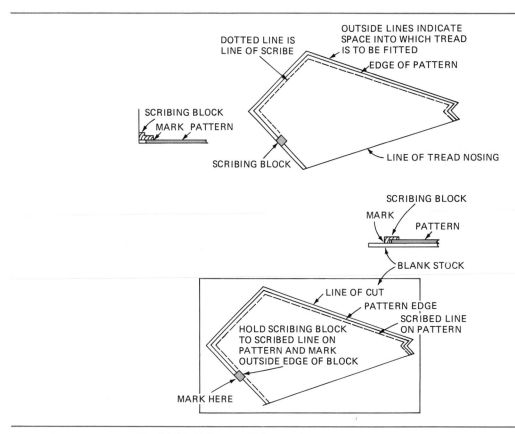

Fig. 28-24 Using a scribing block and pattern to lay out winding treads

sighting the handrail for straightness. Care must be taken when installing balusters not to force the handrail upward, creating a crown in the rail.

INSTALLING TREADS ON WINDERS

Treads on winding steps are especially difficult to fit because of the angles on both ends. A method used by many carpenters involves the use of a pattern and a scribing block.

Cut a thin piece of plywood so it fits in the tread space within 1/2 inch on the ends and back edge. The outside edge should be straight and in line with the nosed edge of the tread when installed.

Tack the plywood pattern in position. Use a 3/4″ block rabbeted on one side the thickness of the pattern to scribe the ends and back edge. Scribe by riding the block against stringers, riser, and post while marking its inside edge on the pattern.

Remove the pattern and tack it on the tread stock. Place the block with its rabbeted side down and inside edge to the scribed lines on the pattern. Mark the tread stock by marking on the outside edge of the block, Figure 28-24.

PROTECTING THE FINISHED STAIRS

When all trim is applied to the stairs, protect the risers and treads by applying a width of rosin paper to them. Unroll a length down the stairway. Hold the paper in position by tacking thin strips of wood to the risers.

REVIEW QUESTIONS

Select the most appropriate answer.

1. The rounded outside edge of a tread that extends beyond the riser is called a
   a. housing.
   b. turnout.
   c. coving.
   d. nosing.

2. Finish boards between the stairway and the wall are called
   a. returns.
   b. balusters.
   c. stringers.
   d. casings.

3. Treads are rabbeted on their back edge to fit into
   a. risers.
   b. housed stringers.
   c. return nosings.
   d. newel posts.

4. An open stringer is
   a. housed to receive risers.
   b. mitered to receive risers.
   c. housed to receive treads.
   d. mitered to receive treads.

5. A volute is part of a
   a. tread.
   b. baluster.
   c. newel post.
   d. handrail.

6. The entire rail assembly on the open side of a stairway is called a
   a. baluster.
   b. balustrade.
   c. guardrail.
   d. finish stringer assembly.

7. In a job-built staircase, the treads and risers are supported by
   a. stair carriages.
   b. housed stringers.
   c. each other.
   d. blocking.

8. One of the first things to do when about to trim a staircase is
   a. check the rough framing for rise and run.
   b. block the staircase so no one can use it.
   c. straighten the stair carriages.
   d. install all the risers.

9. Treads usually project beyond the face of the riser
   a. 3/4 inch.
   b. 1 1/8 inches.
   c. 1 1/4 inches.
   d. 1 3/8 inches.

10. Newel posts are notched around the stair carriage so that their centerline lines up with the
    a. centerline of the stair carriage.
    b. outside face of the stair carriage.
    c. outside face of the open stringer.
    d. centerline of the handrail.

FINISH FLOORS 29

OBJECTIVES

After completing this unit, the student will be able to

- *describe the kinds, sizes, and grades of hardwood finish flooring.*
- *apply strip, plank, and parquet finish flooring.*
- *apply underlayment and resilient tile flooring.*
- *estimate quantities of finish flooring.*

A number of materials are used for finish floors. Wall to wall carpet, applied over an underlayment, is used widely. The carpenter usually applies the underlayment, but not the carpet. Concrete floors, finished in a number of ways, are the responsibility of masons. Carpenters install wood finish floors, long time favorites because of their durability, beauty, and warmth. Resilient sheet and tile floors are widely used in bathrooms and kitchens and are usually installed by specialists. On occasion, however, the carpenter may be required to install a tile floor.

HARDWOOD FINISH FLOORING

Lumber used in the manufacture of hardwood flooring has been air-dried, kiln-dried, cooled, and then accurately machined to exacting standards. It is a fine product that should receive proper care during handling and installation.

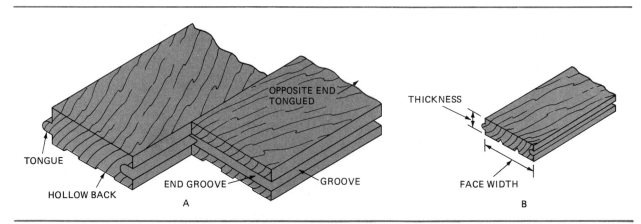

Fig. 29-1 Strip flooring is usually edge and end matched.

If possible, unload flooring in good weather; never unload while it is raining. Flooring should not be delivered to the job site until the building has been closed in — with outside windows and doors in place — and until cement work, plastering, and other materials are thoroughly dry. Leave adequate room around bundles of flooring for good air circulation. Let the bundles become acclimated to the atmosphere of the building for a few days before installation. If the building is heated, temperature should not exceed 72°F. In winter construction, the building should be heated prior to delivery of the flooring, with heat maintained until the floor is installed and finished.

KINDS AND SIZES

Most hardwood finish flooring is made from white or red oak, although beech, birch, hard maple, and pecan are also used commonly. For less expensive finish floors, some softwoods such as Douglas fir, hemlock and southern pine are used.

Hardwood flooring comes in strip, plank, and parquet form. Most finish floors are laid using strips.

Strip Flooring. Strip flooring usually comes with ends and edges matched, Figure 29-1. Some strips are manufactured with square edges. Square-edge flooring is not used frequently, however.

The flooring comes in bundles in lengths of 1 1/4 feet and up. Pieces in the bundle are not of equal lengths. A bundle may include pieces from 6 inches under to 6 inches over the nominal length of the bundle. No pieces shorter than 9 inches are allowed.

Hardwood strip flooring comes in a standard thickness of 3/4 inch and in face widths of 1 1/2, 2, 2 1/4, and 3 1/4 inches. In the 1/2-inch thickness, it is available in face widths of 1 1/2 and 2 inches. A face width is the exposed width when installed. Other thicknesses and widths are available on special order.

Unfinished strip flooring has sharp corners at the intersection of the face and edge. Any unevenness in the faces of adjoining pieces are sanded flush after installation. Because prefinished floor cannot be sanded after installation, a chamfered corner is machined on the flooring. When installed, these chamfered corners form small V-grooves between adjoining pieces, thus obscuring any unevenness of the surface, Figure 29-2.

Plank Flooring. Plank flooring is similar to strip flooring except that it comes in random widths of from 3 to 8 inches. This flooring may have counterbored holes for securing the planks with wood screws. These holes are then filled with wood plugs that are supplied with the flooring, Figure 29-3. Prefinished plank flooring is installed by blind nailing but has plugs of contrasting color already installed to simulate screw fastening. One or more plugs according to the width are used across the face at spaced intervals.

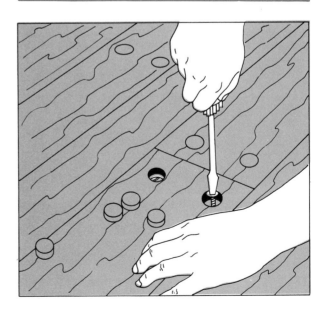

Fig. 29-3 Plank flooring is fastened with screws which are covered by plugs. *(Courtesy of National Oak Flooring Manufacturers' Association)*

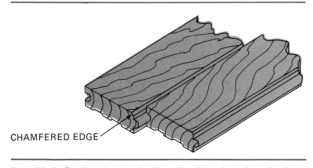

CHAMFERED EDGE

Fig. 29-2 Prefinished strip flooring has chamfered edges.

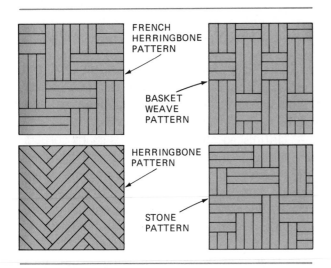

Fig. 29-4 Parquet flooring may be laid in a number of interesting patterns.

Parquet Flooring. Parquet flooring is a type of floor that is laid in a number of intricate patterns and consequently is often referred to as *pattern flooring*, Figure 29-4. This type is manufactured in short pieces with each piece cut to exact lengths to match each other and to match multiples of its width. Each piece is end and edge matched with face lengths running from 6 3/4 to 13 1/2 inches as desired. Widths and thicknesses are generally the same as in standard strip flooring. Herringbone, basketweave, and other interesting patterns are made using parquet strips.

Block flooring is another form of parquet flooring. It differs from pattern flooring in that the strips are assembled into square or rectangu-lar blocks at the factory. Various means are used to hold the pieces together — depending on the manufacturer. The edges of the blocks are tongue-and-grooved and are laid in units. Various sizes are available with 9 x 9-inch squares used frequently, Figure 29-5. Blocks less than 1/2 inch are usually set in a bed of mastic. Thicker blocks are usually blind nailed into place.

GRADES OF HARDWOOD FLOORING

Uniform grading rules have been established for hardwood strip flooring. There are no official grading rules for plank and parquet flooring.

Standards for strip flooring have been established by the National Oak Flooring Manufacturers Association and the Maple Flooring Manufacturers Association. Members of these associations stamp their product with the association trademark to assure consumers of adherence to high quality standards.

Oak. Oak flooring is available *quarter sawed* and *plain sawed*. The grades in declining order are *clear, select, no. 1 common, no. 2 common,* and *1 1/4' shorts.* Grades are based on the number of defects and the length. For instance, bundles of 1 1/4' shorts contain pieces from 9 to 18 inches long. The average length of clear bundles is 3 3/4 feet. Quarter-sawed flooring is available in clear and select grades, Figure 29-6.

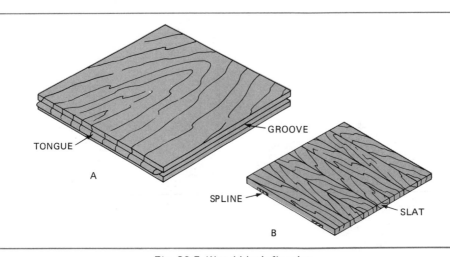

Fig. 29-5 Wood block flooring

Guide to Hardwood Flooring Grades

A brief grade description, for comparison only. NOFMA flooring is bundled by averaging the lengths. A bundle may include pieces from 6 inches under to 6 inches over the nominal length of the bundle. No piece shorter than 9 inches admitted. The percentages under 4 ft. referred to apply on total footage in any one shipment of the item. ¾ inch added to face length when measuring length of each piece.

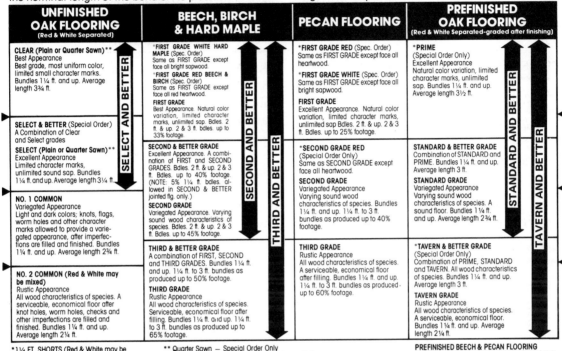

UNFINISHED OAK FLOORING (Red & White Separated)

CLEAR (Plain or Quarter Sawn)
Best Appearance
Best grade, most uniform color, limited small character marks. Bundles 1¼ ft. and up. Average length 3¾ ft.

SELECT & BETTER (Special Order)
A Combination of Clear and Select grades

SELECT (Plain or Quarter Sawn)
Excellent Appearance
Limited character marks, unlimited sound sap. Bundles 1¼ ft. and up. Average length 3¼ ft.

NO. 1 COMMON
Variegated Appearance
Light and dark colors; knots, flags, worm holes and other character marks allowed to provide a variegated appearance, after imperfections are filled and finished. Bundles 1¼ ft. and up. Average length 2¾ ft.

NO. 2 COMMON (Red & White may be mixed)
Rustic Appearance
All wood characteristics of species. A serviceable, economical floor after knot holes, worm holes, checks and other imperfections are filled and finished. Bundles 1¼ ft. and up. Average length 2¼ ft.

SELECT AND BETTER

1¼ FT. SHORTS (Red & White may be mixed)
Unique Variegated Appearance. Lengths 9 inches to 18 inches. Bundles average nominal 1¼ ft. Production limited.

*NO. 1 COMMON & BETTER SHORTS
A combination grade, CLEAR, SELECT, & NO. 1 COMMON 9 inches to 18 inches.

*NO. 2 COMMON SHORTS
Same as No. 2 COMMON, except length 9 inches to 18 inches.

BEECH, BIRCH & HARD MAPLE

*FIRST GRADE WHITE HARD MAPLE (Spec. Order)
Same as FIRST GRADE except face all bright sapwood.
*FIRST GRADE RED BEECH & BIRCH (Spec. Order)
Same as FIRST GRADE except face all red heartwood.

FIRST GRADE
Best Appearance. Natural color variation, limited character marks, unlimited sap. Bundles 2 ft. & up. 2 & 3 ft. bdles. up to 33% footage.

SECOND & BETTER GRADE
Excellent Appearance. A combination of FIRST and SECOND GRADES. Bdles. 2 ft. & up. 2 & 3 ft. Bdles. up to 40% footage. (NOTE: 5% 1¼ ft. bdles. allowed in SECOND & BETTER jointed flg. only.)

SECOND GRADE
Variegated Appearance. Varying sound wood characteristics of species. Bdles. 2 ft. & up. 2 & 3 ft. Bdles. up to 45% footage.

THIRD & BETTER GRADE
A combination of FIRST, SECOND and THIRD GRADES. Bundles 1¼ ft. and up. 1¼ ft. to 3 ft. bundles as produced up to 50% footage.

THIRD GRADE
Rustic Appearance
All wood characteristics of species. Serviceable, economical floor after filling. Bundles 1¼ ft. and up. 1¼ ft. to 3 ft. bundles as produced up to 65% footage.

SECOND AND BETTER

THIRD AND BETTER

** Quarter Sawn — Special Order Only
*Check with supplier for grade and species available.
†NESTED FLOORING: Random length tongued and grooved, end-matched flooring is bundled end to end continuously to form 8 ft. long (nominal) bundles. Regular grade requirements apply.
‡NESTED FLOORING: If put up in 8 ft. nested bundles, 9 to 18 inch pieces will be admitted in ¾" x 2¼" as follows in the species of Beech, Birch & Hard Maple: FIRST GRADE, 4 pcs. per bundle; SECOND GRADE, 8 pcs.; THIRD GRADE, as develops. Average lengths: FIRST GRADE, 42 inches; SECOND GRADE, 33 inches; THIRD GRADE, 30 inches.

PECAN FLOORING

*FIRST GRADE RED (Spec. Order)
Same as FIRST GRADE except face all heartwood.
*FIRST GRADE WHITE (Spec. Order)
Same as FIRST GRADE except face all bright sapwood.

FIRST GRADE
Excellent Appearance. Natural color variation, limited character marks, unlimited sap Bdles. 2 ft. & up. 2 & 3 ft. Bdles. up to 25% footage.

*SECOND GRADE RED (Special Order Only)
Same as SECOND GRADE except face all heartwood.

SECOND GRADE
Variegated Appearance
Varying sound wood characteristics of species. Bundles 1¼ ft. and up. 1¼ ft. to 3 ft. bundles as produced up to 40% footage.

THIRD GRADE
Rustic Appearance
All wood characteristics of species. A serviceable, economical floor after filling. Bundles 1¼ ft. and up. 1¼ ft. to 3 ft. bundles as produced up to 60% footage.

PREFINISHED OAK FLOORING (Red & White Separated-graded after finishing)

*PRIME (Special Order Only)
Excellent Appearance
Natural color variation, limited character marks, unlimited sap. Bundles 1¼ ft. and up. Average length 3½ ft.

STANDARD & BETTER GRADE
Combination of STANDARD and PRIME. Bundles 1¼ ft. and up. Average length 3 ft.

STANDARD GRADE
Variegated Appearance
Varying sound wood characteristics of species. A sound floor. Bundles 1¼ ft. and up. Average length 2¾ ft.

*TAVERN & BETTER GRADE (Special Order Only)
Combination of PRIME, STANDARD and TAVERN. All wood characteristics of species. Bundles 1¼ ft. and up. Average length 3 ft.

TAVERN GRADE
Rustic Appearance
All wood characteristics of species. A serviceable, economical floor. Bundles 1¼ ft. and up. Average length 2¼ ft.

STANDARD AND BETTER

TAVERN AND BETTER

PREFINISHED BEECH & PECAN FLOORING
*TAVERN & BETTER GRADE (Special Order Only)
Combination of PRIME, STANDARD and TAVERN. All wood characteristics of species. Bundles 1¼ ft. and up. Average length 3 ft.

Standard Sizes, Counts & Weights

Nominal	Actual	Counted	Weights M Ft.
TONGUE AND GROOVE-END MATCHED			
** ¾x3¼ in.	¾x3¼ in.	1x4 in.	2210 lbs.
¾x2¼ in.	¾x2¼ in.	1x3 in.	2020 lbs.
¾x2 in.	¾x2 in.	1x2¾ in.	1920 lbs.
¾x1½ in.	¾x1½ in.	1x2¼ in.	1820 lbs.
** ⅜x2 in.	11/32x2 in.	1x2½ in.	1000 lbs.
** ⅜x1½ in.	11/32x1½ in.	1x2 in.	1000 lbs.
** ½x2 in.	15/32x2 in.	1x2½ in.	1350 lbs.
** ½x1½ in.	15/32x1½ in.	1x2 in.	1300 lbs.
SQUARE EDGE			
** 5/16x2 in.	5/16x2 in.	face count	1200 lbs.
** 5/16x1½ in.	5/16x1½ in.	face count	1200 lbs.

Nominal	Actual	Counted	Weights M Ft.
SPECIAL THICKNESSES (T and G, End Matched)			
** 33/32x3¼ in.	33/32x3¼ in.	5/4x4 in.	2400 lbs.
** 33/32x2¼ in.	33/32x2¼ in.	5/4x3 in.	2250 lbs.
** 33/32x2 in.	33/32x2 in.	5/4x2¾ in.	2250 lbs.
JOINTED FLOORING — i.e., SQUARE EDGE			
** ¾x2½ in.	¾x2½ in.	1x3¼ in.	2160 lbs.
** ¾x3¼ in.	¾x3¼ in.	1x4 in.	2300 lbs.
** ¾x3½ in.	¾x3½ in.	1x4¼ in.	2400 lbs.
** 33/32x2½ in.	33/32x2½ in.	5/4x3¼ in.	2500 lbs.
** 33/32x3½ in.	33/32x3½ in.	5/4x4¼ in.	2600 lbs.
**Special Order Only			

Nail Schedule
Tongue and Groove Flooring Must Be Blind Nailed

¾x1½, 2¼ & 3¼ in.	2 in. machine driven fasteners, 7d or 8d screw or cut nail.	10-12 in. apart*
¾x3 in. to 8 in.** Plank	2 in. machine driven fasteners, 7d or 8d screw or cut nail.	8" apart into and between joists.

*If subfloor is ½ inch plywood, fasten into each joist, with additional fastening between.
**Plank Flooring over 4" wide must be installed over a subfloor.

Following flooring must be laid on a subfloor.		
½x1½ & 2 in.	1½ in. machine driven fastener, 5d screw, cut steel or wire casing nail.	10 in. apart
⅜x1½ & 2 in.	1¼ in. machine driven fastener, or 4d bright wire casing nail.	8 in. apart
Square-edge flooring as follows, face-nailed — through top face		
5/16x1½ & 2 in.	1 inch 15 gauge fully barbed flooring brad. 2 nails every 7 inches.	
5/16x1⅓ in.	1 inch 15 gauge fully barbed flooring brad. 1 nail every 5 inches on alternate sides of strip.	

Fig. 29-6 Grades of hardwood flooring *(Courtesy of National Oak Flooring Manufacturers' Association)*

Other. Birch, beech, and hard maple flooring are graded in declining order as *first grade*, *second grade*, *third grade*, and *special grade*. Grades of pecan flooring are *first grade*, *first grade red*, *first grade white*, *second grade*, *second grade red*, and *third grade*. Red grades contain all heartwood. White grades are all bright sapwood.

GRADES OF PREFINISHED FLOORING

Grades of prefinished flooring are determined after the floor has been sanded and finished. The grades of prefinished oak flooring are *prime*, *standard and better*, *standard*, and *tavern*. Prefinished beech and pecan is furnished only in a combination grade called *tavern or better*.

ESTIMATING HARDWOOD FLOORING

To estimate the amount of hardwood flooring material needed, first determine the board feet of flooring needed to cover a given space. To do this, find the area in square feet. Add to the area a percentage of the area that applies to the size flooring to be used. The listed percentages include an additional 5 percent for end matching and normal waste:

- 55 percent for 3/4'' x 1 1/2''
- 42 1/2 percent for 3/4'' x 2''
- 38 1/3 percent for 3/4'' x 2 1/4''

RESILIENT TILE FLOORING

Resilient flooring tiles are made from *linoleum*, *asphalt*, *vinyl*, *vinyl-asbestos*, *rubber*, and *cork*. Perhaps the most widely used is vinyl-asbestos. Many different colors, textures, and patterns are available. Generally, tiles come in 9 x 9-inch and 12 x 12-inch squares and are applied to the floor in a manner similar to that used in applying ceiling tile. The usual tile thicknesses are 1/16 inch and 1/8 inch.

Long strips of the same material, called feature strips, are available from the manufacturer. The strips vary in width from 1/4 inch to 2 inches. They are used between tiles to create unique floor patterns.

LAYING THE FINISH FLOOR

Before laying any type of finish flooring, nail any loose areas and sweep the subfloor clean. Scraping may be necessary to remove all plaster, taping compound, or other materials.

LAYING STRIP FLOORING

Wood strip finish flooring may be laid in either direction on a plywood or diagonal board subfloor. Flooring laid in the direction of the longest dimension of the room gives the best appearance. It must be laid perpendicular to a board subfloor that runs across or at right angles to the joists.

When the subfloor is clean, apply a layer of building paper, lapped 4 inches at the seams. This helps to keep out dust, prevents squeaks in dry seasons, and retards moisture from below which could cause warping of the floor.

Starting Strip. The location and straight alignment of the first course is important. Place a strip of flooring 3/4 inch from the starter wall with the groove side toward the wall. On each side of the room, mark along the edge of the flooring tongue, Figure 29-7. Snap a chalk line between the two points.

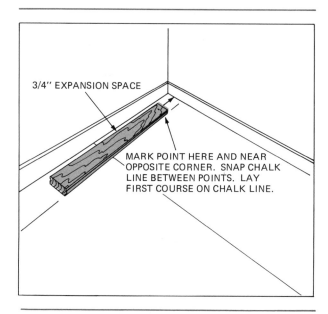

3/4'' EXPANSION SPACE

MARK POINT HERE AND NEAR OPPOSITE CORNER. SNAP CHALK LINE BETWEEN POINTS. LAY FIRST COURSE ON CHALK LINE.

Fig. 29-7 Mark and snap a starting line for strip flooring. *(Courtesy of National Oak Flooring Manufacturers' Association)*

Nail the first course with the tongue of the strips to the chalk line. The gap between the flooring and the wall is needed for expansion and is later covered by the base. Work from left to right with the grooved end of the first piece toward the wall. Left is determined by having the back of the person laying the floor to the wall where the starting strip is laid. When necessary to cut a strip to fit to the right wall, use a strip long enough so that the cut-off piece is 8 inches or longer. Start the next course on the left wall with this piece.

Make sure end joints between strips are driven up tight. The end piece in each course is marked by turning it end for end. In other words, place the grooved end against the wall and mark the other end in line with the end of the last piece installed. Nail the first course by face nailing with finish nails along the grooved edge and then blind nailing into the tongued edge with hardened cut nails.

BLIND NAILING

Flooring is *blind nailed* by driving nails at an angle of about 45° to 50° through the flooring, starting the nail in the corner at the top edge of the tongue. Usually 7d hardened cut nails or hardened wire nails with a spiral shank are used.

For the first two or three courses of flooring, a hammer must be used to drive the fasteners. For floor laying, a 20-ounce hammer is generally used for extra driving power. Care must be taken when using a hammer to fasten

flooring not to let the hammer glance off the nail, thus damaging the edge of the flooring. Care also must be taken that, on the final blows, the hammer head does not crush the top corner of the flooring. To prevent this, raise the hammer slightly on the final blow so that the hammer head hits the nail, but not the corner of the flooring, Figure 29-8.

After the nail is driven home, it must be set slightly so that adjoining strips can come up tight. The head of the next nail to be driven is used as a nail set. When fastening flooring using a hammer in one hand, the carpenter holds a number of nails in the other hand. While driving one nail, the carpenter fingers the next nail to be driven into position to be used as a set. When the nail being driven is home, the next nail is laid on edge with its head on the nail to be set. With one sharp blow, the nail is set, Figure 29-9. The setting nail is then the next nail to be driven.

Note: A nail set should not be used. If used in the usual manner, its tip will be flattened because the flooring nail is harder than the nail set. This ruins the nail set.

Some persons recommend laying the nail set flat along the tongue, on top of the nail, and setting the nail by hitting the side of the nail set with a hammer. This method is slower and may possibly break the nail set.

RACKING THE FLOOR

After the second or third course is in place, lay out seven or eight loose rows of flooring, end to end. Lay out in a staggered pattern with end joints at least 6 inches apart. Find or cut pieces

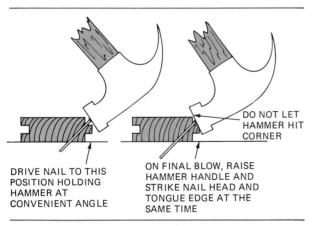

DRIVE NAIL TO THIS POSITION HOLDING HAMMER AT CONVENIENT ANGLE

ON FINAL BLOW, RAISE HAMMER HANDLE AND STRIKE NAIL HEAD AND TONGUE EDGE AT THE SAME TIME

DO NOT LET HAMMER HIT CORNER

Fig. 29-8 Method of driving a blind nail home

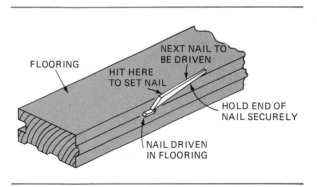

FLOORING

NEXT NAIL TO BE DRIVEN

HIT HERE TO SET NAIL

HOLD END OF NAIL SECURELY

NAIL DRIVEN IN FLOORING

Fig. 29-9 Method of setting a blind nail

511

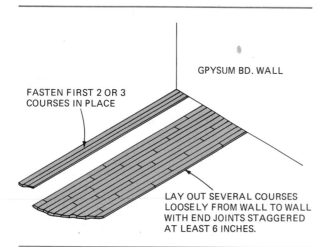

Fig. 29-10 Racking the floor

to fit within 1/2 inch of the end wall. Distribute long and short pieces evenly. Avoid clusters of short strips. Laying out loose flooring in this manner is called racking the floor. This is done to save time, Figure 29-10.

USING THE POWER NAILER

Two or three courses must be laid by hand to provide clearance from the wall before the power nailer can be used. The power nailer drives a special barbed fastener, fed into the machine like staples, through the tongue of the flooring at the proper angle, and sets it. Although it is called a power nailer, a heavy hammer is swung by the operator against a plunger to drive the fastener, Figure 29-11.

The hammer is double-ended. One end is rubber and the other end is steel. The rubber end is used against the flooring edge to bring the strips up tight. The steel end is used against the plunger of the power nailer.

Place the strip in position; drive the edges and end up tight with the rubber end of the hammer. Then use the steel end to drive the fasteners. Slide the power nailer along the tongue edge and drive fasteners about 12 inches apart or as necessary to bring the strip up tight.

Note: When using the power nailer, one and only one heavy blow must be used to drive the fastener. After the first blow, another fastener drops into place ready to be driven. Make sure the first blow is heavy enough to drive the fastener.

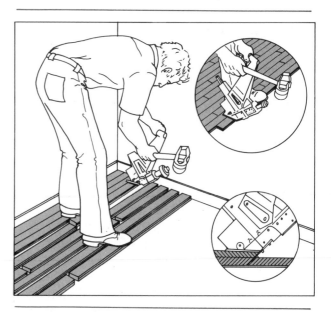

Fig. 29-11 Using a power nailer *(Courtesy of National Oak Flooring Manufacturers' Association)*

ENDING THE FLOORING

Continue across the room, racking seven or eight courses as work progresses. The last three or four courses from the opposite wall must be nailed by hand because of limited room to place the power nailer and swing the hammer.

The next-to-the-last row can be blind nailed if care is taken. However, the flooring must be brought up tightly by prying between the flooring and the wall. Use a gooseneck bar and pry the pieces tight at each nail location. When prying, use a wide board against the wall to avoid puncturing the wall covering, Figure 29-12. The last course is installed in a similar manner except that it must be face-nailed and may need to be ripped to the proper width.

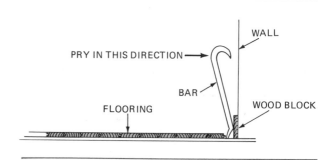

Fig. 29-12 The last two courses of strip flooring must be brought tight by prying.

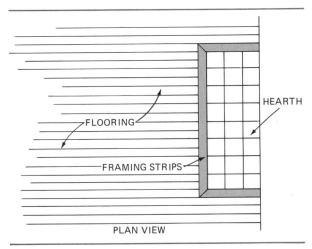

Fig. 29-13 Frame floor obstructions for a professional look.

FRAMING AROUND OBSTRUCTIONS

A much more professional and finished look is given to a strip flooring installation if *hearths* and other obstructions are framed. Use flooring around the obstructions with mitered joints in the corners, Figure 29-13.

CHANGING DIRECTION OF FLOORING

Sometimes it is necessary to change direction of flooring when it extends from a room into another room, hallway, or closet. To do this, face and blind nail the extended piece to a chalk line. Change directions by joining groove edge to groove edge and inserting a *spline* supplied with the flooring, Figure 29-14. For best appearance, always use long flooring strips where most noticeable, such as entrance hallways and living areas. Save most of the shorts for closet areas and apply the rest evenly in the general floor area.

CUSTOM OF FLOOR LAYING IN CERTAIN GEOGRAPHICAL AREAS

It is the custom in certain geographical areas to lay flooring so it butts tightly against a previously installed base. It is felt that by using this method, the finished floor can then be the last interior finish installed. There is less danger of the floor being marred or scratched than if the floor were laid first and then worked upon while other interior finish is being applied.

This method takes longer because starting and ending strips must be scribed to the base, and others fitted around door jambs and casings. This method is not recommended by associations of flooring manufacturers because it does not allow for expansion of the flooring. Finish flooring applied with no room for expansion may buckle or even push exterior walls outward.

To allow for expansion, finish flooring must be applied before the base is installed. Because the base butts against them, door casings must also be applied after the floor is laid. It should be planned, however, that all possible interior finish be applied before the finish floor is laid. After the finish floor is laid, it should be covered with a rosin-type building paper to protect it while other interior trim is applied.

INSTALLING PLANK FLOORING

Plank flooring is installed in a manner similar to that used in installing strip flooring. Alternate the courses by widths. Start with the narrowest pieces, then use increasingly wider courses, and repeat the pattern.

Manufacturers' instructions for fastening the flooring vary and should be followed. Generally, the flooring is blind-nailed through the tongue and one or more counterbored screws, depending on the width, are driven at each end of the plank and at intervals along the plank. The screws are covered with wood plugs and trimmed flush. Prefinished plank flooring has applied plugs and is blind-nailed only.

INSTALLING PARQUET FLOORING

Procedures for the application of parquet flooring vary with the style and the manufacturer.

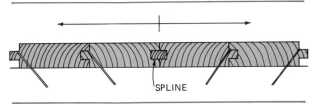

Fig. 29-14 Direction of strip flooring is changed by the use of a spline.

Detailed installation directions are usually provided with the flooring.

Generally, both blocks and individual pieces of parquetry are laid in *mastic.* Use the recommended type and apply with a notched trowel. Allow the mastic to dry solid enough to snap working lines on it. Use blocks of flooring to step on while snapping lines.

METHODS OF LAYING PARQUET BLOCKS

There are two general methods used to lay parquet blocks. The most common is with the edges of the blocks square with the walls of the room. Another method is by applying the blocks in a *diagonal* pattern with their edges at a 45° angle to the walls.

In both patterns, the width of the border blocks must be determined so an equal width is obtained on opposite walls. This is done in the same manner as calculating borders for ceiling tile.

SQUARE PATTERN

When laying parquet blocks in a square pattern, snap a starting line near the hardiest entry door to the room and four or five parquet blocks from the wall. Near the center, snap another line at right angles to the starting line so that border units are equal on opposite walls.

Place one unit at the intersection of the lines. Lay the next units ahead and to one side of the first one and along the lines. Install blocks in a pyramid working from the center outward toward the walls in all directions. Small variations in size are accommodated by making adjustments as installation progresses to prevent misalignment, Figure 29-15.

DIAGONAL PATTERN

To lay parquet blocks in a diagonal pattern, measure an equal distance from one corner of the room along both walls. Snap a base line between the two marks. Near the center of the base line, snap another line at right angles to it. The location of both lines is made with consideration to the width of border units. The diagonal pattern is then laid in the same manner as the square pattern, Figure 29-16.

INSTALLING UNDERLAYMENT

Finish floor material such as resilient sheet or tiles needs a very smooth, solid base under it. Plywood, hardboard, or particleboard is used for underlayment and is installed on top of the subfloor.

Underlayment thickness may range from 1/4 inch to 3/4 inch, depending on the material and job requirements. In many cases, where a

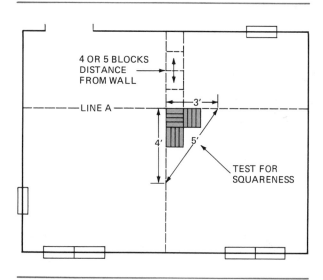

Fig. 29-15 Layout of a square pattern for parquet blocks *(Courtesy of National Oak Flooring Manufacturer's Association)*

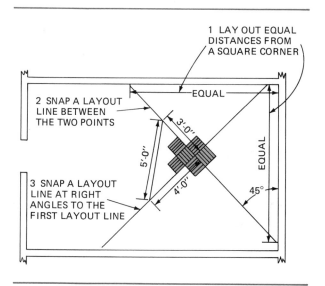

Fig. 29-16 Layout of a diagonal pattern for parquet blocks

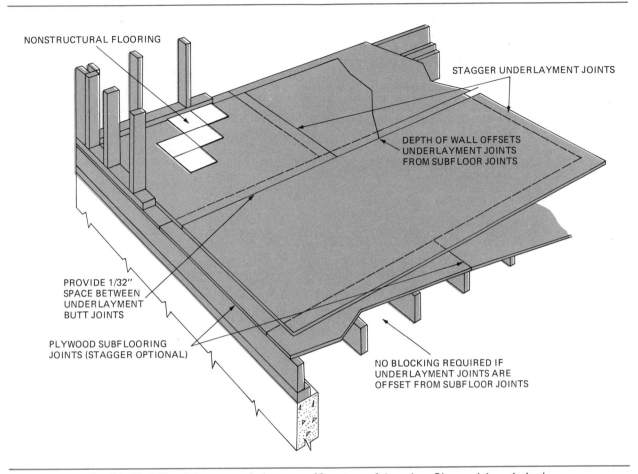

NONSTRUCTURAL FLOORING

STAGGER UNDERLAYMENT JOINTS

DEPTH OF WALL OFFSETS
UNDERLAYMENT JOINTS
FROM SUBFLOOR JOINTS

PROVIDE 1/32"
SPACE BETWEEN
UNDERLAYMENT
BUTT JOINTS

PLYWOOD SUBFLOORING
JOINTS (STAGGER OPTIONAL)

NO BLOCKING REQUIRED IF
UNDERLAYMENT JOINTS ARE
OFFSET FROM SUBFLOOR JOINTS

Fig. 29-17 Offset joints on underlayment *(Courtesy of American Plywood Association)*

finish wood floor meets a tile floor, the underlayment thickness is determined by the difference in the thickness of the two types of finish floor so that both floor surfaces come flush with each other.

Sweep the subfloor as clean as possible and cover with the asphalt felt lapped 4 inches at the edges. Stagger joints between subfloor and underlayment. If installation is started in the same corner as the subfloor, the thickness of the walls will be enough to offset the joints. Leave about 1/32 inch between underlayment panels to allow for expansion.

If the underlayment is to go over a board subfloor, install the underlayment with its face grain across the boards. In all other cases, a stiffer and stronger floor is obtained if the face grain is across the floor joists, Figure 29-17.

Fasten the first row in place with staples or nails. Underlayment requires more nails than subfloor to provide a squeak-free and stiff floor. Nail spacing and size are shown in Figure 29-18 for various thicknesses of plywood underlayment.

Install remaining rows of underlayment with end joints staggered from the previous courses. The last row of panels is ripped to width to fit the remaining space.

If APA Sturd-I-Floor is used, it can double as a subfloor and underlayment. If square-edge panels are used, blocking must be provided under the joints to support the edges. If tongue-and-grooved panels are used, blocking is not required, Figure 29-19. The number of underlayment panels required is found by dividing the area to be covered by the area of a panel.

Plywood Group Number and Grades	Subfloor Type	Plywood Thickness (inch)	Fastener Size (approx.) and Type	Fastener Spacing (inches)	
				Panel Edges	Panel Interior
All Groups:	Plywood	1/4	3d ring-shank nails	3	6 each way
			18 gauge staples	3	6 each way
Groups 1, 2, 3, 4, 5 UNDERLAYMENT INT-APA (with interior or exterior glue), or UNDERLAYMENT C-C Plugged EXT-APA	Lumber or other uneven surface	3/8 or 1/2	3d ring-shank nails	6	8 each way
			16 gauge staples	3	6 each way
		5/8 or 3/4	4d ring-shank nails	6	8 each way
			16 gauge staples	3	6 each way
Group 1 only: Grades as above	Lumber to 4" wide	1/4	3d ring-shank nails	3	6 each way
			18 gauge staples	3	6 each way

Fig. 29-18 Nailing specifications for plywood underlayment

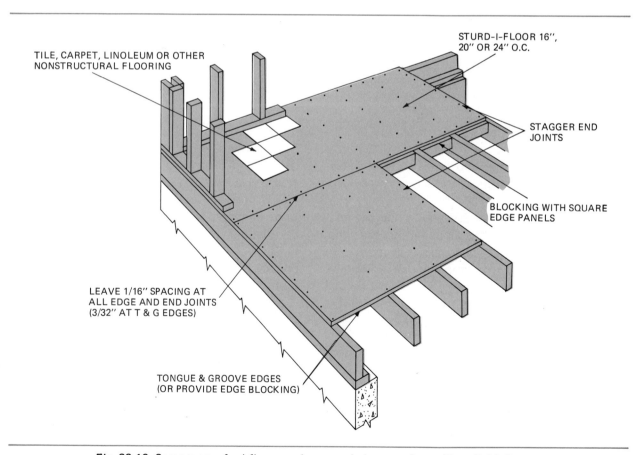

Fig. 29-19 Some types of subfloor require no underlayment for resilient finish floors.

Fig. 29-20 Applying adhesive for a tile floor

Fig. 29-21 Laying resilient floor tile

INSTALLING RESILIENT TILE

When installing resilient tile, make sure underlayment fasteners are flush with or below the surface. Fill any open areas such as splits with a hard, quick-setting filler. Sand off any surface roughness around joints and sweep the floor clean. Snap base lines at right angles to each other on the floor in the proper location for equal width border tiles on opposite walls.

Apply the recommended adhesive with a notched trowel, Figure 29-20. Let the adhesive dry until transparent. The layout lines will show through. If not, snap new lines.

Apply tiles by starting at the intersection of the layout lines and pyramid out from the center in all directions. Make sure the finished face of the tile is up. Lay tiles with edges tight and work toward the walls, Figure 29-21. Watch

the grain pattern. It may be desired to alternate the run of the patterns or to lay the patterns all in one direction. Lay tiles in place. Do not attempt to slide them into position. Apply all tiles except border tiles. Then, cut and fit border tiles in place. Tiles may be cut by scoring with a sharp utility knife and bending or by the use of a special tile cutter.

FITTING BORDER TILES

To fit border tiles, first place the tile to be cut directly on top of and in line with the last tile installed. Place a full tile against the wall and on top of the one to be fitted. Score along the outside edge of the top tile with the knife, Figure 29-22. Bend the tile, break it along the scored line, and fit it into place, Figure 29-23.

Fig. 29-22 Scoring a border tile

Fig. 29-23 Fitting a border tile in place

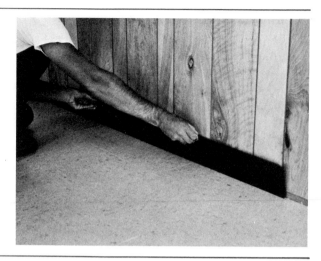

Fig. 29-24 Applying cement to the back of a vinyl cove base and pressing it into place

APPLYING A VINYL COVE BASE

Many times a vinyl cove base is used to trim a tile floor. A special vinyl base cement is applied to its back and the base is pressed into place, Figure 29-24.

ESTIMATING TILE FLOORING

To estimate the amount of tile flooring needed, find the area of the room in square feet. To do this, measure the length and width of the room to the next whole foot. Multiply these figures together to find the area. For 12"x12" tiles, the result is the number of pieces needed. For 9"x9" tiles, multiply the area by 1 1/3.

REVIEW QUESTIONS

Select the most appropriate answer.

1. If hardwood flooring is stored in a heated building, the temperature should not exceed
   a. 72 degrees.   c. 85 degrees.
   b. 78 degrees.   d. 90 degrees.

2. Most hardwood finish flooring is made from
   a. Douglas fir.   c. southern pine.
   b. hemlock.   d. oak.

3. Bundles of strip flooring may contain pieces over and under the nominal length of the bundle by
   a. 4 inches.   c. 8 inches.
   b. 6 inches.   d. 9 inches.

4. No pieces are allowed in bundles of hardwood strip flooring shorter than
   a. 4 inches.
   b. 6 inches.
   c. 8 inches.
   d. 9 inches.

5. The edges of prefinished strip flooring are chamfered to
   a. prevent splitting.
   b. apply the finish.
   c. simulate cracks between adjoining pieces.
   d. obscure any unevenness in the floor surface.

6. The best grade of unfinished oak strip flooring is
   a. prime.
   b. clear.
   c. select.
   d. quarter-sawed.

7. To estimate the amount of 2 1/4-inch face hardwood flooring, add to the area to
   be covered a percentage of the area of
   a. 42 1/2.
   b. 38 1/3.
   c. 29.
   d. 33 1/3.

8. When it is necessary to cut the last strip in a course of flooring, the waste is used
   to start the next course and should be at least
   a. 8 inches long.
   b. 10 inches long.
   c. 12 inches long.
   d. 16 inches long.

9. For floor laying, the hammer weight is generally
   a. 13 ounces.
   b. 16 ounces.
   c. 20 ounces.
   d. 22 ounces.

10. To change direction of strip flooring,
    a. face nail both strips.
    b. turn the extended strip around.
    c. blind nail both strips.
    d. use a spline.

KITCHEN CABINETS 30

OBJECTIVES

After completing this unit, the student will be able to

- *state the sizes and describe the construction of typical base and wall kitchen cabinet units.*
- *plan, order, and install factory-built kitchen cabinets.*
- *build kitchen cabinets on the job site.*
- *make and laminate a countertop.*

Kitchen cabinets may be built piece by piece on the job or may be purchased in preassembled, prefinished units and installed. In either case, the carpenter must be familiar with the construction and sizes of the cabinets and know how to plan, order, and install them, Figure 30-1. Factory-built cabinets come in a wide variety of styles, sizes, and colors. Job-built cabinets are custom-made to the customer's specifications.

KINDS AND SIZES

The two basic kinds of kitchen cabinets are the base unit and the wall unit. The base unit sets on the floor, and the wall unit hangs on the wall.

The distance between the wall and base units is usually 16 to 18 inches. This distance is

Fig. 30-1 Installing and building kitchen cabinets is part of the interior finish.

Fig. 30-2 Section of a preformed kitchen cabinet countertop

enough to accommodate articles such as coffee makers, toasters, blenders, and mixers. The top shelf in the wall unit should not be over 6 feet from the floor if it is to be within easy reach.

BASE UNIT

The standard height of a base unit from the floor to the surface of the countertop is 36 inches. Its depth is 24 inches. The width of the countertop is 25 inches, which allows for a 1-inch overhang. If the countertop has a back-splash, it is usually 4 inches high. Countertops may be built on the job. Preformed counter-tops may be purchased and cut to any desired

length, Figure 30-2. A toe space is provided beneath the base unit. The toe space is usually 2 1/2 inches deep by 2 1/2 inches high.

Usually the base cabinet has one drawer and one door. Some base units contain all drawers. The usual height of the top drawer opening is 5 1/2 inches, Figure 30-3.

WALL UNIT

A standard wall unit is 12 inches deep and 30 inches high. Others are made in heights of 12, 15, 18, and 24 inches. The smaller cabinets are used above sinks, refrigerators, and ranges to keep the tops all the same height. The height of

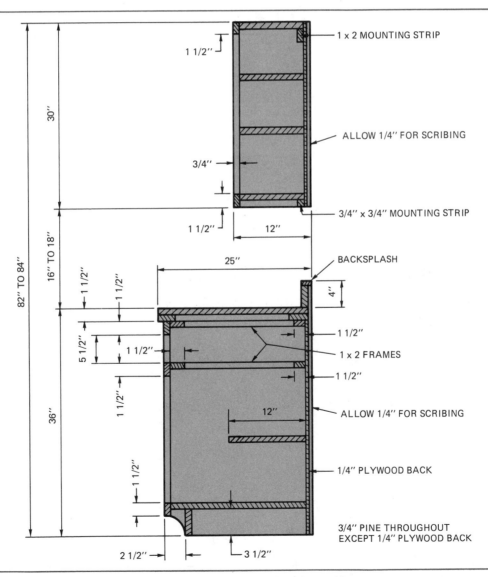

Fig. 30-3 Typical section of kitchen cabinets

521

Specification Sheet

Wall Cabinets
All single door cabinets are reversible for door swing. Single door wall cabinets (except Charter House) have adjustable shelf.

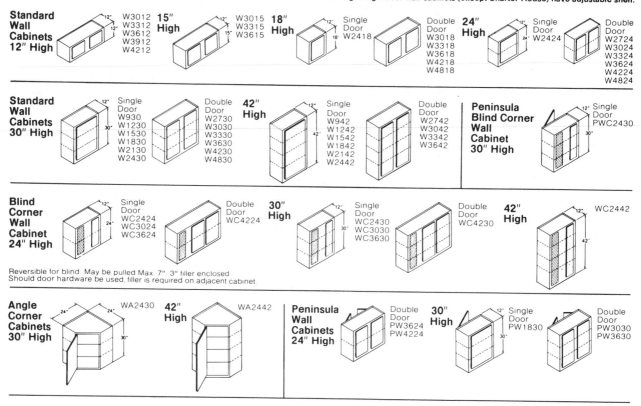

Standard Wall Cabinets 12" High
W3012
W3312
W3612
W3912
W4212

15" High
W3015
W3315
W3615

18" High — Single Door W2418 — Double Door W3018 W3318 W3618 W4218 W4818

24" High — Single Door W2424 — Double Door W2724 W3024 W3324 W3624 W4224 W4824

Standard Wall Cabinets 30" High — Single Door W930 W1230 W1530 W1830 W2130 W2430 — Double Door W2730 W3030 W3330 W3630 W4230 W4830

42" High — Single Door W942 W1242 W1542 W1842 W2142 W2442 — Double Door W2742 W3042 W3342 W3642

Peninsula Blind Corner Wall Cabinet 30" High — Single Door PWC2430

Blind Corner Wall Cabinet 24" High — Single Door WC2424 WC3024 WC3624 — Double Door WC4224

30" High — Single Door WC2430 WC3030 WC3630 — Double Door WC4230

42" High WC2442

Reversible for blind. May be pulled Max. 7". 3" filler enclosed.
Should door hardware be used, filler is required on adjacent cabinet.

Angle Corner Cabinets 30" High WA2430 **42" High** WA2442

Peninsula Wall Cabinets 24" High — Double Door PW3624 PW4224 **30" High** — Single Door PW1830 — Double Door PW3030 PW3630

Base Cabinets
All single door cabinets are hinged left — frames spotted for reverse hinging.

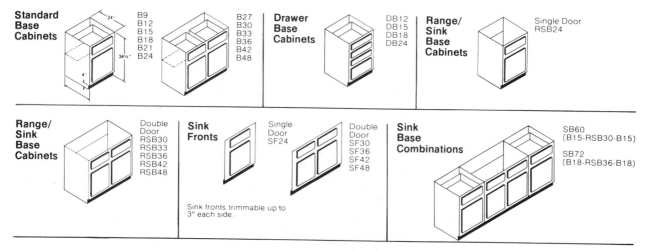

Standard Base Cabinets
B9
B12
B15
B18
B21
B24
B27
B30
B33
B36
B42
B48

Drawer Base Cabinets
DB12
DB15
DB18
DB24

Range/ Sink Base Cabinets — Single Door RSB24

Range/ Sink Base Cabinets — Double Door RSB30 RSB33 RSB36 RSB42 RSB48

Sink Fronts — Single Door SF24 — Double Door SF30 SF36 SF42 SF48

Sink fronts trimmable up to 3" each side.

Sink Base Combinations
SB60 (B15-RSB30-B15)
SB72 (B18-RSB36-B18)

Fig. 30-4 Typical factory-made kitchen cabinets *(Courtesy of Pacific Cabinet Corporation)*

522

Base Cabinets
All single door cabinets are hinged left — frames spotted for reverse hinging.

Blind Corner Base Cabinet (Includes 3" Base Corner Filler)

BC 39/46

BC 42/49

Cabinet must be pulled min. 3" from corner. Max. pull — 7". Reversible by interchanging door and drawer to opposite side.

Should door hardware be used, filler is required on adjacent cabinet.

Blind Corner Base Installation

BC Cabinet

24"

24"

3" Filler.

Peninsula Base Cabinets

Single Door PB18

Double Door PB36 PB42 PB48

Peninsula Blind Corner Base Cabinets (Includes 3" Base Corner Filler)

PBC42 PBC48

Should door hardware be used, filler is required on adjacent cabinet.

Corner Base w/Lazy Susan

LS36

Requires 36" wall space in each direction from corner.

Oven Cabinets and Front

Oven Front Only (Doors applied at factory. Drawer boxes, fronts and hardware shipped to be job applied)

033 Max cut-out width — 30½" height — 65½"
027 Max cut-out width — 24½" height — 65½"
024 Max cut-out width — 21½" height — 65½"
033FF
027FF
024FF
80" high
No toe kick

Utility Cabinets 12" and 24" Deep and Fronts

Order shelf kit if pantry desired. (5 adj. shelves.)

U1812 U1824 UF18

U2412 U2424 UF24

Linen Cabinets and Fronts

Linen front has wide bottom rail and no toe kick. 18" linen cabinet and/or front — single door. Flapper hinged on bottom — opens from top. Secured by chain. Toe kick 5" H.

L1812 L1821 L1824 L2412 L2421 L2424 LF18 LF24

Vanity Base Cabinets

VB12 VB15 VB18

VB24

Toe Kick 5"H

Vanity Sink Cabinets

VS24

VS27 VS30

VS36 VS42 VS48

Vanity Drawer Cabinets

VD12 VD15 VD18 VD24

Vanity Sink Fronts

VSF30 VSF36 VSF42 VSF48

Trimmable up to 3" each side.

Vanity Desk Drawer

ID36

Trimmable up to 6" each side.

Medicine Cabinet

MC1524

Cut-out size 13⅛" W x 22⅛" H. Min. depth recess reqd. = 3½"

Wall Filler

WF 3" x 30"
WF 3" x 42"
WF 6" x 30"
WF 6" x 42"

Wall Corner Turn

WCT 3 x 3 — 30"
WCT 3 x 3 — 42"

Wall Corner Bottom

WCB 15 x 15

Vanity Fillers

VF3 VF6

VF 3 x 84

Toe Kick 5" H.

Base Fillers

BF 3 x 34½" H
BF 6 — 34½" H
BF 3 x 84" H
BF 3 x 96" H

Toe Kick 4" H.

Base Corner Filler

BCF 34½" H

Base Corner Turn

BCT 3 x 3 34½" H

Base Vanity End Panels (One side finished)

BEP/VEP 23⅜₁₆" x 34½"

To be notched on job.

Valance

48" V4
72" V6

Drop-In Front

DIF30

Dishwasher End Panel

DWEP 1½" x 34½"

To be notched on job.

W1812 FF Kit
W2412 FF Kit
W2712 FF Kit
Includes face frame, 2 returns, 4 cleats. Shipped KD.

Mouldings

M 1 ¼" Scribe

M 2 Cove

M 4 Corner

Pacific Cabinet Corporation, Bldg. N-2, Industrial Park, Spokane, WA 99216

Fig. 30-4 (Continued)

a cabinet above a sink or range should be a minimum of 24 inches. The height of a cabinet over a refrigerator depends on the height of the refrigerator.

The number and spacing of shelves depends on the purpose of the cabinet. A standard height wall unit usually contains two equally spaced shelves. However, shelves may be made adjustable and spaced at any convenient height.

SPECIAL UNITS

Some base units, whether built on the job or in the factory, extend from the floor to the top of the wall units. These are used for broom closets, to hold wall ovens, or for general storage. Corner units, with or without lazy Susans, provide easy access. Sink or range fronts consist only of the face and toeboard and are installed between existing cabinets. Filler pieces are also available for wall and base units and are used when no combination of standard factory-built cabinets can fill the existing space. Other special units include spice cabinets, tray and mixer bases, and desk units, Figure 30-4.

BATHROOM VANITIES

Bathroom vanities are similar to base units except for the size. Without the countertop, the unit is usually 30 1/2 inches high by 21 inches deep.

PLANNING LAYOUTS FOR FACTORY-BUILT CABINETS

When planning the layout for factory-built cabinets, measure carefully and accurately the length of the walls on which the cabinets are to be installed. Draw a plan view, to scale, showing the location of all appliances, sinks, windows, and similar items, Figure 30-5. Allow 1/4 to 1/2 inch additional space for dishwasher, range, and trash compactor. Allow a minimum of 3 additional inches for refrigerator space. It may be wise to allow more space as the customer may eventually want to replace the refrigerator with a larger size.

Next, draw elevations of the base cabinets referring to the manufacturer's catalog for

sizes. Always use the largest size cabinets available instead of using two or three smaller ones. This reduces the cost and makes installation easier.

Match up the wall cabinets with the base cabinets, where feasible. If filler strips are necessary place them between a wall and a cabinet or between cabinets in the corner. Identify each unit on the elevations with the manufacturer's model number, Figure 30-6. Make a list of the units in the layout and order from the distributor.

INSTALLING WALL UNITS

Cabinets must be installed in a straight, level, and plumb line. This requires skill because floors and walls are not always level and plumb, especially in older buildings. Many installers prefer to mount the wall units first so that work does not have to be done over base units.

The first step in installing wall units is to locate the bottom of the wall units 52 to 54 inches from the floor. This will leave a 16- to 18-inch space from the base unit countertop.

- Use a level on the floor and find the low point between each end of the line of cabinets to be installed.
- Measure up 52 to 54 inches from the low point and mark the wall.
- Using a level and straightedge, draw a level line through the mark and across the wall. The bottom of the wall units are installed to this line.
- In a wood frame wall, the studs must be located. Screws are driven into the studs to fasten the cabinets to the wall.
- Just above the level line, lightly tap on and across the wall with a hammer. When a solid sound is heard, a stud is located.
- To make sure of the location of a stud, drive a finish nail in at the point the solid sound is heard. Drive the nail above the level line so the nail holes are later covered by the cabinet.
- If a stud is found, mark the location with a pencil. If no stud is found try a little over to one side or the other.

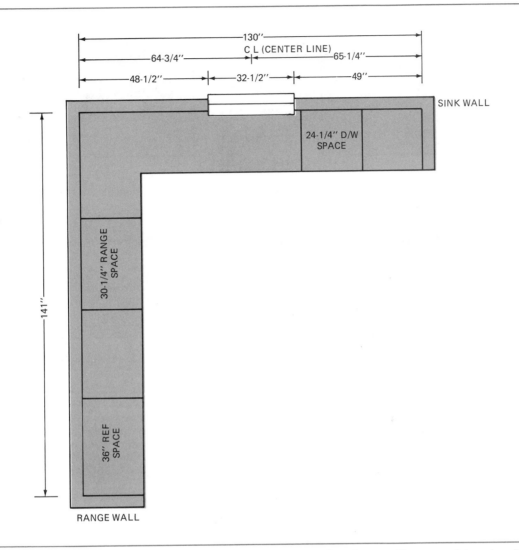

Fig. 30-5 **Plan view of a kitchen cabinet layout showing the location of appliances** *(Courtesy of American Woodmark Cabinets)*

- Measure at 16-inch intervals in both directions from the mark to locate other studs. Drive a finish nail to test for solid wood and mark each stud location. If studs are not found at 16-inch centers, then tap the wall with a hammer to locate each stud.

- At each stud location, draw a plumb line down below the line for the bottom of the cabinets. Projecting below the wall units makes it easier to locate the studs when installing both wall and base units.

- Build, and place the unit on, a stand that holds the unit near the line of installation.

- Shim the unit up with wood shingles until the bottom of the unit is on the layout line.

- Test the face of the unit with a level for plumbness. If the unit is not plumb, shim it between the wall and its back edge with wood shingles until it is plumb.

- Usually the ends of the cabinet project beyond the cabinet back to allow for scribing to the wall. Set the dividers for the farthest distance that the back edge of the unit is away from the wall.

SINK WALL–UPPER (WALL) CABINETS

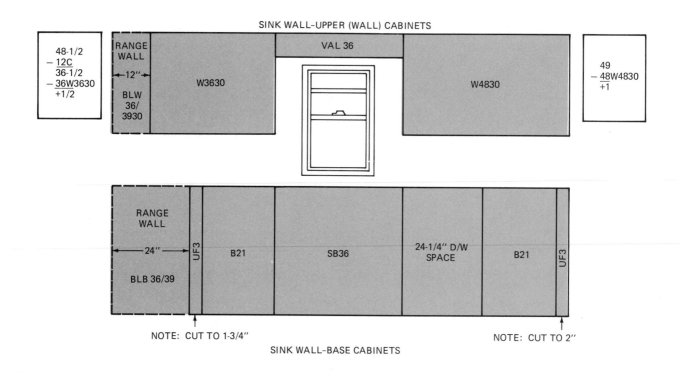

48-1/2
− 12C
36-1/2
− 36W3630
+1/2

RANGE WALL

←12"→

BLW 36/3930

W3630

VAL 36

W4830

49
− 48W4830
+1

RANGE WALL

←24"→

BLB 36/39

UF3

B21

SB36

24-1/4" D/W SPACE

B21

UF3

NOTE: CUT TO 1-3/4"

NOTE: CUT TO 2"

SINK WALL–BASE CABINETS

RANGE WALL — UPPER (WALL) CABINETS

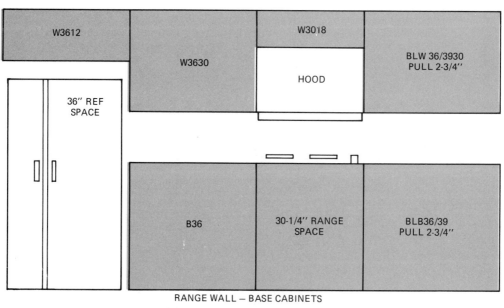

W3612

W3630

W3018

HOOD

BLW 36/3930 PULL 2-3/4"

36" REF SPACE

B36

30-1/4" RANGE SPACE

BLB36/39 PULL 2-3/4"

RANGE WALL — BASE CABINETS

Fig. 30-6 Elevation of a kitchen cabinet layout *(Courtesy of American Woodmark Cabinets)*

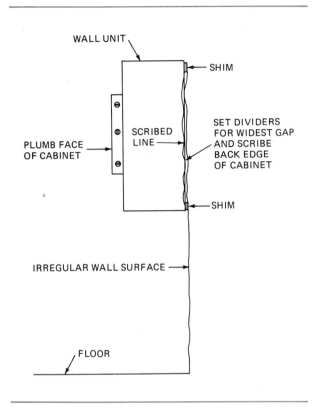

Fig. 30-7 Scribing end panels of
factory-built wall units

- Scribe the back edge by riding the dividers against the wall and marking the back edge of both end panels to the contour of the wall, Figure 30-7.

- Take the cabinet down from the stand. Cut the back edges with a handsaw to the scribed line. A handsaw, rather than a saber saw, is used because a saber saw cuts on the upstroke and splinters out the face side, while a handsaw cuts on the down-stroke.

- Place the cabinet back into position.

- Drill holes through the mounting strips into the studs. A mounting strip is located at the top and bottom of the cabinet.

- Fasten the cabinet in place with wood screws. Usually oval head screws are used with finishing washers. Screws should be of sufficient length to hold the cabinet securely. In most cases, a 2 1/2-inch screw is used.

On masonry walls, first drill holes through the mounting strips, place the cabinet in position and mark the location of the drilled holes on the wall. Remove the cabinet and drill holes into the masonry wall for lead inserts. Replace the cabinet and screw in place.

Adjacent cabinets are installed in the same manner. The back edges of these cabinets are scribed so their face frames are flush with the cabinet previously installed. Adjacent cabinets are fastened to each other by means of bolts through the cabinet ends or screws into the stiles of the face frame.

INSTALLING BASE CABINETS

Before installing the base cabinets, draw a line 16 inches (plus the thickness of the counter-top) below the line previously drawn for the location of the wall units. This line marks the location of the base units without countertops.

- Place the first unit to be installed against the wall in the desired location.

- Shim the bottom with wood shingles until the top is level across its width and depth.

- Adjust the dividers to correspond with the amount the top of the unit is above the line.

- Scribe this amount across both ends and toeboard by riding the dividers along the floor, Figure 30-8.

- Cut both ends and toeboard to the scribed lines.

- Draw a line on the back between the cut ends; cut the back, if necessary.

- Place the unit back in position. The top end should correspond to the layout line. If the back edges do not fit the wall, scribe them in a manner similar to that used for wall units. Fasten the base unit in place.

- Scribe and fit adjacent units the same as you did the first unit. Then, fasten them to the wall and to each other.

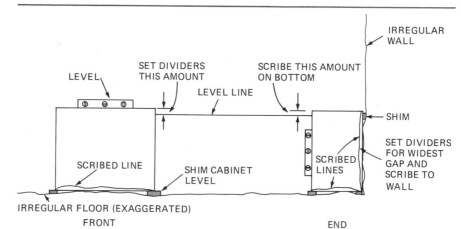

Fig. 30-8 Scribing base units to the floor

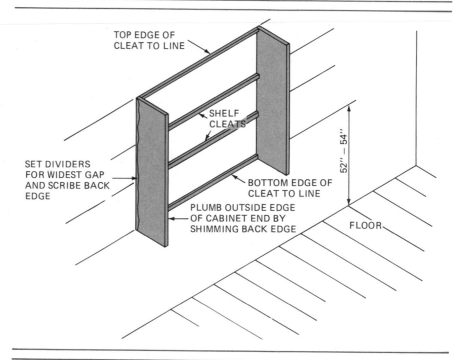

Fig. 30-9 Install cabinet ends by scribing to the wall.

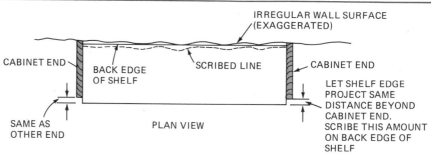

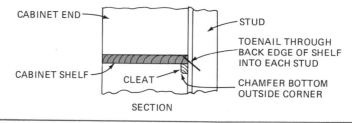

Fig. 30-10 Installing cabinet shelves in the wall unit

INSTALLING
PREFORMED COUNTERTOPS

After the base units are fastened in position, the countertop is laid on top of the units and against the wall.

- Move the countertop, if necessary, so that it overhangs the same amount over the face frame of the base cabinets.

- Adjust the dividers for the difference between the amount of overhang and the desired amount of overhang. Scribe this amount on the top edge of the backsplash if it has a scribing strip.

- Cut the backsplash to the scribed line and fit it to the wall.

- Fasten the countertop to the base cabinets with screws up through the top frame of the base units. Use a stop on the drill bit to prevent drilling through the countertop. Use screws of sufficient length, but not so long as to penetrate the countertop.

Raw ends of preformed countertops are treated by applying specially shaped pieces of plastic laminate to the ends. For interior corners, the countertops are mitered accurately in the factory and joined with special bolts on the underside of the countertop.

Sink cutouts are made by carefully outlining the cutout and cutting with a saber saw. Some masking tape applied to the base of the saber saw will prevent scratching of the countertop. The cutout pattern usually comes with the sink.

JOB-BUILT CABINETS

In many cases, kitchen cabinets are built on the job. Material is conserved because the wall acts as the cabinet backs and there is no need to double cabinet ends as in the case of individual units installed side by side. In addition, the cabinets are custom built without the need of filler blocks, resulting in a neater and more pleasing appearance. The cost, in many cases, may be even less than factory-built cabinets.

WALL UNITS

The wall units are usually constructed first to avoid working over the base units.

- Draw a level line across the wall 52 or 54 inches up from the floor for the bottom of the cabinet.

- Draw level lines for the bottom of each shelf and the cabinet top.

- Draw plumb lines for the inside face of each end of the cabinet.

- Nail 3/4"x1" cleats to the level lines and between the plumb lines. For a better appearance, chamfer the bottom outside corner of each cleat. Fasten the bottom strip with its bottom edge to the line. Fasten the other strips with their top edges to the lines.

- Cut the cabinet ends to length and width. For a standard wall cabinet, the length is 30 inches and the width is 11 1/4 inches.

- Place the cabinet end in position against the ends of the cleats. The bottom end is flush with the bottom edge of the bottom cleat.

- Plumb the outside edge of the cabinet end. Place a shim between it and the wall, if necessary.

- Scribe the inside edge of the cabinet end to the wall to take off as little as possible. Cut to the scribed line.

- Install the cabinet end by fastening it to the ends of the cleats. Install the other end in the same manner, Figure 30-9.

- Draw level lines on the inside faces of the cabinet ends from the top of the cleats for the position of the shelves.

- Cut the bottom, top, and shelves the same length as the cleats.

- Place the bottom in position on top of the cleat and between the cabinet ends.

- Let the ends of the shelf project beyond the outside edge of the cabinet end the same amount on both ends. Scribe this amount from the back edge of the shelf, Figure 30-10. Cut to the scribed line, and fasten in position. Besides nailing through the cabinet ends, toenail through the back edge into each stud. Set all nails.

- Mark and fasten the rest of the shelves in the same manner.

MAKING THE FACE FRAME

Members of the face frame are usually made of 1x2 nominal stock. However, where doors are hinged back to back on the same piece, a 1x3 is needed to provide room for the hinges. In corners, the exposed width of the vertical pieces should be at least 2 1/2 inches to allow for the swing of doors. Smooth all edges with a plane before installing.

- Cut the necessary number of vertical pieces the same length as the cabinet ends. The number needed depends on the width of the cabinet and the number of door openings.

- Apply the stiles flush with the cabinet ends. Apply intermediate pieces with equal spaces between them. Nail to the outside edges of the shelves. Sight between both ends and keep shelves straight. Horizontal members of the face frame are called *rails* and are fitted between the vertical members.

- Fit and fasten the bottom rails so their top edge is flush with the bottom shelf and their bottom edge is flush with the bottom end of the cabinet end.

- Fasten the top rail with its top edge flush with the top end of the cabinet. Fasten the rails into the outside edge of the shelves; toenail their ends into the vertical members.

- Set all nails and sand all joints smooth and flush. Break (round off) all exposed sharp corners, Figure 30-11.

BASE UNITS

Base units must be constructed so their parts lie in level and plumb lines. Care must be taken in the layout, cutting, and installation of the parts.

- Begin by drawing plumb lines for the inside face of the cabinet ends.

- Draw a level line 3 1/2 inches above the floor for the top of the cleat to support the cabinet bottom.

- Install a 1x4 cleat against the wall with its top edge to the line and its ends to the plumb lines.

- Square out on the floor from the ends of the cleat. Snap a line on the floor between the squared lines — 20 3/4 inches away from the wall. Nail a 1x2 cleat on the floor so its outside edge is to the chalk line. This cleat supports the toeboard.

- Fasten a 3 1/2-inch-wide toeboard against the cleat by toenailing into the floor, Figure 30-12. The toeboard is the same length as the cleat. Keep the ends of the toeboard flush with the ends of the cleat. The top edges of the toeboard and the wall cleat must be level with each other.

INSTALLING THE CABINET BOTTOM

The cabinet bottom must fit the wall with its outside edge overhanging the toeboard by the proper amount along its entire length.

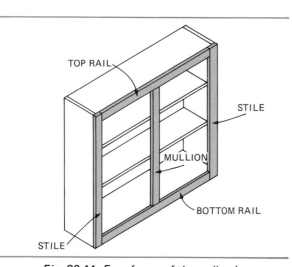

Fig. 30-11 Face frame of the wall unit

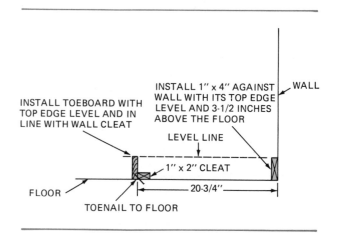

Fig. 30-12 Layout of the toeboard

• Cut a length of 3/4'' plywood the same as the toeboard. Rip it 23 1/2 inches wide. Place it in position on top of the wall cleat and toeboard. Let the outside edge overhang the toeboard the same amount on both ends.

• Set the dividers for the difference between the amount of overhang and 1 3/4 inches. Scribe this amount on the wall edge of the cabinet bottom. Cut to the scribed lines and fasten the bottom in position, keeping its ends flush with the ends of the toeboard and the wall cleat, Figure 30-13.

Other wall cleats are needed to support the countertop, drawer guides, and shelves.

• The level line for the cleats to support drawer guides is determined by the depth of the drawer and the type and method of installation of the drawer guides.

• The position of the cleat to support the countertop is such that the countertop will be 36 inches from the floor when installed.

• Shelf cleats may be installed for any number of desired shelves.

• Cut and fasten these cleats in position, Figure 30-14.

INSTALLING THE CABINET ENDS

The back edge of the cabinet end must fit the wall. Its front edge must be flush with the front edge of the cabinet bottom, and be plumb.

Its bottom end must fit the floor and its top end must be flush with the top edge of the countertop cleat and be level.

• Cut the cabinet ends to a rough length and width with square ends. The length should be about 36 inches and the width about 23 1/2 inches.

• Place one end in position and plumb the outside edge. Use shims between it and the wall and floor, if necessary, to hold it steady.

• Set the dividers for the amount the top end projects above the top cleat and scribe this amount on the bottom. Cut the bottom end to the scribed line to fit to the floor.

• Place the end back in position. Its bottom end should fit the floor, the top end flush and level with the top edge of the top cleat, and its outside edge plumb.

• Set the dividers for the amount the outside edge projects beyond the cabinet bottom. Scribe this amount on the edge against the wall. Cut to the scribed line. Cut out the

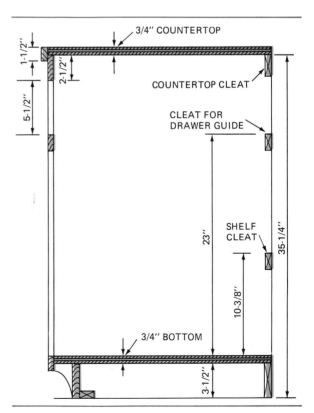

Fig. 30-14 Location of cleats for the base unit

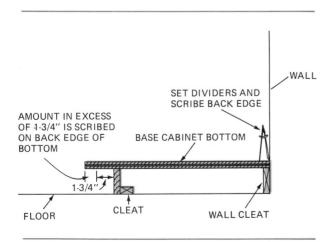

Fig. 30-13 Fitting the bottom of the base unit

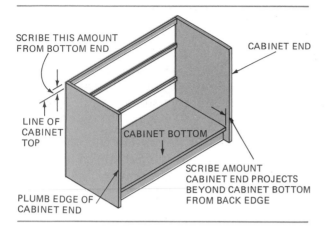

Fig. 30-15 Laying out the base cabinet end panels

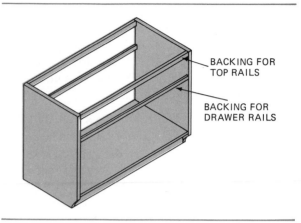

Fig. 30-16 Fasten 1"x2" backing between end panels.

toe space and fasten the end in position. Nail into the end of the cabinet bottom and into the ends of the cleats, Figure 30-15.

- Install the other end in the same manner. Cut and fit a 1x2 between the ends at the top outside corner. It should be the same length as the cleats. Fasten in position with the 1x2 standing on edge, Figure 30-16.

MAKING THE FACE FRAME

The face frame for the base unit is made in the same manner as for wall units except the stiles project 1 1/2 inches below the face of the cabinet bottom and extra rails need to be installed for drawer openings. The top rails are made of 1x3 stock.

- Install vertical pieces as required for the number of door openings. Plan openings so they do not exceed 24 inches in width.

- Cut the bottom rails between the stiles, keeping the top edges flush with the top side of the cabinet bottom. Fasten into the edge of the cabinet bottom.

- The top rails are fastened by clamping and gluing them to the 1x2 strip previously installed or with screws applied through the back side of the strip.

- The drawer rails may be installed in the same manner as the top rails by installing

another 1x2 strip in back of them or by clamping and gluing short blocks of 1x2 behind the joints, Figure 30-17. If wood drawer guides are to be used, *kickers* need to be installed. Kickers are pieces, usually 1-inch stock, installed on edge, centered on the opening, with their bottom edges flush with the opening running between the face frame and the wall. Kickers prevent the drawer from tilting downward when opened.

- Sand all joints flush and smooth.

MAKING THE COUNTERTOP

- Cut a length of plywood to the overall length of the cabinet. Its width should be about 24 1/2 inches. Stock can be conserved by using the waste from the cabinet bottom for the countertop.

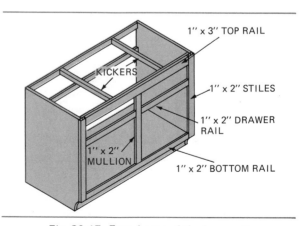

Fig. 30-17 Face frame of the base unit

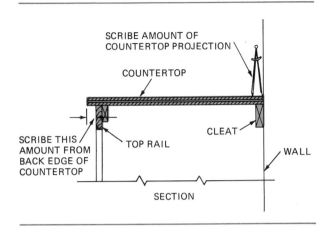

Fig. 30-18 Installation of the countertop

LAMINATING A COUNTERTOP

Before laminating a countertop, make sure all surfaces are flush. There should be no indentations where the pilot of the router bit will ride. Check for protruding nailheads and points. Plane or sand any surfaces that are not flush. Fill in any holes and sand smooth. Always use eye protection when trimming laminates.

USING CONTACT CEMENT

Contact cement is used for bonding plastic laminates and other thin, flexible material. A number of reasons why a contact-bonded piece of laminate may fail to adhere are listed:

- Not enough cement is applied. If the material is porous, like the edge of particleboard or plywood, a second coat is needed after the first coat dries. When enough cement has been applied, a glossy film appears over the entire surface when dry.

- Too little time is allowed for the cement to dry. Both surfaces must be dry before contact is made. To test for dryness, lightly press your finger to the surface. Although it may feel sticky, the cement is dry if no cement remains on the finger.

- The cement is allowed to dry too long. If contact cement dries too long (more than about two hours depending on the humidity), it will not bond properly. To correct this condition, merely apply another coat of cement.

- The surface is not rolled out or tapped after the bond is made. Pressure must be applied to the entire surface using a 3-inch roller or by tapping with a small block of wood and a hammer after the bond is made.

- Place the countertop in position with its outside edges overhanging the face frame by the same amount on both ends.

- Scribe this amount from the back edge, Figure 30-18. Cut to the scribed line and place in position. The ends should be flush with the ends of the cabinet and the front edge flush with the face of the face frame.

- Install 1x2 on edge around the front edge and ends, if an overhang is desired at the ends. Keep the top edge flush with the top side of the countertop.

- Rip a 4-inch-wide length of 3/4-inch plywood for the backsplash, if required, to the same length as the countertop. Fasten the backsplash to the countertop, Figure 30-19.

CAUTION: Some contact cements are flammable. Apply only in a well-ventilated area around no open flame. Avoid inhaling the fumes.

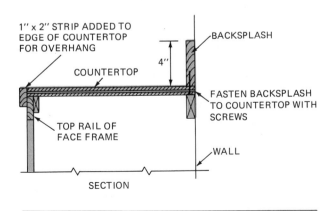

Fig. 30-19 Installation of the backsplash

Fig. 30-20 Applying laminate to countertop

LAMINATING THE COUNTERTOP EDGES

Cut the edge laminate to a rough size (about 1/2 inch wider and longer than needed). Apply a liberal coat of cement to the countertop edges and the back of the laminate with a brush or roller.

After the cement is dry, apply the laminate to the front edge of the countertop, Figure 30-20. Position it so the bottom, top edge, and ends overhang. A permanent bond is made when the two surfaces touch. A mistake in positioning means removing the bonded piece, a time-consuming, frustrating, and difficult job. Roll out or tap the surface.

Apply the laminate to the ends in the same manner as to the front edge piece. Make sure that the square ends butt up firmly against the back side of the overhanging ends of the front edge piece to make a tight joint.

TRIMMING LAMINATED EDGES

Two router bits are used to trim laminates. The *flush trimming bit* is used to cut the first piece of laminate flush with the surface. The *bevel trimming bit* is used to trim the edge of laminate bonded against another, Figure 30-21.

When using a bevel trimming bit, the router base is gradually adjusted to expose the bit so that the laminate is trimmed flush with the first piece but not cutting into it. The bevel of the cutting edge allows the laminate to be trimed without cutting into the adjacent piece. Do not use a flush trimming bit when the pilot rides against another piece of laminate.

Some trimming bits have ball-bearing pilots while others have solid (or dead) pilots that turn with the bit. When bevel trimming with a dead pilot, the laminate that the pilot rides against must be lubricated. Rub wax, solid shortening, or petroleum jelly on the surface to lubricate it and keep it from burning the laminate.

TRIMMING THE EDGES

- Using the bevel trimming bit, trim the ends overhanging the front edge piece.

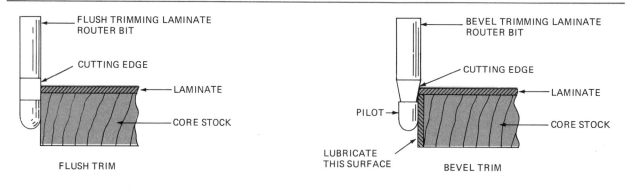

FLUSH TRIMMING LAMINATE ROUTER BIT

CUTTING EDGE

LAMINATE

CORE STOCK

FLUSH TRIM

BEVEL TRIMMING LAMINATE ROUTER BIT

CUTTING EDGE

LAMINATE

PILOT

CORE STOCK

LUBRICATE THIS SURFACE

BEVEL TRIM

ADJUST BEVEL TRIMMING BIT TO CUT FLUSH WITH, BUT NOT INTO EDGE LAMINATE. THE BEVEL KEEPS THE CUTTING EDGE FROM GRAZING THE FIRST LAYER OF LAMINATE.

Fig. 30-21 Flush and bevel trim laminate bits

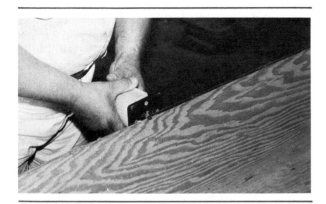

Fig. 30-22 Flush trimming the edge band

- Then, using the flush trimming bit, trim off the bottom and top edges of both front and end edge pieces, Figure 30-22.

- Use a belt sander or a file to make the top and bottom edges flush with the surface.

- Break the sharp edge on the bottom with a file or sandpaper. Keep a sharp edge on the top to assure a tight joint with the countertop laminate.

LAMINATING THE COUNTERTOP SURFACE

- Remove the back splash from the countertop.

- Cut the laminate to a rough size and apply contact bond cement to the countertop and the back side of the laminate.

- To position large pieces such as countertops, use thin strips of wood or metal venetian blind slats with the crowned side up. Place the strips or slats about a foot apart on the surface.

- Lay the laminate to be bonded on the strips or slats and position correctly, Figure 30-23.

- Make contact on one end. Gradually remove the slats one by one until all are removed. The laminate is now positioned correctly with no costly errors.

- Roll out the laminate, Figure 30-24.

- Trim the back edge with a flush trimming bit. Trim the ends and front edge with a bevel trimming bit, Figure 30-25.

- Break any sharp corners with a sanding block and clean excess cement with solvent.

Fig. 30-23 Position the laminate on the countertop using venetian blind slats.

Fig. 30-24 Roll out the laminate to assure a good bond.

Fig. 30-25 Bevel trim the countertop laminate.

LAMINATING A COUNTERTOP WITH TWO OR MORE PIECES

In cases where the countertop must be laminated in two or more pieces, the joint must be made tight to present a good appearance.

- Clamp the two pieces of laminate in a straight line on some strips of 3/4-inch stock.

- Butt the ends together or leave a space less than 1/4 inch between them.

- Using one of the strips as a guide, run the flush trimming bit through the joint. Keep the pilot of the bit against the straightedge to assure making a tight joint, Figure 30-26.

Fig. 30-26 The router, with a flushing trimming bit, is used to make joints between pieces of laminate.

Fig. 30-27 Applying the backsplash

- Bond the sheets as previously described. Press seam-filling compound, especially made for laminates and which comes in various colors, into the joint if necessary. Wipe off excess compound with the recommended solvent.

LAMINATING BACKSPLASHES

Backsplashes are laminated in the same manner as countertops. If possible, laminate the backsplash before attaching it to the countertop. Then fasten it to the countertop by screwing or nailing up through the bottom of the countertop into the bottom edge of the backsplash. Use a little caulking compound in the joint, to prevent any accumulated water from seeping behind the countertop, Figure 30-27.

LAMINATING CURVED SURFACES

If the edge of a countertop has a curve, the laminate can be bent, if necessary, to the desired radius by heating with a heat gun, Figure 30-28. Heat the laminate carefully and bend until the desired radius is obtained. Allow the laminate to cool while holding it in the shape desired. After it cools, it retains the shape in which it was held.

CAUTION: Keep fingers away from the heated area of the laminate. Remember that the laminate retains heat for some time.

Fig. 30-28 Heating and bending laminate with a heat gun

REVIEW QUESTIONS

Select the most appropriate answer.

1. The vertical distance between the base unit and the wall unit is usually
   a. 10 to 12 inches.   c. 16 to 18 inches.
   b. 12 to 14 inches.   d. 18 to 20 inches.

2. The maximum distance, in feet, that the top shelf of a wall unit should be from the floor is
   a. 5.   c. 7.
   b. 6.   d. 8.

3. The distance from the floor to the surface of the countertop is
   a. 30 inches.   c. 36 inches.
   b. 32 inches.   d. 42 inches.

4. In order to accommodate sinks and provide adequate working space, the width of the countertop is
   a. 25 inches.   c. 30 inches.
   b. 28 inches.   d. 32 inches.

5. Standard wall unit height is
   a. 24 inches.   c. 32 inches.
   b. 30 inches.   d. 36 inches.

6. The depth of standard wall units is
   a. 10 inches.   c. 14 inches.
   b. 12 inches.   d. 16 inches.

7. A projection is made on cabinet end panels of factory-built units to allow
   a. for shrinkage.   c. for scribing.
   b. for damage in handling.   d. for mistakes.

8. To find the line of installation for the bottom edge of wall units, measure up from the low point on the floor
   a. 36 to 38 inches.   c. 52 to 54 inches.
   b. 48 to 50 inches.   d. 82 to 84 inches.

9. The amount scribed off the wall edge of job-built base unit end panels is
   a. one inch.
   b. the amount the outside edge projects beyond the cabinet bottom.
   c. the amount the top end extends above the top cleat.
   d. an amount equal to the cabinet end thickness.

10. The cabinet bottom is fitted to the wall by scribing an amount
    a. of 1 3/4 inches.
    b. of the difference between the overhang at the toeboard and 1 3/4 inches.
    c. of the difference between the overhang at the toeboard and 3 1/2 inches.
    d. which is the farthest distance the back edge is away from the wall.

CABINET DOORS AND DRAWERS 31

OBJECTIVES

After completing this unit, the student will be able to

- *classify cabinet doors and drawers according to the method of construction and installation.*
- *make solid and paneled cabinet doors.*
- *install overlay, lipped, and flush cabinet doors.*
- *make overlay, lipped, and flush cabinet drawers.*
- *install drawer guides and fit overlay, lipped, and flush drawers in a cabinet.*
- *apply door and drawer hardware.*

The carpenter may purchase cabinet doors and drawers in a variety of styles and sizes or may have them made to specifications in a mill. However, because of special requirements, the carpenter must often design and construct cabinet doors and drawers on the job.

KINDS OF DOORS AND DRAWERS

Doors and drawers are classified as overlay, lipped, and flush. Classification is based on the method of installation, Figure 31-1.

OVERLAY TYPE

The overlay type of doors and drawers laps over the opening, usually 3/8 inch on all sides

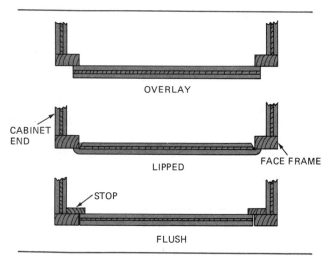

Fig. 31-1 Overlay, lipped, and flush doors

but in many cases may cover the entire face frame. This kind is widely used on plastic laminated kitchen cabinets.

The overlay drawer or door is easy to install because it does not require fitting in the opening. The face frame of the cabinet acts as a stop. The door may have square or reverse bevel edges.

European-style cabinets omit the face frame. Doors and drawers completely overlay the front edges of the cabinet.

LIPPED TYPE

The lipped type of doors and drawers has rabbeted edges that overlap the opening by usually 3/8 inch on all sides. Usually the ends and edges are rounded over to give a more pleasing appearance. Lipped doors and drawers are easy to install because no fitting is required and the rabbeted edges stop against the face frame of the cabinet. However, a little more time is required to shape the edges.

FLUSH TYPE

The *flush type* door or drawer fits into and flush with the opening. These are a little more difficult to install because they have to be fitted in the opening. A fine joint, about the thickness of a dime, must be maintained between the opening and the door or drawer front. Stops must be provided in the cabinet against which to close the drawer or door.

DOOR CONSTRUCTION

Doors are constructed as solid or paneled doors. *Solid doors* are made of plywood, particleboard, or glued-up solid lumber. Designs are often grooved into the face of the door with a router, or molding may be applied to give the door a more attractive appearance, Figure 31-2.

Paneled doors have an exterior framework of solid wood with a panel of solid wood, plywood, hardboard, metal, plastic, glass, or some other material, Figure 31-3. Because of limited woodworking machinery, carpenters make only paneled doors of simple design. More complicated designs are usually manufactured in a mill if custom sizes are needed. If standard sizes are used, these doors may be purchased from a distributor. Both solid doors and paneled doors may be installed in overlay, lipped, or flush fashion.

MAKING SOLID DOORS

OVERLAY DOORS

To make a solid door from plywood or particleboard, simply cut the stock to size. For an overlay door, cut the stock 3/4 inch larger than the opening for a 3/8-inch lap all around. Tilt the table saw blade 30° to cut reverse bevel edges, if desired. Round over the corners slightly.

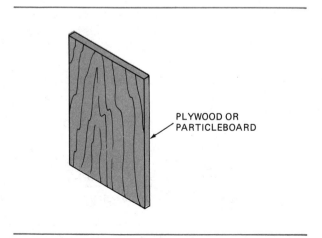

Fig. 31-2 Solid door

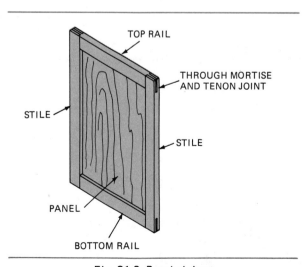

Fig. 31-3 Paneled door

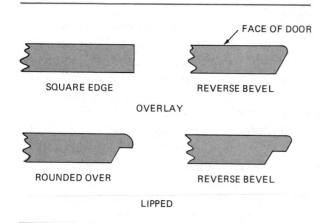

Fig. 31-4 Edge designs on overlay and lipped doors

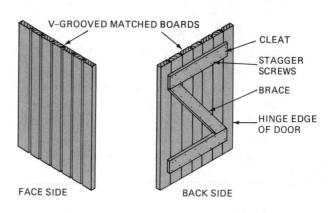

Fig. 31-5 Solid matched board door

LIPPED DOORS

For a lipped door, cut the stock 5/8 inch larger than the opening. Then rabbet the stock 3/8x3/8 on the edges and ends. This gives 1/16-inch clearance between the frame and rabbeted edges and ends of the door when installed. Cutting the rabbet on a slight bevel gives added clearance to swing the door, Figure 31-4.

Usually the edges and ends of a lipped door are rounded over. Use a block plane to round over the edges and ends. Then sand them smooth. A router with a 3/8-inch-radius rounding-over bit may also be used; but, in this case, round over the edges before making the rabbet so that the pilot of the bit will have some wood to ride on. Instead of rounded edges, a reverse bevel may also be made on lipped doors.

FLUSH DOORS

For a flush door, cut the stock slightly oversize. Then, joint the door in the opening with a hand plane.

MAKING A SOLID MATCHED BOARD DOOR

Usually clamping and jointing equipment for edge-gluing boards is not available to the carpenter. To make a door from solid boards on the job, usually V-grooved matched boards are used. Cleats are placed on the back side near the top and the bottom of the door with a diagonal brace between them, Figure 31-5.

Cut the necessary number of pieces to length for the width of the door to be made. Rip square edges on the outside pieces. Tack a straightedge on the workbench or subfloor and place the pieces side by side with their faces down and one edge against the straightedge. Toenail another straightedge to the bench and against the other edge of the door, driving the edges of the pieces tightly against each other, Figure 31-6.

Cut two 1x3 cleats about 1 inch shorter than the width of the door. Fasten them to the back side of the door with screws staggered into each piece. Keep the cleats about 2 inches from the top and bottom edges of the door.

Cut a diagonal brace of the same size material between the cleats. The bottom end of the brace should be toward the hinge edge of the door. Fasten with screws staggered into each piece of the door. Chamfer all exposed edges

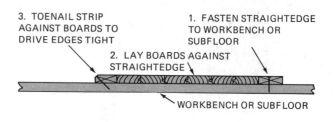

Fig. 31-6 Driving matched board edges tightly together

and ends of the cleats and brace. Shape the edges and ends of the door as desired.

COVERING SOLID DOOR EDGES

On high-quality cabinetwork, the end grain of lumber is completely covered or exposed as little as possible. End grain shows all around plywood doors because of the alternate direction of the plies.

PLASTIC LAMINATES

Door edges and faces can be laminated in the same manner as kitchen countertops. See Unit 30 Kitchen Cabinets. In quality construction the back face of the door is covered with a thin laminate called a *backer* laminate to achieve a balanced core. A balanced core is achieved when both sides are sealed in the same manner to prevent warpage due to moisture entering from one side more than the other.

VENEER STRIPS

Another method used to band door edges is to apply a thin wood veneer of the same material as the door to all the edges, Figure 31-7. These veneers may be purchased in rolls of different kinds of wood. Veneers may also be made on the job by slicing thin strips with the table saw. Make the strips slightly wider than the thickness of the door from scrap material left over after making the doors.

The veneers are applied to door edges with contact cement. Coat both the underside of the strips and the door edges with the cement and let them dry. Apply the strips to the edges of the door, making tight joints at the corners. Trim the excess with a sharp knife and sand smooth.

PLASTIC TAPES

Rolls of plastic tape are available to cover edges of plywood and other core material. These tapes come in a variety of colors and wood grains and are backed with a heat-sensitive adhesive. They are applied by simply using a heating iron to press the tape on the edge and trimming them flush, Figure 31-8.

WOOD STRIPS

Strips of wood are often used to band edges. The strips may be thin, from 1/4 to 1/2 inch, and are applied with glue and brads. The corners are mitered to avoid exposure of any end grain.

TREATING SOLID DOOR FACES

The appearance of solid doors may be improved by routing designs in or applying molding to the door faces.

ROUTING DOOR FACES

A number of designs may be routed in the door faces using a small *veining bit* in the router.

Fig. 31-7 Applying veneer strips to door edges

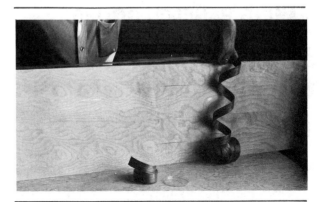

Fig. 31-8 Covering plywood edges with plastic tape

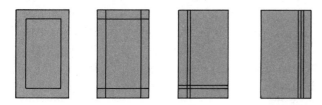

Fig. 31-9 Routed designs on door faces

Straight line cuts, Figure 31-9 may be made by guiding the router base against a straightedge while holding the door securely in a jig.

A French Provincial design may be made by removing the router base and replacing it with an elongated base with rounded ends. Lay the door on a bench and tack some strips a little thicker than the door around it. Start routing about 1/8 inch deep along one side. Bring the router into each corner and rotate it until the cut is complete, Figure 31-10. The radius of the curve in the corners is dependent on the radius of the ends of the base.

APPLYING MOLDING

Small molding that has one flat side, such as screen molding or half-round, may be applied to door faces in a number of interesting designs, Figure 31-11. Lightly mark the location of the molding on the door face. Mark what will be the outside edge of the molding.

Cut the molding to length, mitering the corners, and apply it with a little glue and brads. Apply the molding so the outside edge just covers the line. Wipe off any excess glue with a damp rag and set all brads below the surface.

MAKING PANELED DOORS

Paneled doors of simple design may be made by the carpenter on the job site. Because of limited equipment, through mortise-and-tenon joints are used on the frame. Flat panels of plywood, hardboard (plain or perforated), or decorative plastic may be used.

MAKING THE FRAME

Cut the stiles to the overall height of the door and the rails to the overall width of the

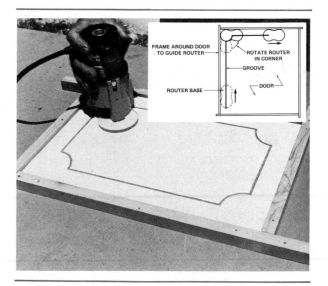

Fig. 31-10 Making a French Provincial design on a door face

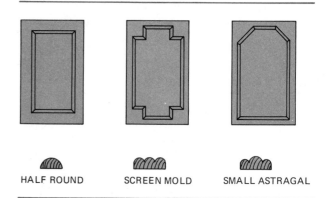

Fig. 31-11 Small molding may be applied to door faces in a number of interesting patterns.

door. Usually 1x2 stock is used for the stiles and rails. Sometimes for appearance and strength, the bottom rail is made wider.

Center a groove on the edge of each piece about 1/4 inch deep and as wide as the thickness of the panel stock. Cut a through mortise on each end of the stiles. This may be done using the table saw. Raise the blade so its height is equal to the width of the rails minus the depth of the groove. Set the fence so that 1/4 inch is removed from the center by standing the piece on end, making a cut, turning the piece around, and making another cut with a single saw blade.

Cut tenons on each end of all the rails. In the same manner, make cuts but leave 1/4 inch in the center between saw cuts. Then set the fence a distance equal to the width of the stiles to the outside of the saw blade. Lower the blade to a height to remove the waste on each side of the tenon. Use the miter gauge as a guide and the rip fence as a stop and cut the waste from the sides of the tenon.

CUTTING THE PANELS

Cut the panels about 1/16 inch undersize to make sure the frame joints will come up tight. Use an auxiliary wood table, if necessary, to prevent thin material like plastic sheets from slipping between the rip fence and the table top.

ASSEMBLING THE DOOR

Assemble the frame and panel, Figure 31-12. Use glue in the joints, but not on the panel edges. Hold each joint tight and square and drive a screw in the center of the joint from the back side of the door. The screw should be as long as possible, but not so long as to penetrate the door frame. Countersink the screw

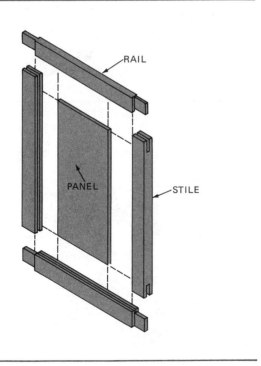

Fig. 31-12 Assembling the frame and panel door

slightly so the head can be filled. Sand the joints and faces smooth, break all sharp corners, and shape the door edges and ends as desired.

TYPES OF HINGES

There are many styles and shapes of cabinet hinges. If the kind of hinge is not specified, select a design that blends well with the cabinet being constructed. Some common types of hinges are surface, offset, overlay, pivot hinge, and butt, Figure 31-13. Many of these are self-closing hinges that eliminate the need for door catches.

SURFACE HINGES

The surface hinge mounts on the exterior surface of the door and frame. The back side of the hinge leaves may lie in a straight line for flush doors or one leaf may be offset for lipped doors, Figure 31-14. The surface hinge is used when it is desirable to show the hardware.

OFFSET HINGES

The offset hinge is used on lipped doors. The semiconcealed type has one leaf bent to a 3/8-inch offset that is screwed to the back of the door. The other leaf screws to the exterior surface of the face frame. A concealed type is designed in which only the pin is exposed, Figure 31-15.

OVERLAY HINGES

The overlay hinge, also available in semiconcealed and concealed types, is used on overlay doors. Special hinges, with one leaf bent at a 30° angle, are used on overlay doors with reverse beveled edges, Figure 31-16.

PIVOT HINGES

The pivot hinge is usually used on overlay doors. It is fastened to the top and bottom of the door and to the inside of the case. It is frequently used when there is no face frame on

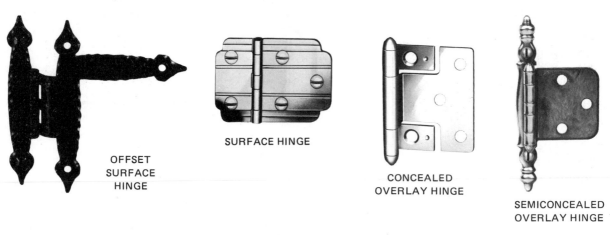

OFFSET
SURFACE
HINGE

SURFACE HINGE

CONCEALED
OVERLAY HINGE

SEMICONCEALED
OVERLAY HINGE

DECORATIVE T HINGE

CONCEALED
OFFSET HINGE

PIVOT HINGE

SEMICONCEALED
OFFSET HINGE

Fig. 31-13 Cabinet door hinges *(Courtesy of Amerock Corporation)*

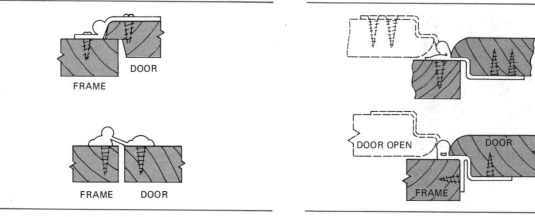

DOOR

FRAME

FRAME   DOOR

Fig. 31-14 Surface hinges

DOOR OPEN

DOOR

FRAME

Fig. 31-15 Offset hinges

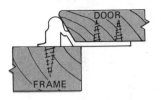
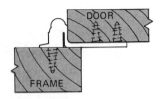
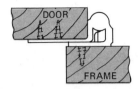

Fig. 31-16 Overlay hinges

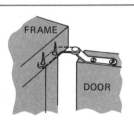

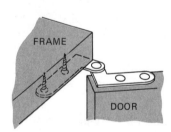

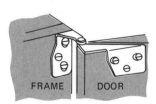

Fig. 31-17 Pivot hinges

the case and when it is desired to have the door completely cover the face of the case, Figure 31-17.

BUTT HINGES

The butt hinge is used on flush doors. The leaves of the hinge are cut into the edges of the frame and the door, in the same manner as for entrance doors. Butt hinges are used when it is desired to conceal most of the hardware; however, they take a little more time to install, Figure 31-18.

HANGING CABINET DOORS

SURFACE HINGES

To hang cabinet doors with surface hinges, first apply the hinges to the door. Then shim the door in the opening so an even joint is obtained all around and screw the hinges to the face frame.

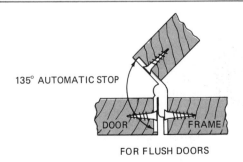

135° AUTOMATIC STOP

FOR FLUSH DOORS

Fig. 31-18 Butt hinges

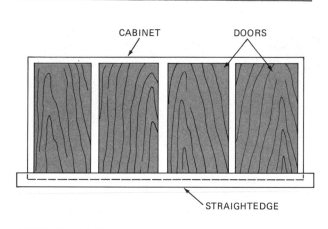

Fig. 31-19 When installing doors, use a straightedge to keep them in line.

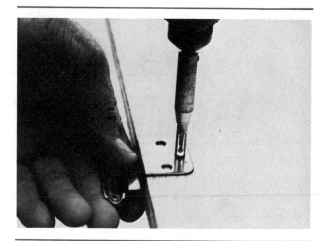

Fig. 31-20 The VIX bit is a self-centering drill stop used for drilling holes for cabinet hinges.

SEMICONCEALED HINGES

For semiconcealed hinges, screw the hinges to the back of the door. Then center the door in the opening and fasten the hinges to the face frame. When more than one door is to be installed side by side, clamp a straightedge to the face frame along the bottom of the openings for the full length of the cabinet. Rest the doors on the straightedge to keep them in line, Figure 31-19.

CONCEALED HINGES

When installing concealed hinges, first screw the hinges on the door. Center the door in the opening and press or tap on the hinge opposite the face frame. Small projections on the hinge mark its location on the face frame. Open the door and screw the hinge to the face frame in the marked location.

BUTT HINGES

Hanging flush cabinet doors with butt hinges is done in the same manner as hanging entrance doors. Drill holes for all screws in hinges centered on the holes in the hinge leaf. Drilling the holes off-center throws the hinge to one side when the screws are driven. This causes the door to be out of alignment. Many carpenters use a tool called a *VIX bit* when drilling

holes for cabinet door hinge screws, Figure 31-20. This tool is attached to the drill to center the hole on the hinge leaf and also acts as a stop to prevent drilling through the door or face frame.

MAKING DRAWERS

Usually drawer fronts are made of 3/4-inch plywood, particleboard, or solid wood. Overlay drawers may have fronts of thinner material. Sides and backs are generally 1/2 inch thick. Often, fiberboard with a printed wood grain is used for drawer sides and backs. The drawer bottom is usually made of 1/4-inch plywood or hardboard. Small drawers may have 1/8-inch hardboard bottoms.

Fig. 31-21 Dovetail joints can be made with a router and a dovetail template. *(Courtesy of Rockwell International)*

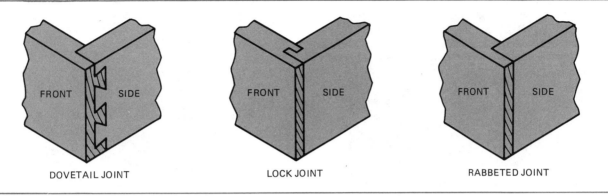

FRONT   SIDE

FRONT   SIDE

FRONT   SIDE

DOVETAIL JOINT

LOCK JOINT

RABBETED JOINT

Fig. 31-22 Typical joints between drawer front and side

DRAWER JOINTS

Typical joints between the front and sides of drawers are the dovetail, lock, and rabbeted joints. The dovetail joint is used in high quality drawer construction. It takes a longer time to make, but is the strongest. Dovetail drawer joints may be made using a router and a dovetail template, Figure 31-21. The lock joint is simpler and can be easily made using a table saw. The rabbeted joint is the easiest to make but must be strengthened with fasteners, Figure 31-22.

Joints used between the sides and back are the dovetail, dado and rabbet, dado, and butt

joints. With the exception of the dovetail joint, the drawer back is usually set in at least 1/2 inch from the ends of the sides for added strength, Figure 31-23.

BOTTOM JOINTS

The drawer bottom is fitted into a groove on all four sides of the drawer, Figure 31-24. An alternate method is to cut the back narrower, assemble the four sides, slip the bottom in, and nail its back edge to the drawer back, Figure 31-25.

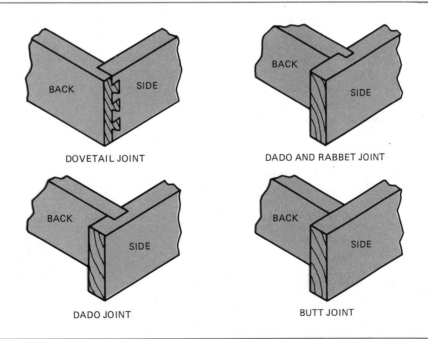

BACK   SIDE

BACK   SIDE

DOVETAIL JOINT

DADO AND RABBET JOINT

BACK   SIDE

BACK   SIDE

DADO JOINT

BUTT JOINT

Fig. 31-23 Typical joints between drawer back and side

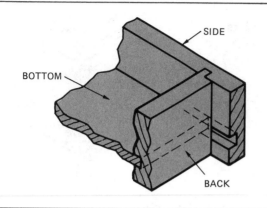

Fig. 31-24 Drawer bottom fitted in groove at drawer back

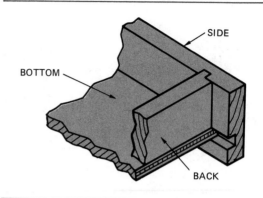

Fig. 31-25 Drawer bottom fastened to bottom edge of drawer back

MAKING AN OVERLAY DRAWER

The overlay drawer consists of the bottom, two sides, a back, a false front, and an overlay front which is fastened to the false front with screws from the inside, Figure 31-26.

Cut the drawer sides to width and length. The width should be at least 1/8 inch less than the height of the drawer opening. This width may be less if metal commercial drawer guides are used. The length of the sides should be slightly less than the depth of the cabinet.

Dado the sides for the back and front of the drawer, 1/4″ deep, Figure 31-27. Make these cuts the same depth to allow the front and back to be cut to the same length.

Cut the false fronts and backs to width and length. The width is the same as the drawer sides. The length determines the overall width of the drawer. The length should be such to allow the proper clearance between the drawer sides and the opening. This depends on the type of drawer guides used. Rabbet the ends of the front to fit the dado in the drawer sides. Round off and smooth the top edges of all sides, fronts, and back.

Make a groove on all parts for the drawer bottom. The width of the groove should be such that the drawer bottom will slide easily into it without being forced.

The depth of the groove should be about 1/4 inch. If the groove is too deep, it will

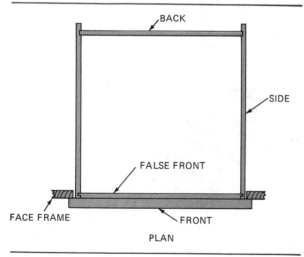

Fig. 31-26 The overlay drawer

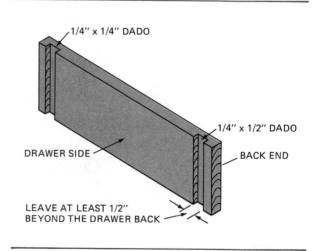

Fig. 31-27 Dado the overlay drawer sides.

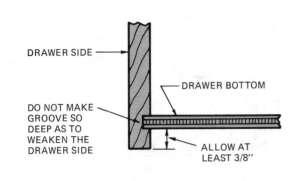

Fig. 31-28 Location of the groove for the drawer bottom

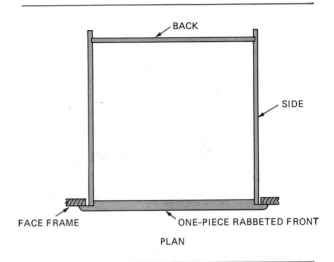

Fig. 31-29 The lipped drawer

weaken the sides. If the groove is too shallow, the bottom may fall out after assembly. The distance from the bottom edge should be about 3/8 inch to the groove, Figure 31-28.

When cutting the grooves for the bottom, alternate the side pieces so that both right and left sides will be obtained. If not carefully done, it is possible to end up with all right-hand sides or all left-hand sides. Cut the bottoms to size about 1/16 inch smaller than measured.

ASSEMBLING THE DRAWER

Glue and fasten the sides to the back. Slip in the bottom. (Do not apply glue to the bottom.) Fasten the false front to the sides. Cut and fasten the overlay front to the drawer with screws driven from the inside.

MAKING A LIPPED DRAWER

The sides, back, and bottom of a lipped drawer are machined in the same way as for an overlay drawer. However, the lipped drawer does not have a false front. Therefore, the front end of the side pieces are not dadoed, Figure 31-29. The sides of the drawer are made up to the lipped drawer front.

Cut the front 5/8 inch over the opening size on both the width and length. Rabbet the top and bottom edges 3/8"x3/8". On each end make rabbets 3/8"x7/8" to allow for the overlap and the thickness of the drawer sides. Shape the edges of the drawer front as desired by rounding over or by cutting a reverse bevel.

The drawer is assembled in a manner similar to that used for the overlay drawer. The

sides are fastened to the rabbeted ends of the front, however.

MAKING A FLUSH DRAWER

Cut out the drawer front to the overall height and width of the drawer opening. Rabbet the two ends of the drawer front to receive the drawer sides plus an allowance for side clearance. Cut out the other parts and assemble in a manner similar to that used for other type drawers. After the drawer is assembled, try it in the opening. Fit the drawer front to the opening by hand planing, if necessary.

Joints other than those described may be used in the making of drawers. To make dovetail joints, follow dovetail template manufacturers' directions.

DRAWER GUIDES

There are many ways of guiding drawers, Figure 31-30. The type of drawer guide selected affects the size of the drawer. The drawer must be guided sideways and vertically and must be kept from tilting down when opened.

WOOD GUIDES

Probably the simplest type wood guide is the center strip. It is installed in the bottom

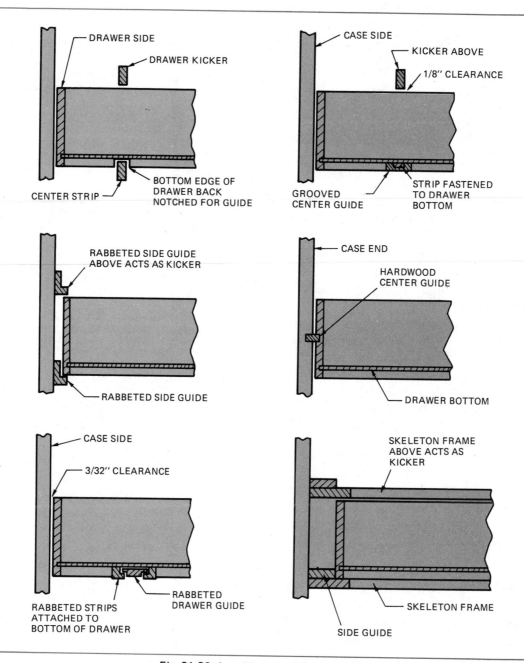

Fig. 31-30 Installing wood drawer guides

center of the opening from front to back, Figure 31-31. The strip projects above the bottom of the opening. The bottom edge of the drawer back is notched to ride in the guide. A kicker is installed above the drawer to keep it from tilting downward when opened.

Another type of wood guide is the grooved center strip, Figure 31-32. The strip is placed in the center of the opening from front to back. A matching strip is fastened to the drawer bottom. In addition to guiding the drawer, this system keeps it from tilting when opened.

Another type of wood guide is a rabbeted strip, Figure 31-33. The drawer sides fit into and slide along the rabbeted pieces. Sometimes these guides are made up of two pieces instead of being rabbeted. A kicker above the drawer is necessary with this type guide.

550

Fig. 31-31 Simple center wood drawer guide (The back of the drawer is notched to run on the guide.)

Fig. 31-32 Grooved center wood drawer guide

Fig. 31-33 Rabbeted wood side guides

METAL DRAWER GUIDES

There are many different types of metal drawer guides. Some have a single track mounted on the bottom center of the opening. Others may be mounted on the top center. Nylon rollers mounted on the drawer ride in the track. Side guides mount either on the top or bottom of the drawer sides and opening, Figure 31-34. Usually a 1/2-inch clearance on each side of the drawer is needed for the installation of side guides.

Instructions for installation differ with each type. When using commercially made drawer guides, read the instructions first, before making the drawer. Allowances must be made for the installation of metal guides. Drawers usually have to be made smaller to accommodate the hardware.

APPLYING DOOR AND DRAWER HARDWARE

PULLS AND KNOBS

Usually doors and drawers are opened by means of pulls or knobs. They come in many styles and designs and are made of metal, plastic, wood, porcelain, or other material, Figure 31-35.

Pulls and knobs are installed by drilling holes in the drawer front and fastening with

Fig. 31-34 Installing metal drawer guides *(Robert Morency, photographer)*

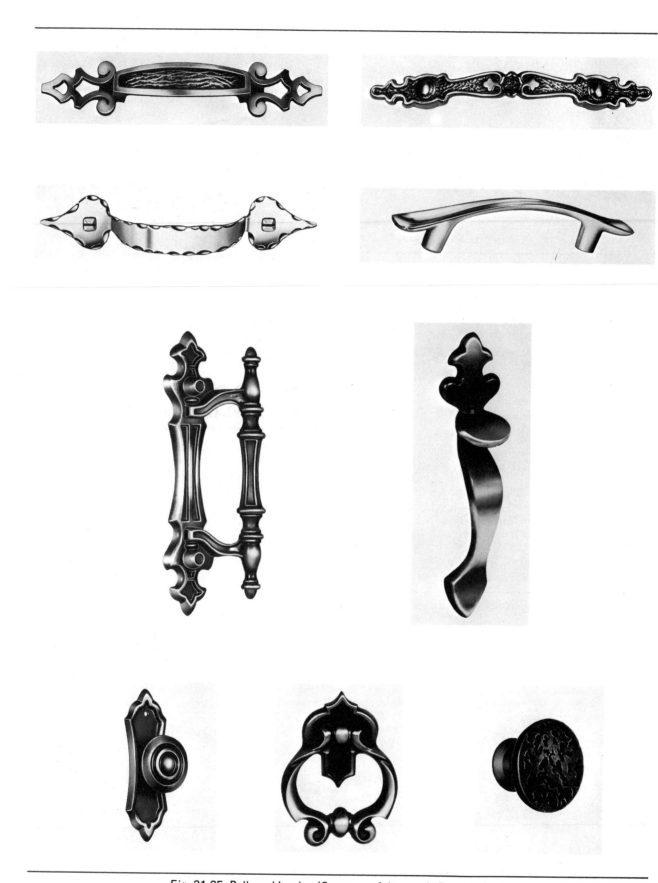

Fig. 31-35 Pulls and knobs *(Courtesy of Amerock Corporation)*

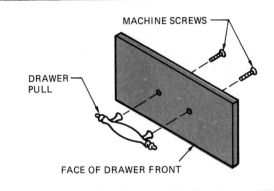

Fig. 31-36 Fastening drawer pulls

BULLET CATCH

ROLLER-TYPE
FRICTION CATCH

DOUBLE MAGNETIC CATCH

machine screws from the inside. When two screws are used to fasten a pull, the holes are drilled slightly oversize in case they are a little off-center. This allows the pulls to be fastened easily without cross-threading the screws, Figure 31-36.

CATCHES

Doors without self-closing hinges need catches to hold the door closed. There are many kinds of catches available, Figure 31-37. Catches should be placed in the most out-of-the-way position possible. For instance, they are placed on the underside of shelves instead of on the top.

Magnetic catches are widely used. They are available in single or double magnets of varying holding power. An adjustable magnet is attached to the inside of the case. A metal plate is attached to the door. First attach the magnet. Then place the plate on the magnet. Close the door and tap it opposite the plate. Projections on the plate mark its location on the door. Attach the plate to the door where marked. Try the door and adjust the magnet, if necessary. When adjusted, a small escutcheon pin (small nail with a round head) is driven in the bracket of the magnet to keep it from moving.

Friction catches are installed in a similar manner to that used for magnetic catches. Fasten the adjustable section to the case and the other section to the door.

Elbow catches are used to hold one door of a double set. They are released by reaching to

FRICTION
CATCH

ELBOW CATCH

DECORATIVE CATCH
FOR FLUSH DOORS

Fig. 31-37 Door catches
(Courtesy of Amerock Corporation)

the back side of the door. These catches are used when one of the doors is locked against the other.

Bullet catches are spring loaded and fit into the edge of the door. When the door is closed, the catch fits into a recessed plate mounted on the frame.

REVIEW QUESTIONS

Select the most appropriate answer.

1. Overlay and lipped doors and drawers are easier to install than flush type because
   a. no catches are necessary.
   b. no jointing is necessary.
   c. they take less material.
   d. simpler joints can be used.

2. A drawer front or door with its edges and ends rabbeted to fit over the opening is called
   a. an overlay type.   c. a lipped type.
   b. a flush type.   d. a rabbeted type.

3. In contrast to other types of doors and drawers, the flush type requires
   a. more time to construct.   c. extra clearance.
   b. no fitting.   d. stops.

4. A lipped door is cut larger than the opening by
   a. 1/4 inch.   c. 1/2 inch.
   b. 3/8 inch.   d. 5/8 inch.

5. Door panels and drawer bottoms are usually cut
   a. undersize.   c. oversize.
   b. exactly to size.   d. with rabbeted edges.

6. The offset hinge is used on
   a. paneled doors.   c. lipped doors.
   b. flush doors.   d. overlay doors.

7. Butt hinges are used on
   a. flush doors.   c. overlay doors.
   b. lipped doors.   d. solid doors.

8. The thickness of drawer sides and backs is usually
   a. 1/4 inch.   c. 5/8 inch.
   b. 1/2 inch.   d. 3/4 inch.

9. The joint used on high-quality drawers is the
   a. dado joint.   c. dovetail joint.
   b. dado and rabbet joint.   d. rabbeted joint.

10. A wood strip installed in the cabinet to prevent a drawer from tilting downward when opened is called a
    a. top guide.   c. sleeper.
    b. kicker.   d. tilt strip.

APPENDIX: SLAB-ON-GRADE CONSTRUCTION

In warm climates, where frost penetration into the ground is not very deep, little excavation is necessary and the first floor may be a concrete slab placed directly on the ground. This is commonly called *slab-on-grade* construction, figure A-1. With improvements in the methods of construction, the need for lower construction costs, and the desire to give the structure a lower profile, floor slabs on grade are being used more often in all climates.

The carpenter's work in slab-on-grade construction is not very different from that in other types of construction. The principles of form construction discussed in Unit 9 are the same for all types of construction. Except for the absence of a framed floor, the framing in slab-on-grade building is the same as in other building types. However, slab-on-grade construction is used extensively in some areas, so this appendix is included to highlight the peculiarities of this type of construction.

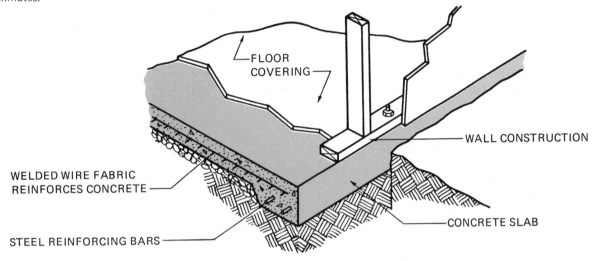

Fig. A-1 Slab-on-grade foundation (From Mark W. Huth, *Understanding Construction Drawings,* © 1983 by Delmar Publishers Inc.)

BASIC REQUIREMENTS

To provide a satisfactory floor, the construction of concrete floor slabs should meet certain basic requirements as follows:

1. The finished floor level must be established high enough so that the finish grade around the slab can be sloped away for good drainage. The top of the slab should be no less than 8 inches above the finish grade.

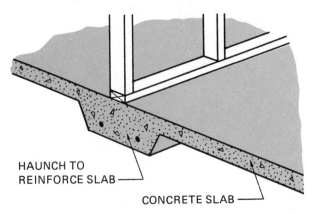

Fig. A-2 The slab is haunched under load-bearing walls. (From Mark W. Huth, *Understanding Construction Drawings,* © 1983 by Delmar Publishers Inc.)

2. All topsoil in the area in which the slab is to be placed must be removed. A base for the slab consisting of 4 to 6 inches of gravel, crushed stone, or other approved material must be well compacted in place.
3. All water and sewer lines, heating ducts, and other utilities that are to run under the slab must be installed.
4. A vapor barrier must be placed under the concrete slab to prevent soil moisture from rising through the slab. The vapor barrier should be a heavy plastic film, such a 6-mil polyethylene or other material having equal or superior resistance to the passage of vapor. It should be strong enough to resist puncturing during the placing of the concrete. Joints in the vapor barrier must be lapped at least 4 inches and sealed.
5. Where necessary, to prevent heat loss through the floor and foundation walls, a waterproof, rigid insulation is installed around the perimeter of the slab.
6. The slab should be reinforced with 6 by 6 inch, #10 welded wire mesh, or by other means to provide equal or superior reinforcing. The concrete slab must be at least 4 inches thick and *haunched* (made thicker) under load bearing walls, figure A-2.

TYPES OF SLABS

In warm climates, where there is little or no frost penetration into the ground, a combined slab and foundation, called a *monolithic slab,* is commonly used, figure A-3. This type of slab is also referred to as a *thickened edge slab.* It consists of a shallow footing around the perimeter which is placed at the same time as the slab, making the slab and footing a one-piece integral unit. The bottom of the footing must be at least 1 foot below the finish grade, unless local building codes dictate otherwise.

The construction of forms for concrete slabs is described in Unit 9 — Concrete Form Construction.

In areas where the ground freezes to any appreciable depth during winter, the footing for the walls of the structure must extend below the frostline. If slab-on-grade construction is desired in these areas, then the concrete slab and foundation wall are separate. This type of slab on grade is called an *independent slab,* and may be constructed in a number of ways according to conditions, figure A-4. The foundation wall acts as a form for the concrete slab. Unit 9 — Concrete Form Construction describes the construction of forms for foundation walls and footings. Rigid insulation is placed between the foundation wall and the slab edge to provide for contraction and expansion of the slab, in addition to preventing heat loss.

INSULATION REQUIREMENTS

As previously stated, the use of perimeter insulation is required to prevent heat loss, except in warm climates. Two general rules to follow when determining insulation requirements are:

1. When average winter low temperatures are 0°F and higher, the R factor should be about 2.0 and the depth of the insulation or the width under the slab should not be less than 1 foot.
2. When average winter temperatures are lower than 0°F, the R factor should be about 3.0 without floor heating, and the depth or width of the insulation should be not less than 2 feet.

TERMITE PROTECTION

Slab-on-grade construction requires certain precautions when protection against termites is necessary. A countersunk opening 1 inch wide and 1 inch deep is made around pipes that pass through the slab and hot tar is poured in the opening. If insulation is used between the foundation wall and the slab, the insulation is kept 1 inch below the top of the slab and the space filled with hot tar. The soil under the slab may also be treated with chemicals for control of termites, but caution is advised. Such treatment should be done only by those thoroughly trained in the use of these chemicals.

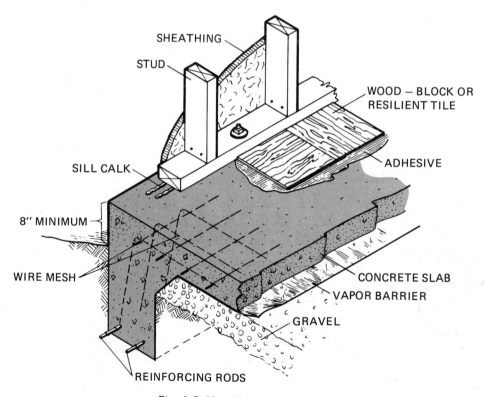

SHEATHING

STUD

WOOD — BLOCK OR RESILIENT TILE

ADHESIVE

SILL CALK

8" MINIMUM

WIRE MESH

CONCRETE SLAB

VAPOR BARRIER

GRAVEL

REINFORCING RODS

Fig. A-3 Monolithic slab on grade

FINISH FLOORS OVER CONCRETE SLABS

Wall-to-wall carpeting is extensively used as a finish floor over concrete slabs in living areas and sleeping quarters. Sometimes wood block or parquet flooring laid in mastic is used. Matched wood strip finish flooring may also be used over pressure-treated sleepers fastened to the concrete slab. In kitchens and baths usually vinyl or vinyl-asbestos sheet or tile floors are laid. See Unit 29 — Finish Floors.

Terrazzo is also used as a finish floor over concrete slabs. Terrazzo is a topping of cement in which small chips of marble or colored stone are embedded. When the topping is set, the surface is ground and highly polished to produce a smooth and decorative surface.

WALL FRAMING

The walls for slab-on-grade construction are framed in the same manner as described in Unit 11 — Exterior Wall Framing and in Unit 12 — Interior Rough Work, with the following exceptions:

1. Any framing in contact with the concrete floor slab or foundation wall, such as the sole plate of the wall, must be pressure-treated with preservative.

2. The sole plates of exterior walls are fastened with anchor bolts to the slab or foundation wall depending on the type of construction. Sole plates of interior partitions may be fastened to the concrete slab with hardened steel nails, or by pins driven by powder-actuated guns, or by other means that are sufficient to hold the plates securely in position.

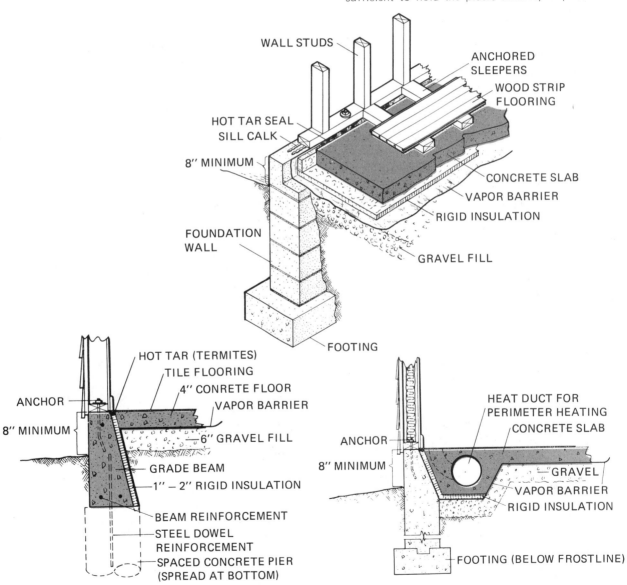

Fig. A-4 Independent slabs are constructed in a number of ways according to conditions.

GLOSSARY

Acoustical board — material used to control or deaden sound

Actual size — the size of lumber after it has been surfaced

Air-dried lumber — lumber which has been seasoned by drying in the air

Anchor bolt — a special type bolt used to fasten a wood member to a concrete wall

Annular ring — the rings seen when viewing a cross section of a log; each ring constitutes one year of tree growth

Apprentice — a beginner who serves for a stated period of time to learn a trade

Arbor — a shaft upon which circular saw blades are inserted

Asphalt felt — a building paper which is saturated with asphalt for waterproofing

Astragal — a semicircular molding often used to cover a joint between doors

Awning window — a type of window in which the sash are hinged at the top and swing outward

Back-bevel — a bevel on the edge or end of stock inclining toward the back side

Backing — strips or blocks installed in walls or ceilings for the purpose of fastening or supporting trim or fixtures

Backing the hip — beveling the top edge of a hip rafter to line it up with adjacent roof surfaces

Back miter — an angle cut starting from the end and coming back on the face of the stock

Backset — the distance an object is set back from an edge, side, or end of stock, such as the distance a hinge is set back from the edge of a door

Backsplash — a raised portion on the back edge of a countertop to protect the wall

Balloon frame — a type of frame in which the studs are continuous from foundation to roof

Baluster — vertical members, usually decorative and spaced closely together, of a stair rail

Balustrade — the entire stair rail assembly including hand rail, balusters and posts

Baseboard — finish board used to cover the joint at the intersection of wall and floor

Base cap — a molding applied to the top edge of the baseboard

Base shoe — a molding applied between the baseboard and floor

Batten — a thin, narrow strip usually used to cover joints between vertical boards

Batter board — a temporary framework erected to hold the stretched lines of a building layout

Bay window — a window, usually three-sided, that projects out from the wall line

Bearer — horizontal members of a wood scaffold that support scaffold plank

Bearing partition — an interior wall that supports the floor above

Bench mark — a reference point for determining elevations during the construction of a building

Bevel — the sloping edge or side of a piece with the angle between not a right angle

Blind joint — a type of joint in which the cuts do not go all the way through

Blind nail — a method of fastening that conceals the nails

Blind stop — part of a window finish applied just inside the exterior casing

Blocking — pieces intalled between studs in a wall usually used to provide fastening

Board — lumber usually eight inches or more in width and less than two inches thick

Board measure — system of designating quantities of lumber in terms of board feet

Bow — a type of warp in which the side of lumber is curved from end to end

Box nail — a thin nail with a head, usually coated with a material to increase its holding power

Brad — a thin, short finishing nail

Break corners — to round sharp exposed corners

Break joints — to stagger joints in adjacent rows

Buck — a rough frame used to form openings in concrete walls

Bullnose — a starting step that has one or both ends rounded

Butt — the placing of one piece against another with a square-cut joint

Cant strip — a thin strip of wood placed under a piece to tilt the piece at a slant

Carbide-tipped — in reference to cutting tools that have small, extremely hard, pieces of carbide steel welded to the tips

Casement window — a type of window in which the sash are hinged at the edge and usually swing outward

Casing — a type of finish lumber or molding used to trim around doors, windows, and other openings

Center punch — a tool used to make an indentation at the centerlines of holes

Chamfer — an edge or end bevel that does not go all the way across the edge or end

Check — lengthwise split in the end of lumber usually resulting from more rapid drying of the end than the rest of the piece

Cheek cut — a beveled cut on the end of certain roof rafters

Cleat — a small strip applied to support a shelf or similar piece

Closed grain — wood in which the pores are small and closely spaced

Closed valley — a roof valley in which the roof covering meets in the center of the valley, completely covering the valley

Collar tie — a horizontal member placed close to the ridge connecting two opposite rafters

Column — a vertical round or rectangular supporting member

Common rafter — a rafter that extends from plate to ridge at right angles to the plate

Compound miter — a bevel cut across the width and also through the thickness of a piece

Condensation — when water, in a vapor form, changes to a liquid due to cooling of the air; the resulting drops of water which accumulate on the cool surface

Conductor — also called *downspout,* a vertical member used to carry water from the gutter downward to the ground

Contact cement — an adhesive used to bond plastic laminates or other thin material; so called, because the bond is made on contact eliminating the need of clamps

Coped joint — a type of joint between moldings in which the end of one piece is cut to fit the molded surface of the other

Corner bead — metal trim used on exterior corners of walls to enforce them

Corner boards — boards used to trim corners on the exterior walls of a building

Corner brace — part of the wall frame used at the corners to stiffen and strengthen the wall

Cornerite — metal lath, cut into strips and bent at right angles, used in interior corners of walls and ceilings, on top of lath to prevent cracks in plaster

Corner post — built-up stud used in the corner of a wall frame

Cornice — the entire finished assembly where the walls of a structure meet the roof

Counterbore — boring a larger hole partway through the stock so that the head of a fastener can be recessed

Countersink — making a flared depression around the top of a hole to receive the head of a flathead screw; also, the tool used to make the depression

Course — a row of brick, siding, roofing, flooring, and similar material

Cove — a concave-shaped molding

Crawl space — a shallow space below the living quarters of a structure without a basement

Cricket — a small, false roof built behind a chimney or other roof obstacle for the purpose of shedding water

Crook — a type of warp in which the edge of lumber is not straight

Cross bridging — small diagonal braces, set in pairs, crossing each other, between floor joists for the purpose of distributing over a wider area the load placed on the floor

Crosscut — a cut made across the grain of lumber

Crown — usually referred to as the high point of the crooked edge of joists, rafters, and other framing members

Cup — a type of warp in which the side of a board is curved from edge to edge

Dado — a relatively wide cut, partway through, and across the grain

Deadman — a T-shaped wood device used to support ceiling drywall panels

Diagonal — at an angle, usually from corner to corner in a straight line

Double acting — doors that swing in both directions or the hinges used on these doors

Double-hung window — a window in which two sash slide vertically by each other

Downspout — see conductor

Drip — that part of a cornice or a course of horizontal siding that projects below another part; also a channel cut in the underside of a windowsill that causes water to drop off instead of drawing back and running down the wall

Drip cap — a molding placed on the top of exterior door and window casings for the purpose of shedding water beyond the outside of the unit

Drip edge — metal edging strips placed on roof edges to provide a support for the overhang of the roofing material

Dropping the hip — increasing the depth of the hip rafter seat cut so that the centerline of its top edge will lie in the plane of adjacent roof surfaces

Drywall — a type of construction usually referred to as the installation of gypsum board

Duplex nail — a double-headed nail used for temporary fastening such as in the construction of wood scaffolds

Dutchman — an odd-shaped piece usually used to fill or cover an opening

Easement — a curved member of a stair handrail

Eaves — that part of a roof that extends beyond the sidewall

Edge — the narrow surface of lumber running with the grain

Edge grain — sometimes called *vertical grain,* refers to boards in which the annular rings are at or near perpendicular to the face

Elevation — a drawing in which the height of the structure or object is shown

Embossed — ornamental designs raised above a surface

End — the extremities of a piece of lumber

Exposure — the amount that courses of siding or roofing are exposed to the weather

Face — the best appearing side of a piece of wood or the side which is exposed when installed, such as finish flooring

Face frame — a framework of narrow pieces on the face of a cabinet containing door and drawer openings

Fascia — a vertical member of the cornice finish installed on the tail end of rafters

Feather edge — the edge of material brought down in a long taper to a very thin edge

Fence — on a table saw, a guide for ripping lumber

Fiber saturation point — the moisture content of wood when the cell cavities are empty but the cell walls are still saturated

Finish carpentry — that part of the carpentry trade involved with the application of exterior and interior finish

Finish nail — a thin nail with a small head designed for setting below the surface of finish material

Finish stringer — the finish board running with the slope of the stairs and covering the joint between the stairs and the wall; also called a *skirt board*

Firecut — an angle cut made on the ends of floor joists bearing on a masonry wall; designed to prevent the masonry wall from toppling in case the joists are burned through and collapse

Firestop — material used to fill air passages in a frame to prevent the spread of fire; in a wood frame, might consist of 2"x4" blocking between studs

First & Seconds — the best grade of hardwood lumber

Fissured — irregular shaped grooves made in material, such as ceiling tile, for acoustical purposes

Flashing — material used at intersections such as roof valleys, dormers, and above windows and doors to prevent the entrance of water

Flat grain — opposite of edge grain, in which the annular rings of lumber lie close to parallel to the sides

Footing — a foundation for a column, wall, or chimney made wider than the object it supports, to distribute the weight over a greater area

Foundation — that part of a wall on which the building is erected

Frieze — a part of the exterior finish applied at the intersection of a overhanging cornice and the wall

Frostline — the depth to which frost penetrates into the ground

Furring strip — strips of lumber spaced at desired intervals for the attachment of wall or ceiling covering

Gable end — the triangular-shaped section on the end of a building formed by the rafters in a common or gable roof and the top plate line

Gable roof — a type of roof that pitches in two directions

Gain — a cutout made in a piece to receive another piece, such as a cutout for a butt hinge

Galvanized — protected from rusting by a coating of zinc

Gambrel roof — a type of roof that has two slopes of different pitches on each side of center

Girder — a heavy timber or beam used to support the ends of floor joists

Glaze — to install glass in a frame

Glazier — a person who installs glass in a frame

Glazing — the act of installing glass

Glazing compound — a soft, plastic-type material, similar to putty, used for sealing lights of glass in a frame

Glazing points — small, triangular or diamond shaped pieces of metal used to secure and hold lights of glass in a frame

Gooseneck — a curved section of handrail used when approaching a landing; also, an outlet in a roof gutter

Grain — in wood, the design on the surfaces caused by the contrast, spacing, and direction of the annular rings

Graphite — a mineral used as pencil lead and also as a lubricant for the working parts of locks and certain tools

Green lumber — lumber which has not been dried to a suitable moisture content

Groove — a relatively wide cut, partway through, and running with the grain of lumber

Ground — strips of wood placed at the base of walls and around openings and used as a guide for the application of an even thickness of plaster

Gusset — a pad of wood or metal used over a joint to stiffen and strengthen it

Gutter — a wood or metal trough used at the roof edge to carry off rain water and water from melting snow

Gypsum board — a sheet product made by encasing gypsum in a heavy paper wrapping

Half round — a molding with its end section in the shape of a semicircle

Handrail — a railing on a stairway intended to be grasped by the hand to serve as a support and guard

Hardboard — a building product made by compressing wood fibers into sheet form

Header — pieces placed at right angles to joists, studs, and rafters to form openings in a wood frame

Hearth — a section of the floor in front of a fireplace usually covered with some type of fireproof material

Heartwood — the wood in the inner part of a tree, usually darker, and containing inactive cells

Heel — the back end of objects, such as a handsaw or hand plane

Hip — the intersection of two surfaces of a hip roof

Hip rafter — extends diagonally from the corner of the plate to the ridge at the intersection of two surfaces of a hip roof

Hip-valley cripple jack rafter — a short rafter, running parallel to common rafters, cut between hip and valley rafters

Hopper window — a type of window in which the sash is hinged at the bottom and swings inward

Horn — an extension of the stiles of doors or the side jambs of window and door frames

Housed stringer — a finished stringer which is dadoed to receive treads and risers of a stairway

Insulated glass — double panes of glass fused together

Insulation — any material used to resist the passage of heat

Jalousie window — a type of window containing movable, horizontal slats of glass

Jamb — the sides and top of window and door frames

Jamb extension — narrow strips of wood fastened to the edge of window jambs to increase their width

Jig — any type of fixture designed to hold pieces or guide tools while work is being performed

Joint — as a verb, denotes straightening the edge of lumber; as a noun, means the place where parts meet and unite

Journeyman — a tradesman who has completed an apprenticeship or who has gained enough experience to perform work without instruction

Kerf — the width of a cut made with a saw

Keyway — a groove made in concrete footings for tying in the concrete foundation wall

Kiln-dried — lumber dried by placing it in huge ovens called kilns

Knot — a defect in lumber caused by cutting through a branch or limb embedded in the log

Laminate — build up with thin layers of wood or other material

Landing — The top of a flight of stairs or an intermediate-level platform

Lath — a base for plaster, usually gypsum board or expanded metal sheets

Lattice — thin strips of wood, spaced apart and applied in two layers at angles to each layer resulting in a kind of grillework

Lazy Susan — used in kitchen cabinets and other places; a set of revolving circular shelves

Ledger — a horizontal member of a wood scaffold that ties the scaffold posts together and supports the bearers

Level — horizontal or perpendicular to the force of gravity

Light — a pane of glass or an opening for a pane of glass

Linear measure — a measurement of length

Lookout — horizontal framing pieces in a cornice, installed to provide fastening for the soffit

Louver — an opening for ventilation consisting of horizontal slats installed at an angle to exclude rain, light, and vision, but to allow the passage of air

Lumber — wood that is cut from the log to form boards, planks, and timbers

Magazine — a container in power nailers and staplers in which the fasteners are placed to be ejected

Mansard — a type of roof that has two different pitches on all sides of the building

Mantel — the ornamental finish around a fireplace, including the shelf above the opening

Masonry — any construction of stone, brick, tile, concrete plaster, and similar materials

Mastic — a thick adhesive

Matched boards — boards which have been finished with tongue-and-grooved edges

Millwork — sometimes called joinery; any wood products which have been machined in large quantities in a factory (ready for use) such as moldings, doors, windows, siding, and others

Miter — the cutting of the end of a piece at any angle other than a right angle

Miter gauge — a guide used on the table saw for making miters and square ends

Miter joint — the joining of two pieces by cutting the end of each piece by bisecting the angle at which they are joined

Modular construction — a method of construction in which parts are preassembled in convenient-sized units

Moisture content — the amount of moisture in wood expressed as a percentage of the dry weight

Moisture equilibrium — the point at which the moisture content of wood is in balance with the humidity of the surrounding air

Moisture meter — a device to determine the moisture content of wood

Molding — decorative strips of wood used for finishing purposes

Mortise — a rectangular cavity cut in a piece of wood to receive a tongue or tenon projecting from another piece

Mullion — a vertical division between windows or panels in a door

Muntin — slender strips of wood between lights of glass in windows or doors

Newel post — an upright post supporting the handrail in a flight of stairs

No. 1 common — a lower grade of hardwood lumber

Nominal size — the stated size of the thickness and width of lumber even though it differs from its actual size; the approximate size of rough lumber before it is surfaced

Nosing — the rounded edge of a stair tread projecting over the riser

O.C. — abbreviation for "on center," meaning the distance from the center of one structural member to the center of the next one

Ogee — a molding with an S-shaped curve

Open valley — a roof valley in which the roof covering is kept back from the centerline of the valley

Panel — a section enclosed by a frame, such as a door panel

Particleboard — a building product made by compressing wood chips and sawdust with adhesives to form sheets

Parting strip — a small strip of wood separating the upper and lower sash of a double-hung window

Partition — an interior wall separating one portion of a building from another

Penny — a term used in designating nail sizes

Perforated — material which has closely spaced holes in a regular or irregular pattern

Perm — a measure of water vapor movement through a material

Phillips head — a type of screw head with a cross-slot

Pier — a column of masonry, usually rectangular in horizontal cross-section, used to support other structural members

Pilot — a guide on the end of edge-forming router bits used to control the amount of cut

Pilot hole — a small hole drilled to receive the threaded portion of a wood screw

Pitch — the amount of slope to a roof expressed as ratio of the total rise to the span

Pitch board — a piece of wood cut in the shape of a right triangle used as a pattern for laying out stair stringers or rafters

Pitch pocket — an opening in lumber between annular rings containing pitch in either liquid or solid form

Pith — the small, soft core at the center of a tree

Pivot — revolving around a point

Plain-sawed — also flat-grain; a method of sawing lumber that produces flat-grain

Planchir (or plancier) — also called *soffit;* the finish member on the underside of a box cornice

Plank — lumber that is 6 or more inches in width and from 1 ½ to 6 inches in thickness

Plan view — a top view of a horizontal section

Plaster — a mixture of Portland cement, sand, and water used for covering walls and ceilings of a building

Plastic laminate — available in a wide choice of colors and designs; a very tough, thin material in sheet form used to cover countertops

Plinth block — a small, decorative block, thicker and wider than a door casing, used as part of the door trim at the base and at the head

Plot plan — also called a *site plan;* a drawing showing a bird's-eye view of the lot, the position of the building, and other pertinent information

Plumb — vertical; at right angles to level

Plumb bob — a pointed weight attached to a line for testing plumbness

Plunge cut — an interior cut made with a portable saw by a method that does not require boring holes before making the cut

Ply — one thickness of several layers of built-up material, such as one of the layers of plywood

Plywood — a building material in which thin sheets of wood are glued together with the grain of adjacent layers at right angles to each other

Pneumatic — powered by compressed air

Pocket — a recess in a wall to receive a piece, such as a recess in a concrete foundation wall to receive the end of a girder

Pocket door — a type of door that when opened slides into a recess in the wall

Polyethylene film — a thin plastic sheet used as a vapor barrier

Preacher — a small piece of wood of the same thickness as the stair risers. It is notched in the center to fit over the finish stringer and rest on the tread cut of the stair carriage. It is used to lay out open finished stringers in a staircase.

Preservative — a substance applied to wood to prevent decay

Pressure-treated — treatment given to lumber that applies preservative under pressure to penetrate the total piece

Primer — the first coat of paint applied to the surface

Quarter round — a type of molding, an end section of which is in the form of a quarter circle

Quartersawed — a method of sawing lumber parallel to the medullary rays to produce edge-grain lumber (See edge grain.)

Rabbet — a cutout along the edge or end of lumber

Rafter — a sloping structural member of a roof frame that supports the roof sheathing and covering

Rail — the horizontal member of a frame

Rake — the sloping portion of the gable ends of a building

Reciprocating — a back-and-forth action, as in certain power tools

Return — a turn and continuation for a short distance of a molding, cornice, or other kind of finish

Return nosing — a separate piece mitered to the open end of a stair tread for the purpose of returning the tread nosing

Reveal — the amount of setback of the casing from the face side of window and door jambs or similar pieces

Ribbon — a narrow board let into studs of a balloon frame to support floor joists

Ridge — the highest point of a roof that has sloping sides

Ridge board — a horizontal member of a roof frame which is placed on edge at the ridge and into which the upper ends of rafters are fastened

Rip — sawing lumber in the direction of the grain

Rise — in stairs, the vertical distance of the flight; in roofs, the vertical distance from plate to ridge; may also be the vertical distance through which anything rises

Riser — the finish member in a stairway covering the space between treads

Rough carpentry — that part of the trade involved with construction of the building frame or other work that will be dismantled or covered by the finish

Rough opening — an opening made in the frame in which to install windows, doors or similar units

Rough stringer — cutout supports for the treads and risers in a staircase; also called *stair carriage* and *stair horse*

Run — the horizontal distance over which rafters, stairs, and other like members travel

R-value — a number given to a material to indicate its resistance to the passage of heat

S2S — surfaced two sides

S4S — surfaced four sides

Saddle — same as cricket

Sapwood — the outer part of a tree just beneath the bark containing active cells

Sash — that part of a window into which the glass is set

Sash balance — a device, usually operated by a spring or tensioned weatherstripping, designed to counterbalance double-hung window sash

S-beam — an I-shaped steel beam

Scab — a length of lumber applied over a joint to stiffen and strengthen it

Scaffold — an elevated, temporary working platform

Scratch coat — the first coat of plastic applied to metal lath

Screed — strips of wood, metal, or pipe secured in position and used as guides to level the top surface of concrete

Scribe — laying out woodwork to fit against an irregular surface

Seasoned lumber — lumber which has been dried to a suitable moisture content

Seat cut (bird's mouth) — the cut on a rafter to allow it to fit and rest on the wall plate

Set — alternate bending of saw teeth to provide clearance in the saw cut

Shake — a defect in lumber caused by a separation of the annular rings

Shank hole — a hole drilled for the thicker portion of a wood screw

Sheathing — boards of sheet material that are fastened to roofs and exterior walls and on which the roof covering and siding is applied

Shed roof — a type of roof that slopes in one direction only

Shim — a thin, wedged-shaped piece of material used behind pieces for the purpose of straightening them, or for bringing their surfaces flush at a joint

Shingle tip — the thin end of a wood shingle

Shortened valley rafter — a valley rafter that runs from the plate to the supporting valley rafter

Side — the wide surfaces of a board, plank, or sheet

Sidelight — a framework containing small lights of glass placed on one or both sides of the entrance door

Siding — exterior sidewall finish covering

Sill — horizontal timbers resting on the foundation supporting the framework of a building; also, the lowest horizontal member in a window or door frame

Skirt board — another name for a finished stringer in a staircase

Sleeper — strips of wood laid over a concrete floor to which finish flooring is fastened

Slump test — a test given to concrete to determine its wetness

Snap tie — a metal device to hold concrete wall forms the desired distance apart

Soffit — the underside trim member of a cornice or any such overhanging assembly

Softwood — wood from coniferous (cone-bearing) trees

Soil stack — part of the plumbing; a vertical pipe extending up through the roof to vent the system

Soleplate — the bottom horizontal member of a wall frame

Solid bridging — solid members of the same thickness and width of floor joists placed between them to distribute the floor load over a wider area

Specifications — written or printed directions of construction details for a building

Spike — a nail, 16d or over

Spline — a thin flat strip of wood inserted in the grooved edges of adjoining pieces

Spreader — a strip of wood used to keep other pieces a desired distance apart

Square — the amount of roof covering that will cover 100 square feet of roof area

Stair carriage — See rough stringer

Stair horse — See rough stringer

Stairwell — an opening in the floor to provide headroom for climbing or descending stairs

Standing cut — a cut made through the thickness of stock at more than a 90° angle between the side and edge or end

Staple — a U-shaped fastener

Starter course — usually used in reference to the first row of shingles applied to a roof or wall

Starting step — the first step in a flight of stairs

Stile — the outside vertical members of a frame, such as in a paneled door

Stool — the bottom horizontal member of a door or window frame; also called *sill*

Stool cap — a horizontal finish piece covering the stool or sill of a window frame on the interior; also called stool

Stop bead — a vertical member of the interior finish of a window against which the sash butts or slides

Storm sash — an additional sash placed on the outside of a window to create a dead air space to prevent the loss of heat from the interior in cold weather

Story — the distance between the upper surface of any floor and the upper surface of the floor above

Story pole — a narrow strip of wood used to lay out the heights of members of a wall frame or courses of siding

Straightedge — a length of wood or metal having at least one straight edge to be used for testing straight surfaces

Strapping — called *stripping* in some locations; refers to furring strips applied at specified spacings for the purpose of attaching wall or ceiling finish

Striated — finish material with random and finely spaced grooves running with the grain

Tack — to fasten temporarily in place

Tail cut — a cut on the extreme lower end of a rafter

Tail joist — short joist running from an opening to a bearing

Taper ground — the inner part of a saw blade ground thinner than the outside edge

Tempered — treated in a special way to be harder and stronger

Tenon — a tongue cut on the end of a piece usually to fit into a mortise

Termite shield — specially shaped pieces of metal, placed at strategic locations, to prevent termites from entering a building

Tile — square or rectangular blocks placed side by side to cover an area

Timber — large pieces of lumber

Toe — the forward end of tools, such as a hand saw and hand plane

Toeboard — a recessed space at the bottom of a base cabinet that provides room for the toes of a person standing close to the cabinet

To the weather — a term used to indicate the exposure of roofing or siding

Tread — horizontal finish members in a staircase upon which the feet of a person are placed ascending or descending the stairs

Trimmer — members of a frame placed between headers in an opening

Truss — in reference to roofs, a triangular combination of members to form a rigid framework for supporting loads over a long span

Undercut — a cut made through the thickness of stock at less than 90 degrees between the side and edges or ends

Underlayment — material placed on the subfloor to provide a smooth, even surface for the application of resilient finish floors

Unit of run — horizontal distance over which rafters travel for each foot of span

Valley — the intersection of two roof slopes at interior corners

Valley cripple jack rafter — a rafter running between two valley rafters

Valley jack rafter — a rafter running between a valley rafter and the ridge

Valley rafter — the rafter placed at the intersection of two roof slopes in interior corners

Vapor barrier — a material used to prevent the passage of vapor

Veneer — a very thin sheet or layer of wood

Vermiculite — a mineral closely related to mica with the ability to expand on heating to form a lightweight material with insulating qualities

Volute — a spiral portion of a handrail

Wainscoting — a wall finish applied part way up the wall from the floor

Waler — horizontal or vertical members of a concrete form used to brace and stiffen the form and to which ties are fastened

Wane — bark, or lack of wood, on the edge of lumber

Warp — any deviation from straightness in a piece of lumber

Water table — finish work applied just above the foundation that projects beyond it and sheds water away from it

W-beam — a wide-flanged I-shaped steel beam

Weatherstripping — narrow strips of thin metal or other material applied to windows and doors to prevent the infiltration of air and moisture

Whet — the sharpening of a tool on a sharpening stone by rubbing the tool on the stone

Wind — a defect in lumber caused by a twist in the stock from one end to the other

Winder — a tread in a stairway, wider on one end then the other, that changes the direction of travel

INDEX